AF443063

Gene Therapy

The HUMAN MOLECULAR GENETICS series

Series Advisors

D.N. Cooper, *Institute of Medical Genetics, University of Wales College of Medicine, Cardiff, UK*

S.E. Humphries, *Division of Cardiovascular Genetics, University College London Medical School, London, UK*

T. Strachan, *Department of Human Genetics, University of Newcastle upon Tyne, Newcstle upon Tyne, UK*

Human Gene Mutation
From Genotype to Phenotype
Functional Analysis of the Human Genome
Molecular Genetics of Cancer
Environmental Mutagenesis
HLA and MHC: Genes, Molecules and Function
Human Genome Evolution
Gene Therapy

Forthcoming titles
Venous thrombosis: from Genes to Clinical Medicine
Molecular Endocrinology
Protein Dysfunction and Human Genetic Disease
Molecular Genetics of Early Human Development

Gene Therapy

Nicholas R. Lemoine
ICRF Oncology Unit, Royal Postgraduate Medical School, Hammersmith Hospital, London, UK

David N. Cooper
Institute of Medical Genetics, University of Wales College of Medicine, Cardiff, UK

βIOS
SCIENTIFIC
PUBLISHERS

BIOS Scientific Publishers Ltd
9 Newtec Place, Magdalen Road, Oxford OX4 1RE, UK.
Tel. +44 (0) 1865 726286. Fax. +44 (0) 1865 246823
World-Wide Web home page: http://www.Bookshop.co.uk/BIOS/

DISTRIBUTORS

Australia and New Zealand
DA Information Services
648 Whitehorse Road, Mitcham
Victoria 3132

India
Viva Books Private Limited
4346/4C Ansari Road
New Delhi 110002

Singapore and South East Asia
Toppan Company (S) PTE Ltd
38 Liu Fang Road, Jurong
Singapore 2262

USA and Canada
BIOS Scientific Publishers
PO Box 605, Herndon,
VA 20172-0605

TO PIERS, LARA, PAUL, CATRIN AND DUNCAN

Typeset by Saxon Graphics Ltd, Derby, UK.
Printed by Biddles Ltd, Guildford, UK.

Contents

Contributors

Alton, E.W.F.W. Ion Transport Unit, National Heart and Lung Institute, Imperial College of Science, Technology and Medicine, Emmanuel Kaye Building, Manresa Road, London SW3 6LR, UK

Banerjee, S. Department of Biochemistry & Biophysics and Lineberger Comprehensive Cancer Center, CB# 7295, School of Medicine, University of North Carolina at Chapel Hill, Chapel Hill, NC 27599-7295, USA

Bartlett, J.S. Gene Therapy Center, CB# 7352, Thurston-Bowles Building, University of North Carolina at Chapel Hill, Chapel Hill, NC 27599-7352, USA

Caplen, N.J. Ion Transport Unit, National Heart and Lung Institute, Imperial College, Manresa Road, London SW3 6LR, UK

Caskey, C.T. Merck & Co., Sumneytown Pike, West Point, PA 19486, USA

Chastain, M. Merck & Co., Sumneytown Pike, West Point, PA 19486, USA

Cheng, S. Genzyme Corporation, One Mountain Road, Framingham, MA 01701, USA

Dorin, J.R. MRC Human Genetics Unit, Western General Hospital, Crewe Road, Edinburgh EH4 2XU, UK

French, B.A. Molecular Cardiology Unit, Department of Medicine – Cardiology, MDR Bldg: Room 521, 511 South Floyd Street, University of Louisville, Louisville, KY 40202, USA

Gäken, J. Department of Molecular Medicine, Rayne Institute, King's College School of Medicine and Dentistry, 123 Coldharbour Lane, London SE5 9NU, UK

Galea-Lauri, J. Department of Molecular Medicine, Rayne Institute, King's College School of Medicine and Dentistry, 123 Coldharbour Lane, London SE5 9NU, UK

Gaspar, H.B. Molecular Immunology Unit, Institute of Child Health, 30 Guilford Street, London WC1N 1EH, UK

Günzburg, W.H. Institut für Molekulaire Virologie, GSF-Forschungszentrum für Umwelt und Gesundheit, D-85764 Oberschleissheim, Germany. Present address: Institute of Virology, VMU, Joseph Baumann Gasse 1, A-86567 Vienna, Austria

Kinnon, C. Molecular Immunology Unit, Institute of Child Health, 30 Guilford Street, London WC1N 1EH, UK

Kurachi, K. Department of Human Genetics, University of Michigan Medical School, 3712 Medical Science II Building, Ann Arbor, MI 48109-0618, USA

Larin, Z. Institute of Molecular Medicine, University of Oxford, John Radcliffe Hospital, Headington, Oxford OX3 9DU, UK

Lemoine, N.R. ICRF Oncology Unit, Royal Postgraduate Medical School, Hammersmith Hospital, Du Cane Road, London W12 0NN, UK

Lowenstein, P.R. Molecular Medicine Unit, Department of Medicine, University of Manchester, Room 1.302, Stopford Building, Oxford Road, Manchester, M13 9PT, UK

Martin, L.-A. ICRF Oncology Unit, Royal Postgraduate Medical School, Hammersmith Hospital, Du Cane Road, London W12 0NN, UK

Middaugh, C.R. Merck & Co., Sumneytown Pike, West Point, PA 19486, USA

Partridge, T.A. Muscle Cell Biology Group, MRC Clinical Sciences Centre, Royal Postgraduate Medical School, Hammersmith Hospital, Du Cane Road, London W12 0NN, UK

Porteous, D.J. MRC Human Genetics Unit, Western General Hospital, Crewe Road, Edinburgh EH4 2XU, UK

Porter, A. MRC Clinical Sciences Centre, Royal Postgraduate Medical School, Hammersmith Hospital, Du Cane Road, London W12 0NN, UK

Ring, C.J.A. ICRF Oncology Unit, Royal Postgraduate Medical School, Hammersmith Hospital, Du Cane Road, London W12 0NN, UK. Present address: Virology Unit, Glaxo Wellcome Research and Development, Medicines Research Centre, Gunnels Wood Road, Stevenage, Herts SG1 2NY, UK

Salmons, B. The Bavarian Nordic Research Institute, Ingolstaedter Landstrasse 1, D-85764 Oberschleissheim, Germany

Samulski, R.J. Gene Therapy Center, CB# 7352, Thurston-Bowles Building, University of North Carolina at Chapel Hill, Chapel Hill, NC 27599-7352, USA

Scheule, R.K. Genzyme Corporation, One Mountain Road, Framingham, MA 01701, USA

Sewry, C.A. Muscle Cell Biology Group, MRC Clinical Sciences Centre, Royal Postgraduate Medical School, Hammersmith Hospital, Du Cane Road, London W12 0NN, UK

Sikora, K. ICRF Oncology Unit, Royal Postgraduate Medical School, Hammersmith Hospital, Du Cane Road, London W12 0NN, UK

Vos, J.-M. Department of Biochemistry & Biophysics and Lineberger Comprehensive Cancer Center, CB# 7295, School of Medicine, University of North Carolina at Chapel Hill, Chapel Hill, NC 27599-7295, USA

Wang, J.-M. Department of Human Genetics, University of Michigan Medical School, 3712 Medical Science II Building, Ann Arbor, MI 48109-0618, USA

Westphal, E.-M. Department of Biochemistry & Biophysics and Lineberger Comprehensive Cancer Center, CB# 7295, School of Medicine, University of North Carolina at Chapel Hill, Chapel Hill, NC 27599-7295, USA

Williamson, R. Murdoch Institute for Research into Birth Defects, Royal Children's Hospital, Flemington Road, Parkville, 3052, Melbourne, Australia

Abbreviations

α-TIF	α-*trans*-inducing factor
AAV	adeno-associated virus
ACS	autonomous consensus sequence
Ad	adenovirus
ADA	adenosine deaminase
AFP	α-fetoprotein
AIDS	acquired immunodeficiency syndrome
ALS	amyotrophic lateral sclerosis
ALV	avian leukaemia virus
APC	*adenomatous polyposis coli* (gene)
APC	antigen-presenting cell
araATP	adenine arabinonucleoside triphosphate
araC	cytosine arabinoside
araM	9-β(-D-arabinofuranosyl)-6-methoxy-9H-purine
ARS	autonomous replicating sequence
ASOR	asialorosomucoid
β-gal	β-galactosidase
bFGF	basic fibroblast growth factor
bGH	bovine growth hormone
BLV	bovine leukaemia virus
BMD	Becker muscular dystrophy
BMT	bone marrow transplantation
c.f.u.	colony-forming units
CAT	chloramphenicol acetyltransferase
CEA	carcinoembryonic antigen
CEN	centromere sequence
CF	cystic fibrosis
CFTR	cystic fibrosis transmembrane conductance regulator
CMD	congenital muscular dystrophy
CMV	cytomegalovirus
CNS	central nervous system
CNTF	ciliary neurotrophic factor
CR	complement receptor
CTL	cytotoxic T lymphocyte
CVS	chorionic villus sampling
dAdo	deoxyadenosine
dATP	deoxyadenosine triphosphate
DBP	DNA-binding protein
DC-Chol	[N-(N',N'-dimethylaminoethane)-carbamoyl] cholesterol chloride
DEAE	diethylaminoethyl
DMD	Duchenne muscular dystrophy
DMRIE	1,2-dimyristyloxypropyl-N,N-dimethyl-hydroxyethyl ammonium bromide
DOGS	dioctadecylamidoglycylspermine tetrafluoroacetic acid
DOPE	dioleoylphosphatidylethanolamine
DOTMA	N-[1-(2,3-dioleyloxy) propyl]-N,N,N-trimethylammonium chloride
DTH	delayed-type hypersensitivity
EBNA	Epstein–Barr nuclear antigen
EBV	Epstein–Barr virus
EGF	epidermal growth factor
ENU	ethylnitrosourea
EPD	electronic pulse delivery
ER	endoplasmic reticulum
ES cells	embryonic stem cells
EtD	ethidium homodimer

5-FC	5-fluorocytosine
5-FU	5-fluorouracil
FACC protein	Fanconi anaemia C complementing protein
FGF	fibroblast growth factor
FH	familial hypercholesterolaemia
FISH	fluorescence *in situ* hybridization
GABA	γ-aminobutyric acid
GCPS	Greig's cephalopolysyndactyly syndrome
GM–CSF	granulocyte–macrophage colony-stimulating factor
GPAT	genetic prodrug activation therapy
GPDH	glyceraldehyde-3-phosphate dehydrogenase
GRP	gastrin-releasing peptide
HA	haemagglutinin
HBV	hepatitis B virus
HAEC	human artificial episomal chromosome
HD	Huntington's disease
HDL	high-density lipoprotein
HFV	human foamy virus
HIV-1	human immunodeficiency virus-1
HLA	human leukocyte antigen
HMG protein	high mobility group protein
HPRT	hypoxanthine guanine phosphoribosyltransferase
HPV	human papilloma virus
HSC	haematopoietic stem cell
HSV	herpes simplex virus
HTLV	human T-cell leukaemia virus
HVS	herpes virus saimiri
ICAM	intercellular cell adhesion molecule
IFN-γ	interferon-γ
Ig	immunoglobulin
IGF	insulin-like growth factor
IL-2	interleukin-2
IRES	internal ribosome entry site
ITR	inverted terminal repeat
i.v.	intravenous
IVF	*in vivo* fertilization
LAK cell	lymphokine activated killer cell
LAT	latency-associated transcript
LCL	lymphoblastoid cell line
LCR	locus control region
LDL	low-density lipoprotein
LFA	leukocyte function-associated antigen
LGMD2A	limb-girdle muscular dystrophy type 2A
LPS	lipopolysaccharide
LTBMC	long-term bone marrow culture
LTCIC	long-term culture initiating cell
LTR	long terminal repeat
luc	luciferase
MA protein	matrix protein
MAC	mammalian artificial chromosome
MCC	mucociliary clearance
MCK	muscle creatine kinase
MEL cells	murine erythroleukaemia cells
MHC	major histocompatibility complex
MLV	murine leukaemia virus
MMTV	mouse mammary tumour virus
MoAb	monoclonal antibody
MOI unit	multiplicity of infection unit
MoMLV	Moloney murine leukaemia virus
MPS	mucopolysaccharidosis
MPTP	1-methyl-4-phenyl-1,2,3,6-tetrahydropyridine
MRI	magnetic resonance imaging

MTX	methotrexate
NDV	Newcastle disease virus
NGF	nerve growth factor
NK cell	natural killer cell
ORC	origin recognition complex
ORF	open reading frame
pDNA	plasmid DNA
PFO	perfringolysin O
p.f.u.	plaque-forming unit
Ph chromosome	Philadelphia chromosome
PBL	peripheral blood lymphocyte
PBS	primer binding site
PCNA	proliferating cell nuclear antigen
PCR	polymerase chain reaction
PEG	polyethylene glycol
PEPCK	phosphoenolpyruvate carboxylase
pLys	poly-L-lysine
PNP	purine nucleoside phosphorylase
Pre-RC	pre-replication complex
ProCon Vector	promoter conversion vector
PRV	pseudorabies virus
PS	phosphatidylserine
rAAV	recombinant adeno-associated virus
RAC	Recombinant DNA Advisory Committee
RBC	red blood cell
RCV	replication-competent virus
RGD	Arg-Gly-Asp
RSV	Rous sarcoma virus
RT–PCR	reverse transcriptase–polymerase chain reaction
SAHH	S-adenosylhomocysteine hydrolase
SCARMD	severe, childhood, autosomal, recessive muscular dystrophy
SCID	severe combined immunodeficiency
SIN vector	self-inactivating vector
SIV	simian immunodeficiency virus
SOD	superoxide dismutase
SP-B	surfactant-associated protein B
SV40	simian virus 40
TACF	telomere-associated chromosome fractionation
TAP	transporter associated with antigen presentation
TcR	T-cell receptor
TFN	transferrin
TGF-β	transforming growth factor-β
TH	tyrosine hydroxylase
Th cell	helper T cell
TIL	tumour-infiltrating lymphocyte
TK	thymidine kinase
TNF	tumour necrosis factor
tPA	tissue-type plasminogen activator
TR	terminal repeat
TRF	telomere repeat factor
ts	temperature-sensitive
uPA	urokinase-type plasminogen activator
VCAM	vascular cell adhesion molecule
VEGF	vascular endothelial growth factor
VHS protein	virion host shut-off protein
VLA	very late antigen
VSMC	vascular smooth muscle cell
VSV	vesicular stomatitis virus
vWF	von Willebrand factor
VZV	varicella zoster virus
WAP	whey acidic protein
YAC	yeast artificial chromosome

Foreword Gene therapy – the hype, the reality and the future

Gene therapy has reached a crossroads during the past year. Up to now, the ratio of hype to data has at times approached infinity. The two journals devoted to the field, *Gene Therapy* and *Human Gene Therapy*, often did not appear to have enough material to fill an issue. Only a very small number of patients had been treated, and even of these only a tiny proportion showed data suggesting success.

There had to be a reassessment of the wild claims made by the media, and by some clinical scientists and biotechnology companies. It is perhaps ironic that the press, which fuelled the hype with outrageous overly positive statements (every experiment, even in cell culture, became a 'cure'), was equally happy to be over-negative ('hopes for cure dashed') when the NIH and MRC provided balanced and realistic assessments. These correctly emphasized that there is indeed long-term promise for gene therapy in many fields of medicine, but that time will be required to develop the safe gene delivery systems required for a clinical product.

The reason why the long-term promise for gene therapy is realistic is now clear. Gene therapy is an enabling technology for delivery of a natural modifier of health and disease. Our health, both in the case of relatively rare single gene disorders and of common diseases such as cancer, heart disease and Alzheimer's, depends on our individual genetic inheritance and the way it interacts with the environment. The Human Genome Project, pulling together all the existing human molecular genetics research in this area, provides the framework for understanding the genetic contribution to human health.

Although understandable in terms of the trials that are taking place, it is a pity that gene therapy has been described primarily in terms of catastrophic diseases of children (such as cystic fibrosis and muscular dystrophy) and of adults (such as terminal cancer). First, most people would rather avoid catastrophic single gene disorders in their children if they can, and choose carrier screening and prenatal diagnosis when it is available, in preference to lifelong therapies of unproven efficacy. Second, and more important, putting gene therapy only in the context of serious disease makes it appear that genes are about illness. Genes are really about health. Using genes for health is an approach which emphasizes their essential usefulness. We should constantly argue that gene therapy is 'green' therapy, using something which most humans have to benefit those who (for some reason or other) have lost the use of a particular DNA sequence.

Because of this, I often emphasize the use of gene therapies for prevention. If it is good to lower blood cholesterol by 20%, perhaps one way to do so is to use genes. Perhaps we could all use an extra dose of the adenomatous polyposis coli (*APC*) gene in our colonic epithelial cells, to provide added resistance to sporadic carcinoma of the colon. But to move in this direction requires safe gene transfecting systems.

This, of course, is clearly the major problem – we do not yet have multifunctional systems to introduce genes into somatic cells efficiently and safely. In our own liposome-mediated gene therapy trial for cystic fibrosis (which was successful as a proof-of-principle) we sacrificed efficiency for extreme safety. The adenoviral cystic fibrosis trials leant in the other direction. However, in both cases the groups made use of 'off-the-shelf' transfecting systems which had been designed for other studies. It is only now that viral, liposome and receptor-targeting systems are being developed specifically for gene therapy, which will not only be more effective but also have higher safety margins. This is essential if the full potential of gene therapy is to be realized.

This book contains 18 chapters, each detailing either a transfection system or a target, written by experts with hands-on experience of the field. It is timely, and I hope that the many readers who are about to consider their own gene therapy approaches read widely through the book. Gene therapy is a multifaceted area of clinical science where we have much to learn from each other, and especially from integrating the different approaches to provide a new unity and imaginative new targets for this powerful technology.

Bob Williamson (*Murdoch Institute for Research into Birth Defects, Melbourne*)

1

Scope and limitations of gene therapy

Karol Sikora

1.1 Introduction

Gene therapy can be defined as the deliberate transfer of DNA for therapeutic purposes. There is a further implication in that it involves only specific sequences containing relevant genetic information; that is, transplantation procedures involving bone marrow, kidney and liver are not considered a form of gene therapy. The concept of transfer of genetic information as a practical clinical tool arose from the gene cloning technology developed during the 1970s. Without the ability to isolate and replicate defined genetic sequences it would be impossible to produce purified material for clinical use. The drive for the practical application of this technology came from the biotechnology industry, with its quest for complex human biomolecules produced by recombinant techniques in bacteria. Within a decade, pharmaceutical-grade insulin, interferon, interleukin-2 (IL-2) and tumour necrosis factor (TNF) were all undergoing clinical trials. The next step was to obtain gene expression *in vivo*.

1.2 Suitable target diseases

Genetic disorders were the obvious first target for such therapies. Abortive attempts were made in the early 1980s to treat two patients with thalassaemia. These experiments were surrounded by controversy as the preclinical evidence of effectiveness was not adequate and full ethical approval had not been given. We now know that thalassaemia – a disorder in which there is transcriptional dysregulation of the globin chains of haemoglobin – may not be an ideal target for gene therapy as we still do not have good methods of regulating the expression level of inserted constructs.

Table 1.1 lists the features of a suitable target disease for gene therapy approaches. Clearly, the disease must be life-threatening so that the potential risk of serious side-effects is ethically acceptable. The gene must be available and its delivery to the relevant tissue feasible. This may involve the *ex vivo* transfection or transduction of cells removed from a patient, which are returned after manipulation. This approach is only possible with a limited range of tissues and most trials so far have used bone marrow. Ideally, a short-term surrogate end-point to demonstrate the physiological benefit of the newly inserted gene should be available. The electrical conductance change in the nasal epithelium after insertion of the cystic fibrosis transmembrane regulator

Gene Therapy, edited by N.R. Lemoine and D.N. Cooper.
© 1996 BIOS Scientific Publishers Ltd, Oxford.

(CFTR) gene is a good example. Other diseases may have more complex, and therefore more difficult to measure, end-points. Finally, there must be some possibility that the disability caused by a disease is reversible. Some of the tragic mental and physical handicaps caused by some genetic metabolic disorders may never be improved by somatic gene therapy, however successful a gene transfer protocol.

Table 1.1. Features of a disease suitable for gene therapy

Life-threatening
Gene cloned
Efficient gene transfer to relevant tissue possible
Precise regulation of gene not required
Proper processing of protein product
Correct subcellular localization of protein
Persistence of gene expression to avoid repetitive dosing
Measurable surrogate end-points
Effects of disease must be potentially reversible

The aim of gene manipulation varies in different diseases. In haemophilia, all that may be required is a suitable 'protein factory' to produce enough circulating coagulation factor to be effective physiologically. In cystic fibrosis, enough CFTR product must be selectively expressed by those cells where its lack causes pathological damage, such as in the lung and gastrointestinal epithelia. A variety of metabolic disorders result from genetic abnormalities in liver proteins. Some form of tissue targeting may be necessary before effective therapeutic strategies can be developed. With cancer, the problem is the requirement to target every single malignant cell. Although a variety of ingenious methods are currently being examined, it would seem more realistic to use systems that do not require the correction of the somatic defect resulting in malignancy.

Over the past seven years, a growing number of protocols have been approved by regulatory authorities throughout the world, the majority in the USA. Protocols for cancer gene therapy now far outstrip all others. This reflects the difficulty in treating patients with advanced cancer by conventional chemotherapy, the low risk–benefit potential of gene therapy and the relatively high level of research funding in this area of biomedical research. *Figure 1.1* outlines gene therapy protocols currently in use from the NIH Recombinant DNA Advisory Committee (RAC) and the European Working Group on Gene Therapy (EWGGT) databases. Out of 163 currently active trials, 119 (65%) are for cancer.

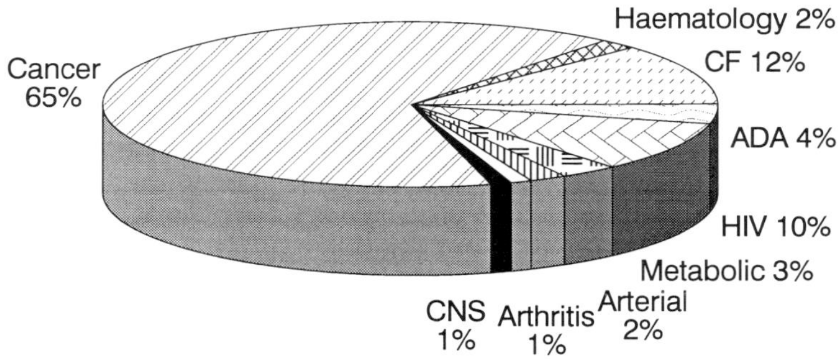

Figure 1.1. Gene therapy: current 163 approved protocols from US and European databases.

1.3 Germ-line gene therapy

So far, no government has seriously considered germ-line gene therapy. The technology is relatively straightforward as the problem of targeting can be solved by direct manipulation. The very sophisticated developments in *in vitro* fertilization together with the growing experience in the generation of transgenic and knockout animals and plants mean that sooner or later we will have to discuss the ethics of such approaches. Techniques for homologous recombination – the targeted replacement of old genes with new – are now possible, at least in mice. Furthermore, the likely explosion in genome information coming both from wholesale sequencing of the human genome and the genetic dissection of functionally related genes in simpler organisms as diverse as nematodes, zebrafish and yeast will almost certainly open up new possibilities for genetic intervention that could transcend generations. It is this permanency that is most frightening to the ethicists, especially if unforeseen problems arise. At some time in the future, a policy of 'genetic cleansing' may become a serious option for governments trying to contain spiralling healthcare costs. Ethics, like beliefs, values and cultures, change with time – so perhaps today's heresy will be tomorrow's routine.

1.4 Ethical and safety considerations

Perhaps the biggest risk from gene manipulation *in vivo* is the possibility of insertional mutagenesis and the activation of oncogenes leading to neoplasia. Such risks are clearly important factors in the consideration of the ethical basis for gene therapy for disorders such as cystic fibrosis, haemophilia and the haemoglobinopathies. For patients with metastatic cancer, the risks are low. Such patients are often desperate for some form of treatment and are already searching for the gene therapy programmes described in the media. Therapies with even minimal potential benefit will be avidly considered. In this situation, the biggest problem is offering false hope. It is unrealistic to expect such new strategies to be effective immediately. The first patients entering trials will provide much information in return for little personal benefit. This must be recognized by both the investigator and patient in order to reduce the 'breakthrough' mentality that surrounds novel treatments (*Table 1.2*).

Table 1.2. Taking gene therapy to the clinic

Safety:	Must demonstrate lack of toxicity, lack of immunological response, lack of viral replication
Efficacy:	Must demonstrate effective delivery and expression leading to phenotype correction
Practicalities:	Issues of scale-up must be addressed Regulatory authorities must be satisfied Therapy must be cost-effective

Various countries have now established regulatory bodies for gene therapy. Most are modelled on the US model of the RAC, which works closely with the country's existing drug regulatory body – the Food and Drug Administration (FDA). The success of the FDA in taking over much of the paperwork has recently led to the disbanding of the RAC. A parallel system has been established in the UK where the Gene

Therapy Advisory Committee (GTAC) together with the Medicines Control Agency (MCA) evaluate proposals from a scientific, ethical and safety standpoint. The creation of the European Medicines Evaluation Agency (EMEA), a single agency for drug and biotechnology product evaluation in Europe based in London, provides an opportunity to standardize approaches across an increasing number of countries. The potential risks that must be considered in each case are detailed in *Table 1.3*.

Table 1.3. Potential risks of gene therapy

Insertional mutagenesis leading to cancer
Recombination of disabled vector resulting in environmental
    pollution by infectious recombinant virus
Toxic shock caused by viraemia
Transfer of non-viral exogenous genetic material
Contamination with other deleterious viruses or organisms
Physiological effects of over-expression

1.5 Milestones of progress

As with any mission-oriented project, certain milestones can be identified without which further progress would have been impossible (*Table 1.4*). However, there are many hurdles to be overcome in the future before this technology can be successfully applied clinically on a routine basis. Perhaps the most significant development yet to come is that of systemically administered, highly selective targeting vectors with high efficiency for stable incorporation in all cells of a chosen population. The delivery problem currently pervades the whole of this exciting field. The following sections consider the problems facing the gene therapist in four diverse clinical situations.

Table 1.4. Milestones in the development of gene therapy

Year	Discovery	Investigators
1928	Transformation of *Diplococcus pneumoniae*	Griffiths
1944	DNA is a transforming substance	Avery
1952	DNA is genetic material	Hershey and Chase
1953	Structure of double helix	Watson and Crick
1963	Genetic code elucidated	Brenner
1968	Transformation by DNA viruses	Sambrook
1976	Cloning of globin genes	Maniatis
1979	Cloned genes expressed	Mulligan and Berg
1981	Clinical thalassaemia studies	Cline
1983	NIH 'Summit' in Banbury, NY	Friedmann
1985	Retroviral vectors developed	Many
1990	RAC 'Points to consider' established	NIH
1990	First trial with the adenosine deaminase (*ADA*) gene	Blaese and Anderson
1990	Gene-marking studies	Rosenberg
1991	Cystic fibrosis protocol	Crystal
1992	Familial hypercholesterolaemia protocol	Wilson
1992	Large number of cancer trials	Many investigators
1993	UK GTAC established	Lloyd

1.6 Severe combined immunodeficiency

The first clinical trial involving a therapeutic gene began on a young girl suffering from severe combined immunodeficiency (SCID) on the 14th of September 1990 at the NIH in Washington, USA. Bone marrow cells were removed and transduced *ex vivo* with a retroviral vector carrying a cDNA copy of the gene for adenosine deaminase (ADA). ADA converts deoxyadenosine to its metabolites. If the enzyme is not present, deoxyadenosine accumulates and is toxic to some cells, especially T lymphocytes. Since these cells are intimately involved in the correct functioning of the immune system, ADA deficiency is usually fatal due to the development of severe combined immunodeficiency of both T- and B-cell lineages.

This disorder was an attractive early target for gene therapy. The gene had been cloned in 1983 and was well characterized. Bone marrow transplantation was an accepted treatment modality if a suitable matching donor could be found. Bone marrow cells could be manipulated in the laboratory and a check for successful transduction carried out after appropriate selection procedures. Furthermore, the patients could be maintained on exogenous ADA administered with polyethylene glycol so that a sudden decline in immune function was unlikely.

Experience with the first two patients at the NIH answered a number of important questions. Safe *ex vivo* manipulation was shown to be possible, although the expression of the transduced *ADA* gene was only transient. This meant that periodic infusions of T lymphocytes were necessary. The first patient showed a dramatic increase in T-cell number and function. She showed improved delayed-hypersensitivity tests and increased levels of isohaemagglutinins – useful short-term surrogate end-points. Since this pioneering study, other groups in different countries set out to repeat and improve the procedure. Several groups have attempted to obtain more long-lasting effects by gene transfer to CD34$^+$ bone marrow stem cells. The disease is extremely rare and very variable in its clinical course, making the assessment of new technology very difficult. However, the principle of successful gene transfer, with at least transient functional improvement, has now been verified several times.

1.7 Cystic fibrosis

There is a vast range of disorders associated with the lung, including malignant disease, asthma and infection. Two hereditary disorders – cystic fibrosis and human α_1-antitrypsin (hAAT) deficiency – are the most obvious targets for gene therapy. Cystic fibrosis is an autosomal recessive disorder resulting in abnormal electrolyte transport of epithelial cells. It is typically characterized by a collection of sticky mucus in the lung, pancreas and liver, which results in chronic inflammation and obstruction. The most common cause of death is respiratory failure caused by sequential bouts of severe chest infection. Cystic fibrosis is caused by a mutation in the *CFTR* gene. This encodes a membrane protein that acts as a chloride channel, pumping Cl$^-$ out of cells in response to an increase in cAMP. It has been shown that when normal *CFTR* cDNA is transferred into epithelial cells isolated from patients with cystic fibrosis, the cells are able to respond to increasing levels and can secrete Cl$^-$. Thus, the abnormalities associated with this disease can essentially be reversed in the laboratory. Can this be achieved *in vivo* in patients?

Two main approaches have been used with encouraging but not dramatic results. The first used an adenovirus-based vector, which has the epithelial cells lining the broncheoalveolar tree as its natural target. In this case, the *CFTR* gene was driven by the adenovirus type-2 major late promoter. The second strategy was to use cationic liposomes, which are positively charged and can therefore bind negatively charged DNA on their outer surface. Initial experiments have used electrical conductance changes in the nasal membranes as a surrogate end-point. Data from two phase-I studies show 20% correction of conductance abnormalities with good patient tolerance. Current studies involve the use of aerosol inhalers to saturate at least the upper part of the respiratory system. Calculations have shown that if the correct form of the *CFTR* gene is expressed in only one in ten cells, then reversal of the pathophysiology is possible. Clearly, optimizing gene delivery, uptake and stable expression are the main goals of investigators. The future therapy of cystic fibrosis may well involve the regular use of gene aerosols from an early age.

1.8 Familial hypercholesterolaemia

Familial hypercholesterolaemia is an autosomal dominant disorder in which there is a defect in the low-density lipoprotein receptor (*LDLR*) gene. The effect is a reduced ability to break down LDL and intermediate-density lipoproteins. Individuals who are heterozygous for this disease have increased serum levels of LDL and are more prone to premature coronary artery disease. Homozygous individuals have very high serum LDL levels and severe atherosclerosis. Current therapeutic approaches include LDL apheresis, ileal bypass and liver transplantation. The success of tissue replacement therapy makes gene replacement the obvious next step if the technological hurdles can be overcome. Furthermore, this type of genetic intervention, if successful, would represent a good paradigm for many other metabolic disorders involving the liver.

The existence of an excellent animal model – the Watanabe heritable hyperlipidaemic rabbit – permitted a series of preclinical experiments to verify the principles of gene therapy in familial hypercholesterolaemia. Newborn rabbit hepatocytes were isolated and transduced with recombinant retrovirus containing the normal LDL receptor. On returning the transduced cells to the animal, near-normal levels of LDL were achieved. The clinical protocol involves the resection of a small wedge of patient's liver and the preparation of a cell suspension using collagenases. The hepatocytes are transduced *ex vivo* with an appropriate retrovirus and returned to the remaining liver via the portal vein. Suggestions of transient clinical benefit have now been obtained and published (see following chapters).

1.9 Cancer

The key problem in the effective treatment of patients with solid tumours is the similarity between tumour and normal cells. Local therapies, such as surgery and radiotherapy, can succeed but only if the malignant cells are confined to the area treated. This is so in around a third of cancer patients. For the majority, some form of systemic selective therapy is required. Although many cytotoxic drugs are available, only a small proportion of patients are actually cured by their use. The success stories of Hodgkin's disease, non-Hodgkin's lymphoma, childhood leukaemia, choriocarcinoma and germ-cell tumours have just not been repeated for the common cancers such as those of the lung, breast and colon. Despite enormous efforts in new drug development, clinical trials of novel drug combinations, the addition of cytokines, high-dose regimens and even bone marrow rescue procedures, the gains have been marginal.

Against this disappointing clinical backdrop, we have seen an explosion of information on the molecular biology of cancer. Although our knowledge of growth control is still rudimentary, we have at last had the first glimpse of its complexity. This has brought a new vision with which to develop novel selective mechanisms to destroy tumours.

The main problem facing the gene therapist is how to get new genes into every tumour cell. If this cannot be achieved, then any malignant cells that remain unaffected will emerge as a resistant clone. Ideal vectors for gene transfer do not as yet as exist. Despite this drawback, there are already over 100 protocols accepted for clinical trial in cancer patients world-wide, the majority in the USA. The ethical issues are fairly straightforward since oncology provides some of the lowest possible risk–benefit ratios. There are several strategies currently under investigation that will be considered in detail later in this volume.

1.9.1 Genetic tagging

There are several situations where the use of a genetic marker to tag tumour cells may help in making decisions on the optimal treatment for an individual patient. The insertion of a foreign marker gene into cells from a tumour biopsy and the replacement of the marked cells into the patient prior to treatment can provide a sensitive new indicator of minimal residual disease after chemotherapy. The most common marker is the gene for neomycin phosphotransferase (the *neo* R gene), an enzyme which metabolizes the aminoglycoside antibiotic G418. When this gene is inserted into an appropriate retroviral vector, it can be stably incorporated into the host cell's genome. Originally detected by antibiotic resistance, it can now be picked up more sensitively by the polymerase chain reaction (PCR). In this way, as few as one tumour cell in one million normal cells can be identified. This procedure can help in the design of aggressive chemotherapy protocols.

1.9.2 Enhancing tumour immunogenicity

The presence of an immune response to cancer has been recognized for many years. The problem is that human tumours seem to be predominantly weakly immunogenic. If ways could be found to elicit a more powerful immune stimulus, then effective immunotherapy could become a reality. Several observations from murine tumours indicate that one reason for weak immunogenicity of certain tumours is the failure to elicit a helper T-cell response. This response in turn releases the necessary cytokines to stimulate the production of cytolytic T cells, which can destroy tumours. The expression of cytokine genes such as those for IL-2, TNF and interferon in tumour cells has been shown to bypass the need for helper T-cells in mice. Similar clinical experiments are now in progress. Melanoma cells have been prepared from biopsies and infected with retrovirus containing the *IL-2* gene. These cells are being used as a vaccine to elicit a more powerful immune response.

1.9.3 Vectoring cytokines to tumours

Cytokines such as the interferons and interleukins have been actively explored for their tumoricidal properties. Although there is evidence of cytotoxicity, their side-effects are profound – limiting the dose that can safely be administered. It is possible to insert cytokine genes into cells that can potentially 'home in' on tumours and so release a high concentration of their protein product locally. TNF genes have been inserted into tumour-infiltrating lymphocytes from patients with melanoma, and given systemically. These experiments are controversial for two reasons. First, it appears from *in vitro* studies that the amount of TNF expressed from such cells is unlikely to be sufficient to cause a significant cytotoxic effect. Secondly, the insertion of a foreign gene limits the ability of the lymphocyte to target into tumour masses. Over 15 patients have so far been treated at the US National Cancer Institute and formal publication of the results is eagerly awaited.

1.9.4 Inserting drug-activating genes

The main problem with existing chemotherapy is its lack of selectivity. If drug-activating genes could be inserted that would only be expressed in cancer cells, then the administration of an appropriate prodrug could be highly selective. There are now many examples of genes preferentially expressed in tumours. In some cases, their promoters have been isolated and coupled to drug-activating enzymes. Examples include α-fetoprotein (AFP) in hepatoma, prostate-specific antigen in prostate cancer and *c-erbB2* in breast cancer. Such promoters can be coupled to enzymes such as cytosine deaminase or thymidine kinase, thereby producing unique retroviral vectors that are able to infect all cells but can only be expressed in tumour cells. These suicide (or Trojan horse) vectors may not have absolute tumour specificity but this may not be essential; it may be possible to perform a genetic prostatectomy or breast ductectomy, so effectively destroying all tumour cells.

1.9.5 Suppressing oncogene expression

The downregulation of abnormal oncogene expression has been shown to revert the malignant phenotype in a variety of tumour lines *in vitro*. It is also possible to develop *in vivo* systems such as the insertion of genes encoding complementary (antisense) mRNA to that produced by the oncogene. Such anti-genes specifically switch off the production of the abnormal protein product. Mutant forms of the *RAS* oncogene family are an obvious target for this approach. Up to 75% of human pancreatic cancers contain a mutation in the 12th amino acid of this protein and reversal of this change in cell lines leads to the restoration of normal growth control.

Clearly, the major problem is to ensure that every single tumour cell gets infected. Any cell that escapes will have a survival advantage and produce a clone of resistant tumour cells. For this reason it may be that future treatment schedules will require the repetitive administration of vectors in a similar way to fractionated radiotherapy or chemotherapy.

1.9.6 Replacing defective tumour suppressor genes

In cell culture, malignant properties can often be reversed by the insertion of normal tumour suppressor genes such as *RB1* (retinoblastoma), *TP53* (encoding p53) and *DCC* (deleted in colorectal cancer). Although tumour suppressor genes were often identified in rare tumour types, abnormalities in their expression and function are abundant in common human cancers. As with anti-gene therapy, the difficulty in this approach lies in the delivery of actively expressed vectors to every single tumour cell *in vivo*. Nevertheless, clinical experiments are in progress in lung cancer where retroviruses which encode *TP53* genes are being administered bronchoscopically.

There have been remarkable advances in our understanding of the molecular biology of cancer. Interventional genetics is now poised to provide new, selective, tumour-destruction mechanisms for patients with widespread cancer. *Figure 1.2* shows the clinical trials currently in progress. There are many hurdles still to overcome: how to transfer genes efficiently and stably into tumour cells *in vivo*; how to ensure safety for both patient and staff; and how best to place genetic approaches alongside more familiar therapies. We are witnessing the beginnings of molecular therapy. Fifty years ago the first alkylating agents were discovered and were about to enter clinical trial as systemic chemotherapy. None of our predecessors could have predicted the successes and disappointments that have led to the practice of modern medical oncology. We are now leaving an era of empiricism and entering an age where our knowledge of genetics and logical molecular design is likely to change cancer treatment radically. Despite all the difficulties in using cancer as a model for gene therapy, a comparative analysis of different disorders (*Table 1.5*) illustrates why it has become the most common disease for the gene therapist to tackle.

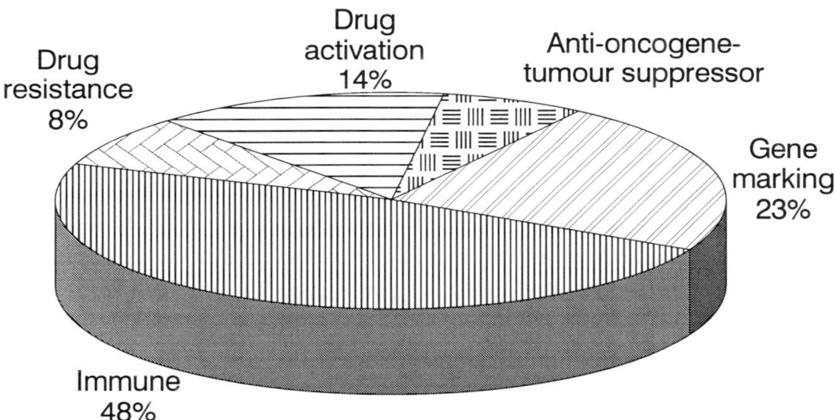

Figure 1.2. Cancer gene therapy: 119 protocols from US and European databases.

Table 1.5. Relative merits of using gene therapy for four target diseases

	SCID	CF	FH	Cancer
Disease common	–	+	–	++++
Life-threatening	++	++	+++	++++
Gene cloned	++++	++++	++++	+
Tight control required	+	+	++	++
Delivery feasible	++++	+	++	+
Surrogate end-point	+++	++++	+++	+
Effects reversible	++++	++	++	++++
Effective therapies	+	++	++	–

SCID, severe combined immunodeficiency; CF, cystic fibrosis; FH, familial hypercholesterolaemia.

1.10 Conclusion

There are many problems yet to be solved before the routine application of gene therapy in clinical practice. Despite the negative views often expressed by critics, the principles of clinical gene transfer have essentially been proven valid. Patient benefit in terms of permanent disease reversion will clearly take much longer to achieve. However, the pace of the genetic age is accelerating. The human genome project; a greater understanding of transcription control; our understanding of gene function from the study of simpler organisms; the use of increasingly sophisticated subtraction techniques to compare DNA and RNA from different cells and tissues; the development of precisely targeting, highly selective vectors; and the uncovering of novel cellular control processes make gene therapy an exciting and promising area of study for the next decade. Furthermore, the dissection of the genetic basis of different diseases will almost certainly result in a much clearer understanding of their molecular pathogenesis, so giving new targets for rational drug design. Classical pharmacology and molecular genetics will become essential partners in the future of therapeutic development.

Tissue-directed gene delivery systems

C.R. Middaugh, M. Chastain and C.T. Caskey

2.1 Introduction

It is generally conceded that an ability to target appropriate DNA constructs spatially will be necessary for gene therapy to provide a practical pharmaceutical approach to disease management. This targeting can occur potentially at a variety of levels within and without cells (*Figure 2.1*). Initially, the aim is to direct a vehicle (e.g. plasmid, plasmid–polymer complex, viral particle) to a particular organ or tissue as well as to individual cell types therein. It is usually thought that subsequent delivery should be aimed at cytoplasmic localization. In many cases, this may involve movement of the gene-containing construct from an endosomal compartment into the cytoplasm since receptor-mediated endocytosis may often be the means by which DNA enters cells. From the cytoplasm, DNA needs to enter the nucleus for transcription to occur. Furthermore, the existence of compartmentalization within the nucleus may require further localization into transcriptionally competent regions. Finally, stable expression may be achieved by integration into the host cell genome, preferably at specifically selected sites. Overall, the question of targeting is complex and we have therefore chosen to focus on targeting vehicles which deliver DNA to specific cell types. The approaches that have been taken are generally based on considerations of endogenous cellular mechanisms of targeting as well as those employed by infectious microorganisms.

There are three distinct approaches that can be envisaged to target DNA to specific tissues and cells. These are: direct injection into a target site; targeting via specific receptor–ligand interactions; and the use of tissue-specific promoters. We shall focus on the first two approaches and only briefly discuss promoters since less has been accomplished in this area. In addition, we shall not review *ex vivo* approaches using tissue harvested from the body and then transfected *in vitro* since we are focusing on the targeting itself. The approaches used to transfect cells *in vitro* are similar to the ones described here.

2.2. Physical delivery into specific tissue

2.2.1 Methods of delivery

The most commonly used method to introduce DNA directly into a tissue employs a conventional needle and syringe. Little shear of DNA is observed for supercoiled

Gene Therapy, edited by N.R. Lemoine and D.N. Cooper.
© 1996 BIOS Scientific Publishers Ltd, Oxford.

plasmids of moderate size (< 10 kb) and injection into multiple sites at relatively well-defined locations can be realized. There has been some suggestion that the damage produced by the needle itself may be important in the transfection process, perhaps mediated by tissue regeneration, but this remains to be conclusively demonstrated. A few attempts have also been made to introduce DNA by needleless injection (Furth *et al.*, 1992). This approach employs some type of gas-driven gun to propel DNA solutions into and through tissue (primarily skin) barriers. Most recently, Vahlsing *et al.* (1994) have

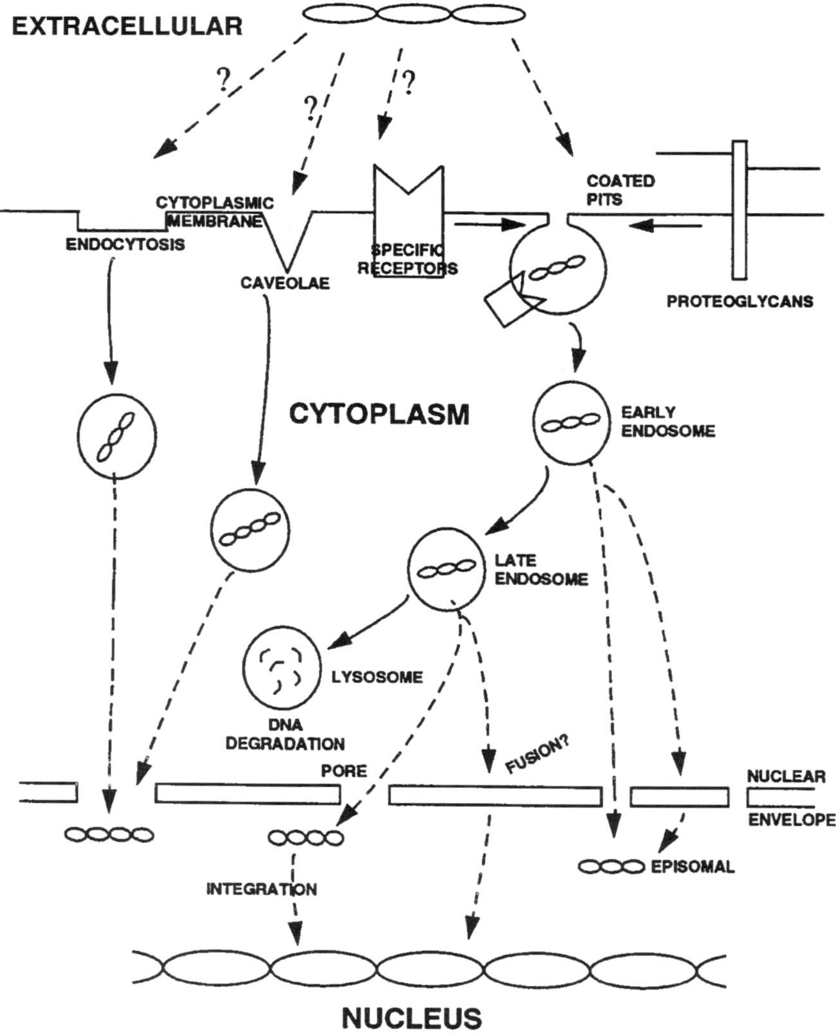

Figure 2.1. Pathways of transport of plasmid DNA into and through cells. In general, the initial recognition site of the DNA or its complexes is unknown, with specific receptors, proteoglycans and caveolae being only a few of many hypothetical targets. Endosomal pathways involving coated pits appear likely in many cases, with entry into the nucleus through pores, direct fusion with the the nuclear envelope, and membrane disruption during cell division all possibilities. Only a few of the potential transport pathways are shown, but the potential complexitiy of the mechanism of DNA transport should be apparent.

demonstrated that supercoiled plasmids can be driven through skin into muscle with significant expression of encoded genes, although some shearing of the plasmid into linear forms could be observed. This method was about one tenth as efficient as needle-mediated injection into muscle (Vahlsing *et al.*, 1994). Nevertheless, potent immune responses to the DNA-encoded influenza A nuclear protein (Vahlsing *et al.*, 1994), the hepatitis B surface antigen (Davis *et al.*, 1994), and a *Cryptosporidium parvum* antigen (Jenkins *et al.*, 1995) have been obtained by this method.

An alternative injection method employs DNA-coated gold particles which are usually prepared by calcium co-precipitation of DNA–gold suspensions (Eisenbraun *et al.*, 1993; Haynes *et al.*, 1996). These are delivered by a 'gene gun' which generates a defined electrostatic discharge event and a subsequent shockwave which mediates delivery. Depth of penetration is a function of complex particle size and the discharge voltage. The efficiency of delivery appears to be controlled primarily by the number of particles introduced and the ratio of the amount of DNA to gold (Eisenbraun *et al.*, 1993). This method produces equivalent transfection at much lower quantities of DNA (two or three orders of magnitude less) than the other direct injection procedures described above (e.g. Fynan *et al.*, 1993). Its use has been primarily limited to skin tissue but more diverse applications are being explored.

In all of the direct injection methods, mechanical energy is employed to overcome the physical barriers to transfection. In fact, at least two other energy sources offer more novel approaches to the direct injection of genes into cells. In electronic pulse delivery (EPD) and related techniques, electromagnetic radiation, usually in the form of well-defined pulses, is used to drive DNA into cells (Zhao, 1995). This technique has so far primarily been used to introduce DNA into isolated cells, but has shown promise in one application to rat brain tissue *in vivo* (Nishi *et al.*, 1996). The method has demonstrated high transfer efficiency (80–90% of cells transfected) with only moderate toxicity (90% cell viability), suggesting further examination is in order. Similarly, ultrasonic radiation has recently been reported to be able to mediate the transdermal delivery of proteins across human skin as well as in living rats (Mitragotri *et al.*, 1995). It seems probable that a similar approach could be used with DNA.

Untargeted intravenous injection of DNA into whole animals results in very little gene expression since 'naked' DNA is rapidly degraded when administered intravenously into an animal (Kawabata *et al.*, 1995). It appears to be primarily eliminated by liver uptake through scavenger receptors on non-parenchymal cells (Kawabata *et al.*, 1995). If plasmid DNA is stabilized by binding to cationic lipids, however, significant expression is seen in many tissues, including vascular endothelial cells, lung, liver, spleen, bone marrow, heart and lymph nodes (Thierry *et al.*, 1995; Zhu *et al.*, 1993). This expression may persist for up to several months (Lew *et al.*, 1995). This route of administration has also been found to result in gene transfer to embryos after intravenous injection of pregnant mice (Tsukamoto *et al.*, 1995). It is possible that systemic delivery could eventually lead to fairly specific tissue targeting since the exact site of injection, the identity of the cationic lipid, and the ratio of lipid to DNA can modulate expression in individual tissues (Philip *et al.*, 1993; Thierry *et al.*, 1995; Zhu *et al.*, 1993).

2.2.2 Direct injection into specific sites

By far, the most commonly employed site of direct injection of plasmid DNA is skeletal muscle. This reflects the initial observations of Wolff *et al.* (1990) who found that direct injection into muscle of DNA and RNA vectors expressing genes for chloramphenicol acetyltransferase (CAT), luciferase (luc) and β-galactosidase (β-gal) resulted in expression comparable to that seen in fibroblasts transfected by the best methods currently available. Although only a small percentage of the total number of cells in a quadriceps muscle and 10–30% of the cells within the injection site appeared to be transfected, this has led to what may turn out to be a revolution in vaccinology. It has now been demonstrated that inclusion of appropriate genes from viruses, bacteria and parasitic organisms, as well as prospective tumour antigens into efficient expression vectors, can be combined with direct injection into muscle to yield both protective humoral and cellular immune responses (e.g. Fynan *et al.*, 1995; Pardoll and Beckerleg, 1995; Waine and McManus, 1995; Wang *et al.*, 1993; Ulmer *et al.*, 1993).

It seems clear that optimization of naked-DNA-associated transgene expression in injected muscle offers one of the more promising approaches to targeted gene therapy. The efficiency of this process is dictated by a number of poorly understood parameters, but the use of hypertonic sucrose to improve distribution of the injected DNA, larger sample volumes, as well as certain local anaesthetics and particular viral and tissue-specific promoters have all been claimed to improve the efficiency of transfection (Davis *et al.*, 1993; Prentice *et al.*, 1994). This phenomenon is not restricted to skeletal muscle, since there are many reports that it works equally well in cardiac muscle (Prentice *et al.*, 1994, 1996). It is clear that problems of inadequate and variable levels of expression will have to be overcome, but its virtues of simplicity and apparent long-term expression make this an extremely attractive approach both for its application to vaccines and gene therapy.

The second most commonly employed site of direct DNA injection is skin. Owing to the ready accessibility of this target, virtually all of the methods mentioned above have been successfully employed, but the gene gun has been especially effective in this regard (Fynan *et al.*, 1993). Again, applications to vaccines have been especially promising since much lower levels of expression appear to be necessary in this application (i.e. the immune system itself supplies internal amplification). In many cases, intradermal immunization appears to produce higher antibody titres than those induced by muscle injection (Fynan *et al.*, 1993; Raz *et al.*, 1994) although the muscle route seems to elicit better cellular immune responses (Ulmer *et al.*, 1994). There is some indication that mucosal routes may also be effective (Fynan *et al.*, 1993) although this has yet to be thoroughly explored.

Balloon catheters have been used to deliver DNA to the endothelial and smooth muscle cells of arteries (reviewed by Nabel *et al.*, 1994). Delivery of genes to arterial walls has promise for treating the proliferation of smooth muscle cells following arterial injury (e.g. restenosis) and for treating limited circulation in the lower extremities. Suicide genes such as that encoding thymidine kinase (tk) can be delivered to halt cell proliferation and growth factors such as vascular endothelial growth factor (VEGF) can be delivered to stimulate the growth of new blood vessels. Delivery of genes to arterial walls has been performed by two approaches using balloon catheters. In the first approach, the balloon is coated with a polymer gel containing the DNA. The balloon is covered in a protective sheath and then routed to the site of delivery. The sheath is removed and the balloon inflated, coating the inside of the vessel with the polymer from which DNA is slowly released (Riessen *et al.*, 1993). The second approach employs two balloons which block the artery while the DNA is released through ports in the catheter between the balloons (Leclerc *et al.*, 1992).

Delivery of DNA to the lungs is also being explored, especially in the search for an effective treatment for cystic fibrosis (CF). Approaches have included the use of naked DNA (Meyer *et al.*, 1995; Tsan *et al.*, 1995), DNA complexed to cationic liposomes (Stribling *et al.*, 1992; Tsan *et al.*, 1995), and permeation enhancers (Freeman and Niven, 1996). Administration by tracheal insufflation or as an aerosol appear to be equally effective, with good expression obtained. Expression appears to be more transient, however, than that observed in skin or muscle. Other sites in which effective direct delivery has been demonstrated include the synovial fluid of joints (Yovandich *et al.*, 1995), the liver (Hickman *et al.*, 1995), tumour nodules (e.g. Nabel *et al.*, 1993) and brain (Abdallah *et al.*, 1995). While not exhaustively reviewing the area of direct introduction of DNA into defined tissues, we can propose a simple generalization. While certain tissues (e.g. muscle) do seem to have properties particularly amenable to this approach, in many if not most cases, it seems DNA is able to enter cells and express encoded genes. A few years ago this would have been a surprising observation and it remains one that is poorly understood. Nevertheless, it serves as the basis for a very simple approach to gene targeting which should see increasing application.

2.2.3 Mechanism of DNA transport into cells

For direct introduction methods to be used confidently, it is essential that some understanding of the mechanism by which DNA enters cells be obtained. Unfortunately, uncomplexed DNA transfects isolated cells in culture with very low efficiency, making this phenomenon very difficult to study experimentally. Although occasional reports have appeared suggesting the presence of specific receptors for DNA on cell surfaces, the ubiquity and specificity of these entities are unclear. It should be noted that bacterial DNA molecules possess multiple copies of short recognition sequences that facilitate DNA uptake (Smith *et al.*, 1995). In *Escherichia coli*, uptake may be mediated by a complex of polyhydroxybutyrate, calcium and inorganic polyphosphate (Castuma *et al.*, 1995). One might postulate the formation of analogous complexes on mammalian cell surfaces with heparan sulphates playing the role of polyphosphate, but there is little evidence for this at present. DNA can be envisaged to enter cells in a number of different ways including through pores, direct fusion, phagocytosis or some type of receptor-mediated endocytosis. Some attempt has been made to address these possibilities in the case of cationic lipid–DNA and calcium phosphate–DNA complexes. Whereas initial studies suggested that calcium phosphate-precipitated DNA enters cells through phagocytosis (Loyter *et al.*, 1982a, b), more recent work suggests that an active endocytotic process may be involved with intermediary endosomes transferring DNA directly to the nucleus (Orrantia and Chang, 1990). Similarly, while cationic lipid–DNA complexes were originally suspected to deliver DNA to cells by lipid-mediated fusion events, more recent studies argue that endocytosis is involved (Felgner *et al.*, 1994; van der Woude *et al.*, 1995, Wu-Pong *et al.*, 1992). However, a pore-mediated delivery cannot yet be excluded. Delivery by either method results in transfection that seems to be primarily limited by the rate of nuclear translocation rather than cell entry (Dowty *et al.*, 1995; Loyter *et al.*, 1982b; Zabner *et al.*, 1995). Whatever the mechanism, it seems probable that interactions between DNA or DNA complexes and cell-surface molecules on target cells are important to DNA entry. Identification of these entities, whether they be protein, lipid or carbohydrate, is a high priority to understand targeting by direct injection methods. In the case of naked DNA injection into muscle, a role for specialized structures such as caveolae and T-tubules has been suggested (Wolff *et al.*, 1992) but this hypothesis requires further investigation.

We conclude this section by noting that the precise nature of the actual transfecting species is unknown. In the case of naked DNA introduced directly into a biological system, the polynucleotide undoubtedly becomes coated with endogenous ions as well as basic proteins. These substances may mediate a targeted receptor-mediated process which is the method most often selected for targeting when employing a more systemic delivery approach. This does not mean, however, that alternative mechanisms cannot be used to enhance DNA uptake into cells. For example, fusion peptides can be complexed to DNA to facilitate translocation across membranes (e.g. Derossi and Prochiantz, 1995; Plank *et al.*, 1994). Similarly, proteins from the surface of numerous microorganisms which enter cells through poorly defined, protein-mediated mechanisms might be employed to the same end. Such artificial vectors are under active development in a number of laboratories, but present significant problems of efficacy, structural complexity and immunogenicity (*Figure 2.2*).

2.3 Ligand–receptor-mediated entry

2.3.1 Non-viral receptor-mediated gene delivery

Introduction. Endocytosis is a collection of processes by which cells bring materials from the external milieu into cellular interiors. Incoming molecules are then distributed through highly regulated but poorly understood mechanisms to different locations such as the cytoplasm, the lysosome, the trans-Golgi network, or back to the cell surface (Trowbridge *et al.*, 1993). The best-characterized form of endocytosis occurs at clathrin-coated pits on the cell surface. Small clathrin-coated vesicles, 100–150 nm in diameter, constitutively invaginate from the plasma membrane, trapping extracellular fluid inside the vesicles. In addition to the extracellular fluid, certain extracellular molecules can bind tightly to specific cell-surface receptor proteins located in coated pit regions. These ligands are then taken up much more efficiently by cells because of their high affinity for cellular receptors. This process is known as receptor-mediated endocytosis. There is also considerable evidence in favour of separate endocytic pathways through uncoated vesicles (Lamaze and Schmid, 1995; Sandvig and van Deurs, 1994), although there remains some question over the existence of these pathways (Watts and Marsh, 1992). One such uncoated vesicle pathway is thought to involve specialized sites on the cell surface known as caveolae. Caveola-rich regions in the membrane are distinguished by their uncoated, invaginated appearance and high concentration of glycosylphosphatidylinositol-linked cell-surface proteins (Anderson, 1993). Another well-characterized endocytic pathway is that of phagocytosis; however, since this process normally occurs only in specialized cell types such as macrophages, we will not consider it further here.

Endocytic vesicles pinch off from the cell surface carrying extracellular fluid and ligands within the vesicles. The vesicles then enter the endosomal pathway, a series of tubular and vesicular membrane structures (Thilo, 1994). Endosomal vesicles fuse with each other and with larger membrane compartments to form intermediate sorting compartments. The pH in these compartments drops, usually dissociating extracellular ligands from their receptors. Receptors are then sorted into vesicles which return to the cell surface, and dissociated ligands are carried by other vesicles to lysosomes where they are degraded to provide raw materials for the cell. There is some controversy over whether vesicles that form at caveolae follow this same pathway or whether they translocate to alternative regions of the cell (Leamon and Low, 1991; Mislick *et al.*, 1995; Turek *et al.*, 1993).

Receptor-mediated endocytosis for gene delivery. Wu and Wu (1987) were the first to employ receptor-mediated endocytosis to deliver plasmids into cells. They used the protein asialoorosomucoid prepared by enzymatic asialyation of the native serum protein to target the asialoglycoprotein receptor, which is uniquely expressed on the surface of liver cells. Asialoorosomucoid was conjugated to polylysine which

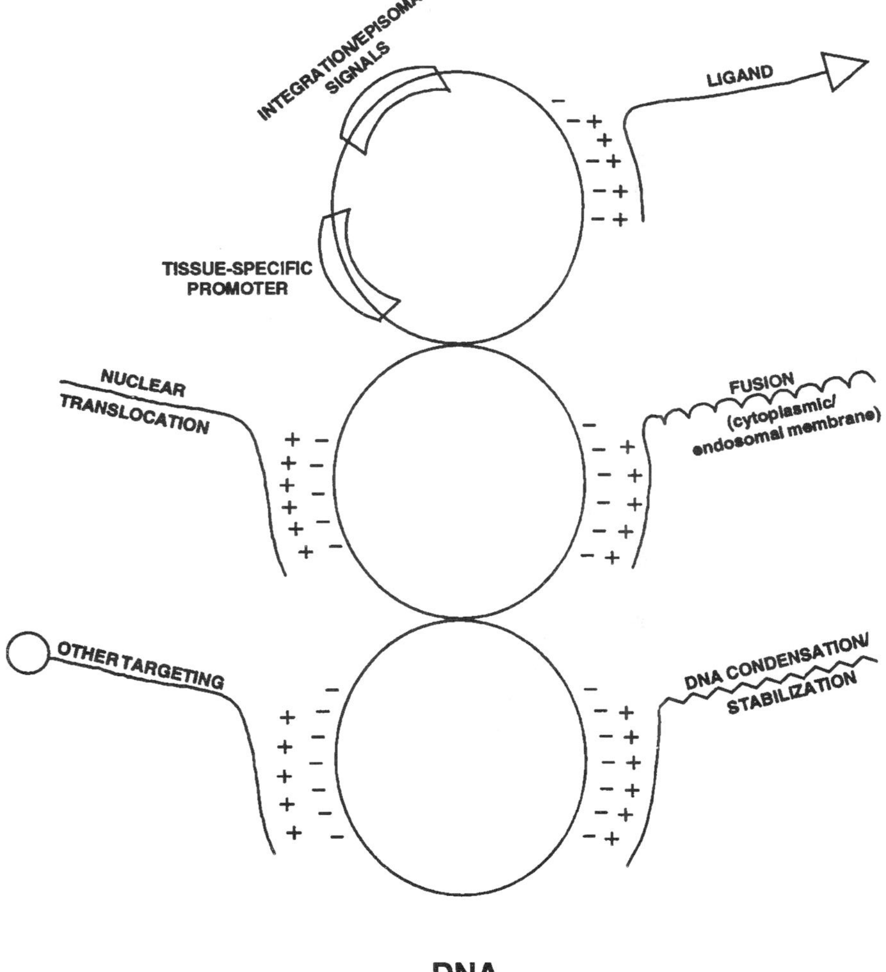

Figure 2.2. A generic, hypothetical, non-viral DNA delivery vehicle. Various molecular entities such as peptides, polysaccharides, synthetic organic polymers or amphiphiles (including lipids) can be complexed to the DNA. The complex is shown to be electrostatic in nature, but other types of interaction can also be employed. It is generally assumed that such complexes, usually with supercoiled DNA, are collapsed into small particles (typically 10–100 nm in diameter).

induced the protein conjugate to bind to plasmid DNA through electrostatic interactions. This conjugate was bound to plasmid DNA encoding the *CAT* reporter gene and added to cells expressing the asialoglycoprotein receptor. This resulted in significant CAT expression in cells expressing the receptor, while expression was undetectable in cells lacking the receptor. Additional evidence that transfection was due to the specific interaction between the asialoorosomucoid and the receptor was shown by the ability of free asialoorosomucoid to prevent gene expression by the protein conjugate complexed to DNA.

Receptor-mediated gene delivery as currently practised. Since the pioneering work of Wu and Wu established that ligands bound to plasmid DNA can effectively deliver DNA to cells, many groups have applied this concept to a variety of ligands and receptors. A representative sampling of these studies is summarized in *Table 2.1*. Rather than discuss in detail individual examples (reviewed by Cotten *et al.*, 1993; Cristiano and Roth, 1995; Findeis *et al.*, 1993; Guy *et al.*, 1995; Michael and Curiel, 1994; Perales *et al.*, 1994a), we will consider this approach generally, with particular attention to the targeting of complexes (*Figure 2.2*) to specific tissues. As indicated above, receptor-mediated

Table 2.1. Receptor-targeted non-viral vectors

Receptor	Tissue	Ligand	Reference
Asialoglycoprotein receptor	Liver	Asialoorosomucoid	Wu and Wu (1987)
Asialoglycoprotein receptor	Liver	Galactosylated polylysine	Perales *et al.* (1994b)
Asialoglycoprotein receptor	Liver	Lactosylated polylysine	Midoux *et al.* (1993)
Transferrin receptor	Lung	Transferrin	Harris *et al.* (1993)
Transferrin receptor	Gastrointestinal	Transferrin epithelium	Batra *et al.* (1994)
Folate receptor	Tumours	Folate	Gottschalk *et al.* (1994)
Insulin receptor	Liver	Insulin	Huckett *et al.* (1990)
Insulin receptor	Liver	Insulin	Rosenkranz *et al.* (1992)
Immunoglobulin receptor	Airway epithelium	Antibody	Ferkol *et al.* (1993, 1995)
Unknown receptor	Liver	Malaria circumsporozoite protein	Ding *et al.* (1995)
Mannose receptor	Macrophages	Mannose	Ferkol *et al.* (1996)
Integrins	Epithelial	RGD peptide	Hart *et al.* (1995)
EGF receptor	Tumour	EGF	Cristiano and Roth (1996)
EGF receptor	Tumour	Antibody	Chen *et al.* (1994)
CD5 receptor	T lymphocytes	Antibody	Merwin *et al.* (1995)
AM Fc receptor	Alveolar macrophages	Antibody	Rojanasakul *et al.* (1994)
Thrombomodulin	Lung endothelial	Antibody	Trubetskoy *et al.* (1992b)
Unknown receptor	Airway epithelium	Pulmonary surfactant protein A	Ross *et al.* (1995)
Unknown receptor	Airway epithelium	Pulmonary surfactant protein B	Baatz *et al.* (1994)
c-kit receptor	Haematopoietic stem cells	Steel factor	Schwarzenberger *et al.* (1996)
Tn antigen	Leukaemic T cells	Antibody	Thurnher *et al.* (1994)

endocytosis delivers ligands to vesicles which normally end up fusing with lysosomes. In the case of gene delivery, this is unacceptable since DNA delivery to lysosomes would result in DNA degradation and consequently no gene expression. The precise mechanism by which some DNA escapes this degradative pathway and results in gene expression is currently not well understood. Curiel *et al.* (1991) first demonstrated that adding adenovirus particles, which are known to disrupt endosomal vesicles, to the DNA–ligand complexes resulted in a 2000-fold enhancement of gene expression. Unfortunately, adding a virus, even an inactivated one, to the gene transfer vehicle creates a variety of problems including potential toxicity and immunogenicity. To minimize these problems, a number of groups have replaced the intact adenovirus particle with various peptides which can disrupt vesicles, usually in a pH-dependent fashion. Plank *et al.* (1994) demonstrated that the ability of various peptides to enhance DNA delivery correlated with their ability to lyse erythrocytes. The best peptides enhance DNA delivery almost as well as intact adenovirus (Plank *et al.*, 1994).

Receptor-mediated targeting has also been applied to cationic liposome-associated delivery of DNA (e.g. Remy *et al.*, 1995; Trubetskoy *et al.*, 1992a). Cationic lipid-induced delivery of DNA is highly variable with regard to cell specificity both *in vitro* and *in vivo* (reviewed in Chapter 6). It is thought to be mediated by the excess of positive charge in the DNA–lipid complexes which permit binding to the anionic surface of cells followed by endocytosis of the complexes. Adding ligands to the surfaces of the DNA–lipid complexes has been shown to target the complexes to specific cell types, but results in less efficient transfection than traditional cationic lipids without targeting ligands (Remy *et al.*, 1995). It remains to be seen whether ligands can be added to cationic lipids in such a way as to give more efficient cell-specific transfection agents. Furthermore, the possibility that cationic lipid–DNA complexes bind endogenous proteins or other substances upon their introduction into biological systems that ultimately provide some type of ligand-specificity remains to be explored.

In vivo results. Receptor-mediated gene transfer complexes are very efficient transfection agents *in vitro*, resulting in >90% of cells being transfected (Cristiano *et al.*, 1993; Ledley, 1995). *In vivo*, their success has been more limited. Several groups have found low-level, short-lived expression *in vivo*, usually lasting no more than 3 or 4 days (Chowdhury *et al.*, 1993; Gao *et al.*, 1993; Wilson *et al.*, 1992; Wu *et al.*, 1989). Even the low level and short-term gene expression typically seen in these studies was sufficient to give therapeutic benefits in a rabbit model of hypercholesterolaemia (Wilson *et al.*, 1992). Much longer *in vivo* gene expression, up to 140 days, has been found after injection of a complex encoding the factor IX gene targeting the asialoglycoprotein receptor on liver cells (Perales *et al.*, 1994b). Neither adenovirus nor other agents were used to enhance release from endosomes in this study. Perales *et al.* (1994b) did find that preparing the DNA–polylysine complexes in high salt while vortexing resulted in the formation of very small particles 10–12 nm in diameter. Previous methods of preparation have generally resulted in much larger (80–100 nm) particles (Cristiano *et al.*, 1993; Wagner *et al.*, 1991). The persistent gene expression seen *in vivo* with the smaller complexes suggests that receptor-mediated gene delivery could achieve real therapeutic benefit and raises the important question of the relationships between the size and properties of the complexed DNA and its ability to enter cells.

Size of particles. Unfortunately, there have been no systematic studies of the effect of particle size on the efficiency of gene delivery to various tissues. Several groups have claimed that 80–100 nm particles are normally formed when ligand–polylysine conjugates bind to DNA, and that this size is compatible with the available volume of endocytic vesicles (Cristiano *et al.*, 1993; Wagner *et al.*, 1991). As described in the previous section, however, Perales *et al.* (1994b) found that condensing the particles to a diameter of 10–12 nm enhanced uptake by the liver *in vivo*. Previous work has shown that the size of the particles can have a dramatic effect on cellular uptake in the liver (Bijsterbosch and Van Berkel, 1991; Bijsterbosch *et al.*, 1989; Schlepper-Schafer *et al.*, 1986). Schlepper-Schafer *et al.* (1986) used gold particles coated with glycoproteins to target the asialoglycoprotein receptor. Beads up to about 10 nm in size were removed by hepatocytes, but larger beads were only taken up by macrophages and Kupffer cells. Similar results were obtained using lipoprotein complexes of varying sizes (Bijsterbosch and Van Berkel, 1991; Bijsterbosch *et al.*, 1989). Furthermore, it was found that the density of ligands on the particle could influence the site of uptake in addition to particle size (Bijsterbosch and Van Berkel, 1990). *In toto*, these studies strongly suggest that the size of the DNA-containing particles can have a large effect on the efficiency of delivery. More detailed studies need to address the relationship between the size of the complex and the efficiency of gene delivery in a variety of tissues and receptors. For example, two studies using the smaller 10–12 nm particles that resulted in persistent gene expression in the liver did not give persistent expression when targeting macrophages or airway epithelial cells (Ferkol *et al.*, 1995, 1996).

Targeting cell-surface proteins. What kind of cellular receptor should one target when designing a receptor-mediated gene delivery agent? Although to some extent this must be dictated by the molecules present on the cells to be targeted, very little of the research needed to answer this question has yet been performed. Most of the cell-surface proteins which have so far been selected for receptor-mediated gene delivery have been actively recycling receptors that associate with coated pits. Such proteins get taken up into endosomes much more frequently than most cell-surface proteins. Actively internalized receptors, however, are just a subset of available proteins on cell surfaces (reviewed by Roth, 1993). Some proteins are anchored to the cytoskeleton and rarely, if ever, undergo endocytosis. A second class appear to be specifically excluded from coated pits and are taken up approximately 100 times less often than bulk membrane. A third class of surface proteins are endocytosed at the same rate as bulk membrane, a temporal process resulting in translocation of about 1% of the cell-surface protein per minute (Almond and Eidels, 1994; Roth, 1993). By comparison, proteins which preferentially localize to coated pits are taken up at a rate of about half that of the cell-surface protein per minute (Brown and Goldstein, 1986; Rothenberger *et al.*, 1987). Receptors associated with coated pits contain an internalization signal which is composed of a β-turn containing an exposed tyrosine within the cytoplasmic domain of the protein (reviewed by Roth, 1993; Trowbridge *et al.*, 1993). Transfer of the internalization sequence to cell-surface proteins which normally internalize slowly greatly increases the rate at which they are internalized (Collawn *et al.*, 1991).

The hypothesis that rapidly internalized cell-surface proteins would make the best receptors seems reasonable, but it has not been tested. In fact, a study which examined the toxicity of diphtheria toxin targeted either to an actively recycling or non-recycling receptor found that the toxin exhibited greater toxicity when added to cells expressing the non-recycling receptor (Almond and Eidels, 1995). More extensive studies of this type are crucial to obtaining an understanding of the basic cell biology involved in optimal receptor-targeted gene delivery. One of the most direct ways to target a cell-surface protein is to employ an antibody which recognizes it (see *Table 2.1*). However, if only cell-surface proteins which contain an internalization signal efficiently deliver DNA, it will be necessary to identify appropriate tissue-specific cell-surface proteins which contain internalization signals before generating antibodies.

An additional issue concerning the choice of a receptor is that no studies have been performed to determine how the kinetic and thermodynamic properties of receptor–ligand binding influence DNA delivery. It seems reasonable to postulate that high-affinity ligand–receptor systems would be optimal, but a ligand which binds too tightly to its receptor might not dissociate from the receptor in the endosome and thus the complex could be unproductively returned to the cell exterior. Thus, it might be better for a ligand to have a more rapid 'on-rate' for the receptor than for it to have a large equilibrium binding constant. A more rapid on-rate should reduce removal of the complex by circulatory processes (Williams, 1991), and a higher 'off-rate' could enhance the release of uncomplexed intracellular DNA.

Once a receptor has been chosen for targeting, either its natural ligand or an antibody to the receptor can be conjugated to polylysine or some other polycation or coupling agent to deliver DNA. More often, however, a unique receptor in the tissue of interest is unknown. A powerful new approach to finding ligands which bind to specific cell types has recently been described (Barry *et al.*, 1996; Pasqualini and Ruoslahti, 1996). The cells of interest are used to select phage containing unique peptide sequences from a phage display library. Several rounds of selection and amplification can be performed resulting in peptide sequences which bind specifically to the cell type of interest. Negative selection can be included to ensure that the peptide does not bind to other cell types.

Prospects for the future. Receptor-mediated gene delivery is probably the most promising non-viral delivery technique for gene therapy, considering its combination of high transfection efficiency, tissue-specific targeting, and presumably limited immunogenicity. However, many fundamental studies still need to be undertaken to make this a practical technique. As mentioned above, the size of the particles and the effect of size on *in vivo* delivery needs to be established. Particle size can potentially be manipulated by the nature of the complexing process and solution conditions in which they are prepared. In addition, the types and mode of entry of cell-surface proteins which can be effectively targeted to deliver DNA needs to be further explored.

The efficiency with which DNA can be delivered could be augmented in several ways. As discussed above, agents which disrupt endosomes greatly increase the efficiency of gene transfer. It is quite possible that more potent endosomal release agents than the limited selection of peptides so far examined can be developed. The presence of a nuclear localization signal might also increase the efficiency with which DNA reaches the nucleus after it escapes from endosomes into the cytoplasm. Furthermore, it might be possible to add signals which would deliver the DNA to transcriptionally active regions of the nucleus (Bregman *et al.*, 1995). Efficiency could also be enhanced by delivering DNA which could actively replicate in dividing cells either maintained as an episome or stably integrated into chromosomes. Delivering DNA which will persist in cells raises many safety issues but it may be necessary to correct certain types of genetic diseases. The basic concept of the flexible gene delivery vehicle is illustrated in *Figure 2.2*.

A final concern about the use of this method for gene therapy in humans is the possibility that the complexes used will be immunogenic. If these complexes do generate a strong immune response, this will clearly limit their effectiveness upon repeat administration. In addition, they might induce anti-DNA antibodies. A precedent for this possibility comes from studies in mice using either peptide–DNA complexes (Desai *et al.*, 1993) or simian virus 40 (SV40) T antigen bound to DNA (Moens *et al.*, 1995). In both cases, the DNA–protein complexes induced autoantibodies against DNA. Kidney pathology associated with the presence of such complexes is a possibility.

2.3.2 Receptor-mediated viral gene delivery

Most viruses bind to a specific receptor or receptors on the surface of cells and are subsequently brought into the cell through endocytosis at clathrin-coated pits (reviewed by Marsh and Helenius, 1989; Weiss and Tailor, 1995). The two viral vectors used most widely for gene therapy, adenovirus (reviewed in Chapter 4) and murine leukaemia virus (MLV), infect a wide variety of cell types. Changing the proteins on the surface of these viruses might allow the tissue tropism of a given virus to be modified in order to deliver genes to particular tissues. Presumably, it will be more difficult to change the tropism of a virus than to engineer tissue specificity into non-viral gene vectors as described in Section 2.3.1, since modifying the outer surface proteins of a virus could also interfere with packaging or uncoating and result in a non-functional vector.

It may be easier to change the surface proteins of enveloped viruses such as retroviruses than non-enveloped particles like adenovirus. It has been known for some time that most enveloped viruses will incorporate surface proteins from other enveloped viruses if two different viruses infect the same cell (Zavada, 1982). Thus, it is at least in principle possible to replace a retroviral envelope protein with an envelope protein of another virus to achieve tissue specificity. Altering a retroviral envelope has generally been performed to alter the species specificity of the virus, rather than its tissue specificity. Replacing retroviral envelope proteins with other viral proteins has also been shown to enhance several properties of the resultant vectors. One example is the incorporation of the vesicular stomatitis virus (VSV) G glycoprotein into retroviral vectors; this increased the titre from the normal range of 10^5–10^6 (a significant limitation in achieving high levels of infection *in vivo*) to > 10^9 (Burns *et al.*, 1993). A second example is the modification of a human immunodeficiency virus (HIV) vector which uses the MLV envelope protein and is capable of infecting non-dividing cells unlike other retroviral vectors in use for gene therapy (Naldini *et al.*, 1996).

Instead of replacing retroviral envelope proteins with another viral protein, it may be possible to incorporate a host protein into the virus which would confer tissue specificity to the resulting altered virus. Although many cellular proteins are actively excluded from the virion membrane (Zavada, 1982), there are examples of cellular membrane proteins incorporated into virus membranes. The cellular proteins β_2-microglobulin and human leukocyte antigen (HLA) DR are present at very high amounts in HIV virions (Arthur *et al.*, 1992). Young *et al.* (1990) demonstrated the feasibility of incorporating novel cellular proteins into virions. They found that human CD4 is efficiently incorporated into avian leukaemia virus (ALV) when both are expressed in quail cells. The use of such systems for targeting viruses to specific cell types has also been demonstrated. Matano *et al.* (1995) incorporated CD4–envelope chimeras into MLV virions (in addition to the wild-type MLV envelope proteins) and found that such virions could specifically infect HIV-infected human cells. In general, it may be possible to introduce targeting peptide ligands into retroviruses by modifying cellular proteins which are incorporated into the virion or by adding modified envelope proteins in addition to wild-type surface molecules.

The most success in modifying retroviruses to target specific tissue types has been accomplished by inserting peptide ligands into wild-type retroviral envelope proteins. The first demonstration that the receptor specificity of viruses could be altered was performed by inserting a 16 amino acid sequence specific for integrin receptors at different positions in the envelope protein of ALV and showing that one of the resulting virus stocks was capable of infecting a cell line resistant to infection by the wild-type virus (Valsesia-Wittmann *et al.*, 1994). Since then, a number of groups have modified retroviruses in similar ways by inserting antibodies as well as peptide ligands (summarized in *Table 2.2*). The difficulty with this approach is that there are no general rules for the types of insertions that might be tolerated nor are the best sites within the envelope into which to insert peptide sequences known. It has been shown, however, that for ligand insertions at the amino terminus of the envelope protein, optimizing the spacing between the envelope and the ligand can increase infection efficiency by up to 100 times (Valsesia-Wittmann *et al.*, 1996). In general, the titres of the viruses bearing modified envelope proteins are lower than those of wild-type retroviruses, which are themselves relatively low.

There has been much more work published on changing the surface proteins of retroviruses to achieve tissue specificity than on analogous alterations of other

Table 2.2. Receptor-targeted viral vectors

Virus	Ligand	Receptor	Reference
MLV	Erythropoietin	Erythropoietin receptor	Kasahara *et al.* (1994)
ALV	RGD peptide	Integrins	Valsesia-Wittmann *et al.* (1994)
MLV	Heregulin	HER-2	Nan *et al.* (1995
Spleen necrosis virus	Antibody to DNP	DNP conjugated to surface	Chu *et al.* (1994)
MLV	Antibody	LDL receptor	Somia *et al.* (1995)
MLV	Lactose	Asialoglycoprotein receptor	Neda *et al.* (1991)
MLV	Antibody to MHC I	MHC I	Roux *et al.* (1989)
MLV	Antibody to EGF receptor	EGF receptor	Etienne-Julan *et al.* (1992)
MLV	Antibody to insulin receptor	Insulin receptor	Etienne-Julan *et al.* (1992)
MLV	Amphotrophic envelope	Ram-1 phosphate transporter	Cosset *et al.* (1995)
MLV	EGF	EGF receptor	Cosset *et al.* (1995)
Spleen necrosis virus	Single chain antibody	Carcinoma antigen	Chu and Dornburg (1995)
MLV	CD4-envelope chimera	HIV envelope (HIV-infected cells)	Matano *et al.* (1995)

viruses. There is, however, significant research in progress involving modification of the tissue specificity of adenovirus by altering either the fibre or penton capsid proteins. The tissue specificity of the adenovirus penton base protein was changed by replacing an RGD with an LDV sequence. This changed the binding specificity of the penton protein from endothelial and epithelial cells expressing the $\alpha_v\beta_3$ and $\alpha_v\beta_5$ integrins to lymphocyte and monocyte cells expressing the $\alpha_4\beta_1$ integrin (Wickham *et al.*, 1995). Whether this strategy will result in a virus with altered tissue tropism is an unanswered question since no results on viruses containing the mutated penton protein have yet been reported.

Viruses can also be targeted to specific tissues by other modifications of their surface. Lactose residues were added to the surface of MLV, resulting in a virus which specifically infected cells expressing the asialoglycoprotein receptor (Neda *et al.*, 1991). In addition to modifying the surface of the virus chemically, combinations of biotinylated antibodies which recognize cell-surface proteins and viral surface proteins linked by streptavidin were used to bind viruses to specific cell types, resulting in low levels of infection (Etienne-Julan *et al.*, 1992; Roux *et al.*,1989). In these studies, major histocompatibility complexes (MHCs) I and II, the epidermal growth factor (EGF) receptor and the insulin receptor all served as targets for viral entry, whereas the transferrin, high-density lipoprotein and galactose receptors were unsuccessful (Etienne-Julan *et al.*, 1992; Roux *et al.*, 1989).

The hypothesis that any complementary receptor–ligand pair of proteins on the surface of a virus and a cell will lead to infection seems unlikely. Various aspects of the specific geometry of the interaction or of induced conformational changes as a result of the interaction may well be important to the recognition and cell entry process. On the other hand, in the case of transcriptional activation in the yeast two-hybrid system, binding seems sufficient in itself and relative geometry does not seem required for biological activity (Fields and Song, 1989). It will probably not be this simple for virus infection. One example of a modified virus which can bind but does not lead to productive infection is a retrovirus containing 53 amino acids of EGF at the amino terminus of the envelope protein. This virus binds to cells expressing the EGF receptor, but is not infectious (Cosset *et al.*, 1995). A surprising number of modified viruses are capable of infecting their target cell types, as shown by the examples in *Table 2.2*, but no general rules for how to construct such viruses have yet emerged.

2.4 Tissue-specific expression

Tissue specificity can also be achieved by non-specific delivery of DNA to tissues followed by tissue-specific expression (reviewed by Hart, 1996). Tissue-specific expression of genes is controlled by promoter and enhancer elements. These have been extensively characterized in a wide variety of tissues (reviewed by Dillon, 1993). Tissue-specific expression could in principle be combined with many of the techniques described above, such as direct injection into specific tissues or receptor-mediated cell entry to achieve maximal levels of tissue-specific expression. Restricting expression to desired tissues could be particularly important in several gene therapy applications. For example, the success of delivering suicide genes to tumour cells will depend critically on expression of those genes exclusively in tumour cells and not in normal tissues because of potential toxicity problems (Vile, 1994). Furthermore, in gene delivery applications to tissues which undergo further differentiation such as haematopoietic cells, specific promoters could restrict gene expression to cells in a particular stage of differentiation.

Tissue-specific promoters have most commonly been used with retroviral delivery systems, largely because control of expression is much more important when the gene is integrated into host chromosomes and consequently could be expressed for very long periods of time. Retroviral delivery potentially creates a problem with genes under control of tissue-specific promoters since the viral long terminal repeat (LTR) promoter often interferes with internal promoters (Emerman and Temin, 1986). An additional difficulty with retroviral delivery is that expression can be influenced by the position of integration of the provirus in the genome. DNA regions which insulate promoters from position effects have been identified and are termed locus control regions (Grosveld *et al.*, 1987). These regions are an area of significant research, but they are often too large to be practically incorporated into retroviral vectors. Gene delivery using tissue-specific promoters has been attempted in a wide variety of tissues including lung, muscle, liver, neurons and haematopoietic cells (e.g. Bauer *et al.*, 1994; Cheng *et al.*, 1993; Couture *et al.*, 1994; Einerhand *et al.*, 1995; Ferrari *et al.*, 1995; Huber and Richards, 1996; Vile, 1994; Vile *et al.*, 1994). So far, use of tissue-specific promoters has often led to only low levels of expression *in vivo*. Continued research should improve expression from such promoters as well as increase the amount of DNA which can be delivered, making it easier to introduce larger, more sophisticated DNA promoter and enhancer sequences.

Another strategy to obtain tissue-specific gene expression is to employ drug-regulated gene expression. This strategy uses genetically engineered promoters and transcriptional activators (Baim *et al.*, 1991; Delort and Capecchi, 1996; Wang *et al.*, 1994). The most promising example of such a system employs the tetracycline repressor protein from *E. coli* which can be mutated so that it binds to DNA only in the presence of tetracycline (Gossen *et al.*, 1995; Shockett *et al.*, 1995). This mutated protein is fused to the transcriptional activation domain VP16 from herpes simplex virus (HSV). In the presence of tetracycline, the fusion protein binds to DNA and activates transcription 1000-fold over the level seen in the absence of the drug. The major problem with systems of this type is that three agents have to be delivered to cells – the gene of therapeutic benefit, the transcriptional activation gene, and the drug used to induce transcription.

2.5 Conclusion

It is already possible to obtain significant tissue-specific expression using a variety of different approaches. Our ability to mimic the many lessons nature has provided in this regard has resulted in several promising approaches to obtain specificity. Nevertheless, this aspect of gene therapy must still be considered to be in its infancy. While the basic ideas outlined above are clearly technically feasible, there is still a major gap between the effects seen and application of these methods to pharmaceutically practical gene therapy. The following three major issues need to be much more extensively addressed.

(i) As specificity is increased, this is often at the expense of increased complexity. For example, targeting by antibodies or antibody fragments is feasible, but the need for this recombinantly reagent greatly increases the cost, difficulty of manufacture and analytical complexity of any gene delivery vehicle containing such a protein. The ultimate extent of this problem should not be underestimated.

Although cost and complexity may be acceptable for a limited number of rare genetic disorders, applications of gene therapy to more widespread disorders such as heart disease, diabetes or cancer, or use in vaccination, will be limited if relatively simple solutions to the targeting problem are not found.

(ii) As the DNA-delivery vehicle becomes more complex, potential problems of immunogenicity are enhanced. This has already led to serious problems with the attempted use of viral vectors in gene therapy. One of the major potential advantages of non-viral methods has always been thought to be their low immunogenicity. Specificity at the expense of increased immunogenicity is obviously problematic. Note, however, that targeting ligands of endogenous origin are at least in principle of low immunogenicity and merit continued investigation. As discussed previously (Section 2.3.1), the presence of additional components, especially peptides or proteins, could also enhance an anti-DNA immune response with possible negative consequences. Viruses have evolved many defenses against the immune system (Barinaga, 1992; Katze, 1993), and several viral proteins are promising candidates to mitigate immune responses to gene delivery vehicles (Fruh *et al.*, 1995; Hill *et al.*, 1995; Levitskaya *et al.*, 1995).

(iii) The use of tissue-specific expression implies a degree of understanding of both the biology of the pathological process as well as that of the normal tissue that, in fact, is often lacking. The more we know about the basic biochemistry involved, the better the chance that targeting the tissue will have potential consequences reasonably anticipated. In other words, we can target genes to tissues of interest, but to do it in a safe, pharmaceutically acceptable and efficacious manner will require many more years of basic research and development before real clinical success is achieved.

References

Abdallah B, Sachs L, Demeneix BA. (1995) Non-viral gene transfer: applications in developmental biology and gene therapy. *Biol. Cell* 85: 1–7.

Almond BD, Eidels L. (1994) The cytoplasmic domain of the diptheria toxin receptor (HB-EGF precursor) is not required for receptor-mediated endocytosis. *J. Biol. Chem.* 269: 26635–26651.

Almond BD, Eidels L. (1995) The effect of receptor rapid-internalization signals on diptheria toxin endocytosis and cell sensitivity. *Mol. Microbiol.* 18: 623–630.

Anderson RGW. (1993) Caveolae: where incoming and outgoing messengers meet. *Proc. Natl Acad. Sci. USA* 90: 10909–10913.

Arthur LO, Bess JW, Jr, Sowder RC, II, Benveniste RE, Mann DL, Chermann JC, Henderson LE. (1992) Cellular proteins bound to immunodeficiency viruses: implications for pathogenesis and vaccines. *Science* 258: 1935–1938.

Baatz JE, Bruno MD, Ciraolo PJ, Glasser, SW, Stripp BR, Smyth KL, Korfhagen TR. (1994) Utilization of modified surfactant-associated protein B for delivery of DNA to airway cells in culture. *Proc. Natl Acad. Sci. USA* 91: 2547–2551.

Baim SB, Labow MA, Levine AJ, Shenk T. (1991) A chimeric mammalian transactivator based on the lac repressor that is regulated by temperature and isopropyl-β-thiogalactopyranoside. *Proc. Natl Acad. Sci. USA* 88: 5072–5076.

Barinaga M. (1992) Viruses launch their own 'star wars'. *Science* 258: 1730–1731.

Barry MA, Dower WJ, Johnston SA. (1996) Toward cell-targeting gene therapy vectors: selection of cell-binding peptides from random peptide-presenting phage libraries. *Nature Med.* 2: 299–305.

Batra RK, Berschneider H, Curiel DT. (1994) Molecular conjugate vectors mediate efficient gene transfer into gastrointestinal epithelial cells. *Cancer Gene Ther.* 1: 185–192.

Bauer TR, Jr, Osborne WRA, Kwok WW, Hickstein DD. (1994) Expression from leukocyte integrin promoters in retroviral vectors. *Hum. Gene Ther.* 5: 709–716.

Bijsterbosch MK, Van Berkel TJC. (1990) Uptake of lactosylated low-density lipoprotein by galactose-specific receptors in rat liver. *Biochem. J.* **270**: 233–239.

Bijsterbosch MK, Van Berkel TJC. (1991) Lactosylated high density lipoprotein: a potential carrier for the site-specific delivery of drugs to parenchymal liver cells. *Mol. Pharmacol.* **41**: 404–411.

Bijsterbosch MK, Ziere GJ, Van Berkel TJC. (1989) Lactosylated low density lipoprotein: a potential carrier for the site-specific delivery of drugs to Kupffer cells. *Mol. Pharmacol.* **36**: 484–486.

Bregman DB, Du L, van der Zee S, Warren SL. (1995) Transcription-dependent redistribution of the large subunit of RNA polymerase II to discrete nuclear domains. *J. Cell Biol.* **129**: 287–298.

Brown MS, Goldstein JL. (1986) A receptor-mediated pathway for cholesterol homeostasis. *Science* **232**: 34–47.

Burns JC, Fiedmann T, Driever W, Burrascano M, Yee JK. (1993) Vesicular stomatitis virus G glycoprotein pseudotyped retroviral vectors: concentration to very high titer and efficient gene transfer into mammalian and nonmammalian cells. *Proc. Natl Acad. Sci. USA* **90**: 8033–8037.

Castuma CE, Huang R, Kornberg A, Reusch RN. (1995) Inorganic polyphosphates in the acquisition of competence in *Eschericia coli*. *J. Biol. Chem.* **270**: 12980–12983.

Chen J, Gamou S, Takayanagi A, Shimizu N. (1994) A novel gene delivery system using EGF receptor-mediated endocytosis. *FEBS Lett.* **338**: 167–169.

Cheng L, Ziegelhoffer PR, Yang NS. (1993) *In vivo* promoter activity and transgene expression in mammalian somatic tissues evaluated by using particle bombardment. *Proc. Natl Acad. Sci. USA* **90**: 4455–4459.

Chowdhury NR, Wu CH, Wu GY, Yerneni PC, Bommineni VR, Chowdhury JR. (1993) Fate of DNA targeted to the liver by asialoglycoprotein receptor-mediated endocytosis *in vivo*. *J. Biol. Chem.* **268**: 11265–11271.

Chu THT, Dornburg R. (1995) Retroviral vector particles displaying the antigen-binding site of an antibody enable cell-type-specific gene transfer. *J. Virol.* **69**: 2659–2663.

Chu THT, Martinez I, Sheay WC, Dornburg R. (1994) Cell targeting with retroviral vector particles containing antibody-envelope fusion proteins. *Gene Ther.* **1**: 292–299.

Collawn JF, Kuhn LA, Liu LFS, Tainer JA, Trowbridge, IS. (1991) Transplanted LDL and mannose-6-phosphate receptor internalization signals promote high-efficiency endocytosis of the transferrin receptor. *EMBO J.* **10**: 3247–3253.

Cosset FL, Morling FJ, Takeuchi Y, Weiss RA, Collins MKL, Russell SJ. (1995) Retroviral retargeting by envelopes expressing an N-terminal binding domain. *J. Virol.* **69**: 6314–6322.

Cotten M, Wagner E, Birnstiel ML. (1993) Receptor-mediated transport of DNA into eukaryotic cells. *Methods Enzym.* **217**: 618–644.

Couture LA, Mullen CA, Morgan RA. (1994) Retroviral vectors containing chimeric promoter/enhancer elements exhibit cell-type-specific gene expression. *Hum. Gene Ther.* **5**: 667–677.

Cristiano RJ, Roth JA. (1995) Molecular conjugates: a targeted gene delivery vector for molecular medicine. *J. Mol. Med.* **73**: 479–486.

Cristiano RJ, Roth JA. (1996) Epidermal growth factor mediated DNA delivery into lung cancer cells via the epidermal growth factor receptor. *Cancer Gene Ther.* **3**: 4–10.

Cristiano RJ, Smith LC, Kay MA, Brinkley BR, Woo SLC. (1993) Hepatic gene therapy: efficient gene delivery and expression in primary hepatocytes utilizing a conjugated adenovirus–DNA complex. *Proc. Natl Acad. Sci. USA* **90**: 11548–11552.

Curiel DT, Agarwal S, Wagner E, Cotten M. (1991) Adenovirus enhancement of transferrin–polylysine-mediated gene delivery. *Proc. Natl Acad. Sci. USA* **88**: 8850–8854.

Davis HL, Whalen RG, Demeneix BA. (1993) Direct gene transfer into skeletal muscle *in vivo*: factors affecting efficiency of transfer and stability of expression. *Hum. Gene Ther.* **4**: 151–159.

Davis HL, Michel ML, Mancini M, Schleef M, Whalen RG. (1994) Direct gene transfer in skeletal muscle: plasmid DNA-based immunization against the hepatitis B virus surface antigen. *Vaccine* **12**: 1503–1507.

Delort JP, Capecchi MR. (1996) TAXI/UAS: A molecular switch to control expression of genes *in vivo*. *Hum. Gene Ther.* **7**: 809–820.

Derossi D, Prochiantz A. (1995) Internalization of macromolecules by live cells. *Restor. Neurol. Neurosci.* **8**: 7–10.

Desai DD, Krishnan MR, Swindle JT, Marion, TN. (1993) Antigen-specific induction of antibodies against native mammalian DNA in nonautoimmune mice. *J. Immunol.* **151**: 1614–1626.

Dillon N. (1993) Regulating gene expression in gene therapy. *Trends Biotechnol.* **11**: 167–173.

Ding ZM, Cristiano RJ, Roth JA, Takacs B, Kuo MT. (1995) Malarial circumsporozoite protein is a novel gene delivery vehicle to primary hepatocyte cultures and cultured cells. *J. Biol. Chem.* **270**: 3667–3676.

Dowty ME, Williams P, Zhang G, Hagstrom JE, Wolff JA. (1995) Plasmid DNA entry into postmitotic nuclei of primary rat myotubes. *Proc. Natl Acad. Sci. USA* **92**: 4572–4576.

Einerhand MPW, Antoniou M, Zolotukhin S, Muzyczka N, Berns KI, Grosveld F, Valerio D. (1995) Regulated high-level human β-globin gene expression in erythroid cells following recombinant adeno-associated virus-mediated gene transfer. *Gene Ther.* **2**: 336–343.

Eisenbraun MD, Fuller DH, Haynes JR. (1993) Examination of parameters affecting the elicitation of humoral immune responses by particle bombardment-mediated genetic immunization. *DNA Cell Biol.* **12**: 791–797.

Emerman M, Temin HM. (1986) Comparison of promoter suppression in avian and murine retrovirus vectors. *Nucleic. Acids Res.* **14**: 9381–9386.

Etienne-Julan M, Roux P, Carillo S, Jeanteur P, Piechaczyk M. (1992) The efficiency of cell targeting by recombinant retroviruses depends on the nature of the receptor and the composition of the artificial cell-virus linker. *J. Gen. Virol.* **73**: 3251–3255.

Felgner JH, Kumar R, Sridhar CN, Wheeler CJ, Tsai YJ, Border R, Ramsey P, Martin M, Felgner PL. (1994) Enhanced gene delivery and mechanism studies with a novel series of cationic lipid formulations. *J. Biol. Chem.* **269**: 2550–2561.

Ferkol T, Kaetzel CS, Davis PB. (1993) Gene transfer into respiratory epithelial cells by targeting the polymeric immunoglobulin receptor. *J. Clin. Invest.* **92**: 2394–2400.

Ferkol T, Perales JC, Eckman E, Kaetzel CS, Hanson RW, Davis PB. (1995) Gene transfer into the airway epithelium of animals by targeting the polymeric immunoglobulin receptor. *J. Clin. Invest.* **95**: 493–502.

Ferkol T, Perales JC, Mularo F, Hanson RW. (1996) Receptor-mediated gene transfer into macrophages. *Proc. Natl Acad. Sci. USA* **93**: 101–105.

Ferrari G, Salvatori G, Rossi C, Cossu G, Mavilio F. (1995) A retroviral vector containing a muscle-specific enhancer drives gene expression only in differentiated muscle fibers. *Hum. Gene Ther.* **6**: 733–742.

Fields S, Song O. (1989) A novel genetic system to detect protein–protein interactions. *Nature* **340**: 245–246.

Findeis MA, Merwin JR, Spitalny GL, Chiou HC. (1993) Targeted delivery of DNA for gene therapy via receptors. *Trends Biotechnol.* **11**: 202–205.

Freeman DJ, Niven RW. (1996) The influence of sodium glycocholate and other additives on the *in vivo* transfection of plasmid DNA in the lungs. *Pharm. Res.* **13**: 202–209.

Fruh K, Ahn K, Djaballah H, Sempe P, van Endert PM, Tampe R, Peterson PA, Yang Y. (1995) A viral inhibitor of peptide transporters for antigen presentation. *Nature* **375**: 415–418.

Furth PA, Shamay A, Wall RJ, Hennighausen L. (1992) Gene transfer into somatic tissues by jet injection. *Anal. Biochem.* **205**: 365–368.

Fynan EF, Webster RG, Fuller DH, Haynes JR, Santoro JC, Robinson HL. (1993) DNA vaccines: protective immunizations by parenteral, mucosal, and gene-gun inoculations. *Proc. Natl Acad. Sci. USA* **90**: 11478–11482.

Fynan EF, Webster RG, Fuller DH, Haynes JR, Santoro JC, Robinson HL. (1995) DNA vaccines: a novel approach to immunization. *Int. J. Immunopharmacol.* **17**: 79–83.

Gao L, Wagner E, Cotten M, Agarwal S, Harris C, Romer M, Miller L, Hu PC, Curiel D. (1993) Direct *in vivo* gene transfer to airway epithelium employing adenovirus–polylysine–DNA complexes. *Hum. Gene Ther.* **4**: 17–24.

Gossen M, Freundlieb S, Bender G, Muller G, Hillen W, Bujard H. (1995) Transcriptional activation by tetracyclines in mammalian cells. *Science* **268**: 1766–1769.

Gottschalk S, Cristiano RJ, Smith LC, Woo SLC. (1994) Folate receptor-mediated DNA delivery into tumor cells: potosomal disruption results in enhanced gene expression. *Gene Ther.* **1**: 185–191.

Grosveld F, van Assendelft GB, Greaves DR, Kollias G. (1987) Position-independent, high-level expression of the human beta-globin gene in transgenic mice. *Cell* **51**: 975–985.

Guy J, Drabek D, Antoniou M. (1995) Delivery of DNA into mammalian cells by receptor-mediated endocytosis and gene therapy. *Mol. Biotechnol.* **3**: 237–248.

Han X, Kasahara N, Kan YW. (1995) Ligand-directed retroviral targeting of human breast cancer cells. *Proc. Natl Acad. Sci. USA* **92**: 9747–9751.

Harris CE, Agarwal S, Hu PC, Wagner E, Curiel DT. (1993) Receptor-mediated gene transfer to airway epithelial cells in primary culture. *Am. J. Respir. Cell. Mol. Biol.* **9**: 441–447.

Hart IR. (1996) Tissue specific promoters in targeting systemically delivered gene therapy. *Semin. Oncol.* **23**: 154–158.

Hart SL, Harbottle RP, Cooper R, Miller A, Williamson R, Coutelle C. (1995) Gene delivery and expression mediated by an integrin-binding peptide. *Gene Ther.* **2**: 552–554.

Haynes JR, McCabe DE, Swain WF, Widera G, Fuller JT. (1996) Particle-mediated nucleic acid immunization. *J. Biotechnol.* **44**: 37–42.

Hickman MA, Malone RW, Sih TR, Akita GY, Carlson DM, Powell JS. (1995) Hepatic gene expression after direct DNA injection. *Adv. Drug Deliv. Rev.* **17**: 265–271.

Hill A, Jugovic P, York I, Russ G, Bennink, J, Yewdell J, Ploegh H, Johnson D. (1995) Herpes-simplex virus turns off the TAP to evade host immunity. *Nature* **375**: 411–415.

Huber BE, Richards CA. (1996) Regulated expression of artificial chimeric genes contained in retroviral vectors: implications for virus-directed enzyme prodrug therapy (VDEPT) and other gene therapy applications. *J. Drug Target.* **3**: 349–356.

Huckett B, Ariatti M, Hawtrey AO. (1990) Evidence for targeted gene transfer by receptor-mediated endocytosis. *Biochem. Pharmacol.* **40**: 253–263.

Jenkins M, Kerr D, Fayer R, Wall R. (1995) Serum and colostrum antibody responses induced by jet-injection of sheep with DNA encoding a *Cryptosporidium parvum* antigen. *Vaccine* **13**: 1658–1664.

Kasahara N, Dozy AM, Kan YW. (1994) Tissue-specific targeting of retroviral vectors through ligand-receptor interactions. *Science* **266**: 1373–1375.

Katze MG. (1993) Games viruses play – a strategic initiative against the interferon-induced dsRNA activated 68,000 Mr protein-kinase. *Semin. Virol.* **4**: 259–268.

Kawabata K, Takakura Y, Hashida M. (1995) The fate of plasmid DNA after intravenous injection in mice: involvement of scavenger receptors in its hepatic uptake. *Pharm. Res.* **12**: 825–830.

Lamaze C, Schmid SL. (1995) The emergence of clathrin-independent pinocytic pathways. *Curr. Opin. Cell Biol.* **7**: 573–580.

Leamon CP, Low PS. (1991) Delivery of macromolecules into living cells: a method that exploits folate receptor endocytosis. *Proc. Natl Acad. Sci. USA* **88**: 5572–5576.

Leclerc G, Gal D, Takeshita S, Nikol S, Weir L, Isner JM. (1992) Percutaneous arterial gene transfer in a rabbit model. *J. Clin. Invest.* **90**: 936–944.

Ledley FD. (1995) Nonviral gene therapy: the promise of genes as pharmaceutical products. *Hum. Gene Ther.* **6**: 1129–1144.

Levitskaya J, Coram M, Levitsky V, Imreh S, Steigerwald-Mullen PM, Klein G, Kurilla MG, Masucci MG. (1995) Inhibition of antigen processing by the internal repeat region of the Epstein-Barr virus nuclear antigen-1. *Nature* **375**: 685–688.

Lew D, Parker SE, Latimer T, Abai AM, Kuwahara-Rundell A, Doh SG, Yang ZY, Laface D, Gromkowski SH, Nabel GJ, Manthorpe M, Norman J. (1995) Cancer gene therapy using plasmid DNA: pharmacokinetic study of DNA following injection in mice. *Hum. Gene Ther.* **6**: 553–564.

Loyter A, Scangos G, Juricek D, Keene D, Ruddle FH. (1982a) Mechanisms of DNA entry into mammalian cells. *Exp. Cell Res.* **139**: 223–234.

Loyter A, Scangos GA, Ruddle FH. (1982b) Mechanisms of DNA uptake by mammalian cells: fate of exogenously added DNA monitored by the use of fluorescent dyes. *Proc. Natl Acad. Sci. USA* **79**: 422–426.

Marsh M, Helenius A. (1989) Virus entry into animal cells. *Adv. Virus. Res.* **36**: 107–151.

Matano T, Odawara T, Iwamoto A, Yoshikura H. (1995) Targeted infection of a retrovirus bearing a CD4-Env chimera into human cells expressing human immunodeficiency virus type 1. *J. Gen. Virol.* **76**: 3165–3169.

Merwin JR, Carmichael EP, Noell GS, DeRome ME, Thomas WL, Robert N, Spitalny G, Chiou HC. (1995) CD5-mediated specific delivery of DNA to T lymphocytes: compartmentalization augmented by adenovirus. *J. Immunol. Methods* **186**: 257–266.

Meyer KB, Thompson MM, Levy MY, Barron LG, Szoka FC, Jr. (1995) Intratracheal gene delivery to the mouse airway: characterization of plasmid DNA expression and pharmacokinetics. *Gene Ther.* **2**: 450–460.

Michael SI, Curiel DT. (1994) Strategies to achieve targeted gene delivery via the receptor-mediated endocytosis pathway, *Gene Ther.* **1**: 223–232.

Midoux P, Mendes C, Legrand A, Raimond J, Mayer R, Monsigny M, Roche AC. (1993) Specific gene transfer mediated by lactosylated poly-L-lysine into hepatoma cells. *Nucleic Acids Res.* **21**: 871–878.

Mislick KA, Baldeschwieler JD, Kayyem JF, Meade TJ. (1995) Transfection of folate–polylysine DNA complexes: evidence for lysosomal delivery. *Bioconjugate Chem.* **6**: 512–515.

Mitragotri S, Blankschtein D, Langer R. (1995) Ultrasound-mediated transdermal protein delivery. *Science* **269**: 850–853.

Moens U, Seternes OM, Hey AW, Silsand Y, Traavik T, Johansen B, Rekvig OP. (1995) *In vivo* expression of a single viral DNA-binding protein generates systemic lupus erythematosus-related autoimmunity to double-stranded DNA and histones. *Proc. Natl Acad. Sci. USA* **92**: 12393–12397.

Nabel EG, Pompili VJ, Plautz GE, Nabel GJ. (1994) Gene transfer and vascular disease. *Cardiovasc. Res.* **28**: 445–455.

Nabel GJ, Nabel EG, Yang ZY, Fox BA, Plautz GE, Gao X, Huang L, Shu S, Gordon D, Chang AE. (1993) Direct gene transfer with DNA–liposome complexes in melanoma: expression, biological activity, and lack of toxicity in humans. *Proc. Natl Acad. Sci. USA* **90**: 11307–11311.

Naldini L, Blomer U, Gallay P, Ory D, Mulligan R, Gage FH, Verma IM, Trono D. (1996) *In vivo* gene delivery and stable transduction of nondividing cells by a lentiviral vector. *Science* **272**: 263–267.

Neda H, Wu CH, Wu GY. (1991) Chemical modification of an ecotropic murine leukemia virus results in redirection of its target cell specificity. *J. Biol. Chem.* **266**: 14143–14146.

Nishi T, Yoshizato K, Yamashiro S, Takeshima H, Sato K, Hamada K, Kitamura I, Yoshimura T, Saya H, Kuratsu J, Ushio Y. (1996) High-efficiency *in vivo* gene transfer using intraarterial plasmid DNA injection following *in vivo* electroporation. *Cancer Res.* **56**: 1050–1055.

Orrantia E, Chang PL. (1990) Intracellular distribution of DNA internalized through calcium phosphate precipitation. *Exp. Cell Res.* **190**: 170–174.

Pardoll DM, Beckerleg AM. (1995) Exposing the immunology of naked DNA vaccines. *Immunity* **3**: 165–169.

Pasqualini R, Ruoslahti E. (1996) Organ targeting *in vivo* using phage display peptide libraries. *Nature* **380**: 364–366.

Perales JC, Ferkol T, Molas M, Hanson RW. (1994a) An evaluation of receptor-mediated gene transfer using synthetic DNA–ligand complexes. *Eur. J. Biochem.* **226**: 255–266.

Perales JC, Ferkol T, Beegen H, Ratnoff OD, Hanson RW. (1994b) Gene transfer *in vivo*: sustained expression and regulation of genes introduced into the liver by receptor-targeted uptake. *Proc. Natl Acad. Sci. USA* **91**: 4086–4090.

Philip R, Liggitt D, Philip M, Dazin P, Debs R. (1993) Efficient transfection of T lymphocytes in adult mice. *J. Biol. Chem.* **268**: 16087–16090.

Plank C, Oberhauser B, Mechtler K, Koch C, Wagner E. (1994) The influence of endosome-disruptive peptides on gene transfer using synthetic virus-like gene transfer systems. *J. Biol. Chem.* **269**: 12918–12924.

Prentice H, Kloner RA, Prigozy T, Christensen T, Newman L, Li Y, Kedes L. (1994) Tissue restricted gene expression assayed by direct DNA injection into cardiac and skeletal muscle. *J. Mol. Cell. Cardiol.* **26**: 1393–1401.

Prentice H, Kloner RA, Li Y, Newman L, Kedes L. (1996) Ischemic/reperfused myocardium can express recombinant protein following direct DNA or retroviral injection. *J. Mol. Cell. Cardiol.* **28**: 133–140.

Raz E, Carson DA, Parker SE, Parr TB, Abai AM, Aichinger G, Gromkowski SH, Singh M, Lew D, Yankauckas MA, Baird SM, Rhodes GH. (1994) Intradermal gene immunization: the possible role of DNA uptake in the induction of cellular immunity to viruses. *Proc. Natl Acad. Sci. USA* **91**: 9519–9523.

Remy JS, Kichler A, Mordvinov V, Schuber F, Behr, JP. (1995) Targeted gene transfer into hepatoma cells with lipopolyamine-condensed DNA particles presenting galactose ligands: a stage toward artificial viruses. *Proc. Natl Acad. Sci. USA* **92**: 1744–1748.

Riessen R, Rahimizadeh H, Blessing E, Takeshita S, Barry JJ, Isner JM. (1993) Arterial gene transfer using pure DNA applied directly to a hydrogel-coated angioplasty balloon. *Hum. Gene Ther.* **4**: 749–758.

Rojanasakul Y, Wang LY, Malanga CJ, Ma JKH, Liaw J. (1994) Targeted gene delivery to aveolar macrophages via Fc receptor-mediated endocytosis. *Pharm. Res.* **11**: 1731–1736.

Rosenkranz AA, Yachmenev SV, Jans DA, Serebryakova NV, Murav'ev VI, Peters R, Sobolev AS. (1992) Receptor-mediated endocytosis and nuclear transport of a transfecting DNA construct. *Exp. Cell Res.* **199**: 323–329.

Ross GF, Morris RE, Ciraolo G, Huelsman K, Bruno M, Whitsett JA, Baatz JE, Korfhagen TR. (1995) Surfactant protein A–polylysine conjugates for delivery of DNA to airway cells in culture. *Hum. Gene Ther.* **6**: 31–40.

Roth MG. (1993) Endocytic receptors. In: *Advances in Cell and Molecular Biology of Membranes*. (eds B Storrie, RF Murphy). JAI Press, Greenwich, CT, pp. 19–50.

Rothenberger S, Iacopetta BJ, Kuhn LC. (1987) Endocytosis of the transferrin receptor requires the cytoplasmic domain but not its phosphorylation site. *Cell* **49**: 423–431.

Roux P, Jeanteur P, Piechaczyk M. (1989) A versatile and potentially general approach to the targeting of specific cell types by retroviruses: application to the infection of human cells by means of major histocompatibility complex class I and class II antigens by mouse ecotropic murine leukemia virus-derived viruses. *Proc. Natl Acad. Sci. USA* **86**: 9079–9083.

Sandvig K, van Deurs B. (1994) Endocytosis without clathrin. *Trends Cell Biol.* **4**: 275–277.

Schwarzenberger P, Spence SE, Gooya JM, Michiel D, Curiel DT, Ruscetti FW, Keller JR. (1996) Targeted gene transfer to human hematopoietic progenitor cell lines through the c-kit receptor. *Blood* **87**: 472–478.

Schlepper-Schafer J, Hulsmann D, Djovkar A, Meyer, HE, Herbertz L, Kolb H, Kolb-Bachofen V. (1986) Endocytosis via galactose receptors *in vivo*. *Exp. Cell Res.* **165**: 494–506.

Shockett P, Difilippantonio M, Hellman N, Schatz DG. (1995) A modified tetracycline-regulated system provides autoregulatory, inducible gene expression in cultured cells and transgenic mice. *Proc. Natl Acad. Sci. USA* **92**: 6522–6526.

Smith HO, Tomb JF, Dougherty BA, Fleischmann RD, Venter JC. (1995) Frequency and distribution of DNA uptake signal sequences in the *Haemophilus influenzae* Rd genome. *Science* **269**: 538–540.

Somia NV, Zoppe M, Verma IM. (1995) Generation of targeted retroviral vectors by using single-chain variable fragment: an approach to *in vivo* delivery. *Proc. Natl Acad. Sci. USA* **92**: 7570–7574.

Stribling R, Brunette E, Liggitt D, Gaensler K, Debs R. (1992) Aerosol gene delivery *in vivo*. *Proc. Natl Acad. Sci. USA* **89**: 11277–11281.

Thierry AR, Lunardi-Iskandar Y, Bryant JL, Rabinovich P, Gallo RC, Mahan LC. (1995) Systemic gene therapy: biodistribution and long-term expression of a transgene in mice. *Proc. Natl Acad. Sci. USA* **92**: 9742–9746.

Thilo L. (1994) Endocytosis: aspects of organellar processing. *Ann. NY Acad. Sci.* **710**: 209–216.

Thurnher M, Wagner E, Clausen H, Mechtler K, Rusconi S, Dinter A, Birnstiel, ML, Berger EG, Cotten M. (1994) Carbohydrate receptor-mediated gene transfer to human T leukaemic cells. *Glycobiology* **4**: 429–435.

Trowbridge IS, Collawn JF, Hopkins CR. (1993) Signal-dependent membrane protein trafficking in the endocytic pathway. *Annu. Rev. Cell Biol.* **9**: 129–161.

Trubetskoy VS, Torchilin VP, Kennel S, Huang L. (1992a) Cationic liposomes enhance targeted delivery and expression of exogenous DNA mediated by N-terminal modified poly(L-lysine)–antibody conjugate in mouse lung endothelial cells. *Biochim. Biophys. Acta* **1131**: 311–313.

Trubetskoy VS, Torchilin VP, Kennel SJ, Huang L. (1992b) Use of N-terminal modified poly(L-lysine)–antibody conjugate as a carrier for targeted gene delivery in mouse lung endothelial cells. *Bioconjugate Chem.* **3**: 323–327.

Tsan MF, White JE, Shepard B. (1995) Lung-specific direct *in vivo* gene transfer with recombinant plasmid DNA. *Am. J. Physiol.* **268**: L1052-L1056.

Tsukamoto M, Ochiya T, Yoshida S, Sugimura T, Terada M. (1995) Gene transfer and expression in progeny after intravenous DNA injection into pregnant mice. *Nature Genet.* **9**: 243–248.

Turek JJ, Leamon CP, Low PS. (1993) Endocytosis of folate–protein conjugates: ultrastructural localization in KB cells. *J. Cell Sci.* **106**: 423–430.

Ulmer JB, Donnelly JJ, Parker SE, Rhodes GH, Felgner PL, Dwarki VJ, Gromkowksi SH, Deck RR, DeWitt CM, Friedman A, Hawe LA, Leander KR, Martinez D, Perry HC, Shiver JW, Montgomery DL, Liu MA. (1993) Heterologous protection against influenza by injection of DNA encoding a viral protein. *Science* **259**: 1745–1749.

Ulmer JB, Deck RR, DeWitt CM, Friedman A, Donnelly JJ, Liu MA. (1994) Protective immunity by intramuscular injection of low doses of influenza virus DNA vaccines. *Vaccine* **12**: 1541–1544.

Vahlsing HL, Yankauckas MA, Swadey M, Gromkowski SH, Manthorpe M. (1994) Immunization with plasmid DNA using a pneumatic gun. *J. Immunol. Methods* **175**: 11–22.

Valsesia-Wittmann S, Drynda A, Deleage G, Aumailley M, Heard JM, Danos O, Verdier G, Cosset FL. (1994) Modifications in the binding domain of avian retrovirus envelope protein to redirect the host range of retroviral vectors. *J. Virol.* **68**: 4609–4619.

Valsesia-Wittmann S, Morling FJ, Nilson BHK, Takeuchi Y, Russell SJ, Cosset FL. (1996) Improvement of retroviral retargeting by using amino acid spacers between an additional binding domain the N terminus of Moloney murine leukemia virus SU. *J. Virol.* **70**: 2059–2064.

van der Woude I, Visser HW, ter Beest MBA, Wagenaar A, Ruiters MHJ, Engberts JBFN, Hoekstra D. (1995) Parameters influencing the introduction of plasmid DNA into cells by the use of synthetic amphiphiles as a carrier system. *Biochim. Biophys. Acta* **1240**: 34–40.

Vile RG. (1994) Tumor-specific gene expression. *Semin. Cancer Biol.* **5**: 429–436.

Vile R, Miller N, Chernajovsky Y, Hart I. (1994) A comparison of the properties of different retroviral vectors containing the murine tyrosinase promoter to achieve transcriptionally targeted expression of the HSVtk or IL-2 genes. *Gene Ther.* **1**: 307–316.

Wagner E, Zenke M, Cotten M, Beug H, Birnstiel ML. (1990) Transferrin–polycation conjugates as carriers for DNA uptake into cells. *Proc. Natl Acad. Sci. USA* **87**: 3410–3414.

Wagner E, Cotten M, Foisner R, Birnstiel ML. (1991) Transferrin–polycation–DNA complexes: the effect of polycations on the structure of the complex and DNA delivery to cells. *Proc. Natl Acad. Sci. USA* **88**: 4255–4259.

Waine GJ, McManus DP. (1995) Nucleic acids: vaccines of the future. *Parasitol. Today* **11**: 113–116.

Wang B, Ugen KE, Srikantan V, Agadjanyan MG, Dang K, Refaeli Y, Sato AI, Boyer J, Williams WV, Weiner DB. (1993) Gene inoculation generates immune responses against human immunodeficiency virus type 1. *Proc. Natl Acad. Sci. USA* **90**: 4156–4160.

Wang Y, O'Malley JR, Tsai SY, O'Malley BW. (1994) A regulatory system for use in gene transfer. *Proc. Natl Acad. Sci. USA* **91**: 8180–8184.

Watts C, Marsh M. (1992) Endocytosis: what goes in and how? *J. Cell Sci.* **103**: 1–8.

Weiss RA, Tailor CS. (1995) Retrovirus receptors. *Cell* **82**: 531–533.

Wickham TJ, Carrion ME, Kovesdi I. (1995) Targeting of adenovirus penton base to new receptors through replacement of its RGD motif with other receptor-specific peptide motifs. *Gene Ther.* **2**: 750–756.

Williams AF. (1991) Cellular interactions: out of equilibrium. *Nature* **352**: 473–474.

Wilson JM., Grossman M, Wu CH, Chowdhury NR, Wu GY, Chowdhury JR. (1992) Hepatocyte-directed gene transfer *in vivo* leads to transient improvement of hypercholesterolemia in low density lipoprotein receptor-deficient rabbits. *J. Biol. Chem.* **267**: 963–967.

Wolff JA, Malone RW, Williams P, Chong W, Acsadi G, Jani A, Felgner PL. (1990) Direct gene transfer into mouse mucle *in vivo*. *Science* **247**: 1465–1468.

Wolff JA, Dowty ME, Jiao S, Repetto G, Berg RK, Ludtke JJ, Williams P. (1992) Expression of naked plasmids by cultured myotubes and entry of plasmids into T tubules and caveolae of mammalian skeletal muscle. *J. Cell Sci.* **103**: 1249–1259.

Wu CH, Wilson JM, Wu GY. (1989) Targeting genes: delivery and persistent expression of a foreign gene driven by mammalian regulatory elements *in vivo*. *J. Biol. Chem.* **264**: 16985–16987.

Wu GY, Wu CH. (1987) Receptor-mediated *in vitro* gene transformation by a soluble DNA carrier system. *J. Biol. Chem.* **262**: 4429–4432.

Wu-Pong S, Weiss TL, Hunt CA. (1992) Antisense c-*myc* oligodeoxyribonucleotide cellular uptake. *Pharm. Res.* **9**: 1010–1017.

Young JAT, Bates P, Willert K, Varmus HE. (1990) Efficient incorporation of human CD4 protein into avian leukosis virus particles. *Science* **250**: 1421–1423.

Yovandich J, O'Malley B, Jr, Sikes M, Ledley FD. (1995) Gene transfer to synovial cells by intra-articular administration of plasmid DNA. *Hum. Gene Ther.* **6**: 603–610.

Zabner J, Fasbender AJ, Moninger T, Poellinger KA, Welsh MJ. (1995) Cellular and molecular barriers to gene transfer by a cationic lipid. *J. Biol. Chem.* **270**: 18997–19007.

Zavada J. (1982) The pseudotypic paradox. *J. Gen. Virol.* **63**: 15–24.

Zhao X. (1995) EPD, a novel technology for drug delivery. *Adv. Drug Deliv. Rev.* **17**: 257–262.

Zhu N, Liggitt D, Liu Y, Debs R. (1993) Systemic gene expression after intravenous DNA delivery into adult mice. *Science* **261**: 209–211.

3

Retroviral vectors

Walter H. Günzburg and Brian Salmons

3.1 Introduction

The ability to be able to deliver and express genes in specific cells is central to the concept of gene therapy. Many methods have been devised to accomplish gene transfer to mammalian cells over the past 25 years. Of these, the use of recombinant viruses may be the most ideal, and a number of viruses have or are being modified to create vehicles for gene delivery. Retroviruses were one of the first types of virus to be engineered as such vehicles or vectors, and were used for the first gene transfer studies in patients. This chapter reviews the development of retroviral vectors, as well as some of the problems associated with their use and how these problems are currently being tackled.

3.2 Retroviruses and tumorigenicity

Retroviruses are found in many different species and became a particular focus of interest in the 1970s as a result of their association with tumour formation in susceptible animals. Although the genetic information carried by retroviruses is in the form of RNA, this becomes reverse-transcribed (*Figure 3.1*) into DNA in the infected cell. The DNA form of the retroviral genome must then integrate into the genome of the infected cell (provirus) for the retroviral life cycle to proceed. This obligatory integration event ensures that the retroviral genome becomes a part of the host-cell genetic information and every daughter of the original infected cell inherits the provirus. It is ironic that this property of retroviruses that makes them so useful for stable long-term gene delivery and expression is also their Achilles heel. This is because the mechanism by which retroviruses are implicated in tumorigenesis involves their essentially random integration into, or in the vicinity of, a cellular gene involved in growth control (proto-oncogene or tumour suppressor gene; reviewed in Fan, 1994). Since the chance of a single retroviral integration event occurring in such a gene locus is extremely low, multiple successive integrations are required before such an event is even likely to occur. This necessitates a retrovirus capable of integration, production of new retroviral particles followed by their integration, another phase of virus production, and so on. Normally, this only occurs with an unmodified replication-competent or wild-type retrovirus.

The possibility that a similar mechanism may cause malignancies in patients treated with retroviral vectors carrying therapeutic genes intended to treat other pre-existing medical conditions, has posed a recurring ethical problem. However, as will become apparent in the following sections, recombinant retroviruses, designed to

Gene Therapy, edited by N.R. Lemoine and D.N. Cooper.

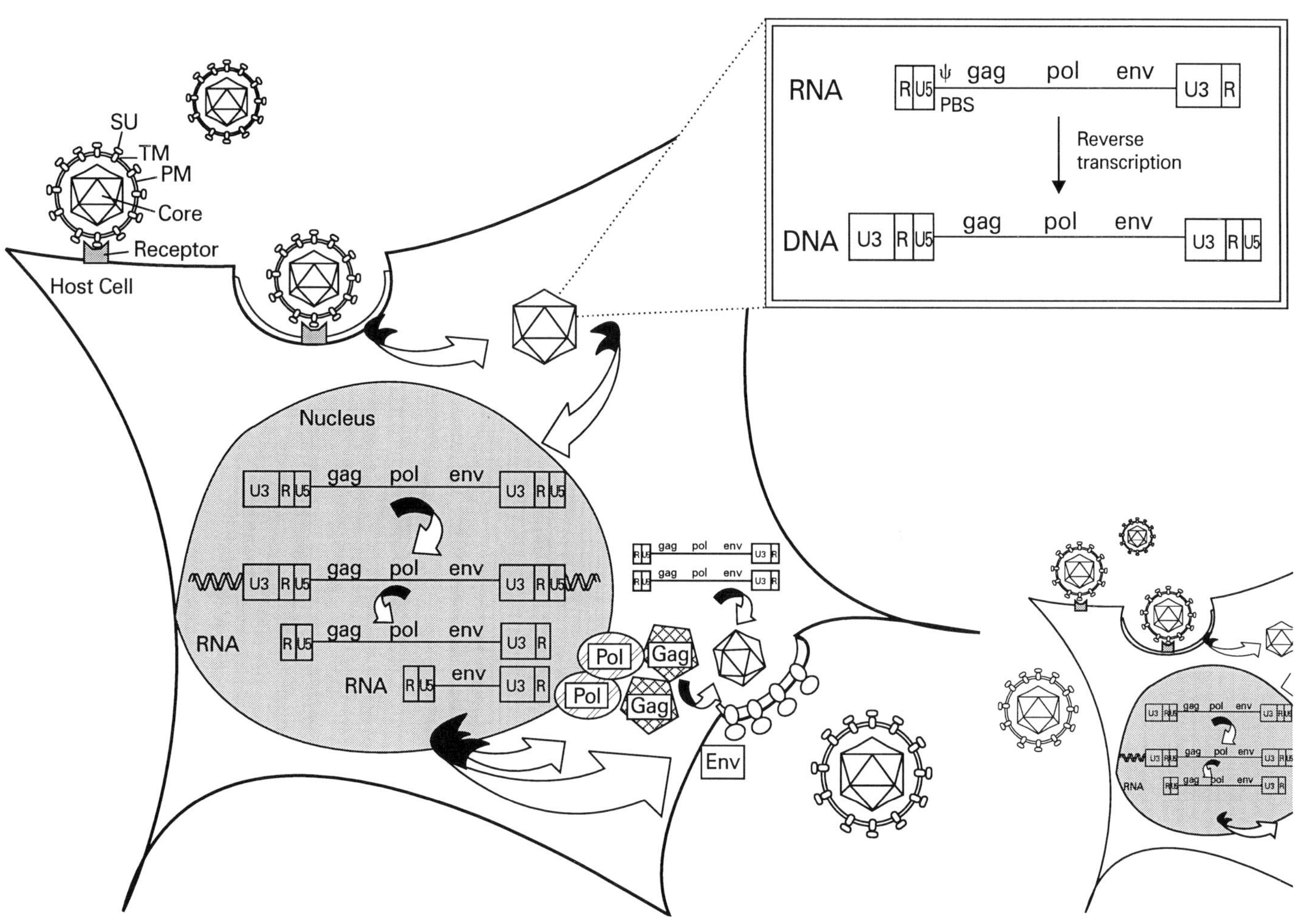
RNA
R U5
ψ
PBS
gag pol env
U3 R
Reverse transcription
DNA
U3 R U5
gag pol env
U3 R U5
SU
TM
PM
Core
Receptor
Host Cell
Nucleus
U3 R U5
gag pol env
U3 R U5
U3 R U5
gag pol env
U3 R U5
RNA
R U5
gag pol env
U3 R
RNA
R U5
env
U3 R
R U5
gag pol env
U3 R
R U5
gag pol env
U3 R
Pol
Pol
Gag
Gag
Env

carry therapeutic genes for gene therapy, do so at the expense of viral gene information. Thus, the resulting viruses are replication-defective, are only able to integrate once, and are *unable* to produce any new retroviral particles. Most researchers would agree that the probability of such a replication-defective virus integrating into or near a cellular gene involved in controlling cell proliferation is vanishingly small. Only if the replication-defective virus *recombined* with other retroviral sequences to produce a new replication-competent virus could there be a problem. The explosive expansion of a population of replication-competent retroviruses from a single event would eventually provide enough integration events to make phenotypic integration a very real possibility.

It has been well documented that recombination events between components of the retroviral vector system can lead to the generation of potentially pathogenic, replication-competent virus (RCV) and several generations of vector systems have now been constructed to minimize this risk of recombination (reviewed in Salmons and Günzburg, 1993; *Figure 3.2*). Unfortunately, little is known about the finite probabilities both of recombination occurring and of the RCV integrating at a specific gene locus. Thus, it becomes very important to determine empirically the chance of (i) insertional disruption or activation of single genes by retrovirus integration and (ii) the risk of generation of RCV by recombination in current generations of packaging cell lines (see Section 3.3.2). The frequency of retroviral vector-mediated gene inactivation of the

---

Figure 3.1. The life cycle of a typical retrovirus (such as MLV). Retrovirus particles (virions) consist of a core containing two identical single-stranded viral RNAs and enzymes (reverse transcriptase, integrase) surrounded by host-cell plasma membrane (PM) into which are inserted the viral surface (SU) and transmembrane (TM) envelope (Env) proteins. The infection process is initiated by a specific interaction between the SU envelope protein and specific receptors expressed on the host-cell surface. This leads to uptake of the virus particle and removal of the envelope by the host cell. The liberated virus core is then the site of the reverse transcription of the viral RNA into a double-stranded DNA form, using a tRNA specifically bound to the primer binding site (PBS) as a primer (see inset box). This reaction is performed by the virus reverse transcriptase (RT) enzyme present as part of a viral nucleic acid–enzyme/protein complex. During reverse transcription, sequences located at either end of the viral RNA (U5 and U3; see inset box) become duplicated and placed at the opposite ends of the newly synthesized viral DNA. This forms two identical terminally repeated sequences known as long terminal repeats (LTRs) with the structure U3-R-U5 (see inset box). After or during synthesis of the double-stranded DNA, the whole viral nucleic acid–enzyme/protein complex re-locates to the nucleus. The viral integrase protein (IN) associated with this complex then inserts the viral DNA into the host-cell DNA in a reaction that is specific from the point of view of the provirus, in other words, maintaining the structure and order U3-R-U5-gag-pol-env-U3-R-U5. However, there appears to be a multitude of sites in host-cell DNA into which the viral DNA can integrate. The integrated DNA form of the virus genome, known as the provirus, is then transcribed like any cellular gene by the host-cell transcription machinery from the viral promoter located in the U3 region of the LTR. The provirus carries at least three coding regions (*gag, pol* and *env*) encoding the viral core proteins, enzymatic activities (RT and IN) and envelope proteins. Transcription from the LTR gives rise to viral RNAs that are used for translation of the virus proteins as well as the genomic RNA for virus particles. The virus proteins and RNA assemble into new cores which interact with regions of the host-cell membrane that contain the newly synthesized virus envelope proteins. Newly produced virus particles then bud out of the infected cell where they undergo the final stages of maturation to form fully infectious virus particles. These newly synthesized virions can then infect more host cells.

1st Generation vector system

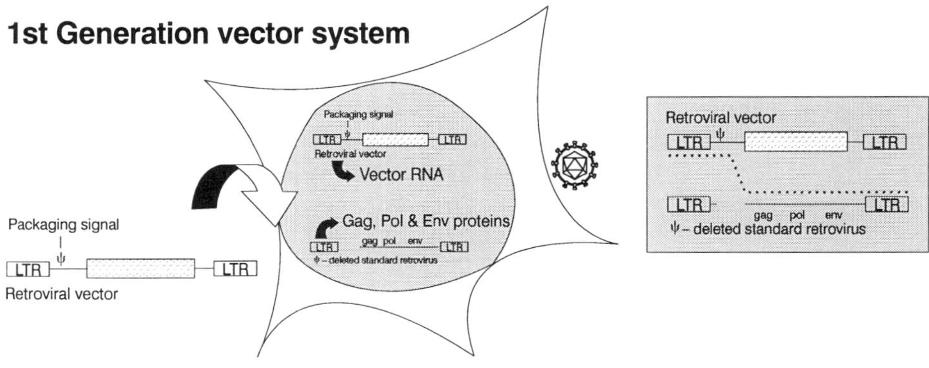

2nd Generation vector system

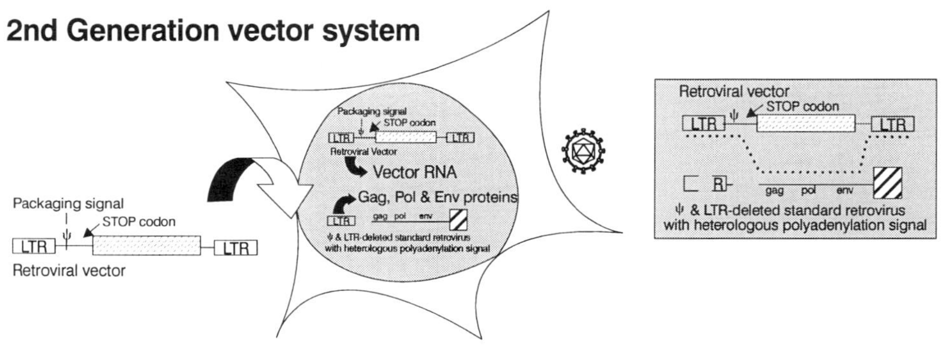

3rd Generation vector system

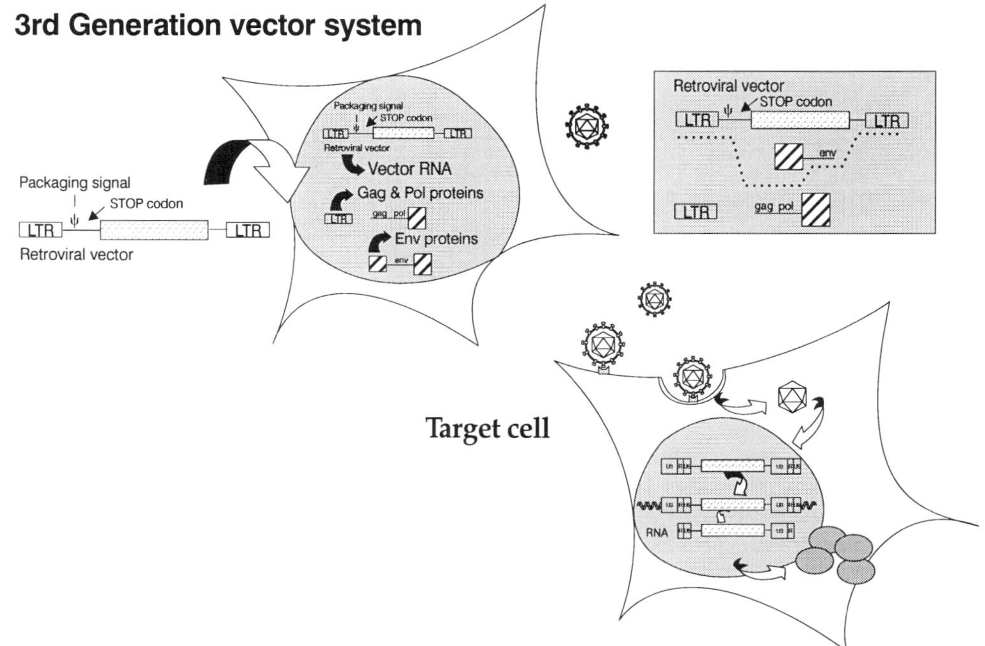

thymidine kinase (*tk*) gene used as a model gene target, has been estimated at 2×10^{-5}; that is, five times above background inactivation (Grosovsky *et al.*, 1993). A detailed examination of the mechanism by which these events occur will also allow the construction of new types of system designed to limit these events or even to target integration to non-essential regions of the genome. Current generations of retroviral vectors are very safe, but since it will never be possible to reduce the risk of this system to zero, an informed risk–benefit decision will always have to be taken.

3.3 Retroviral vectors as vehicles for gene transfer

Although retroviral vectors based on a number of different retroviruses (including those listed in *Table 3.1*) have been constructed, the vectors that are presently being used for clinical gene therapy trials are derived exclusively from murine leukaemia virus (MLV). This is because the biology of MLV is relatively well understood (*Figure 3.1*), the vector systems give relatively high titres of recombinant virus, and also because of the availability of retroviral variants (*amphotropic* or *xenotropic* MLVs; see Section 3.7.1) that efficiently infect human cells. Almost all retroviral vector systems consist of two components. The first component is the recombinant retroviral vector, an expression vector that carries the therapeutic gene(s) and that encodes the RNA that will form the genome of the retroviral vector particle. The therapeutic gene is carried in place of the retroviral coding information (*Figure 3.1*). The *replacement* of retroviral genes with therapeutic genes rather than the additional *insertion* of such genes is necessary to allow efficient incorporation of the recombinant retroviral genome into the vector virus particle, since only RNA up to a maximum size of about 9 kb can be efficiently packaged. However, the loss of retroviral coding information renders the vector replication-defective and vector virus particles cannot be produced without the second component of the system, the packaging cell line. These cells are engineered to produce the missing retroviral structural proteins from one or more

---

Figure 3.2. Generations of retroviral vector systems. In the first generation system, a retroviral vector construct carrying the necessary packaging signals (Ψ) and a therapeutic and/or marker gene(s) (stippled box) is introduced into a cell line that carries a retroviral provirus lacking the Ψ packaging sequences. This construct directs the synthesis of the viral structural proteins and enzymes but the RNA that is used for the translation of these proteins cannot be efficiently packaged since it lacks the Ψ signal. Initially, almost all of the virus particles produced from this system carry the retroviral vector genome. However, one recombination event between the vector and the packaging construct (inset box) can lead to standard, replication-competent virus production. The second generation of packaging cell line carries a modified retroviral provirus lacking, in addition to the Ψ, part of the 5′LTR and the 3′LTR which is replaced by heterologous polyadenylation signals. At least two recombination events between the vector and the packaging construct (inset box) must occur before standard, replication-competent viruses can be produced from this system. Third generation: the latest safety-conscious packaging cell lines carry two independent constructs, one of which expresses the *gag* and *pol* gene products and a second that directs the expression of the Env proteins. Consequently, at least three recombination events between the vector and the two packaging constructs are required for the production of standard replication-competent virus. Retroviral vector particles produced from any of these systems are used to infect target cells. The vector RNA genome carrying the therapeutic gene is reverse-transcribed into a double-stranded DNA which then integrates into the genome of the target cell. The gene is then expressed like any cellular gene. No further virus production is possible from the infected cell because the genetic information encoding the viral structural and enzymatic proteins is not present in these cells.

Table 3.1. Mammalian retroviruses used as a basis for the construction of vector systems

Retrovirus	Reference
MLV family	
MoMLV	Reviewed by McLachlin *et al.* (1990)
FrMLV	Cohen-Haguenauer *et al.* (1995)
AKV	Paludan *et al.* (1989)
SL-3	Paludan *et al.* (1989)
MuMPSV	Laker *et al.* (1987)
HaMSV	Pastan *et al.* (1988)
MMTV	Günzburg and Salmons (1986); Morris *et al.* (1989); Salmons *et al.* (1989); Shackleford and Varmus (1988)
HIV-1	Buchschacher and Panganiban (1992); Carroll *et al.* (1994); Page *et al.* (1990); Parolin *et al.* (1994); Pozansky *et al.* (1991); Richardson *et al.* (1993)
SIV	Rizvi and Panganiban (1992)
BLV	Boris-Lawrie and Temin (1995); Derse and Martarano (1990); Milan and Nicolas (1991)
Human foamy virus	Russell (1996); Rethwilm (1996)

AKV, endogenous ectotropic MLV; SL-3, highly leukaemic MLV; MuMPSV, murine myeloproliferative sarcoma virus; HaMSV, Harvey murine sarcoma virus.

expression construct. The introduction of the vector into the packaging cells allows the production of recombinant viral particles that are able to infect (transduce) target cells (*Figure 3.2*). Much effort has been devoted to the production of vector constructs and packaging cells that produce high titres of retroviral vector but that also minimize the chance of production of RCV.

3.3.1 Retroviral vector constructs

Although much of the retroviral genomic sequence information can be removed to create a vector, a number of regions of the viral genome are apparently indispensable for successful gene transfer. These include the outermost sequences at each end of the long terminal repeats (LTRs) required for integration of the proviral DNA (Bushman and Craigie, 1990), the primer binding site (PBS) which binds the cellular tRNA used as a primer for reverse transcription, as well as the sequences involved in the specific interaction between the virus genome and the virus proteins that leads to the incorporation of the genome into the newly synthesized virus particle. This latter event is also known as encapsidation or packaging. The packaging signals for Moloney MLV (MoMLV) have been defined as the sequence (ψ) between the splice donor site immediately downstream of the 5′LTR and the AUG start codon for the retroviral Gag protein (Mann *et al.*, 1983) (*Figure 3.2*). Retroviral RNAs lacking these sequences cannot be encapsidated into virus particles. Other facultative signals for packaging have been identified that are located within the *gag* gene (Armentano *et al.*, 1987; Bender *et al.*, 1987) and the U5 region of the LTR (Murphy and Goff, 1989).

In addition to carrying a therapeutic gene, many retroviral vectors carry a marker gene, allowing the easy positive selection or identification of infected cells. The use of such a gene may not be necessary or even desirable in a clinical situation since (i) selection will usually not be possible in a patient and (ii) an immune response may be generated against the product of the marker gene. The genes that are carried by the retroviral vector can be expressed in a number of different ways (*Figure 3.3*). The retroviral promoter within the LTR can drive heterologous gene expression when such genes are cloned into the positions formerly occupied by the *gag*, *pol* and *env* genes, the gene placed in the *env* position then being expressed from the subgenomic viral RNA (*Figure 3.1*). A second heterologous gene can also be expressed from a heterologous promoter introduced along with the gene construct as an expression cassette (*Figure 3.3*).

Generally, the use of such an additional heterologous internal promoter proves to be problematic since promoter interference between heterologous promoters and the retroviral promoter has been documented (Li *et al.*, 1992; McLachlin *et al.*, 1993; Wu *et al.*, 1996; Xu *et al.*, 1989). Two recent reports suggest that expression of a therapeutic gene from an internal heterologous promoter gives poor and relatively short-term expression in comparison with that obtained from the retroviral promoter after infection of haematopoietic cells (Mavilio *et al.*, 1994; Riviere *et al.*, 1995). This has led to the retroviral LTR promoter being used to drive expression in many clinical applications in the absence of a second heterologous promoter. In retroviral vectors carrying only one promoter, bi-cistronic transcription units can be used to allow the expression of two genes (*Figure 3.3*). These vectors carry an internal ribosome entry site (IRES), allowing ribosomes to recognize and translate the second protein from one RNA molecule (Dirks *et al.*, 1993; Koo *et al.*, 1992; Levine *et al.*, 1991; Morgan *et al.*, 1992; Sugimoto *et al.*, 1994). The inclusion of an IRES in retroviral vectors also permits the introduction of larger genes since a second promoter is not required, and results in stable gene expression. Finally, the titre of virus (see Section 3.4) obtained with bi-cistronic retroviral vectors virus has been reported to be higher than that obtained from two-gene, two-promoter vectors (Levine *et al.*, 1991). Retroviral vectors carrying the human glucocerebrosidase cDNA and the CD24 surface antigen as a bi-cistronic unit have been used to deliver these gene products to fibroblasts derived from patients with Gaucher disease (Migita *et al.*, 1995).

These vectors rely on transcription being directed from the retroviral promoter within the LTR. It has been repeatedly observed that this promoter can be shut down or silenced after a variable period in the target cell (Naviaux and Verma, 1992). Methylation status of the provirus, host-cell factors and negatively acting regulatory elements, as well as site of integration effects, have been proposed as culprits for this shut-off (Challita and Kohn, 1994; Challita *et al.*, 1995; Naviaux and Verma, 1992). Efforts have been made to alter both the retroviral LTR and downstream sequences to prevent transcriptional silencing in certain cell types. The silencing of gene expression in haematopoietic stem cells has been especially intensively studied since these cells represent a particularly attractive target for gene therapy in a variety of metabolic disorders and leukaemias. The data available suggest that both the MLV U3 enhancer and the PBS are involved in repressing expression (Baum *et al.*, 1995) in an analogous fashion to the repression of retroviral expression described in embryonic stem cells (Challita *et al.*, 1995). Much higher levels of expression can be achieved if the PBS of murine embryonic stem cell virus and the U3 regions of either Friend mink cell focus-forming virus or myeloproliferative sarcoma virus are used in the context of an MLV retroviral vector (Baum *et al.*, 1995). Alternatively, the complete MLV LTR can be replaced by the LTR of myeloproliferative sarcoma virus, resulting in improved

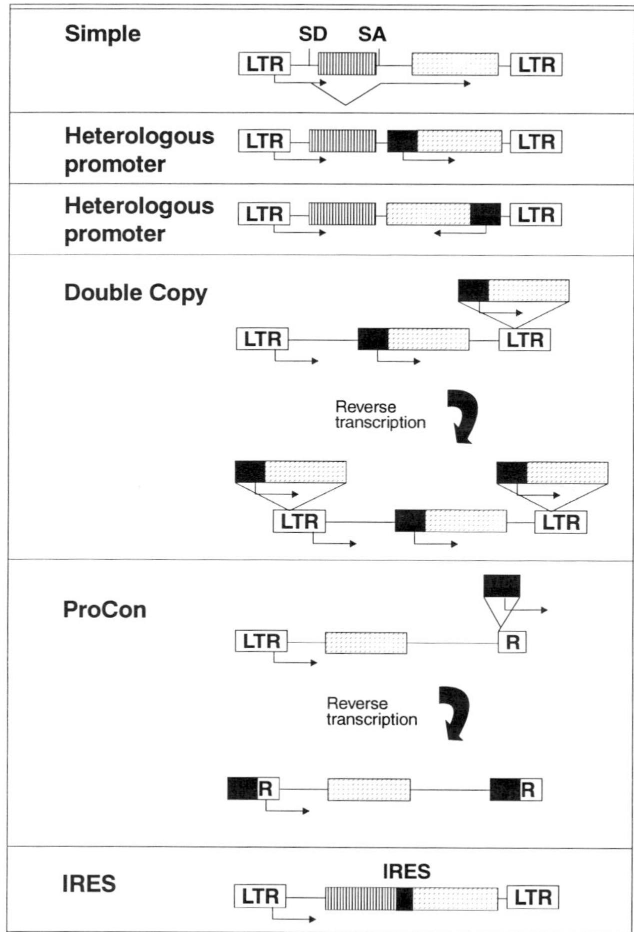

Figure 3.3. Retroviral vector constructs. A number of different types of retroviral vectors are shown. The simple type carry one or two heterologous genes (vertically striped box and stippled box) in place of the *gag, pol* and *env* genes. Both of the genes are under the transcriptional control of the *MLV* promoter contained within the retroviral LTR. The first gene is expressed from full-length transcripts (analogous to the genomic *gag-pol* transcript) and the second gene is expressed as a spliced transcript (analogous to the *env* transcript) using the splice donor (SD) and splice acceptor (SA). Alternatively, the second gene can be expressed from an internal heterologous promoter (black box), either in the sense or antisense orientation. Double copy retroviral vectors carry a gene often linked to a promoter in the U3 or R regions of the 3′LTR of the vector. After packaging of the RNA form of the vector genome into retroviral vector particles and infection of cells, a double-stranded DNA form is synthesized by reverse transcription, which results in the duplication of the gene so that the second copy is present in the 5′LTR. ProCon retroviral vectors carry a heterologous promoter (black box), often from a gene that is expressed in certain cell types for targeting purposes, in place of the entire U3 normally present in 3′LTR. After infection and reverse transcription, the heterologous retroviral promoter is duplicated and placed in the 5′ LTR as the only promoter that directs expression of the retrovirally carried gene. IRES or bi-cistronic vectors carry two genes expressed from the *MLV* promoter as one transcript. Both genes are translated from the transcript, the translation of the second gene being facilititated by an internal ribosome entry site (IRES).

expression in bone marrow (Riviere *et al.*, 1995), embryonic carcinoma and stem cells (Challita *et al.*, 1995). Recently, a vector has been constructed based on Friend murine leukaemia virus (FrMLV) (Cohen-Haguenauer *et al.*, 1995) and this may also allow good expression in stem cells. Sequences within one therapeutic gene itself (clotting factor VIII) have also been shown to decrease the titre of vector virus and result in poor expression levels of the therapeutic protein in transduced cells (Lynch *et al.*, 1993), and it seems likely that the coding sequences of other genes may exert similar, though perhaps less dramatic, effects. Clearly some of these problems can also arise when using other viral or even non-viral methods of gene transfer since any expression vector in which two or more heterologous sequences are linked can yield unpredictable effects.

Silencing of expression or interference between promoters in retroviral vectors may be avoided by modifing or replacing the retroviral promoter. A number of approaches have been taken that exploit a unique feature of the retroviral life cycle, the reverse transcription of viral genomic RNA into a double-stranded form (*Figure 3.1*). During this process, sequences from the 5′ end of the RNA (U5) are duplicated and placed additionally at the 3′ end of the DNA, while sequences from the 3′ end of the RNA (U3) are copied on to the 5′ end of the DNA. The process generates the identical LTR structures (U3-R-U5) which flank the viral genome (*Figure 3.1*). The first vectors to exploit this feature of the retroviral life cycle were the self-inactivating (SIN) vectors, in which the enhancer carried in the U3 region of the 3′LTR is deleted in the vector construct (Yu *et al.*, 1986). After reverse transcription and resultant translocation of the deleted U3 region to the 5′ LTR, this resulted in loss of retroviral promoter activity in infected cells. This strategy has been employed to alleviate the inhibitory effects of the retroviral promoter on internal promoters (Soriano *et al.*, 1991). If, instead of deletion, heterologous sequences are *inserted* within the U3 or R region of the 3′LTR they become *duplicated* in the infected cell after reverse transcription. Thus, a promoter–gene cassette inserted in this way will be present twice in each LTR (i.e. in double copy) in the target cell (Hantzopoulos *et al.*, 1989). In practice, it has been shown that when cDNAs encoding the myogenic determination gene (MyoD) and purine nucleoside phosphorylase (PNP) are inserted into the R region of double-copy retroviral vectors, slightly reduced titres are obtained. However, the expression levels of MyoD (but not PNP) in vector-infected cells is increased (Adam *et al.*, 1995).

A similar strategy has been used to create a series of vectors which carry heterologous promoters or enhancer elements within their LTRs, so as to transcribe viral RNA from the normal start site but under the control of this element. Promoters or enhancers of interest include those that are tissue-specific or that can be regulated conditionally (see Section 3.7.2). Tissue-specific promoters have been successfully inserted in place of, or in addition to, the MLV regulatory elements and these vectors appear to direct expression of genes carried by the retroviral vector to the expected cell types (Ferrari *et al.*, 1995; Vile *et al.*, 1995). Replacement of the entire U3 region (carrying the retroviral promoter–enhancer) with heterologous promoters from genes that are expressed in a tissue-specific manner has also been successfully employed for targeting of expression (Günzburg *et al.*, 1995; Saller, 1994; Salmons *et al.*, 1995; Mrochen *et al.*, manuscript submitted). These promoter conversion (ProCon; *Figure 3.3*) vectors may also be advantageous from a safety point of view since (i) the frequency of recombination with viral sequences in the producer or target cell should be reduced due to the lack of viral U3 sequences in the vector and (ii) it is not clear whether a promoter from a cellular gene can activate or inactivate cellular genes in the context of a retrovirus. The strategy of tissue-specific targeting, either at the level of expression or of infection, will be of paramount importance for future protocols for *in vivo* gene therapy (see Section 3.7). A combination of the bi-cistronic vector with the ProCon vector would remove any possibility of retro-viral promoter interference, while at the same time ensuring adequate expression of multiple IRES-separated genes from the heterologous promoter inserted in the LTR.

3.3.2 Packaging cell lines

These engineered cell lines contain retrovirus-derived constructs that give rise to transcripts which direct the synthesis of authentic viral Gag, Pol and Env proteins. However, the Ψ signal has been deleted in these constructs, which precludes the packaging of the RNA transcripts into virion particles. Upon transfection of the retroviral vector construct into the packaging cell line, the RNA transcribed from the vector, which carries the Ψ signal, is packaged by the retroviral proteins and recombinant virus is released. After infection of a target cell, the recombinant retroviral vector RNA is reverse-transcribed into a proviral DNA that integrates into the DNA of the host cell. Therapeutic genes carried by the vector are then expressed in the infected cell. However, in the absence of Gag, Pol and Env protein, further virus particles cannot be generated (*Figure 3.2*).

Currently, most retroviral vector systems are based upon the third-generation concept, in which the *gag* and *pol* gene products are expressed from one construct and the *env* products from another in the packaging cell line (*Figure 3.2*). In these 'safe' vector systems, at least three recombination events are required for production of RCV (see Section 3.2). The best-known examples of these third-generation packaging cell lines are the GP+E-86 line producing ecotropic virus, the GP+envAm12 producing amphotropic virus (Markowitz *et al.*, 1988a, b) and PG13, in which the *env* gene is from gibbon ape leukaemia virus, a retrovirus that can infect cells of many species but not mouse (Miller *et al.*, 1991). The RNA from the modified retroviral constructs in the packaging cell line cannot be packaged since the Ψ signal has been deleted. However, other sequences are present in the host-cell genome (including endogenous retroviruses and retrovirus-like sequences such as VL30 elements; Keshet *et al.*, 1980) that can also be packaged, suggesting that these elements may even compete for packaging, after introduction of the retroviral vector, with the vector RNA genome (Hatzoglou *et al.*, 1990a; Muenchau *et al.*, 1990; Purcell *et al.*, 1996; Scarpa *et al.*, 1991).

Most of these endogenous elements have yet to be associated with disease. However, primates have developed T-cell lymphomas after receiving cells that had been genetically modified by infection with a retroviral vector stock that was contaminated with RCV (Donahue *et al.*, 1992). Endogenous retroviruses and VL30 elements as well as recombinant MLV, generated as a result of recombination between the packaging and vector constructs, could be detected in DNA isolated from these lymphomas (Purcell *et al.*, 1996). The use of non-murine cells as a basis for retroviral vector packaging cells, although carrying other endogenous viral sequences, would at least reduce the possibility of the production of RCV as a result of recombination with endogenous retroviral elements, since retroviruses of differing species show little or no sequence homology (Leib-Mösch *et al.*, 1990). Human embryonic kidney 293 cells have been used as a basis for the ecotropic retroviral packaging cell line BOSC 23 (Pear *et al.*, 1993). These cells carry two constructs, one encoding the Gag and Pol proteins and the other encoding the Env proteins (Danos and Mulligan, 1988), and give titres of $>10^6$ infectious particles per millilitre (Pear *et al.*, 1993).

3.4 Retroviral vector titre

In comparison to vector systems based upon adenovirus or adeno-associated virus (see Chapters 4 and 5), standard retroviral vector systems usually yield a lower number of infectious virus particles per unit volume (titre). The issue of titre has been deemed important, particularly when the target cells (such as stem or progenitor cells) represent a small and difficult to isolate proportion of a larger pool of cells, all of which are potentially infectable, or when vectors are to be used for *in vivo* gene transfer. However, recent data suggest that the titre measurement *in vitro* may not be relevant for the success of clinical gene therapies. Indeed, assays of virus titre may be inaccurate since it has been recently shown that end-point dilution estimates of virus titre do not predict gene transfer efficiency but rather tend to overestimate it. The dilutions of virus used in these assays remove an inhibitory activity, probably due to the presence of free virus envelope protein or empty particles (Forestell *et al.*, 1995). Moreover, the conditions that have been previously used to estimate infection efficiency in cell culture systems may not have been optimal (Morgan *et al.*, 1995). The ability to target vector virus to cells of interest would contribute to greater gene transfer efficiency to the relevant cell type, even if titres were not increased. However, since targeting is almost invariably associated with reduced titres (see Section 3.7), considerable effort has been devoted to improving retroviral vector titres.

The most dramatic increase in titre has been obtained by replacing the MLV Env protein with that of another enveloped virus, vesicular stomatitis virus (VSV), a process known as pseudotyping (see Section 3.7.1). This was achieved by establishing packaging cells that express the Gag and Pol proteins of MLV and the surface protein of VSV. A very broad range of cell types from all kinds of species can be efficiently infected by VSV and this property is conferred upon MLV/VSV pseudotypes (Yee *et al.*, 1994). Further, the pseudotyped virions are more stable than MLV virions, facilitating the concentration of virus without appreciably affecting infectability, and resulting in titres of more than 10^9 colony-forming units (c.f.u.) ml^{-1} (Burns *et al.*, 1993).

The amount of virus produced by packaging cells also can be increased by treating these cells with agents such as sodium butyrate, which enhances expression of the vector and packaging construct, leading to a 10–1000-fold increase in virus production (Olsen and Sechelski, 1995; Pages *et al.*, 1995; Soneoka *et al.*, 1995). Alternatively, treatment of vector virus particles with dNTPs has been shown to increase their infectivity by a factor of 10. The relatively high concentration of dNTPs required (5 mM) appears to initiate and drive reverse transcription *within the virion* before the virus infects a cell, suggesting that reverse transcription may normally be inefficient and thus a limiting factor for infection efficiency (Zhang *et al.*, 1995).

Other strategies have also been used for improving the titres of retroviral vectors. Concentration of virus by centrifugation results in more virions per unit volume but a relative reduction of infectivity due to physical damage. Recently, concentration of vector virus has been achieved, without such loss of infectivity, by filtration through hollow fibres (Kotani *et al.*, 1994; Paul *et al.*, 1993). In order to increase the efficiency of the subsequent infection, the vector virus can be centrifuged on to the target cells (Kotani *et al.*, 1994). Even though virus production appears to be optimal at 37°C, it has been reported that the virus particles are more stable at 32°C, due to a retarded rate of inactivation (Forestell *et al.*, 1995). Thus, incubation of vector-producing packaging cells at 32°C has been used to enhance virus titre (Bunnell *et al.*, 1995; Kotani *et al.*, 1994). Phosphate starvation of the target cells to be infected has also been used to increase the expression of the phosphate transporter used as the amphotropic MLV receptor (see Section 3.7.1), and this has led to enhanced transduction (Bunnell *et al.*, 1995; Kavanaugh *et al.*, 1994).

3.5 Retroviral vector inactivation by complement

Murine retroviruses are directly inactivated by human serum complement (Welsh *et al.*, 1975). Retroviral vectors suffer the same fate *in vivo*. Complement-mediated inactivation is non-lytic and depends both on the retrovirus and on the cells from which the virus has been produced. For example, MLV produced from mouse cells is much more readily inactivated than MLV produced from human or mink cells (Takeuchi *et al.*, 1994), and this appears to be at least in part due to the expression of α-1-3 galactosyl tranferase in murine cells which modifies the retroviral vector envelope proteins. Human cells lack a functional α-1-3 galactosyl transferase. Furthermore, human serum contains antibodies directed against this Gal (α1-3) Gal modification (Takeuchi *et al.*, 1996). Human immunodeficiency virus-1 (HIV-1) and human T-cell leukaemia virus (HTLV) are not inactivated by human serum complement (Banapour *et al.*, 1986; Hoshino *et al.*, 1984). Thus, it should be possible to use combinations of virus and packaging cells that show high resistance to complement inactivation. Even though the construction of retroviral vector systems based on retroviruses with more complex regulation mechanisms such as HIV is difficult (see Section 3.7.1), elements from these viruses could be exploited in conjunction with MLV-based vector systems to give hybrid vectors resistant to human complement. Since complement resistance is a property of the viral surface proteins, it may be possible to incorporate the surface or Env proteins from HIV or HTLV (also known as pseudotyping; see Section 3.7.1) into the envelope of conventional MLV retroviral vectors (Wilson *et al.*, 1989). Human foamy viruses are also not inactivated by the human complement system and vectors are presently being constructed based upon this retrovirus (Russell and Miller, 1996; Schmidt and Rethwilm, 1995). Alternatively, the administration of monoclonal antibodies that block specific components of the complement pathway may prevent inactivation, at least for the duration of the gene therapy (Rother *et al.*, 1995). Despite these problems of complement inactivation, local rather than systemic administration of the vector, for example by direct intra-tumoral injection (Culver *et al.*, 1992; Oldfield *et al.*, 1993) or by injection of vector-producing packaging cells in immunoprotective capsules (Saller et al., 1995), should also be successful.

3.6 Infection of dividing cells

A prerequisite for the infection of target cells with MLV is that these cells are dividing (Miller *et al.*, 1990), since nuclear membrane breakdown is necessary for the movement of the viral core into the nucleus (Lewis and Emerman, 1994; Roe *et al.*, 1993). By contrast, it has been reported that HIV is able to infect non-dividing cells, a property that is conferred by the Gag and Vpr proteins of this virus (reviewed in Stevenson, 1996). Although HIV-based vectors (see Section 3.7.1) may offer an alternative to classic MLV-based retroviral vectors for the infection of non-dividing cells, the construction of such vectors has been fraught with difficulties due to the virus's complex regulatory machinery. Even if these problems (and the psychological problems) associated with the use of HIV-based vectors can be overcome, it seems clear that HIV still preferentially (and much more efficiently) infects dividing cells

(Dobrescu *et al.*, 1995). Most probably, other strategies involving the incorporation of subcomponents, (such as HIV Gag and Vpr proteins, possibly linked to proteins that can push a quiescent cell through one round of cell division) into conventional MLV retroviral systems will prove superior.

In the meantime, this specific property of retroviral vectors to infect rapidly proliferating cells preferentially, often cited as a disadvantage, has been used to target rapidly proliferating cells, such as tumour cells, on a background of quiescent, end-differentiated normal cells. One of the first studies in which this property of retroviral vectors was exploited, targeted the delivery of the thymidine kinase gene of herpes simplex virus (*HSV-tk*) to rapidly dividing brain tumour cells in clinical trials (Culver *et al.*, 1992; Oldfield *et al.*, 1993) or in animal models (Ram *et al.*, 1993). *HSV-tk* gene expression allows the selective conversion of the non-toxic prodrug ganciclovir to the phosphorylated form that is toxic for replicating cells. It was reasoned that non-transformed brain cells do not undergo cell division and therefore should be refractory both to infection with the retroviral vector and to the lethal toxic effects of phosphorylated ganciclovir. Thus, only the tumour cells should be infected and become susceptible to the killing effects of phosphorylated ganciclovir. In practice, it turned out that the vector-producing cells transmitted the toxic effect to the brain cells by a so-called bystander effect (Bi *et al.*, 1993; Freeman *et al.*, 1993; Ram *et al.*, 1993), in which phosphorylated ganciclovir is thought to pass via gap junctions from the initial retrovirus-infected cell to adjoining non-infected cells (Bi *et al.*, 1993; Elshami *et al.*, 1996) rather than by infection. Targeting of *HSV-tk* gene delivery to proliferating tumour cells has also been attempted in mice after systemic injection of vector particles via the tail vein (Vile *et al.*, 1994). Interestingly, it appears that the immune system also plays a major role in eliminating tumour cells in this model, presumably being stimulated by the release of previously hidden tumour antigens from destroyed cells (Vile *et al.*, 1994).

3.7 Targeting of retroviral vector-mediated gene delivery and expression

Many of the current gene therapy protocols involve an *ex vivo* approach in which cells are removed from the patient, modified in the laboratory and then returned to the individual. The success of such an approach for the treatment of severe combined immunodeficiency resulting from adenosine deaminase (ADA) deficiency was recently documented (Blaese *et al.*, 1995; Bordignon *et al.*, 1995; Kohn *et al.*, 1995; see Chapter 13). These investigators were able to establish long-term correction of haematopoetic cells after *ex vivo* gene therapy by using natural selection for the *function* of the transferred gene to maintain corrected haematopoetic cells expressing the introduced *ADA* gene. Expression of this gene protects lymphocytes against the destructive effects of the ADA substrate, which otherwise results in devastation of the immune system and immunodeficiency. *Ex vivo* correction of haematopoietic cells with a functional *ADA* gene copy resulted in long-term (up to 3 years has so far been measured) re-population by *ADA*-expressing lymphocytes.

The most interesting target cells for such *ex vivo* approaches are haematopoetic cells, which are easily harvested and readily re-infused; also, most of the applications of haematopoetic cell modification are for inborn disease, and thus require long-term correction. Retroviral vectors, in contrast to adenovirus and adeno-associated virus vectors, are ideally suited for long-term correction since they integrate into the genome of the target cells and, becoming part of the genome, are not lost with time.

However, many inborn and acquired diseases manifest themselves in cells or organs which cannot be removed, brought into culture, modified and then re-introduced. These situations call for *in vivo* gene therapy. In such cases, the retroviral vector carrying a therapeutic gene should be locally or systemically introduced into the patient and the infection event allowed to occur *in vivo*. Ideally, in such an *in vivo* therapeutic approach, the vector should only infect the cells in which the defect manifests itself and/or the expression of the introduced gene should be controlled by regulatory elements that target the expression to the relevant cell type.

Retroviral vectors that are replication-competent may have to be used with this approach, particularly if the cells of a multicellular organ or dense tumour mass are to be targeted or if very high virus titres are required. To prevent unsolicited replication of the retrovirus during or after gene therapy, vectors should be designed that can be conditionally silenced or inactivated. The remainder of this article will deal with some of the potential strategies addressing these issues and will also review progress being made in these areas.

3.7.1 Restriction at the level of the infection event

The most commonly used retroviral vectors are based on MoMLV, which is able to infect a wide range of cell types. One obvious way in which infection may be targeted is by limiting the types of cell that the retroviral vector can infect. Alternatively, retroviral vectors based upon retroviruses other than MoMLV or on combinations of components from different types of retrovirus may also target the infection.

Modification of the MoMLV infection spectrum. MLVs can be classified according to their host infection spectrum. Isolates that are able to infect only rodent cells are called *ecotropic*, whereas MLVs that infect exclusively cells from species other than rodents are termed *xenotropic*. *Amphotropic* MLVs are able to infect cells of most species including rodents and humans. This observed tropism is determined by the Env protein (also known as SU for surface protein) of the MLV together with the availability of the corresponding receptor on the target cell. The Env protein of amphotropic MoMLV (gp70), which is commonly used for gene therapy, interacts with a phosphate transporter which serves as the host-cell receptor and is expressed in many tissues and on many cell types (Miller and Miller, 1994; van Zeijl *et al.*, 1994). It is the amphotropic Env that is normally provided in retroviral vectors used for human gene therapy.

Nevertheless, it has recently been shown that it is possible to limit the infection spectrum of this virus and consequently of MoMLV-based vectors. In many of these studies, ecotropic MLV vectors were modified to ensure that infection targeting had taken place since these retroviral vectors cannot normally infect human or other non-rodent cells. The earliest attempts to achieve infection targeting involved the coupling of antibodies, directed against known proteins that are expressed on the surface of the target cell, to antibodies specific for the virus Env protein via streptavidin. Using antibodies directed against class I and class II major histocompatibility complex (MHC) antigens (Roux *et al.*, 1989) or against the receptor for epidermal growth factor (EGF; Etienne-Julan *et al.*, 1992), it was possible to target infection to cells expressing these molecules on their surface. However, the efficacy of targeting of retroviral vector infection depends on the type of cell-encoded protein that is chosen as a target. For example, antibodies to transferrin attached to virus in this fashion resulted in binding and internalization of the virus but integration was not observed (Goud *et al.*, 1988). Thus, some potential target proteins may not be suitable for this approach of antibody-mediated targeting of infection (Etienne-Julan *et al.*, 1992). A second problem with this strategy is that infection is inefficient and titres are severely reduced compared with non-targeted amphotropic retroviral vectors.

Chemical modification of the retroviral Env protein has also been used to target the infection spectrum of retroviral vectors. This approach has been used to enable a chemically modified ecotropic MoMLV to infect human hepatocytes. Hepatocytes are unique in that they carry receptors that are able to internalize asialoglycoproteins. Viral Env proteins can be artificially converted to asialoglycoproteins by coupling them to lactose (Neda *et al.*, 1991). After this chemical modification, the ecotropic virus behaved more like an amphotropic virus in that it was able to infect hepatocytes of human origin efficiently and was less able to infect previously susceptible host rodent cells (Neda *et al.*, 1991). However, although this may be useful for hepatocytes, such an approach may not be generally applicable for other cell types since it relies on specialized receptors and the ability to modify the viral Env chemically to suit the receptors.

In the past few years, the most popular strategy for modifying the infection spectrum of retroviral vectors has involved the genetic engineering of the viral *env* gene carried in the retroviral packaging cell line. This has been made possible by the identification of the regions of the gp70SU Env protein of MLV involved in receptor recognition (Battini *et al.*, 1992; Morgan *et al.*, 1993; Ott and Rein, 1992). These regions have been replaced with gene segments encoding epitopes that would recognize other receptors, thereby allowing the selective control of receptor targeting of the resultant chimeric Env protein. In these approaches, the ligand–Env protein is produced in the packaging cell line *in addition* to the normal, non-modified retrovirus envelope protein which is presumably required for stability. Very often, the ligand that is used is the variable domain of a single-chain antibody specific for a defined receptor or cell-surface protein. A summary of these experiments is given in *Table 3.2*. Even though re-targeting of the infection event by modification of the retroviral envelope protein has been successfully achieved by a number of groups, it is invariably associated with reduced titres. Clearly, more knowledge is required about the mechanisms that govern the normal functioning

Table 3.2. Re-targeting retroviral vector infection spectrum by envelope protein modification

Virus	Ligand	Receptor	Reference
MLV	scvf	NIP hapten	Russell *et al.* (1993)
ALV	RGD peptide	Integrin	Valsesia-Wittmann *et al.* (1994)
MoMLV	Erythropoeitin (EPO)	EPO receptor	Kasahara *et al.* (1994)
MoMLV	EGF	EGF receptor	Cosset *et al.* (1995)
MoMLV	Amphotropic domain	Ram-1	Cosset *et al.* (1995)
MoMLV	scvf	LDLR	Somia *et al.* (1995)
MoMLV	Heregulin	HER-3/HER-4	Han *et al.* (1995)
SNV	scvf	DNP hapten	Chu *et al.* (1994)
FrMLV	CD4	gp120	Matano *et al.* (1995)

ALV, avian leukosis virus; DNP, dinitrophenol; NIP, 4-hydroxy-5-iodo-3-nitrophenacetyl caproate; RGD, Arg-Gly-Asp; scvf, single chain variable fragment; SNV, spleen necrosis virus.

of retroviral envelope proteins before chimeric envelopes can be constructed that retain the ability of these proteins to recognize the receptor *and* initiate infection efficiently.

Recently, phage display libraries which facilitate rapid screening have been used to select 12 or 20 amino acid peptides that bind to cell receptors present on specific cell types. Peptides of 20 amino acids generally bind cell receptors more efficiently than dodecapeptides (12 amino acids). Peptides identified using this system could be incorporated into any gene delivery system, including retroviral vectors, to achieve targeting (Barry *et al.*, 1996).

The chimeric Envs that have been used for targeting are expressed in packaging cell lines that also produce the standard retroviral Env protein. This raises the possibility that another approach to re-directing retroviral vector infection specificity could be the co-expression of other ligands on the virus surface along with the normal amphotropic Env SU proteins. This strategy should still allow normal viral internalization and may also result in favoured uptake by cells expressing the receptor for the co-expressed ligand.

Retroviral vectors based on other retroviruses. Some retroviruses such as bovine leukaemia virus (BLV), simian immunodeficiency virus (SIV), mouse mammary tumour virus (MMTV) and HIV show restricted *in vivo* tissue tropism, which is due, at least in part, to restrictions at the level of the infection event. Retroviral vectors based upon these retroviruses may likewise be expected to deliver genes specifically to the same limited number of cell types as the parental virus. Retroviral vector systems are being developed that are based on a number of retroviruses (*Table 3.1*) including HIV, SIV and MMTV.

Vector systems based upon HIV would have a number of advantages including the ability to deliver genes preferentially to CD4-bearing T cells, a cell type that is particularly relevant for gene therapy. There is evidence that other receptors may also be used by HIV, thereby extending the possible target cell range for HIV-based retroviral vectors. Such a vector system would be resistant to the effects of human complement and one pathway of entry may actually even use the complement system (Thielens *et al.*, 1994). Moreover, HIV, unlike MLV, is able to infect non-dividing cells, a property that seems to be conferred by the matrix (MA) protein of the *gag* gene and Vpr (reviewed in Stevenson, 1996). Retroviral vectors based upon HIV may also incorporate virus regulatory mechanisms such as those mediated by the viral proteins Tat and Rev. This would allow the expression of therapeutic genes carried by HIV-based retroviral vectors only in cells that are infected by HIV and expressing Tat and Rev. Although retroviral vectors based upon HIV have been constructed (see *Table 3.1*), the more complex genome structure and gene regulatory mechanisms of HIV (reviewed by Levy, 1993) have made their construction more problematic than that of retroviral vectors based on MLV. To date, titres of only 10–100 c.f.u. ml^{-1} have been obtained with HIV-derived vector systems, although this can be increased to 10^5 c.f.u. ml^{-1} by pseudotyping the HIV vector with either the MLV or the VSV envelope (Naldini *et al.*, 1996). Further, the stigma associated with acquired immunodeficiency syndrome (AIDS) and the possibility of generating novel human retroviruses by recombination may preclude the use of such vectors for clinical gene therapy. However, if HIV-based vectors were to gain public acceptance, HIV-2 might be more suitable as a starting point for vector construction since it is generally less pathogenic than HIV-1 (Garzino-Demo *et al.*, 1995).

MMTV is also an interesting virus for vector construction since it preferentially infects and is expressed in mammary epithelial cells and B-lymphocytes (Acha-Orbea and MacDonald, 1995; Günzburg and Salmons, 1992). Further, MMTV (like the amphotropic MLVs) is able to infect cells of non-rodent origin (Howard and Schlom, 1978). However, the genome organization and regulatory mechanisms that control virus expression and production of MMTV are again much more complex than those of the MLVs (Acha-Orbea *et al.*, 1991; Choi *et al.*, 1991; Günzburg *et al.*, 1993; Salmons *et al.*, 1990; Wintersperger *et al.*, 1995). These features are probably responsible for the poor titres that have been obtained when using retroviral vector systems based upon this virus (Salmons *et al.*, 1989). Greater insight into the biology of MMTV should provide the basis for the construction of further generations of improved vector systems. Nevertheless, components of both HIV and MMTV may be useful for the construction of hybrid gene delivery systems or retroviral vectors based on MLV but carrying the Env proteins of more infection-restricted retroviruses.

Pseudotyped vectors. It has long been known that cells simultaneously infected with two enveloped viruses are able to produce phenotypically mixed virus particles, also known as pseudotyped viruses (Zavada, 1972). These virus particles carry the genetic information and core of one virus and an envelope containing the Env proteins of either the second virus or of both viruses. Thus, pseudotyped viruses have an extended infection spectrum since they are able to infect cells that can be infected by the second virus (reviewed by Weiss, 1993). Pseudotyped retroviral vectors have been produced by a number of groups using packaging cell lines that produce only the Gag and Pol proteins from one virus and only the Env proteins from a second virus (*Table 3.3*). This ensures that the retroviral vector shows the infection spectrum of only the Env-providing virus. Pseudotyped retroviral vectors based on MLV and carrying the surface protein of VSV have also been created (see Section 3.4) but these vectors, although giving very high titres, are non-targeted gene delivery vehicles since they can infect many different cell types from a variety of animals including insects, fish and amphibians (Burns *et al.*, 1993; Yee *et al.*, 1994). Targeted gene transfer should be achievable using the envelope from a virus that shows a narrow infection spectrum (e.g. HIV, HTLV or MMTV) and preliminary data suggest that this is indeed the case (Salmons *et al.*, 1995).

Table 3.3. Pseudotyped retroviral vectors

Virus providing genome and core	Virus providing Env	Reference
MoMLV	HTLV-I	Wilson *et al.* (1989)
HIV	MLV (amphotropic domain)	Page *et al.* (1990)
MoMLV	GaLV	Miller *et al.* (1991); Wilson *et al.* (1989)
HIV	HTLV-I	Landau *et al.* (1991)
RSV	influenza HA	Dong *et al.* (1992)
MLV	RSV	Landau and Littman (1992)
MLV	VSV	Burns *et al.* (1993); Yee *et al.* (1994)
MLV	MMTV	Salmons *et al.* (1995)

GaLV, gibbon ape leukaemia virus; HA, haemagglutinin; RSV, Rous sarcoma virus.

3.7.2 Restriction at the level of expression

Even if a therapeutic gene is delivered promiscuously to both target and non-target cell types, it is theoretically possible to ensure that the therapeutic gene product is only made in the desired target cells. This can be achieved by limiting gene expression with the use of transcriptional control elements or tissue-specific promoters. Gene expression is regulated by the availability (e.g. presence or absence) of multiple transcription factors that recognize specific regulatory elements present in the promoter region of genes. Thus, if a factor required for expression is not present in a cell, then this cell will not express efficiently a gene under the transcriptional control of that factor. Likewise, the inclusion of the specific regulatory element that mediates the ability of the factor to drive expression of a gene that is not normally regulated by this factor, results in gene expression in cells that contain the factor. Such regulatory elements are useful for targeting gene expression, whatever gene transfer method is used, and are commonly used in transgenic animal studies.

The promoter of MLV is relatively promiscuous in that it directs expression of the virus in most cell types. Regulatory elements that confer tissue-specific expression can be included in retroviral vectors either in the body of the vector (*Figure 3.3*) or in addition to, or in place of, the retroviral promoter or regulatory elements (see Section 3.3.1). Regulation of expression as a means of targeting has been attempted in tumour, liver and mammary gland tissue in different model systems.

Expression targeting to tumours. The ability to target the expression of therapeutic genes becomes an issue of particular importance when the therapeutic gene encodes a toxic product or if its expression confers sensitivity to a prodrug (suicide gene) upon a cell (Moolten, 1994). These therapeutic genes are especially useful for gene therapy approaches directed towards the treatment of cancers. Several examples of tissue-selective and tumour-selective targeting are cited in Chapter 16.

A number of mammary-specific promoters have been defined that may be useful for targeting the expression of therapeutic genes to breast cancer cells. Retroviral vectors have been constructed in which the promoter region of MLV has been completely replaced by a putative mammary-specific regulatory region from the gene for murine whey acidic protein (WAP) (Kolb *et al.*, 1994, 1995), a gene that is expressed exclusively in pregnant and lactating mammary glands, but also in some mammary tumours (Öztürk, manuscript in preparation). Although such ProCon retroviral vectors infect many different cell types, the expression of an indicator gene carried in the retroviral vector under the transcriptional control of the WAP promoter is detectable only in primary mammary gland cells isolated from pregnant mice but not in fibroblasts or normal mammary cells (Saller, 1994; Salmons *et al.*, 1995). Further, indicator gene expression could be detected after injection of the virus into the mammary gland of pregnant but not non-pregnant mice (Günzburg *et al.*, 1995; Saller, 1994).

Expression targeting to specific cell types and organs. Expression targeting of retroviral vectors to normal organs has been mainly studied using the liver or hepatocytes as a target. The liver is of particular interest with respect to gene therapy since many metabolic disorders are the result of genetic defects in liver-specific enzyme-encoding genes (Horwitz, 1991; Strauss, 1994). Also, recent evidence suggests that genetically modified hepatocytes can re-populate the liver if they express genes that confer a selective advantage upon them (Overturf *et al.*, 1996). There are a number of promoters available which have been isolated from genes specifically or preferentially expressed in the liver.

The promoter of the phosphoenolpyruvate carboxylase (*PEPCK*) gene contains regulatory elements that can direct the expression of heterologous genes to the liver and kidney. Expression of these heterologous genes is subsequently controlled by hormones and diet, analogously to the normal *PEPCK* gene (McGrane *et al.*, 1988). This promoter has been independently coupled to both the neomycin resistance gene and the bovine growth hormone (*bGH*) gene as indicators and placed internally into a MoMLV-based retroviral vector. Recombinant vector virus was injected into the peritoneal cavity of fetal rats or into the portal vein of hepatectomized adult rats. Whereas 38–50% of the rats were successfully infected if the virus was injected *in utero*, only 10% of adult rats were positive for proviral DNA in hepatic cells after infection via the portal vein. The indicator genes were expressed in the livers of more than half of the rats that were successfully infected. Moreover, the expression of the chimeric constructs was regulated by diet and hormones as expected (Hatzoglou *et al.*, 1990b).

A second liver-specific promoter is that of the α-fetoprotein (*AFP*) gene. This promoter has been linked to the *tk* gene of varicella-zoster virus (VZV) and inserted into a retroviral vector. The VZV-tk enzyme converts the non-toxic prodrug 6-methopurine arabinonucleoside (araM) to the cytotoxic compound adenine arabinonucleoside triphosphate (araATP). Infection of hepatoma cell lines with the retroviral vector resulted in the expression of VZV-tk, and cell death when araM was administered. By contrast, cell lines derived from other organs did not allow expression of VZV-tk upon infection with the vector and these cells were thus unaffected by araM (Huber *et al.*, 1991).

A number of hepatocyte-specific promoters have been directly compared for their ability to direct a high level of expression of an indicator gene in the context of a retroviral vector, after *in vivo* infection of hepatocytes. The human α_1-antitrypsin (*hAAT*) promoter gave rise to the highest levels of expression, followed by the murine albumin (*mAlb*) promoter, whereas the rat *PEPCK* was weaker than the *MLV* promoter (Hafenrichter *et al.*, 1994). Although these vectors were not tested for *specific* or even *selective* expression, the evaluation of promoter activity from genes that are expressed in one particular cell type in the context of a retroviral vector had previously not been undertaken. In a more recent study from the same group, retroviral vectors carrying as internal promoters either the *hAAT* or *mAlb* promoters, as well as the constitutive promoter of the RNA *PolII* gene, were examined for their ability to drive expression of the *hAAT* reporter gene after infection of fibroblasts or hepatoma-derived cells. Whereas the liver-specific *hAAT* promoter was only active in hepatoma cells, the *mAlb* promoter was active in fibroblasts as well as hepatoma cells. The activation of the liver-specific *mAlb* promoter in fibroblasts could be shown to be due to an interaction between the *mAlb* and *MLV* promoters (Wu *et al.*, 1996). However, the coupling of *hAAT* coding sequences to its own promoter could conceivably have played a role in determining the absolute liver specificity observed, since transcriptional regulatory elements are also known to be located within coding regions of certain genes.

One problem associated with this approach is that, depending on the site of integration of the retroviral vector in the host-cell genome, the tissue-specific regulation conferred upon the retroviral promoter may be overridden by strong cellular regulatory elements located in the vicinity of the integration site. DNA sequences, termed locus control region (LCR) sequences, confer position-independent, high-level expression upon genes (Dillon and Grosveld, 1993). To ensure appropriate levels of therapeutic gene expression, it has been proposed to include LCR sequences along with the therapeutic gene. The inclusion of a 36 bp core subsequence of the human β-globin LCR in a retroviral vector carrying a human β-globin gene was shown to enhance the expression of human β-globin in mouse erythroleukaemia cells, although expression levels were still therapeutically suboptimal (Chang et al., 1992). More recently, a similar retroviral vector construct was shown to give widely varying levels of expression (4–146%) in diffferent infected cell clones (Sadelin et al., 1995), suggesting either that more of the LCR region is necessary to obtain position-independent expression or that LCR sequences are not able to function in the context of retroviral vectors.

An alternative strategy to ensure position-independent expression of genes in retroviral vectors is to shield them from the effect of enhancers or repressors located in the vicinity of the integration site (Duch et al., 1994). A number of such insulators have been identified in *Drosophila* (Cai and Levine, 1995; Gerasimova et al., 1995; Jupe et al., 1995; Roseman et al., 1995). Similar elements exist in mammalian cells (Kalos and Fournier, 1995) and these could be incorporated into future vector designs.

3.7.3 Inducible vectors

The ability to regulate the expression of genes delivered by retroviral vectors may be useful in certain situations as well as for targeting (Salmons and Günzburg, 1993). Vector systems based on other retroviruses such as HIV or MMTV can be constructed to maintain conditional expression that is characteristic of these viruses. Unfortunately, the efficiency of gene transfer obtained using these vectors is at present troublesome (see Section 3.7.1).

Nevertheless, retrovirus-encoded regulatory factors may be used both to target and to control the expression of retroviral vectors, if the relevant regulatory elements are included in the vector. The best example of this is HIV, in which the virally encoded transcriptional activator Tat is required for efficient expresssion from the HIV promoter (reviewed by Levy, 1993). In theory, therapeutic genes under the control of the HIV promoter or the transactivating response (TAR) element that mediates Tat-induced transcription should be targeted to HIV-infected cells, since only these cells express Tat. Inducibility through the Tat/TAR activation system has been employed in studies designed to direct the expression of the gene for the diphtheria toxin A (DT-A) chain to HIV-infected cells. The gene for the DT-A chain was inserted into an amphotropic MoMLV vector under the transcriptional control of the HIV-LTR carrying the TAR element. Following transduction of the human H9 cell line with the resultant recombinant vector virus, these cells were assayed for their ability to support HIV production after superinfection or transfection with HIV. The H9 cells transduced with the retrovirus vector were selectively eliminated, presumably due to the expression of Tat by the incoming HIV provirus which led to activation of the expression of DT-A which is toxic for the cells (Harrison et al., 1992).

The TAR element from the HIV-LTR has also been coupled to the cytomegalovirus (CMV) immediate–early enhancer/promoter and inserted into the U3 region of the 3'LTR of a retroviral vector. In the presence of the HIV transactivator Tat, higher titres of this vector could be obtained than from the parental MLV vector, suggesting that the Tat/TAR enhancer could upregulate expression from the MLV promoter in the 5′LTR. Furthermore, the basal expression levels of the chloramphenicol acetyltransferase (CAT) indicator gene, carried in the retroviral vector, observed in cells infected with the modified virus were higher than those from the parental, vector-infected cells and expression could be further enhanced when Tat was present (Robinson *et al.*, 1995).

The glucocorticoid-regulated promoter of MMTV has proved to be a useful inducible promoter for a variety of purposes (Günzburg and Salmons, 1992) including the regulatable expression of potentially therapeutic genes (Sparmann *et al.*, 1994). This promoter has been inserted into MLV ProCon vectors (see Section 3.3) and, in the presence of glucocorticoids, can drive maximal expression of a marker gene (Salmons *et al.*, 1995) and of the p21^{SDI-1} (SDI, senescent cell-derived inhibitor) cell-cycle inhibitor (Noda *et al.*, 1994; Stinchcomb, 1995), which is a possible therapeutic gene (Mrochen, submitted). However, even though the hormone inducibility of the MMTV promoter is useful in *in vitro* models, it is questionable, in view of the levels of endogenous hormone, whether this system can be used to regulate expression in patients.

The promoters of a number of cellular genes also carry regulatory elements that confer inducibility upon them. One example of such a promoter that has been inserted into an MLV-based vector and linked to an indicator gene is the promoter of the L-type pyruvate kinase gene. Transcription from this promoter can be regulated by glucose. Hepatocytes infected with this retroviral vector showed concentration-dependent, glucose-responsive expression of the marker gene (Chen *et al.*, 1995). Artificial promoters regulatable by tetracycline (Gossen and Bujard, 1992) have also been inserted into retroviral vectors. The expression of indicator genes in cells infected with these vectors can be repressed by tetracyclin (Paulus *et al.*, 1996). Recently, this system has also been adapted to allow tetracyclin induction of linked heterologous genes (Gossen *et al.*, 1995) and this represents a very real application for the regulation of gene expression in gene therapy. However, the efficiency of these systems is to some extent dependent on the cell type in which they will be used (Gossen and Bujard, 1995) and, depending on the integration site, basal levels of expression may be detectable (see Section 3.7.2).

3.8 Summary

Retroviral vectors are ideal gene transfer systems for efficient and stable gene transfer and expression. Approaches have been taken to target either therapeutic gene delivery or expression to any cell type for which it is known that a gene exists that is specifically or preferentially expressed. A plethora of genes are now being analysed with respect to the regulation of expression, since an understanding of the control of gene expression is a major goal of molecular biology. These studies are and will continue to be useful for the isolation and identification of promoter and regulatory elements that can be used to target the expression of therapeutic genes, not only for retroviral vectors but also for other gene transfer systems. Furthermore, regulatable promoters may also be of value, even though these elements may not result in absolutely strict on–off situations, and it seems likely that many applications will still require such targeted and/or regulated vectors to be used for gene transfer to physically isolated cells, possibly by local application of vector rather than by systemic delivery.

References

Acha-Orbea H, MacDonald HR. (1995) Superantigens of mouse mammary tumor virus. *Annu. Rev. Immunol.* **13**: 459–486.

Acha-Orbea H, Shakhov AN, Scarpellino L, Kolb E, Mueller V, Vessaz-Shaw A, Fuchs R, Bloechlinger K, Rollini P, Billotte J, Sarfidou M, Macdonald HR, Diggelmann H. (1991) Clonal deletion of V beta 14-bearing T cells in mice transgenic for mammary tumour virus. *Nature* **350**: 207–211.

Adam MA, Osbourne WRA, Miller AD. (1995) R-region cDNA inserts in retroviral vectors are compatible with virus replication and high level protein synthesis from the insert. *Hum. Gene Ther.* **6**: 1169–1176.

Armentano D, Yu SF, Kantoff P, von Ruden T, Anderson W, Gilboa E. (1987) Effect of internal viral sequences on the utility of retroviral vectors. *J. Virol.* **61**: 1647–1650.

Banapour B, Sernatinger J, Levy JA. (1986) The AIDS-associated retrovirus is not sensitive to lysis or inactivation by human serum. *Virology* **152**: 268–271.

Barry MA, Dower WJ, Johnston SA. (1996) Toward cell-targeting gene therapy vectors: selection of cell-binding peptides from random peptide-presenting phage libraries. *Nature Med.* **2**: 299–305.

Battini J-L, Heard JM, Danos O. (1992) Receptor choice determinants in the envelope glycoproteins of amphotropic, xenotropic and polytropic murine leukemia viruses. *J. Virol.* **66**: 1468–1475.

Baum C, Hegewischbecker S, Eckert HG, Stocking C, Ostertag W. (1995) Novel retroviral vectors for efficient expression of the multidrug-resistance (MDR-1) gene in early hematopoietic cells. *J. Virol.* **69**: 7541–7547.

Bender MA, Palmer TD, Gelinas RE, Miller AD. (1987) Evidence that the packaging signal of moloney murine leukaemia virus extends into the gag region. *J. Virol.* **61**: 1639–1646.

Bi WL, Parysek LM, Warnick R, Stambrook PJ. (1993) *In vitro* evidence that metabolic cooperation is responsible for the bystander effect observed with HSV tk retroviral gene therapy. *Hum. Gene Ther.* **4**: 725–731.

Blaese RM, Culver KW, Miller AD, Carter CS, Fleisher T, Clerici M, Shearer G, Chang L, Chiang Y, Tolstoshev P, Greenblatt JJ, Rosenberg SA, Klein H, Berger M, Mullen CA, Ramsey WJ, Muul L, Morgan RA, Anderson WF. (1995) T lymphocyte-directed gene therapy for ADA SCID: initial trial results after 4 years. *Science* **270**: 475–480.

Bordignon C, Notarangelo N, Nobili N, Ferrari G, Casorati G, Panina P, Mazzolari E, Maggioni D, Rossi C, Servida P, Ugazio AG, Mavilio F. (1995) Gene therapy in peripheral blood lymphocytes and bone marrow for ADA-immunodeficient patients; gene transfer to hematopoietic progenitor cells using a retrovirus vector. *Science* **270**: 470–475.

Boris-Lawrie K, Temin HM. (1995) Genetically simpler bovine leukemia virus derivatives can replicate independently of Tax and Rex. *J. Virol.* **69**: 1920–1924.

Buchschacher GL, Panganiban AT. (1992) Human immunodeficiency virus vectors for inducible expression of foreign genes. *J. Virol.* **66**: 2731–2739.

Bunnell BA, Muul LM, Donahue RE, Blaese RM, Morgan RA. (1995) High-efficiency retroviral-mediated gene transfer into human and nonhuman primate peripheral-blood lymphocytes. *Proc. Natl Acad. Sci. USA* **92**: 7739–7743.

Burns JC, Friedmann T, Driever W, Burrascano M, Yee JK. (1993) Vesicular stomatitis virus G glycoprotein pseudotyped retroviral vectors: concentration to very high titer and efficient gene transfer into mammalian and nonmammalian cells. *Proc. Natl Acad. Sci. USA* **90**: 8033–8037.

Bushman FD, Craigie R. (1990) Sequence requirements for integration of Moloney Murine Leukemia Virus DNA *in vitro*. *J. Virol.* **64**: 5645–5648.

Cai H, Levine V. (1995) Modulation of enhancer–promoter interactions by insulators in the *Drosophila* embryo. *Nature* **376**: 533–536.

Carroll R, Lin J-T, Dacquel EJ, Mosca JD, Burke DS, St Louis DC. (1994) A human immunodeficiency virus type 1 (HIV-1)-based retroviral vector system utilizing stable HIV-1 packaging cell lines. *J. Virol.* **68**: 6047–6051.

Challita P-M, Kohn DB. (1994) Lack of expression from a retroviral vector after transduction of murine hematopoietic stem cells is associated with methylation *in vivo*. *Proc. Natl Acad. Sci. USA* **91**: 2567–2571.

Challita PM, Skelton D, El-Khoueiry A, Yu XJ, Weinberger K, Kohn DB. (1995) Multiple modification in *cis* elements of the long terminal repeat of retroviral vectors lead to increased expression and decreased DNA methylation in embryonic carcinoma cells. *J. Virol.* **69**: 748–755.

Chang JC, Liu D, Kan YW. (1992) A 36-base-pair core sequence of locus control region enhances retrovirally tranferred human beta-globin gene expression. *Proc. Natl Acad. Sci. USA* **89**: 3107–3110.

Chen R, Doiron B, Kahn A. (1995) Glucose responsiveness of a reporter gene transduced into hepatocytic cells using a retroviral vector. *FEBS Lett.* **365**: 223–226.

Choi Y, Kappler JW, Marrack P. (1991) A superantigen encoded in the open reading frame of the 3′ long terminal repeat of mouse mammary tumour virus. *Nature* **350**: 203–207.

Chu T-HT, Martinez I, Sheay WC, Dornburg R. (1994) Cell targeting with retroviral vector particles containing antibody-envelope fusion proteins. *Gene Ther.* **1**: 292–299.

Cohen-Haguenauer O, Restrepo L-M, Upegui-Gonzalez L-C, Masset M, Boiron M, Marty M. (1995) Design of an original Friend virus derived retroviral construct and use to confer resistance to *cis*-platinum derivatives and alkylating agents to human haematopoietic CD34⁺ cells. *Gene Ther.* **2**: 570–571.

Cosset F-L, Morling FJ, Takeuchi Y, Weiss RA, Collins MKL, Russell SJ. (1995) Retroviral retargeting by envelopes expressing an N-terminal binding domain. *J. Virol.* **69**: 6314–6322.

Culver KW, Ram Z, Wallbridge S, Ishii H, Oldfield EH, Blaese RM. (1992) *In vivo* gene transfer with retroviral vector producer cells for treatment of experimental brain tumors. *Science* **256**: 1550–1552.

Danos O, Mulligan RC. (1988) Safe and efficient generation of recombinant retroviruses with amphotropic and ecotropic host ranges. *Proc. Natl Acad. Sci. USA* **85**: 6460–6464.

Derse D, Martarano L. (1990) Construction of a recombinant bovine leukemia virus vector for analysis of virus infectivity. *J. Virol.* **64**: 401–405.

Dillon N, Grosveld F. (1993) Transcriptional regulation of multigene loci: multilevel control. *Trends Genet.* **9**: 134–137.

Dirks W, Wirth M, Hauser H. (1993) Dicistronic transcription units for gene expression in mammalian cells. *Gene* **129**: 247–249.

Dobrescu D, Ursea B, Pope M, Asch AS, Posnett DN. (1995) Enhanced HIV-I replication in Vbeta12 T cells due to human cytomegalovirus in monocytes: evidence for a putative herpesvirus superantigen. *Cell* **82**: 753–763.

Donahue RE, Kessler SW, Bodine D, McDonagh K, Dunbar C, Goodman S, Agricola B, Byrne E, Raffeld M, Moen R, Bacher J, Zsebo KM, Nienhuis A. (1992) Helper virus induced T cell lymphoma in nonhuman primates after retroviral mediated gene transfer. *J. Exp. Med.* **176**: 1125–1135.

Dong J, Roth MG, Hunter E. (1992) A chimeric retrovirus containing the influenza virus hemagglutinin gene has an expanded host range. *J. Virol.* **66**: 7374–7382.

Duch M, Paludan K, Jorgensen P, Pedersen FS. (1994) Lack of correlation between basal expression levels and susceptibility to transcriptional shutdown among single-gene murine leukemia virus vector proviruses. *J. Virol.* **68**: 5596–5601.

Elshami AA, Saavedra A, Zhang H, Kucharczuk JC, Spray DC, Fishman GI, Amin KM, Kaiser LR, Albelda SM. (1996) Gap junctions play a role in the 'bystander effect' of the herpes simplex virus thymidine kinase/ganciclovir system *in vitro*. *Gene Ther.* **3**: 85–92.

Etienne-Julan M, Roux P, Carillo S, Jeanteur P, Piechaczyk M. (1992) The efficiency of cell targeting by recombinant retroviruses depends on the nature of the receptor and the composition of the artificial cell-virus linker. *J. Gen. Virol.* **73**: 3251–3255.

Fan H. (1994) Retroviruses and their role in cancer. In: *The Retroviridae* (Vol. 3) (ed. JA Levy). Plenum Press, New York, pp. 313–362.

Ferrari G, Salvatori G, Rossi C, Cossu G, Mavilio F. (1995) A retroviral vector containing a muscle-specific enhancer drives gene expression only in differentiated muscle fibers. *Hum. Gene Ther.* **6**: 733–742.

Forestell SP, Böhnlein E, Rigg RJ. (1995) Retroviral end-point titre is not predictive of gene transfer efficiency: implications for vector production. *Gene Ther.* **2**: 723–730.

Freeman SM, Abboud CN, Whartenby KA, Packman CH, Koeplin DS, Moolten FL, Abraham GN. (1993) The 'bystander effect': tumor regression when a fraction of the tumor mass is genetically modified. *Cancer Res.* **53**: 5274–5283.

Garzino-Demo A, Gallo RC, Arya SK. (1995) Human immunodeficiency virus type 2 (HIV-2): packaging signal and associated negative regulatory element. *Hum. Gene Ther.* **6**: 177–184.

Gerasimova TI, Gdula DA, Gerasimov DV, Simonova O, Corces VG. (1995) A *Drosophila* protein that imparts directionality on a chromatin insulator is an enhancer of position-effect variegation. *Cell* **82**: 587–597.

Gossen M, Bujard H. (1992) Tight control of gene expression in mammalian cells by tetracycline-responsive promoters. *Proc. Natl Acad. Sci. USA* **89**: 5547–5551.

Gossen M, Bujard H. (1995) Efficacy of tetracycline-controlled gene expression is influenced by cell-type. *Bio/Techniques* **19**: 213–216.

Gossen M, Freundlieb S, Bender G, Muller G, Hillen W, Bujard H. (1995) Transcriptional activation by tetracyclines in mammalian cells. *Science* **268**: 1766–1769.

Goud B, Legrain P, Buttin G. (1988) Antibody-mediated binding of a murine ecotropic moloney retroviral vector to human cells allows internalization but not the establishment of the proviral state. *Virology* **163**: 251–254.

Grosovsky AJ, Skandalis A, Hasegawa L, Walter BN. (1993) Insertional inactivation of the tk locus in a human B lymphoblastoid cell line by a retroviral shuttle vector. *Mutat. Res.* **289**: 297–308.

Günzburg WH, Salmons B. (1986) Mouse mammary tumour virus mediated transfer and expression of neomycin resistance to infected cultured cells. *Virology* **155**: 236–248.

Günzburg WH, Salmons B. (1992) Factors controlling the expression of mouse mammary tumour virus. *Biochem. J.* **283**: 625–632.

Günzburg WH, Heinemann F, Wintersperger S, Miethke T, Wagner H, Erfle V, Salmons B. (1993) Endogenous superantigen expression controlled by a novel promoter located in the MMTV long terminal repeat. *Nature* **364**: 154–158.

Günzburg WH, Saller R, Salmons B. (1995) Retroviral vectors directed to predefined cell types for gene therapy. *Biologicals* **23**: 5–12.

Hafenrichter DG, Wu XY, Rettinger SD, Kennedy SC, Flye MW, Ponder KP. (1994) Quantitative evaluation of liver-specific promoters from retroviral vectors after *in vivo* transduction of hepatocytes. *Blood* **84**: 3394–3404.

Han X, Kasahara N, Kan YW. (1995) Ligand-directed retroviral targeting of human breast cancer cells. *Proc. Natl Acad. Sci. USA* **92**: 9747–9751.

Hantzopoulos PA, Sullenger BA, Ungers G, Gilboa E. (1989) Improved gene expression upon transfer of the adenosine deaminase minigene outside the transcriptional unit of a retroviral vector. *Proc. Natl Acad. Sci. USA* **86**: 3519–3523.

Harrison GS, Long CJ, Curiel TJ, Maxwell F, Maxwell IH. (1992) Inhibition of human immunodeficiency virus-1 production resulting from transduction with a retrovirus containing an HIV-regulated diphtheria toxin A chain. *Hum. Gene Ther.* **3**: 461–469.

Hatzoglou M, Hodgson CP, Mularo F, Hanson RW. (1990a) Efficient packaging of a specific VL30 retroelement by psi2 cells which produce MoMLV recombinant retroviruses. *Hum. Gene Ther.* **1**: 385–397.

Hatzoglou M, Lamers W, Bosch F, Wynshaw-Boris A, Clapp DW, Hanson RW. (1990b) Hepatic gene transfer in animals using retroviruses containing the promoter from the gene for phosphoenolpyruvate carboxykinase. *J. Biol. Chem.* **265**: 17285–17293.

Horwitz AL. (1991) Inherited hepatic enzyme defects as candidates for liver-directed gene therapy. *Curr. Top. Microbiol. Immunol.* **168**: 185–200.

Hoshino H, Tanaka H, Miwa M, Okada H. (1984) Human T-cell leukaemia virus is not lysed by human serum. *Nature* **310**: 324–325.

Howard DK, Schlom J. (1978) Isolation of host-range variants of mouse mammary tumor viruses that efficiently infect cells *in vitro. Proc. Natl Acad. Sci. USA* **75**: 5718–5722.

Huber BE, Richards CA, Krenitsky TA. (1991) Retroviral-mediated gene therapy for the treatment of hepatocellular carcinoma: an innovative approach for cancer therapy. *Proc. Natl Acad. Sci. USA* **88**: 8039–8043.

Jupe ER, Sinden RR, Cartwright IL. (1995) Specialized chromatin structure domain boundary elements flanking a *Drosophila* heat-shock gene locus are under torsional strain *in vivo. Biochemistry* **34**: 2628–2633.

Kalos M, Fournier REK. (1995) Position-independent transgene expression mediated by boundary elements from the apolipoprotein-B chromatin domain. *Mol. Cell. Biol.* **15**: 198–207.

Kasahara N, Dozy AM, Kan YW. (1994) Tissue-specific targeting of retroviral vectors through ligand–receptor interactions. *Science* **266**: 1373–1376.

Kavanaugh MP, Miller DG, Zhang W, Law W, Kozak SL, Kabat D, Miller AD. (1994) Cell-surface receptors for gibbon ape leukemia virus and amphotropic murine retrovirus are inducible sodium-dependent phosphate symporters. *Proc. Natl Acad. Sci. USA* **91**: 7071–7075.

Keshet E, Shaull Y, Kaminchik J, Aviv H. (1980) Heterogeneity of 'virus-like' genes encoding retrovirus associated 30S RNA and their organization within the mouse genome. *Cell* **20**: 431–439.

Kohn DB, Weinberg KI, Nolta JA, Heiss LN, Lenarsky C, Crooks GM, Hanley ME, Annett G, Brooks JS, El-Khoreiy A, Lawrence K, Wells S, Moen RC, Bastian J, Williams-Herman DE, Elder M, Wara D, Bowen T, Hershfield MS, Mullen CA, Blaese RM, Parkman R. (1995) Engraftment of gene-modified umbilical cord blood cells in neonates with adenosine deaminase deficiency. *Nature Med.* **1**: 1017–1023.

Kolb AF, Günzburg WH, Albang R, Brem G, Erfle V, Salmons B. (1994) A negative regulatory element in the mammary specific whey acidic protein promoter. *J. Cell. Biochem.* **55**: 245–261.

Kolb AF, Albang R, Brem G, Erfle V, Günzburg WH, Salmons B. (1995) Characterization of a protein that binds a negative regulatory element in the mammary-specific whey acidic protein promoter. *Biochem. Biophys. Res. Commun.* **217**: 1045–1052.

Koo HM, Brown AMC, Kaufman RJ, Prorock CM, Ron Y, Dougherty JP. (1992) A spleen necrosis virus-based retroviral vector which expresses two genes from a dicistronic mRNA. *Virology* **186**: 669–675.

Kotani H, Newton PB, Zhang S, Chiang YL, Otto E, Weaver L, Blaese RM, Anderson WF, McGarrity GJ. (1994) Improved methods of retroviral vector transduction and production for gene therapy. *Hum. Gene Ther.* **5**: 19–28.

Laker C, Stocking C, Bergholz U, Hess N, De Lamarter JF, Ostertag W. (1987) Autocrine stimulation after transfer of the granulocyte–macrophage colony stimulating factor gene and autonomous growth are distinct by interdependent steps in the oncogenic pathway. *Proc. Natl Acad. Sci. USA* **84**: 8458–8462.

Landau NR, Littman DR. (1992) Packaging system for rapid production of murine leukemia virus vectors with variable tropism. *J. Virol.* **66**: 5110–5113.

Landau NR, Page KA, Littman DR. (1991) Pseudotyping with human T-cell leukemia virus type I broadens the human immunodeficiency virus host range. *J. Virol.* **65**: 162–169.

Leib-Mösch C, Brack-Werner R, Werner T, Bachmann M, Faff O, Erfle V, Hehlmann R. (1990) Endogenous retroviral elements in human DNA. *Cancer Res.* **50**: 5636–5642.

Levine F, Yee JK, Friedmann T. (1991) Efficient gene expression in mammalian cells from a dicistronic transcriptional unit in an improved retroviral vector. *Gene* **108**: 167–174.

Levy JA. (1993) Pathogenesis of human immunodeficiency virus infection. *Microbiol. Rev.* **57**: 183–289.

Lewis PF, Emerman M. (1994) Passage through mitosis is required for oncoretroviruses but not for the human immunodeficiency virus. *J. Virol.* **68**: 510–516.

Li M, Hantzopoulos PA, Banerjee SC, Zhao SC, Scweitzer BI, Gilboa E, Bertino JR. (1992) Comparison of the expression of a mutant dihydrofolate reductase under control of different internal promoters in retroviral vectors. *Hum. Gene Ther.* **3**: 381–390.

Lynch CM, Israel DI, Kaufmann RJ, Miller D. (1993) Sequences in the coding region of clotting factor VIII act as dominant inhibitors of RNA accumulation and protein production. *Hum. Gene Ther.* **4**: 259–272.

Mann R, Mulligan RC, Baltimore D. (1983) Construction of a retrovirus packaging mutant and its use to produce helper-free defective retrovirus. *Cell* **33**: 153–159.

Markowitz D, Goff S, Bank A. (1988a) Construction and use of a safe and efficient amphotropic packaging cell line. *Virology* **167**: 400–406.

Markowitz D, Goff S, Bank A. (1988b) A safe packaging line for gene transfer: separating viral genes on two different plasmids. *J. Virol.* **62**: 1120–1124.

Matano T, Odawara T, Iwamoto A, Yoshikura H. (1995) Targeted infection of a retrovirus bearing a CD4-Env chimera into human cells expressing human immunodeficiency virus type 1. *J. Gen. Virol.* **76**: 3165–3169.

Mavilio F, Ferrari G, Rossini S, Nobili N, Bonini C, Casorati G, Traversari C, Bordgnon C. (1994) Peripheral blood lymphocytes as target cells of retroviral vector-mediated gene transfer. *Blood* **83**: 1988–1997.

McGrane M, de Vente J, Yun J, Bloom J, Park E, Wynshaw-Boris A, Wagner T, Rottman FM, Hanson RW. (1988) Tissue-specific expression and dietary regulation of a chimeric phosphoenolpyruvate carboxykinase/bovine growth hormone gene in transgenic mice. *J. Biol. Chem.* **263**: 11443–11451.

McLachlin JR, Cornetta K, Eglitis MA, Anderson WF. (1990) Retroviral-mediated gene transfer. *Prog. Nucleic Acid Res. Mol. Biol.* **38**: 91–135.

McLachlin JR, Mittereder N, Daucher MB, Kadan M, Eglitis MA. (1993) Factors affecting retroviral vector function and structural integrity. *Virology* **195**: 1–5.

Migita M, Medin JA, Pawliuk R, Jacobson S, Nagle JW, Anderson S, Amiri M, Humphries RK, Karlsson S. (1995) Selection of transduced CD34⁺ progenitors and enzymatic correction of cells from Gaucher patients, with bicistronic vectors. *Proc. Natl Acad. Sci. USA* **92**: 12075–12079.

Milan D, Nicolas J-F. (1991) Activator dependent and activator independent defective recombinant retroviruses from bovine leukemia virus. *J. Virol.* **65**: 1938–1945.

Miller AD, Garcia JV, von Suhr N, Lynch CM, Wilson C, Eiden MV. (1991) Construction and properties of retrovirus packaging cells based on gibbon ape leukemia virus. *J. Virol.* **65**: 2220–2224.

Miller DG, Miller AD. (1994) A family of retroviruses that utilize related phosphate transporters for cell entry. *J. Virol.* **68:** 8270–8276.

Miller DG, Adam MA, Miller AD. (1990) Gene transfer by retrovirus vectors occurs only in cells that are actively replicating at the time of infection. *Mol. Cell. Biol.* **10:** 4139–4242.

Moolten FL. (1994) Drug sensitivity ('suicide') genes for selective cancer chemotherapy: use of suicide gene e.g. herpes simplex virus thymidine-kinase, for prodrug activation and selective tumor killing. *Cancer Gene Ther.* **1:** 279–287.

Morgan JR, LeDoux JM, Snow RG, Tompkins RG, Yarmush ML. (1995) Retrovirus infection: effect of time and target cell number. *J. Virol.* **69:** 6994–7000.

Morgan RA, Couture L, Elroy-Stein O, Ragheb J, Moss B, Anderson WF. (1992) Retroviral vectors containing putative internal ribosome entry sites: development of a polycistronic gene transfer system and applications to human gene therapy. *Nucleic Acids Res.* **20:** 1293–1299.

Morgan RA, Nussbaum O, Muenchau DD, Shu L, Couture L, Anderson WF. (1993) Analysis of the functional and host range-determining regions of the murine ecotropic and amphotropic retrovirus envelope proteins. *J. Virol.* **67:** 4712–4721.

Morris DW, Bradshaw HD, Billy HT, Munn RJ, Cardiff RD. (1989) Isolation of a pathogenic clone of mouse mammary tumor virus. *J. Virol.* **63:** 148–158.

Muenchau DD, Freeman SM, Cornetta K, Zwiebel JA, Anderson WF. (1990) Analysis of retroviral packaging lines for generation of replication-competent virus. *Virology* **176:** 262–265.

Murphy JE, Goff SP. (1989) Construction and analysis of deletion mutations in the U5 region of Moloney Murine Leukemia Virus: effects on RNA packaging and reverse transcription. *J. Virol.* **63:** 319–327.

Naldini L, Blömer U, Gallay P, Ory D, Mulligan R, Gage FH, Verma IM, Trono D. (1996) In vivo delivery and stable transduction of nondividing cells by a lentiviral vector. *Science* **272:** 263–267.

Naviaux RK, Verma IM. (1992) Retroviral vectors for persistent expression *in vivo. Curr. Biol.* **3:** 540–547.

Neda H, Wu CH, Wu GY. (1991) Chemical modification of an ecotropic murine leukemia virus results in redirection of its target cell specificity. *J. Biol. Chem.* **266:** 14143–14146.

Noda A, Ning Y, Venable SF, Periera-Smith OM, Smith JR. (1994) Cloning of senescent cell-derived inhibitors of DNA synthesis using an expression screen. *Exp. Cell. Res.* **211:** 90–98.

Oldfield EH, Ram Z, Culver KW, Blaese RM, Devroom HL, Anderson WF. (1993) Gene therapy for the treatment of brain tumors using intra-tumoral transduction with the thymidine-kinase gene and intravenous ganciclovir; using mouse retrovirus vector sensitizes tumor to ganciclovir. *Hum. Gene Ther.* **4:** 39–69.

Olsen JC, Sechelski J. (1995) Use of sodium-butyrate to enhance production of retroviral vectors expressing *CFTR* cDNA. *Hum. Gene Ther.* **6:** 1195–1202.

Ott D, Rein A. (1992) Basis for receptor specificity of nonecotropic murine leukemia virus surface glycoprotein gp70SU. *J. Virol.* **66:** 4632–4638.

Overturf K, Al-Dhalimy M, Tanguay R, Brantly M, Ou C-N, Finegold M, Grompe M. (1996) Hepatocytes corrected by gene therapy are selected *in vivo* in a murine model of hereditary tyrosinaemia type I. *Nature Genet.* **12:** 266–273.

Page KA, Landau NR, Littman DR. (1990) Construction and use of a human immunodeficiency virus vector for analysis of virus infectivity. *J. Virol.* **64:** 5270–5276.

Pages JC, Loux N, Farge D, Briand P, Weber A. (1995) Activation of Moloney murine leukemia-virus LTR enhances the titer of recombinant retrovirus in Psi-Grip packaging cells. *Gene Ther.* **2:** 547–551.

Paludan K, Dai HY, Duch M, Jorgensen P, Kjeldgaard NO, Pedersen FS. (1989) Different relative expression from two murine leukemia virus long terminal repeats in unintegrated transfected DNA and in integrated retroviral vector proviruses. *J. Virol.* **63:** 5201–5207.

Parolin C, Dorfman T, Palu G, Goettlinger H, Sodroski J. (1994) Analysis in human immunodeficiency virus type 1 vectors of *cis*-acting sequences that affect gene transfer into human lymphocytes. *J. Virol.* **68:** 3888–3895.

Pastan I, Gottesman M, Ueda K, Lovelace E, Rutherford AV, Willingham MC. (1988) A retrovirus carrying a MDR 1 cDNA confers multidrug resistance and polarized expression of p-glycoprotein in MDCK cells. *Proc. Natl Acad. Sci. USA* **85:** 1388–1393.

Paul RW, Morris D, Hess B, Dunn J, Overall RW. (1993) Increased viral titer through concentration of viral harvests from retroviral packaging lines. *Hum. Gene Ther.* **4:** 609–615.

Paulus W, Baur I, Boyce FM, Breakefield XO, Reeves SA. (1996) Self-contained, tetracyclin-regulated retroviral vector system for gene delivery to mammalian cells. *J. Virol.* **70:** 62–67.

Pear WS, Nolan GP, Scott ML, Baltimore D. (1993) Production of high-titer helper-free retroviruses by transient transfection. *Proc. Natl Acad. Sci. USA* **90:** 8392–8396.

Pozansky M, Lever AM, Bergeron L, Haseltine W, Sodroski J. (1991) Gene transfer into human lymphocytes by a defective human immunodeficiency virus type 1 vector. *J. Virol.* **65**: 532–536.

Purcell DFJ, Broscius CM, Vanin EF, Buckler CE, Nienhuis AW, Martin MA. (1996) An array of murine leukemia virus-related elements is transmitted and expressed in a primate recipient of retroviral gene transfer. *J. Virol.* **70**: 887–897.

Ram Z, Culver KW, Walbridge S, Blaese RM, Oldfield EH. (1993) *In situ* retroviral-mediated gene transfer for the treatment of brain tumors in rats. *Cancer Res.* **53**: 83–88.

Richardson JH, Child LA, Lever AML. (1993) Packaging of human immunodeficiency virus type 1 RNA requires *cis*-acting sequences outside the 5′ leader region. *J. Virol.* **67**: 3997–4005.

Riviere I, Brose K, Mulligan RC. (1995) Effects of retroviral vector design on expression of human adenosine deaminase in murine bone marrow transplant recipients engrafted with genetically modified cells. *Proc. Natl Acad. Sci. USA* **92**: 6733–6737.

Rizvi TA, Panganiban AT. (1992) Propagation of SIV vectors by genetic complementation with a heterologous *env* gene. *AIDS Res. Hum. Retrovir.* **8**: 89–95.

Robinson D, Elliott JF, Chang LJ. (1995) Retroviral vector with a CMV-IE/HIV-TAR hybrid LTR gives high basal expression levels and is up-regulated by HIV-1 Tat. *Gene Ther.* **2**: 269–278.

Roe T, Reynolds TC, Yu G, Brown PO. (1993) Integration of murine leukaemia virus DNA depends on mitosis. *EMBO J.* **12**: 2099–2108.

Roseman RR, Swan JR, Geyer PK. (1995) A *Drosophila* insulator protein facilitates dosage compensation of the X-chromosome mini-white gene located at autosomal insertion sites. *Development* **121**: 3573–3582.

Rother RP, Squinto SP, Mason JM, Rollins SA. (1995) Protection of retroviral vector particles in human blood through complement inhibition. *Hum. Gene Ther.* **6**: 429–435.

Roux P, Jeanteur P, Piechaczyk M. (1989) A versatile and potentially general approach to the targeting of specific cell types by retroviruses: application to the infection of human cells by means of major histocompatibility complex class I and class II antigens by mouse ecotropic murine leukemia virus-derived viruses. *Proc. Natl Acad. Sci. USA* **86**: 9079–9083.

Russell DW, Miller AD. (1996) Foamy virus vectors. *J. Virol.* **70**: 217–222.

Russell SJ, Hawkins RE, Winter G. (1993) Retroviral vectors displaying functional antibody fragments. *Nucleic Acids Res.* **21**: 1081–1085.

Sadelin M, Wang CHJ, Antoniou M, Grosfeld F, Mulligan RC. (1995) Generation of a high-titer retroviral vector capable of expressing high levels of the human beta-globin gene. *Proc. Natl Acad. Sci. USA* **92**: 6728–6732.

Saller RM. (1994) Design von locus- und gewebespezifische retroviralen Vektoren für eine *in vivo* Gentherapie. Doctoral Thesis, Biology Faculty, Ludwig-Maximilian University, Munich.

Saller RM, Günzburg WH, Salmons B. (1995) Microcapsules provide a novel alternative for systemic virus release. *Gene Ther.* **2 (Suppl. 1)**: 12.

Salmons B, Günzburg WH. (1993) Targeting of retroviral vectors for gene therapy. *Hum. Gene Ther.* **4**: 129–141.

Salmons B, Erfle V, Brem G, Günzburg WH. (1990) *naf*, a *trans*-regulating negative-acting factor encoded within the Mouse Mammary Tumour Virus open reading frame region. *J. Virol.* **64**: 6355–6359.

Salmons B, Moritz-Legrand S, Garcha I, Günzburg WH. (1989) Construction and characterization of a packaging cell line for MMTV-based retroviral vectors. *Biochem. Biophys. Res. Commun.* **159**: 1191–1198.

Salmons B, Saller RM, Baumann J, Günzburg WH. (1995) Construction of retroviral vectors for targeted delivery and expression of therapeutic genes. *Leukemia* **9 (Suppl. 1)**: 53–60.

Scarpa M, Cournoyer D, Muzny DM, Moore KA, Belmont JW, Caskey CT. (1991) Characterization of recombinant helper retroviruses from Moloney-based vectors in ecotropic and amphotropic packaging cell lines. *Virology* **80**: 849–852.

Schmidt M, Rethwilm A. (1995) Replicating foamy virus-based vectors directing high expression of foreign genes. *Virology* **210**: 167–178.

Shackleford GM, Varmus HE. (1988) Construction of a clonable, infectious, and tumorigenic mouse mammary tumor virus provirus and a derivative genetic vector. *Proc. Natl. Acad Sci. USA* **85**: 9655–9659.

Slamon DJ, Clark GM, Wong SG, Levin WJ, Ullrich A, McGuire WL. (1987) Human breast cancer: correlation of relapse and survival with amplification of the HER-2/neu oncogene. *Science* **235**: 177–182.

Somia NV, Zoppe M, Verma IM. (1995) Generation of targeted retroviral vectors by using single-chain variable fragment: an approach to *in vivo* gene delivery. *Proc. Natl Acad. Sci. USA* **92**: 7570–7574.

Soneoka Y, Cannon PM, Ramsdale EE, Griffiths JC, Romano G, Kingsman SM, Kingsman AJ. (1995) A transient 3-plasmid expression system for the production of high-titer retroviral vectors. *Nucleic Acids Res.* **23**: 628–633.

Soriano P, Friedrich G, Lawinger P. (1991) Promoter interactions in retrovirus vectors introduced into fibroblasts and embryonic stem cells. *J. Virol* **65**: 2314–2319.

Sparmann G, Walther W, Günzburg WH, Uckert W, Salmons B. (1994) Conditional expression of human TNF-α: a system for inducible cytotoxicity. *Int. J. Cancer* **59**: 103–107.

Stevenson M. (1996) Portals of entry: uncovering HIV nuclear transport pathways. *Trends Cell Biol.* **6**: 9–15.

Stinchcomb DT. (1995) Constraining the cell cycle: Regulating cell division and differentiation by gene therapy. *Nature Med.* **1**: 1004–1006.

Strauss M. (1994) Liver-directed gene therapy: prospects and problems. *Gene Ther.* **1**: 156–164.

Sugimoto Y, Aksentijevich I, Gottesman MM, Pastan I. (1994) Efficient expression of drug-selectable genes in retroviral vectors under control of an internal ribosome entry site. *Bio/Technology* **12**: 694–698.

Takeuchi Y, Cosset F-LC, Lachmann PJ, Okada H, Weiss RA, Collins MKL. (1994) Type C retrovirus inactivation by human complement is determined by both the viral genome and the producer cell. *J. Virol.* **68**: 8001–8007.

Takeuchi Y, Porter CD, Strahan KM, Preece AF, Gustafsson K, Cosset FL, Weiss RA, Collins MKL. (1996) Sensitization of cells and retroviruses to human serum by (alpha 1–3) galactosyltransferase. *Nature* **379**: 85–88.

Thielens N, Bally I, Ebenbichler C, Dierich M, Arlaud G. (1994) Characterization of the interaction between the C1q subcomponent of human C1 and the transmembrane envelope glycoprotein gp41 of HIV-1. *AIDS Res. Hum. Retrovir.* **10**: S64.

Valsesia-Wittmann S, Drynda A, Deleage G, Aumailly M, Heard JM, Danos O, Verdier G, Cosset FL. (1994) Modifications in the binding domain of avian retrovirus envelope protein to redirect the host-range of retroviral vectors. *J. Virol.* **68**: 4609–4619.

van Zeijl M, Johann SV, Closs E, Cunningham J, Eddy R, Shows T, O'Hara B. (1994) A human amphotropic retrovirus receptor is a second member of the gibbon ape leukemia virus receptor family. *Proc. Natl Acad. Sci. USA* **91**: 1168–1172.

Vile RG, Nelson JA, Castleden S, Chong H, Hart I. (1994) Systemic gene therapy of murine melanoma using tissue-specific expression of the HSVtk gene involves an immune component. *Cancer Res.* **54**: 6228–6234.

Weiss RA. (1993) Cellular receptors and viral glycoproteins involved in retrovirus entry. In: *The Retroviridae* (Vol. 2) (ed. JA Levy). Plenum Press, New York, pp. 1–108.

Welsh RM, Cooper NR, Jensen FC, Oldstone MBA. (1975) Human serum lyses RNA tumour viruses. *Nature* **257**: 612–614.

Wilson C, Reitz MS, Okayama H, Eiden MV. (1989) Formation of infectious hybrid virions with gibbon ape leukemia virus and human T cell leukemia virus retroviral envelope glycoproteins and the Gag proteins and Pol proteins of Moloney murine leukemia virus. *J. Virol.* **63**: 2374–2378.

Wintersperger S, Salmons B, Miethke T, Erfle V, Wagner H, Günzburg WH. (1995) Negative acting factor and superantigen are separable activities encoded by the mouse mammary tumor virus long terminal repeat. *Proc. Natl Acad. Sci. USA* **92**: 2745–2749.

Wu X, Holschen J, Kennedy SC, Parker-Ponder K. (1996) Retroviral vector sequences may interact with some internal promoters and influence expression. *Hum. Gene Ther.* **7**: 159–171.

Xu L, Yee JK, Wolff JA, Friedmann T. (1989) Factors affecting long-term stability of moloney murine leukemia virus-based vectors. *Virology* **171**: 331–341.

Yee JK, Friedmann T, Burns JC. (1994) Generation of high-titer pseudotyped retroviral vectors with very broad host range. *Methods Cell Biol.* **43**: 99–112.

Yu SF, von Ruden T, Kantoff PW, Garber C, Seiberg M, Ruther U, Anderson WF, Wagner EF, Gilboa E. (1986) Self-inactivating retroviral vectors designed for transfer of whole genes into mammalian cells. *Proc. Natl Acad. Sci. USA* **83**: 3194–3198.

Zavada J. (1972) Pseudotypes of vesicular stomatitis virus with the coat of murine leukaemia and of avian myeloblastosis viruses. *J. Gen. Virol.* **15**: 183.

Zhang H, Duan L-X, Dornadula G, Pomeranz RJ. (1995) Increasing transduction efficiency of recombinant murine retrovirus vectors by initiation of endogenous reverse transcription: potential utility for genetic therapies. *J. Virol.* **69**: 3929–3932.

4

Adenovirus vectors

Christopher J.A. Ring

4.1 Introduction

Adenoviruses possess a number of features making them attractive candidates for the transfer and expression of therapeutic genes. They have a very broad host range including quiescent and terminally differentiated cells. Between 10^4 and 10^5 virions are produced per infected cell; these remain cell-associated long after yields have reached maximum levels, thereby facilitating concentration of virus. Indeed, adenoviral vectors can be prepared typically at titres as high as 10^{11} infectious units ml^{-1}. Adenoviruses do not possess an envelope and are therefore more stable and less susceptible to complement-mediated inactivation than enveloped viruses. Adenovirus types 2 and 5 (Ad2, Ad5), upon which most of the available adenoviral vectors are based, are non-oncogenic in rodents (Ginsberg, 1984; Horwitz, 1990). Adenovirus genomes generally remain episomal and therefore vectors based on these viruses may only prove useful for directing the short-term expression of transgenes. However, this may make them a safer alternative than retroviruses which have the disadvantage of random chromosomal integration that can at least theoretically result in insertional mutagenesis of host-cell genes.

Although wild-type Ad2 and Ad5 commonly cause mild upper respiratory tract illness, they are not associated with severe disease, even in immunocompromised individuals. The natural affinity of adenoviruses for respiratory epithelium led initially to the development of adenoviral vectors for use in the genetic treatment of cystic fibrosis (Crystal *et al.*, 1994; Rosenfeld *et al.*, 1992) and emphysema resulting from α_1-antitrypsin deficiency (Rosenfeld *et al.*, 1991; Setoguchi *et al.*, 1994). However, strategies using recombinant adenoviruses are also being developed for the treatment of many other conditions including haemophilias A and B (Connelly *et al.*, 1996; Fang *et al.*, 1995; Kay *et al.*, 1994), Duchenne muscular dystrophy (Acsadi *et al.*, 1996; Petrof *et al.*, 1995), familial hypercholesterolaemia (Herz and Gerard, 1993; Kozarsky *et al.*, 1993), arterial restenosis following angioplasty (Chang *et al.*, 1995; French *et al.*, 1994), deficiencies in surfactant protein B (Yei *et al.*, 1994a) and erythropoietin (Descamps *et al.*, 1994), and various forms of cancer (Colak *et al.*, 1995; Eastam *et al.*, 1995; Elshami *et al.*, 1996; Kaneko *et al.*, 1995; O'Malley *et al.*, 1995; Rosenfeld *et al.*, 1995; Vincent *et al.*, 1996).

4.2 Virus biology

Adenovirions are non-enveloped particles, 80–90 nm in diameter, with a spiked icosohedral morphology. The three major structural proteins are the hexon, forming the faces of the icosahedron, the penton base, located at each of the 12 vertices of the

Gene Therapy, edited by N.R. Lemoine and D.N. Cooper.

capsid, and the fibre, protruding from each penton base protein. Serologically, human adenoviruses have been classified into nearly 50 distinct types. Adenoviruses have been documented to cause several clinical illnesses including minor upper respiratory tract infections, kerato-conjunctivitis, gastroenteritis, pneumonia, bronchitis, hepatitis and cystitis. Importantly, while adenoviral infections are very common, they only cause mild, self-limiting illnesses in most individuals (Straus, 1984). The problem of acute epidemic outbreaks of respiratory illness in military personnel has served as an incentive for the development of an adenovirus vaccine. Over the last 20 years, more than 10 million subjects have received live oral virus tablets with good success and no detectable toxicity. Currently, these vaccines are approved for human use, but restricted to military personnel in Canada and the USA (Rubin and Rorke, 1988).

4.3 The adenoviral genome

The adenoviral genome is a linear, double-stranded DNA molecule approximately 36 kb in length and bears a 55 kDa protein at both ends (terminal protein, TP). The genome is traditionally divided into 100 map units (mu), 1 mu representing 360 bp (Ginsberg, 1984; Horwitz, 1990). The genome possesses inverted terminal repeat (ITR) sequences approximately 100 bp in length. The sequences necessary for encapsidation of the viral genome are located close to the left-hand ITR (Hearing *et al.*, 1987). The genome is organized into several early and late transcriptional regions, each of which plays a specific role in the viral life cycle. There are four early regions (E1–E4) and one major late region with five principal coding units (L1–L5) (see *Figure 4.1*). In addition, there are several minor intermediate and/or late regions.

The E1 region is active immediately upon entry of the viral genome into the host-cell nucleus and encodes proteins that regulate all the other early functions (Grand, 1987; Jones and Shenk, 1979). The E2 region encodes proteins involved in viral DNA replication, namely the protein bound to the termini of viral DNA (i.e. the TP), a DNA-binding protein (DBP) and the DNA polymerase (Ginsberg, 1984; Horwitz, 1990). The DBP appears also to be involved in transcriptional control. The E3 region encodes polypeptides that are involved in counteracting the anti-viral immune response of the host. For example, an E3-encoded 19 kDa glycoprotein associates with the class I antigens of the major histocompatability complex (MHC) and inhibit their transport to the surface of infected cells (Anderson *et al.*, 1985; Burgert and Kvist, 1985). As a consequence, recognition of infected cells by cytotoxic T cells is inhibited. Furthermore, the 10.4 kDa, 14.5 kDa and the 14.7 kDa E3-encoded proteins have

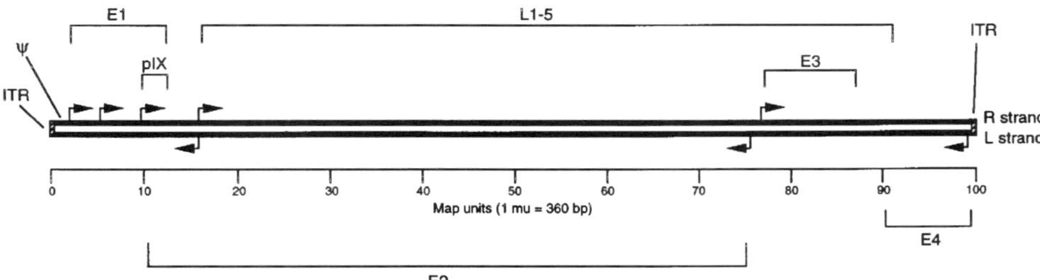

Figure 4.1. Schematic representation of the adenoviral genome showing the principal transcriptional regions (E1–E4, L1–L5), the ITRs and the genome-packaging signal (ψ). The location of the pIX gene is also shown. Promoters for the early genes are depicted as arrows and denote the direction of transcription.

been shown to inhibit the lysis of infected cells by tumour necrosis factor-α (TNFα) (Gooding *et al.*, 1988, 1991). The E3 region can be deleted from the genome with no apparent effect on the infectivity of the virus (Kelly and Lewis, 1973). During lytic infection, E4-encoded proteins are involved in several aspects of the regulation of viral and cellular gene expression, namely viral DNA replication, late viral mRNA accumulation, viral protein synthesis and the shut-off of host protein synthesis (Halbert *et al.*, 1985; Weinberg and Ketner, 1986). The major late region codes for most of the polypeptides that make up the capsid (Ginsberg, 1984; Horwitz, 1990). A minor structural protein, designated pIX, which has been shown to be required for the encapsidation of genomes greater than 34 kb in size is encoded by a transcriptional unit within the E1 region (Ghosh-Choudhury *et al.*, 1987).

4.4 Virus life cycle

The entry of adenoviruses into host cells appears to occur via two separate receptors, one mediating attachment and the other mediating internalization. The virus attaches to cells by way of the fibre protein which interacts with an, as yet unknown, receptor (Defer *et al.*, 1990). The secondary receptors, responsible for the internalization of the virus into endosomes, have recently been identified as α_v integrins (Wickham *et al.*, 1993). Acidification of the endosome triggers a conformational change within the viral capsid proteins which leads to escape of the virion through endosomolysis (Seth *et al.*, 1984). The virion, now free in the cytoplasm, is transported via nuclear targeting signals in the capsid proteins into the cell nucleus where the various transcriptional regions are expressed. Proteins encoded by the E1 region upregulate their own expression and activate expression of the other early regions. After about 8 h, viral DNA replication begins and late proteins are subsequently expressed, leading to the assembly of progeny virions within the infected cell nucleus. Adenovirus dominates the protein synthetic machinery of the host cell, promoting translation of its own transcripts, while suppressing that of the host (Schneider and Shenk, 1987). Approximately 30–40 h after infection, cell death occurs, releasing the progeny virions (Ginsberg, 1984; Horwitz, 1990).

4.5 The uses of adenoviruses in the transfer of therapeutic genes

The bulk of this review concerns the use of adenoviruses whose genomes have been manipulated so that they contain sequences that encode proteins of potential therapeutic benefit. A number of reports have shown that adenoviruses can also be used to enhance the delivery of genes by other means. For example, Curiel *et al.* (1991) reported that the efficiency of gene transfer by DNA–transferrin–polylysine conjugates is greatly enhanced upon exposure of cells to a replication-defective adenovirus. This augmentation was shown to be dependent upon the endosome-disrupting properties of adenovirus (see Section 4.4). A number of investigators have coupled DNA directly to adenoviral capsids. Cristiano *et al.* (1993) have used asialoorosomucoid–polylysine conjugated with DNA and adenovirus to target hepatocytes *in vitro*. Curiel and colleagues have further reported successful transfer of DNA to a variety of cells by coupling polylysine–DNA complexes to adenoviral capsids (Curiel *et al.*, 1992; Wagner *et al.*, 1992). Exposure of cells to adenoviruses has also been shown to enhance the efficiency of liposome-mediated gene transfer (Raja-Walia *et al.*, 1995; Yoshimura *et al.*, 1993), a phenomemon also likely to be due to the endosome-disrupting properties of adenovirus.

The use of adenovirus as an endosome-disrupting agent rather than a gene-packaging system allows the transduction of cells with DNA molecules that are too large to be accommodated within the adenoviral capsid (approximately 38 kb). Indeed, Cotten *et al.* (1992) reported that cells could be transduced with a 48 kb cosmid molecule using adenovirus-enhanced, receptor-mediated delivery. It should be emphasized, however, that transduction is still more efficient when the genes of interest are incorporated within the capsid of an infectious virus. For this reason, recombinant adenoviruses encoding foreign genes are likely to be more useful for gene therapy.

Adams *et al.* (1995) showed that, in the presence of adenovirus, several different ecotropic retroviral vectors that are usually only able to infect murine and rat cells were able to infect human cells. This observation may have implications for the development of safer retroviral vectors for gene therapy. The use of ecotropic vectors, in conjunction with adenovirus to enhance the entry of the retrovirus into human cells, would limit the risks associated with the subsequent generation of replication-competent retrovirus since such vectors, in the absence of an adenovirus 'helper', would be incapable of infecting human cells.

4.6 Construction of adenoviral vectors for use in gene therapy

Adenovirus gene expression has long been the subject of intense investigation as a model of eukaryotic gene expression. Ad2 and Ad5 have been most widely studied in terms of their genomic organization and the pattern of gene expression. Indeed, complete nucleotide sequences are available for both types (Chroboczek *et al.*, 1992; Van Ormondt and Galibert, 1984). As a consequence, most of the adenoviral vectors constructed to date are based upon these two viral serotypes.

At present, only replication-defective viruses are deemed suitable for use in gene therapy. Since E1-encoded functions are required for the efficient expression of all the other viral genes and can be provided in *trans* by cell lines such as 293 (Graham *et al.*, 1977), substitution of E1 sequences by foreign DNA is the most common strategy for generating conditionally replication-defective adenoviral vectors. Since deletion of the E3 region of the adenoviral genome has no apparent effect upon viral infectivity, sequences within the E3 region may also be deleted in order to make more room for foreign DNA. Further, since the maximum amount of DNA that can be packaged into adenoviral capsids is 105% of the wild-type genome (Bett *et al.*, 1993) (i.e. about 2 kb of extra DNA), combined deletions in E1 and E3 allow the insertion of approximately 8 kb of foreign DNA into vectors that are able to replicate in 293 cells (Bett *et al.*, 1993, 1994). The need to clone larger fragments of foreign DNA requires deletion of other viral sequences, with the missing functions provided in *trans* either by a complementing cell line or a helper virus.

Recombinant adenoviral genomes are generated either by ligation of subgenomic fragments *in vitro* or, more commonly, by homologous recombination *in vivo* between overlapping fragments following co-transfection into a complementing cell line. The initial step in the construction of an E1-replacement adenoviral vector is the insertion of the foreign DNA into a bacterial plasmid such as pΔE1sp1A (Bett *et al.*, 1994), containing sequences from the left-hand end of the adenoviral genome including the ITR, packaging signal and a partially deleted E1 region. These plasmids containing the foreign DNA (termed transfer plasmids) are then co-transfected into the E1-complementing cell line together with DNA representing the remainder of the genome. This DNA is provided either in the form of virion DNA that has been digested with a restriction endonuclease in order to reduce its infectivity, or bacterial plasmids such as pJM17 (McGrory *et al.*, 1988) or those in the pBHG series (Bett *et al.*, 1994). The pJM17 DNA, being over 40 kb in size, is packaged very inefficiently. Therefore, infectious adenovirus is generally only recovered when recombination between E1 sequences substitute the pJM17 plasmid replication origin and antibiotic resistance genes for the expression cassette, thereby reducing the size of the genome so that it can be efficiently encapsidated. The pBHG plasmids, on the other hand, have been rendered non-infectious by deletion of the viral packaging signal. Consequently, infectious virus can only be produced when the expression cassette and a functional packaging signal are introduced into the pBHG backbone by recombination with the transfer plasmid (see *Figure 4.2*). The presence or absence of E3 sequences in the resulting vectors is governed by the pBHG plasmid used for recombination. Use of the pBHG3 plasmid allows the construction of Ad5 vectors with a wild-type E3 region and inserts of up to 5.2 kb in E1, whereas use of pBHG11 allows the construction of E3⁻ viruses with inserts of up to 8.3 kb in E1. Furthermore, ligation of foreign DNA into the unique *Pac*I site of pBHG11 allows the construction of vectors with expression cassettes replacing the E3 region (Bett *et al.*, 1994).

4.7 Improved methods for vector construction

The use of bacterial plasmids such as pJM17 and those in the pBHG series has greatly facilitated the generation of recombinant adenoviruses by significantly reducing and even eliminating the background of parental virus. However, the use of digested virion DNA, despite yielding this background, is more efficient at generating recombinants. Two recent reports have described methods that improve the selection and recovery of E1 recombinants generated by co-transfecting transfer plasmids and virion DNA. Schaack *et al.* (1995a) described the use of DNA prepared from Ad5 mutants containing β-galactosidase genes in the E1 region. Recombinants will therefore generate white plaques when the complementing cells are overlayed with X-gal-containing agarose, whereas the parental virus will generate blue ones. Imler *et al.* (1995) used DNA prepared from an adenovirus in which the E1 region has been replaced by the herpes simplex virus thymidine kinase (*HSV-tk*) gene. This allows the positive selection of recombinants by the addition of the nucleoside analogue ganciclovir to the cell culture medium.

Two reports have described methods of manipulating adenoviral genomes by homologous recombination in *Saccharomyces cerevisiae* (Ketner *et al.*, 1994) and *Escherichia coli* (Chartier *et al.*, 1996). Such methods offer several advantages over conventional methods for the generation of recombinant adenoviruses. They allow the manipulation of any region of the viral genome, they are independent of the positioning of restriction sites in the viral DNA and, since cloning is carried out in the yeast or bacterial host, they eliminate the time-consuming purification of recombinant viruses in cell culture.

When transfected into permissive cells, adenoviral DNA bearing TP has been shown to produce 100 times more plaques than naked adenoviral DNA (Sharp *et al.*, 1976). Miyake and colleagues have exploited this increased transfection efficiency in order to generate recombinant adenoviruses (Miyake *et al.*, 1996). These researchers

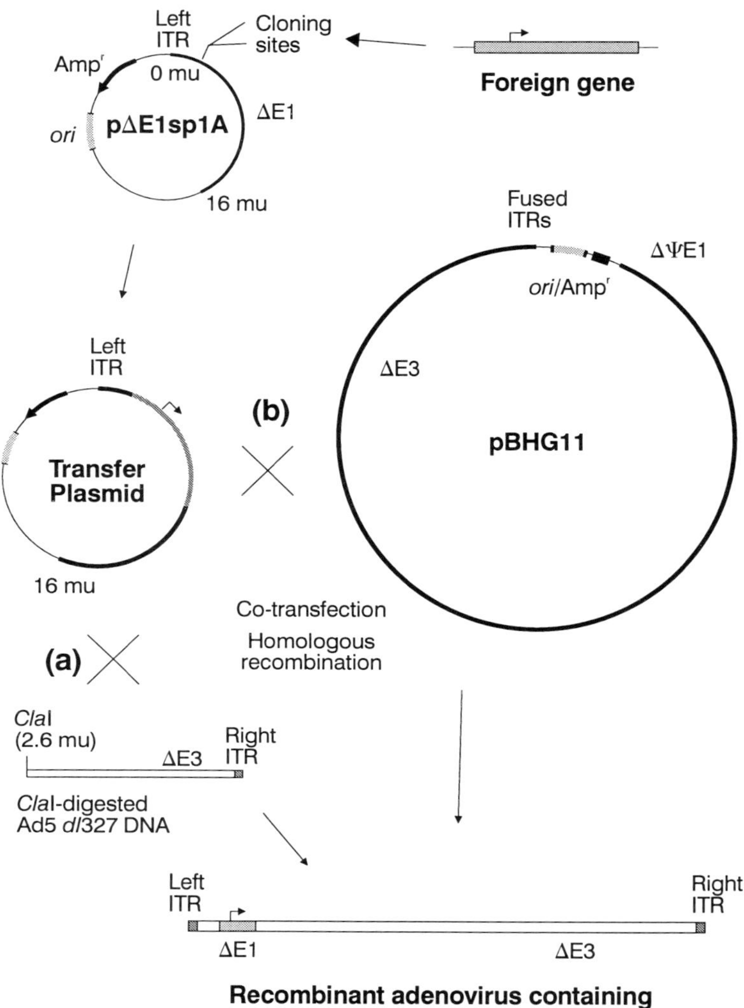

Figure 4.2. Construction of E1⁻/E3⁻ adenovirus vectors by homologous recombination *in vivo*. The foreign gene is initially cloned into a plasmid such as pΔE1sp1A (Bett *et al.*, 1994), which contains the left end of the adenoviral genome up to 16 mu. The resulting transfer plasmid is then transfected into a complementing cell line together with either (a) *Cla*I-digested Ad5 *dl*327 DNA (Shenk and Williams, 1984) or (b) a plasmid such as pBHG11 (Bett *et al.*, 1994). Both Ad5 *dl*327 and pBHG11 possess deletions in the E3 region. Plasmid pBHG11 also posseses deletions in E1 and the packaging signal (ψ). Since pBHG lacks a packaging signal, it is non-infectious when transfected alone. Homologous recombination between the E1 sequences in the transfer plasmid and those in the Ad5 *dl*327 DNA results in the incorporation of the extreme left-hand adenoviral DNA sequences and the gene of interest into the *dl*327 backbone. The Ad5 *dl*327 DNA is pre-digested with *Cla*I in order to reduce its infectivity in the complementing cell line. Similarly, homologous recombination between E1 sequences in the transfer plasmid and pBHG11 results in the substitution of the *ori* sequence and the gene for ampicillin resistance (Amp^r) in pBHG11 by the gene of interest and a functional packaging signal. Acquisition of a packaging signal enables the recombinant viral genome to be encapsidated.

have developed cosmid vectors bearing an Ad5 genome with cloning sites in either the E1 or E4 regions. Following ligation of an expression cassette into the cloned adenoviral genome, the cosmid is co-transfected into 293 cells with adenoviral DNA–TP complex (Ad DNA–TPC) which has been digested at numerous sites by restriction endonucleases in order to reduce its infectivity. Recombinant adenoviruses are generated by homologous recombination between overlapping viral sequences in the cosmid and the adenoviral DNA. Using this protocol, Miyake *et al.* (1996) claim that the efficiency of obtaining a desired recombinant is improved by a factor of almost 100 compared with conventional methods, and that the production of parental virus is very rare.

The most widely used complementing cell line for recombinant adenovirus production is the 293 line (Graham *et al.*, 1977) but recently Fallaux *et al.* (1996) described a novel human embryonic retinoblast-derived cell line (designated 911) and claimed that it possessed certain advantages over 293 cells. Adenoviral plaques were reported to appear more rapidly on 911 cells than on 293 cells, thereby reducing the time required for plaque assays. Furthermore, yields of E1-deleted viruses on 911 cells were claimed to be three times greater than those produced in 293 cells. This cell line may facilitate the propagation and titration of adenoviral vectors. However, it should be stressed that overlap does exist between the viral sequences present in 911 cells and currently used adenovirus vectors. Thus, 911 cells can potentially yield replication-competent virus (RCV) by homologous recombination.

4.8 Improving the design of adenoviral vectors

The adenoviral vectors constructed to date show great potential for the transfer and expression of therapeutic genes. Adenoviral vectors are able to transduce a large number of cell types with high efficiency. However, expression of transgenes is usually short-lived. Transient expression may not be a problem in instances such as the expression of 'suicide genes' in tumours, when the aim is to eliminate the target population. Since the majority of diseases are likely to require long-term expression of transgenes for effective treatment, improvements in vectors have to be made to extend the duration of transgene expression. In addition, the need to express very large genes (e.g. *dystrophin*) requires the cloning capacity of current vectors to be increased. Finally, since inflammatory responses have been observed in non-human primates (Bout *et al.*, 1994; Brody *et al.*, 1994; Yei *et al.*, 1994b) and cystic fibrosis patients (Crystal *et al.*, 1994) following administration of large doses of adenoviral vectors, current adenoviral vectors need to be modified in order to increase their safety.

4.8.1 Prolonging the duration of transgene expression

Since adenoviral genomes are not generally integrated into the chromosomes of infected cells, the expression of transgenes from adenoviral vectors is likely to be short-lived. Furthermore, the presence of viral genes in the current adenoviral vectors means that there is always the potential for viral gene expression and a subsequent immune response directed toward the transduced cells, leading to loss of transgene expression. Indeed, current replication-deficient adenoviral vectors have been reported to express viral proteins and induce cellular immune responses leading to the destruction of the infected cells and subsequent re-population of the organ with non-transduced cells (Yang *et al.*, 1994a).

One approach which can be taken to reduce immune-mediated destruction of trans-duced cells is to inhibit viral gene expression. A number of attempts have been made to reduce the ability of E1-deleted adenoviruses to express viral proteins. A mutation has been introduced into the *E2a* gene which results in the expression of a tempera-ture-sensitive (ts) DBP. At non-permissive temperatures, these viruses fail to encode late viral gene products. Furthermore, infection of mouse liver and lung with lacZ- and cystic fibrosis transmembrane regulator (CFTR)-encoding *E2a* ts vectors, respectively, were shown to be associated with less inflammation and substantially longer transgene expression than when first-generation vectors (encoding a normal DBP) were used (Yang *et al.*, 1994b). Since revertants have been detected in populations of *E2a* ts virus (Kruijer *et al.*, 1983), however, such ts vectors are unlikely to be approved for gene ther-apy. Cell lines expressing the Ad5 TP precursor (Schaack *et al.*, 1995b) or the Ad2 DNA polymerase (Amalfitano *et al.*, 1996) have recently been reported. Since viruses that are unable to encode a functional TP or DNA polymerase are also defective for viral DNA replication, the introduction of TP or DNA polymerase mutations in vectors would be expected to decrease the ability of vectors to express late proteins.

An alternative way of reducing immune-mediated destruction of transduced cells may be to exploit the mechanisms used by adenoviruses to diminish the recognition and lysis of infected cells. Such functions have been attributed to proteins encoded by E3, a region deleted in many of the existing adenoviral vectors. It has been demon-strated that, upon intranasal inoculation, adenoviruses bearing deletions in the E3 region stimulate more intense inflammation in the lungs of cotton rats and mice than viruses with intact E3 regions (Ginsberg *et al.*, 1989; Sparer *et al.*, 1996). Furthermore, the introduction of the gp19 gene into E3-deleted adenoviruses has been shown to sig-nificantly inhibit cytotoxic T-cell lysis of infected cells *in vitro* (Lee *et al.*, 1995). These observations suggest that adenoviral vectors possessing an intact E3 region can inhibit the antiviral immune response and thus may be able to direct the expression of foreign genes for a longer period of time. Such vectors may also be safer than E3-deleted ones.

Perhaps a more obvious approach to tackle the problem of the immune-mediated destruction of transduced cells is to immunosuppress the recipient of the vector directly. A number of researchers have indeed demonstrated a significant increase in the duration of transgene expression when their recombinant adenoviral vectors were co-administered with the immunosuppressive agent cyclosporin A (Dai *et al.*, 1995; Fang *et al.*, 1995; Gilgenkrantz *et al.*, 1995). Co-administration of vector and agents such as cyclophosphamide, anti-CD4 and interleukin-4 (IL-4) has also been sug-gested as a way of inhibiting immune cytolysis of transduced cells and thus increas-ing the duration of transgene expression (Wilson, 1995).

Even if the problem of immune-mediated destruction of transduced cells can be solved, the episomal nature of the adenoviral genome is likely to necessitate repeated vector administration for effective long-term treatments. It is possible that humoral immunity may prevent adenoviral infection in individuals with pre-existing anti-adenoviral immunity or following repeated administration. One strategy that has been suggested to circumvent such a problem is to use adenoviruses of different sub-types for each administration so the vector is not neutralized by antibodies stimulated by the previous dose (Kass-Eisler *et al.*, 1996; Mastrangeli *et al.*, 1996). This strategy would require the construction of a number of adenoviral vectors, each containing the same transgene but encoding different fibre proteins. Perhaps an easier approach to take would be to inhibit the synthesis of neutralizing antibodies. Indeed, co-administration of interferon-γ or IL-12 with recombinant adenovirus into the airways of mice has been shown to diminish neutralizing antibody production, allowing efficient re-administration of recombinant virus (Yang *et al.*, 1995).

4.8.2 Increasing the capacity of adenoviral vectors for foreign DNA

Current adenovirus vectors with deletions in E1 and E3 have a maximum cloning capacity of approximately 8 kb (Bett *et al.*, 1993, 1994). However, it may be advantageous to use vectors with intact E3 regions (Section 4.8.1), in which case the maximum capacity would be reduced to about 5 kb. Increasing the cloning capacity requires either the deletion of sequences that are not required for viral replication, or the deletion of sequences whose functions can be provided by complementing cell lines. Many E4 deletion mutants are defective for viral replication. However, those that retain the ability to express the E4 open reading frame 6 (ORF6), or to a lesser extent ORF3, are able to replicate *in vitro* (Bridge and Ketner, 1989). Recently, Armentano *et al.* (1995) constructed an adenovirus vector with a modified E4 region so that it only expresses the ORF6 protein. Deletion of the other E4 ORFs had little effect on the ability of the virus to replicate *in vitro* and increases the capacity for foreign DNA by 1.9 kb. Three groups have recently described the construction of 293-based cell lines that express all the E4 proteins, allowing the replication of E4-deficient viruses (Krougliak and Graham, 1995; Wang *et al.*, 1995; Yeh *et al.*, 1996). Deletion of the entire E4 region potentially increases the cloning capacity by 2.8 kb over those possessing E4. Since E4-encoded proteins upregulate viral gene expression, deletion of E4 is likely to reduce the level of viral gene expression, thereby increasing both the duration of transgene expression and vector safety. In addition to complementing E4 functions, the cell lines reported by Krougliak and Graham (1995) also conditionally express the minor structural protein pIX. This potentially allows the construction of E1⁻/E3⁻/pIX⁻/E4⁻ vectors with cloning capacities of approximately 11 kb. Even if adenoviral vectors retaining intact E3 regions are required, the cloning capacity of E1⁻/pIX⁻/E4⁻ vectors would be no smaller than the capacity of the current E1⁻/E3⁻ vectors.

In order to guarantee that no viral proteins are expressed in transduced cells, deletion of all the viral genes from the vector is required so that only the ITRs and the packaging signal remain. This would enable over 30 kb of foreign DNA to be cloned but such a strategy would necessitate the use of a helper virus, or the construction of packaging cell lines, in order to provide all the viral proteins in *trans*. The production of cell lines constitutively expressing adenoviral proteins has been notoriously difficult, probably due to their toxic nature. The construction of adenovirus packaging lines is therefore unlikely, leaving the use of a helper virus as the only method of providing viral proteins. For safety reasons, such a helper should not be packaged into virions or, if packaged, should be easily separable from the vector virions. Two recent reports have described adenoviral vectors in which all of the viral genes have been deleted (Fisher *et al.*, 1996; Kochanek *et al.*, 1996). Foreign DNA was cloned in plasmids between the sequences constituting the left ITR and genome packaging signal and the right ITR. Recombinant adenoviruses were then generated either by co-transfecting the resulting plasmid into 293 cells with DNA from a helper adenoviral mutant (Kochanek *et al.*, 1996) or by transfecting the plasmid into 293 cells that had been previously infected with a E1-deleted helper adenovirus (Fisher *et al.*, 1996). In both reports, the progeny virus was harvested and the vector virions were separated from helper virions by buoyant density centrifugation. Such methods have been used to generate adenoviral vectors encoding CFTR (Fisher *et al.*, 1996) and dystrophin proteins (Kochanek *et al.*, 1996). The reported yields of purified vector viruses were low in comparison to yields obtained from the infection of 293 cells with wild-type or E1/E3-deleted viruses necessitating the development of improved production and purification strategies before these new vectors can be evaluated *in vivo*.

Caravokyri and Leppard (1995) recently reported the construction of a 293-based cell line that constitutively expresses large amounts of the pIX structural protein that is required for packaging of adenoviral genomes greater than 34 kb in length (Ghosh-Choudhury *et al.*, 1987). This cell line was used to construct a novel pIX⁻ virus (designated CE5), which is too large to be packaged in unmodified 293 cells. It may therefore be possible, providing that vector constructs are less than 34 kb in length, to generate helper-free vector stocks in 293 cells by using the CE5 virus to provide replication and packaging proteins.

4.8.3 Increasing the safety of adenoviral vectors

Prior to the administration of any viral vector, it must be certain that the preparation does not contain RCV. Since current adenoviral vectors share sequences that are present in 293 cells, there is a theoretical possibility of generating E1-encoding, and therefore replication-competent, virus by homologous recombination. Indeed, the presence of RCV has been detected in adenoviral vector stocks during multiple passage in 293 cells (Lochmuller *et al.*, 1994). It is only when complementing cell lines lacking homology with vector sequences are constructed that this possibility of recombination will be minimized. Such cell lines, constructed by stable transfection of the A549 human lung carcinoma line with E1 expression vectors, have recently been reported (Imler *et al.*, 1996b). Krougliak and Graham (1995) recently described vectors and complementing cell lines for the construction of E4⁻ vectors. The cell lines are 293-based and conditionally express E4 functions. Since no overlap exists between the E4 sequences in the vectors and cell lines, the re-introduction of E4 sequences into the vectors is minimized. Thus, even if the vectors regain E1 sequences by recombining with E1 in the 293 cells, the viruses will remain E4⁻ and will thus be severely compromised for replication.

A recent study in which either infectious or inactivated virus was instilled intratracheally into mice demonstrated that the inflammatory responses induced by inactivated virus was quantitatively similar to that induced by replication-competent virus (McCoy *et al.*, 1995). This suggests that the inflammation reported previously in recipients of adenoviral vectors (Bout *et al.*, 1994; Brody *et al.*, 1994; Crystal *et al.*, 1994; Yei *et al.*, 1994b) is not stimulated solely by the expression of viral proteins within the transduced cells but also by virion components in the vector preparation. Indeed, Boudin and colleagues demonstrated nearly 20 years ago that the adenoviral penton base is directly cytotoxic (Boudin *et al.*, 1979). In the short term, the only way to overcome the problem of direct virion toxicity may be to develop vectors which express the therapeutic gene at a sufficiently high level so that a therapeutic effect is achieved using lower doses of vector. Recently, Connelly *et al.* (1996) described a modification to a previously reported adenoviral vector that resulted in increased potency allowing the administration of lower, less toxic doses of vector.

4.8.4 Development of targeted adenoviral vectors

Adenoviruses are capable of infecting a wide variety of cell types including those that are normally quiescent. This is generally considered an advantage of the adenoviral system. However, there are situations where transgene expression may need to be targeted to certain cell populations. One approach which may achieve this is the use of viruses containing the transgene under the control of cell-type-specific promoters. Indeed, liver-specific expression of the gene for human factor VIII has been achieved using the *albumin* promoter (Connelly *et al.*, 1996) and expression of marker genes has been shown to be restricted to certain neuronal cell types by the use of the *Purkinje cell protein 2* and *myelin basic protein* promoters (Hashimoto *et al.*, 1996). Moreover, transgene expression selective for hepatic and breast cancer cells has also been achieved when the *α-fetoprotein* and *MUC1* promoters, respectively, have been used (Chen *et al.*, 1995; Kaneko *et al.*, 1995; Wills *et al.*, 1995). Using an adenovirus vector containing the gene for LacZ under the control of the *CFTR* promoter, Imler *et al.* (1996a) failed to demonstrate a cell-type specificity of expression strictly paralleling that of endogenous *CFTR*. This suggests that cell-specific expression may only be achieved with certain transcriptional regulatory elements.

An alternative approach to achieve cell-type-specific expression is to target the infection of defined cell populations by manipulating the virion proteins that interact with cellular receptors. As discussed in Section 4.4, adenovirions interact with two distinct cellular receptors, the primary one binding to the fibre protein and the secondary receptor interacting with the penton base. Development of a targeted adenoviral infection system thus requires manipulation of the fibre or penton base proteins. Stevenson *et al.* (1995) have shown that it is possible to change the receptor specificity of the fibre by manipulation of sequences in the carboxy terminal head domain of the fibre protein. Recently, the receptor tropism of Ad5 has been changed by substituting its fibre protein for that encoded by Ad7 (Gall *et al.*, 1996). As an initial step towards developing a targeted infection system, Michael *et al.* (1995) placed the terminal decapeptide of the gastrin-releasing peptide (GRP) at the 3′ end of the Ad5 fibre gene. When expressed in a T7 vaccinia system, the fibre–GRP fusion protein was shown to possess a quaternary structure indistinguishable from the wild-type protein and to be correctly transported to the nucleus of HeLa cells following synthesis. Furthermore, the fusion protein was accessible to binding by an anti-GRP antibody. In order to demonstrate whether such a fibre modification results in cell-specific infection, the fusion protein must be re-introduced into the adenovirus capsid. However, the preliminary data reported by Michael *et al.* (1995) does suggest that novel binding specificities may be introduced into adenovirus by genetically manipulating the fibre protein.

In an attempt to modify the tropism of adenovirus by manipulating the protein that interacts with the secondary receptor, Wickham *et al.* (1995) constructed Ad5 penton base chimeras that recognize tissue-specific integrin receptors. This was achieved by replacing the wild-type RGD peptide motif with $\alpha_v\beta_3$- or $\alpha_4\beta_1$-specific peptide motifs. In one chimera, the wild-type HAIRGDTFA amino acid sequence motif was replaced with EILDVPST, which mediated binding to the integrin $\alpha_4\beta_1$ that is expressed at high levels on lymphocytes and monocytes but not expressed on epithelial or endothelial cells. In a second chimera, the sequences flanking the RGD motif were altered in order to abolish its interaction with $\alpha_v\beta_5$ while retaining its specificity for $\alpha_v\beta_3$. The integrin $\alpha_v\beta_5$ is expressed primarily on epithelial cells whereas integrin

$\alpha_v\beta_3$ is normally expressed on endothelial cells. The integrin $\alpha_v\beta_3$ is also aberrantly expressed on certain metastatic melanomas and glioblastomas. It is hoped that future investigations will show that these fibre and penton base chimeras, when incorporated into adenovirions, will indeed narrow the tropism of adenovirus to defined cell types.

4.9 Conclusion

The ability to infect a wide range of cell types and the absence of a requirement for dividing cells has made adenovirus an attractive candidate for use as a gene therapy vector. However, a low background level of viral gene expression leads to immune-mediated destruction of transduced cells. This, together with the episomal nature of the viral genome, leads to a waning of transgene expression that necessitates the re-administration of vector for long-term treatment. Attempts to abolish viral gene expression and inhibit anti-viral immune responses are currently being made in order to increase the efficacy and safety of adenoviral vectors. Results to date indicate that recombinant adenoviruses only mediate toxic effects when administered in very high doses. This supports the continued development of adenoviruses as vectors for use in gene therapy.

References

Acsadi G, Lochmuller H, Jani A, Huard J, Massie B, Prescott S, Simoneau M, Petrof BJ, Karpati G. (1996) Dystrophin expression in muscles of mdx mice after adenovirus-mediated *in vivo* gene transfer. *Hum. Gene Ther.* **7**: 129–140.

Adams RM, Wang M, Steffen D, Ledley FD. (1995) Infection by retroviral vectors outside of their host range in the presence of replication-defective adenovirus. *J. Virol.* **69**: 1887–1894.

Amalfitano A, Begy CR, Chamberlain JS. (1996) Improved adenovirus packaging cell lines to support the growth of replication-defective gene-delivery vectors. *Proc. Natl Acad. Sci. USA* **93**: 3352–3356.

Anderson M, Paabo S, Nilsson T, Peterson P. (1985) Impaired intracellular transport of class I MHC antigens as a possible means for adenovirus to evade immune surveillance. *Cell* **43**: 215–222.

Armentano D, Sookdeo CC, Hehir KM, Gregory RJ, St George JA, Prince GA, Wadsworth SC, Smith AE. (1995) Characterization of an adenovirus gene transfer vector containing an E4 deletion. *Hum. Gene Ther.* **6**: 1343–1353.

Bett AJ, Prevec L, Graham FL. (1993) Packaging capacity and stability of human adenovirus type 5 vectors. *J. Virol.* **67**: 5911–5921.

Bett AJ, Haddara W, Prevec L, Graham FL. (1994) An efficient and flexible system for construction of adenovirus vectors with insertions or deletions in early regions 1 and 3. *Proc. Natl Acad. Sci. USA* **91**: 8802–8806.

Boudin, M-L, Moncany M, D'Halluin J-C, Boulanger PA. (1979) Isolation and characterization of adenovirus type 2 vertex capsomer (penton base). *Virology* **92**: 125–138.

Bout A, Imler J-L, Schultz H, Perricaudet M, Zurcher C, Herbrink P, Valerio D, Pavirani A. (1994) *In vivo* adenovirus-mediated transfer of human CFTR cDNA to Rhesus monkey airway epithelium: efficacy, toxicity and safety. *Gene Ther.* **1**: 385–394.

Bridge E, Ketner G. (1989) Redundant control of late gene expression by early region 4. *J. Virol.* **63**: 631–638.

Brody SL, Metzger M, Danel C, Rosenfeld MA, Crystal RG. (1994) Acute responses of non-human primates to airway delivery of an adenovirus vector containing the human cystic fibrosis transmembrane conductance regulator cDNA. *Hum. Gene Ther.* **5**: 821–836.

Burgert H, Kvist S. (1985) An adenovirus type 2 glycoprotein blocks cell surface expression of human histocompatibility class I antigens. *Cell* **41**: 987–997.

Caravokyri C, Leppard KN. (1995) Constitutive episomal expression of polypeptide IX (pIX) in a 293-based cell line complements the deficiency of pIX mutant adenovirus type 5. *J. Virol.* **69**: 6627–6633.

Chang MW, Ohno T, Gordon D, Lu MM, Nabel GJ, Nabel EG, Leiden JM. (1995) Adenovirus-mediated transfer of the herpes simplex virus thymidine kinase gene inhibits vascular smooth muscle cell proliferation and neointima formation following balloon angioplasty of the rat carotid artery. *Mol. Med.* **1**: 172–182.

Chartier C, Degryse E, Gantzer M, Dieterle A, Pavirani A, Mehtali M. (1996) Efficient generation of recombinant adenoviral vectors by homologous recombination in *Escherichia coli. J. Virol.* **70**: 4805–4810.

Chen L, Chen D, Manome Y, Dong Y, Fine HA, Kufe DW. (1995) Breast cancer selective gene expression and therapy mediated by recombinant adenoviruses containing the DF3/MUC1 promoter. *J. Clin. Invest.* **96**: 2775–2782.

Chroboczek J, Bieber F, Jacrot B. (1992) The sequence of the genome of adenovirus type 5 and its comparison with the genome of adenovirus type 2. *Virology* **186**: 280–285.

Colak A, Goodman JC, Chen S-H, Woo SLC, Grossman RG, Shine HD. (1995) Adenovirus-mediated gene therapy in an experimental model of breast cancer metastatic to the brain. *Hum. Gene Ther.* **6**: 1317–1322.

Connelly S, Gardner JM, McClelland A, Kaleko M. (1996) High-level tissue-specific expression of functional human factor VIII in mice. *Hum. Gene Ther.* **7**: 183–195.

Cotten M, Wagner E, Zatloukal K, Phillips S, Curiel DT, Birnstiel ML. (1992) High-efficiency receptor-mediated delivery of small and large (48 kilobase) gene constructs using the endosome-disruption activity of defective or chemically inactivated adenovirus particles. *Proc. Natl. Acad. Sci. USA* **89**: 6094–6098.

Cristiano RJ, Smith LC, Woo SLC. (1993) Hepatic gene therapy: adenovirus enhancement of receptor-mediated gene delivery and expression in primary hepatocytes. *Proc. Natl Acad. Sci. USA* **90**: 2122–2126.

Crystal RG, McElvaney NG, Rosenfeld MA, Chu C-S, Mastrangeli A, Hay JG, Brody SL, Jaffe HA, Eissa NT, Danel C. (1994) Administration of an adenovirus containing the human CFTR cDNA to the respiratory tract of individuals with cystic fibrosis. *Nature Genet.* **8**: 42–51.

Curiel DT, Agarwal S, Wagner E, Cotten M. (1991) Adenovirus enhancement of transferrin–polylysine-mediated gene delivery. *Proc. Natl Acad. Sci. USA* **88**: 8850–8854.

Curiel DT, Wagner E, Cotten M, Birnstiel ML, Agarwal S, Li C-M, Loechel S, Hu P-C. (1992) High efficiency *in vitro* gene transfer mediated by adenovirus coupled to DNA–polylysine complexes. *Hum. Gene Ther.* **3**: 147–154.

Dai Y, Schwarz EM, Gu D, Zhang W-W, Sarvetnick N, Verma IM. (1995) Cellular and humoral immune responses to adenoviral vectors containing factor IX gene: tolerization of factor IX and vector antigens allows for long-term expression. *Proc. Natl Acad. Sci. USA* **92**: 1401–1405.

Defer C, Belin M-T, Caillet-Boudin M-L, Boulanger P. (1990) Human adenovirus–host cell interactions: comparative study with members of subgroups B and C. *J. Virol.* **64**: 3661–3673.

Descamps V, Blumenfeld N, Villeval J-L, Vainchenker W, Perricaudet M, Beuzard Y. (1994) Erythropoietin gene transfer and expression in adult normal mice: use of an adenovirus vector. *Hum. Gene Ther.* **5**: 979–985.

Eastam JA, Hall SJ, Sehgal I, Wang J, Timme TL, Yang G, Connell-Crowly L, Elledge SJ, Zhang W-W, Harper JW, Thompson TC. (1995) *In vivo* gene therapy with *p53* or *p21* adenovirus for prostate cancer. *Cancer Res.* **55**: 5151–5155.

Elshami AA, Kucharczuk JC, Zhang HB, Smythe WR, Hwang HC, Litzky LA, Kaiser LR, Albelda SM. (1996) Treatment of pleural mesothelioma in an immunocompetent rat model utilizing adenoviral transfer of the herpes simplex virus thymidine kinase gene. *Hum. Gene Ther.* **7**: 141–148.

Fallaux FJ, Kranenburg O, Cramer SJ, Houweling A, Van Ormondt H, Hoeben RC, Van Der Eb AJ. (1996) Characterization of 911: a new helper cell line for the titration and propagation of early region 1-deleted adenoviral vectors. *Hum. Gene Ther.* **7**: 215–222.

Fang B, Eisensmith RC, Wang H, Kay MA, Cross RE, Landen CN, Gordon G, Bellinger DA, Read MS, Hu PC, Brinkhous KM, Woo SLC. (1995) Gene therapy for haemophilia B: Host immunosuppression prolongs the therapeutic effect of adenovirus-mediated factor IX expression. *Hum. Gene Ther.* **6**: 1039–1044.

Fisher KJ, Choi H, Burda J, Chen S-J, Wilson JM. (1996) Recombinant adenovirus deleted of all viral genes for gene therapy of cystic fibrosis. *Virology* **217**: 11–22.

French BA, Mazur W, Ali NM, Geske RS, Finnigan JP, Rodgers GP, Roberts R, Raizner AE. (1994) Percutaneous transluminal *in vivo* gene transfer by recombinant adenovirus in normal porcine coronary arteries, atherosclerotic arteries and two models of coronary restenosis. *Circulation* **90**: 2402–2413.

Gall J, Kass-Eisler A, Leinwand L, Falck-Pedersen E. (1996) Adenovirus type 5 and 7 capsid chimera: Fibre replacement alters receptor tropism without affecting primary immune neutralization epitopes. *J. Virol.* **70**: 2116–2123.

Ghosh-Choudhury G, Haj-Ahmed Y, Graham FL. (1987) Protein IX, a minor component of the human adenovirus capsid, is essential for the packaging of full-length genomes. *EMBO J.* **6**: 1733–1739.

Gilgenkrantz H, Duboc D, Juillard V, Couton D, Pavirani A, Guillet JG, Briand P, Kahn A. (1995) Transient expression of genes transferred *in vivo* into heart using first-generation adenoviral vectors: Role of the immune response. *Hum. Gene Ther.* **6**: 1265–1274.

Ginsberg HS. (1984) *The Adenoviruses.* Plenum Press, New York.

Ginsberg HS, Lundholm-Beauchamp U, Horswood RL, Pernis B, Wold WSM, Chanock RM, Prince GA. (1989) Role of early region 3 (E3) in pathogenesis of adenovirus disease. *Proc. Natl Acad. Sci. USA* **86**: 3823–3827.

Gooding LR, Elmore LE, Tollefson AE, Brody HA, Wold WSM. (1988) A 14,700 MW protein from the E3 region of adenovirus inhibits cytolysis by tumour necrosis factor. *Cell* **53**: 341–346.

Gooding LR, Ranheim TS, Tollefson AE, Aquino L, Duerksen-Hughes P, Horton TM, Wold WSM. (1991) The 10,400- and 14,500-dalton proteins encoded by region E3 of adenovirus function together to protect many but not all mouse cell lines against lysis by tumour necrosis factor. *J. Virol.* **65**: 4114–4123.

Graham FL, Smiley J, Russell WC, Nairn R. (1977) Characteristics of a human cell line transformed by DNA from human adenovirus type 5. *J. Gen. Virol.* **36**: 59–74.

Grand RJA. (1987) The structure and functions of the adenovirus early region 1 proteins. *Biochem. J.* **241**: 25–38.

Halbert DN, Cutt JR, Shenk T. (1985) Adenovirus early region 4 encodes functions required for efficient DNA replication, late gene expression and host cell shut-off. *J. Virol.* **56**: 250–257.

Hashimoto M, Aruga J, Hosoya Y, Kanegae Y, Saito I, Mikoshiba K. (1996) A neural cell-specific expression system using recombinant adenovirus vectors. *Hum. Gene Ther.* **7**: 149–158.

Hearing P, Samulski RJ, Wishart WL, Shenk TJ. (1987) Identification of a repeated sequence element required for efficient encapsidation of the adenovirus type 5 chromosome. *J. Virol.* **61**: 2555–2558.

Herz J, Gerard RD. (1993) Adenovirus-mediated transfer of low density lipoprotein receptor gene acutely accelerates cholesterol clearance in normal mice. *Proc. Natl Acad. Sci. USA* **90**: 2812–2816.

Horwitz MS. (1990) Adenoviridae and their replication. In: *Virology* (ed. BN Fields, DM Knipe). Raven Press Ltd, New York, pp. 1679–1721.

Imler J-L, Chartier C, Dieterele A, Dreyer D, Mehali M, Pavirani A. (1995) An efficient procedure to select and recover recombinant adenovirus vectors. *Gene Ther.* **2**: 263–268.

Imler J-L, Dupuit F, Chartier C, Accart N, Dieterele A, Schultz H, Puchelle E, Pavirani A. (1996a) Targeting cell-specific gene expression with an adenovirus vector containing the lacZ gene under the control of the CFTR promoter. *Gene Ther.* **3**: 49–58.

Imler J-L, Chartier C, Dreyer D, Dieterele A, Sainte-Marir M, Faure T, Pavirani A, Mehtali M. (1996b) Novel complementation cell lines derived from lung carcinoma A549 cells support the growth of E1-deleted adenoviral vectors. *Gene Ther.* **3**: 75–84.

Jones N, Shenk T. (1979) An adenovirus type 5 early gene function regulates expression of other early viral genes. *Proc. Natl Acad. Sci. USA* **76**: 3665–3669.

Kaneko S, Hallenbeck P, Kotani T, Nakabayashi H, McGarrity G, Tamaoki T, Anderson WF, Chiang YL. (1995) Adenovirus-mediated gene therapy of hepatocellular carcinoma using cancer-specific gene expression. *Cancer Res.* **55**: 5283–5287.

Kass-Eisler A, Leinwand L, Gall J, Bloom B, Falck-Pedersen E. (1996) Circumventing the immune response to adenovirus-mediated gene therapy. *Gene Ther.* **3**: 154–162.

Kay MA, Landen CN, Rothenberg SR, Taylor LA, Leland F, Wiehle S, Fang B, Bellinger D, Finegold M, Thompson AR, Read M, Brinkhous KM, Woo SLC. (1994) *In vivo* hepatic gene therapy: Complete albeit transient correction of factor IX deficiency in haemophilia B dogs. *Proc. Natl Acad. Sci. USA* **91**: 2353–2357.

Kelly TJ, Lewis AM. (1973) Use of non-defective adenovirus-simian virus 40 hybrids for mapping the SV40 genome. *J. Virol.* **12**: 643–652.

Ketner G, Spencer F, Tugendreich S, Connelly C, Hieter P. (1994) Efficient manipulation of the human adenovirus genome as an infectious yeast artificial chromosome clone. *Proc. Natl Acad. Sci. USA* **91**: 6186–6190.

Kochanek S, Clemens PR, Mitani K, Chen H-H, Chan S, Caskey CT. (1996) A new adenoviral vector: replacement of all viral coding sequences with 28 kb of DNA independently expressing both full-length dystrophin and β-galactosidase. *Proc. Natl Acad. Sci. USA* **93**: 5731–5736.

Kozarsky K, Grossman M, Wilson JM. (1993) Adenovirus-mediated correction of the genetic defect in hepatocytes from patients with familial hypercholesterolaemia. *Somatic Cell Mol. Genet.* **19**: 449–458.

Krougliak V, Graham FL. (1995) Development of cell lines capable of complementing E1, E4 and protein IX defective adenovirus type 5 mutants. *Hum. Gene Ther.* **6**: 1575–1586.

Kruijer W, Nicolas J-C, Van Schaik FMA, Sussenbach, JS. (1983) Structure and function of DNA binding proteins from revertants of adenovirus type 5 mutants with a temperature-sensitive DNA replication. *Virology* **124**: 425–433.

Lee MG, Abina MA, Haddada H, Perricaudet M. (1995) The constitutive expression of the immunomodulatory gp19 k protein in E1-, E3- adenoviral vectors strongly reduces the host cytotoxic T cell response against the vector. *Gene Ther.* **2**: 256–262.

Lochmuller H, Jani A, Huard J, Prescott S, Simoneau M, Massie B, Karpati G, Acsadi G. (1994) Emergence of early region 1-containing replication-competent adenovirus in stocks of replication-defective adenovirus recombinants (ΔE1+ΔE3) during multiple passages in 293 cells. *Hum. Gene Ther.* **5**: 1485–1491.

Mastrangeli A, Harvey B-G, Yao J, Wolff G, Kovesdi I, Crystal RG, Falck-Pedersen E. (1996) 'Sero-switch' adenovirus-mediated *in vivo* gene transfer: Circumvention of anti-adenovirus humoral immune defenses against repeat adenovirus vector administration by changing the adenovirus serotype. *Hum. Gene Ther.* **7**: 79–87.

McCoy RD, Davidson BL, Roessler BJ, Huffnagle GB, Janich SL, Laing TJ, Simon RH. (1995) Pulmonary inflammation induced by incomplete or inactivated adenoviral particles. *Hum. Gene Ther.* **6**: 1553–1560

McGrory WJ, Bautista DS, Graham FL. (1988) A simple technique for the rescue of early region I mutations into infectious human adenovirus type 5. *Virology* **163**: 614–617.

Michael SI, Hong JS, Curiel DT, Engler JA. (1995) Addition of a short peptide ligand to the adenovirus fiber protein. *Gene Ther.* **2**: 660–668.

Miyake S, Makimura M, Kanegae Y, Harada S, Sato Y, Takamori K, Tokuda C, Saito I. (1996) Efficient generation of recombinant adenoviruses using adenovirus DNA-terminal protein complex and a cosmid bearing the full-length virus genome. *Proc. Natl. Acad. Sci. USA* **93**: 1320–1324.

O'Malley BW, Chen S-H, Schwarzt MR, Woo SLC. (1995) Adenovirus-mediated gene therapy for human head and neck sqamous cell cancer in a nude mouse model. *Cancer Res.* **55**: 1080–1085.

Petrof BS, Acsadi G, Jani A, Massie B, Bourdon J, Matusiewicz N, Yang L, Lochmuller H, Karpati G. (1995) Efficiency and functional consequences of adenovirus-mediated *in vivo* gene transfer to normal and dystrophic (mdx) mouse diaphragm. *Am. J. Respir. Cell Mol. Biol.* **13**: 508–517.

Raja-Walia R, Webber J, Naftilan J, Chapman GD, Naftilan AJ. (1995) Enhancement of liposome-mediated gene transfer into the vascular tissue by replication-deficient adenovirus. *Gene Ther.* **2**: 521–530.

Rosenfeld MA, Siegfried W, Yoshimura K, Yoneyama K, Fukayama M, Stier LE, Paakko PK, Gilardi P, Stratford-Perricaudet LD, Parricaudet M, Jallat S, Pavirani A, Lecocq J-P, Crystal RG. (1991) Adenovirus-mediated transfer of a recombinant α1-antitrypsin gene to the lung epithelium *in vivo*. *Science* **252**: 431–434.

Rosenfeld MA, Yoshimura K, Trapnell BC, Yoneyama K, Rosenthal ER, Dalemans W, Fukayama M, Bargon J, Stier LE, Stratford-Perricaudet LD, Parricaudet M, Guggino WB, Pavirani A, Lecocq J-P, Crystal RG. (1992) *In vivo* transfer of the human cystic fibrosis transmembrane conductance regulator gene to the airway epithium. *Cell* **68**: 143–155.

Rosenfeld ME, Feng M, Michael SI, Siegal GP, Alvarez RD, Curiel DT. (1995) Adenoviral-mediated delivery of the herpes simplex virus thymidine kinase gene selectively sensitizes human ovarian carcinoma cells to ganciclovir. *Clin. Cancer Res.* **1**: 1571–1580.

Rubin BA, Rorke LB. (1988) Adenovirus vaccines. In: *Vaccines* (eds SA Plotkin and EA Mortimer). W.B. Saunders, Philadelphia, PA, pp. 492–512.

Schaack J, Langer S, Guo X. (1995a) Efficient selection of recombinant adenoviruses by vectors that express β-galactosidase. *J. Virol.* **69**: 3920–3923.

Schaack J, Guo X, Ho WY-W, Karlok M, Chen C, Ornelles D. (1995b) Adenovirus type 5 precursor terminal protein-expressing 293 and HeLa cell lines. *J. Virol.* **69**: 4079–4085.

Schneider RJ, Shenk T. (1987) Impact of virus infection on host cell protein synthesis. *Annu. Rev. Biochem.* **56**: 317–332.

Seth P, Fitzgerald D, Willingham M, Pastan I. (1984) Pathway of adenovirus entry into cells. *Mol. Cell. Biol.* **4**: 1528–1533.

Setoguchi Y, Jaffe HA, Chu C-S, Crystal RG. (1994) Intraperitoneal *in vivo* gene ther.apy to deliver α1-antitrypsin to the systemic circulation. *Am. J. Respir. Cell Mol. Biol.* **10**: 369–377.

Sharp PA, Moore C, Haverty JL. (1976) The infectivity of adenovirus 5 DNA–protein complex. *Virology* **75**: 442–456.

Shenk T, Williams J. (1984) Genetic analysis of adenoviruses. *Curr. Top. Microbiol. Immunol.* **111**: 1–39.

Sparer TE, Tripp RA, Dillehay DL, Hermiston TW, Wold WSM, Gooding LR. (1996) The role of human adenovirus early region 3 proteins (gp19K, 10.4K, 14.5K and 14.7K) in a murine pneumonia model. *J. Virol.* **70**: 2431–2439.

Stevenson SC, Rollence M, White B, Weaver L, McClelland A. (1995) Human adenovirus serotypes 3 and 5 bind to two different cellular receptors via the fibre head domain. *J. Virol.* **69**: 2850–2857.

Straus SE. (1984) Adenovirus infections in humans. In: *The Adenoviruses* (ed. HS Ginsberg). Plenum Press, New York, pp. 451–496.

Van Ormondt H, Galibert F. (1984) Nucleotide sequences of adenovirus DNAs. *Curr. Top. Microbiol. Immunol.* **110**: 73–142.

Vincent AJPE, Vogels R, Someren GV, Esandi MC, Noteboom JL, Avezaat CJJ, Vecht C, Bekkum DWV, Valerio D, Bout A, Hoogerbrugge PM. (1996) Herpes simplex virus thymidine kinase gene therapy for rat malignant brain tumours. *Hum. Gene Ther.* **7**: 197–205.

Wagner E, Zatloukal K, Cotten M, Kirlappos H, Mechtler K, Curiel DT, Birnstiel ML. (1992) Coupling of adenovirus to transferrin–polylysine/DNA complexes greatly enhances receptor-mediated gene delivery and expression of transfected genes. *Proc. Natl Acad. Sci. USA* **89**: 6099–6103.

Wang Q, Jia X-C, Finer MH. (1995) A packaging cell line for propagation of recombinant adenovirus vectors containing two lethal gene-region deletions. *Gene Ther.* **2**: 775–783.

Weinberg DH, Ketner G. (1986) Adenoviral early region 4 is required for efficient viral DNA replication and for late gene expression. *J. Virol.* **57**: 833–838.

Wickham TJ, Mathias P, Cheresh DA, Nemerow GR. (1993) Integrins $\alpha_v\beta_3$ and $\alpha_v\beta_5$ promote adenovirus internalization but not virus attachment. *Cell* **73**: 309–319.

Wickham TJ, Carrion ME, Kovesdi I. (1995) Targeting of adenovirus penton base to new receptors through replacement of its RGD motif with other receptor-specific peptide motifs. *Gene Ther.* **2**: 750–756.

Wills KN, Huang W-M, Harris MP, Machemer T, Maneval DC, Gregory RJ. (1995) Gene therapy for hepatocellular carcinoma: Chemosensitivity conferred by adenovirus-mediated transfer of the HSV-1 thymidine kinase gene. *Cancer Gene Ther.* **3**: 191–197.

Wilson JM. (1995) Gene therapy for cystic fibrosis: Challenges and future directions. *J. Clin. Invest.* **96**: 2547–2554.

Yang Y, Nunes FA, Berencsi K, Furth EE, Gonczol E, Wilson JM. (1994a) Cellular immunity to viral antigens limits E1-deleted adenoviruses for gene therapy. *Proc. Natl Acad. Sci. USA* **91**: 4407–4411.

Yang Y, Nunes FA, Berencsi K, Gonczol E, Engelhardt JF, Wilson JM. (1994b) Inactivation of *E2a* in recombinant adenoviruses improves the prospect for gene therapy in cystic fibrosis. *Nature Genet.* **7**: 362–369.

Yang Y, Trinchieri G, Wilson JM. (1995) Recombinant IL-12 prevents formation of blocking IgA antibodies to recombinant adenovirus and allows repeated gene therapy to mouse lung. *Nature Med.* **1**: 890–893.

Yeh P, Dedieu J-F, Orsini C, Vigne E, Denefle P, Perricaudet M. (1996) Efficient dual transcomplementation of adenovirus E1 and E4 regions from a 293-derived cell line expressing a minimal E4 functional unit. *J. Virol.* **70**: 559–565.

Yei S, Bachurski CJ, Weaver TE, Wert S, Trapnell BC, Whitsett JA. (1994a) Adenoviral-mediated gene transfer of human surfactant protein B to respiratory epithelial cells. *Am. J. Respir. Cell Mol. Biol.* **11**: 329–336.

Yei S, Mittereder N, Wert S, Whitsett JA, Wilmott RW, Trapnell BC. (1994b) *In vivo* evaluation of the safety of adenovirus-mediated transfer of the human cystic fibrosis transmembrane conductance regulator cDNA to the lung. *Hum. Gene Ther.* **5**: 731–744.

Yoshimura K, Rosenfeld MA, Seth P, Crystal RG. (1993) Adenovirus-mediated augmentation of cell transfection with unmodified plasmid vectors. *J. Biol. Chem.* **268**: 2300–2303.

5

Adeno-associated virus vectors for human gene therapy

Jeffrey S. Bartlett and Richard J. Samulski

5.1 Introduction

Presently, human gene therapy is limited by the efficiency of stable gene transfer. To overcome this obstacle, adeno-associated virus (AAV) is being developed as a vector. This unique member of the parvovirus family possesses several properties which distinguish it from other gene transfer vectors. Its advantages include stable and efficient integration of viral DNA into the host genome (Berns *et al.*, 1975; Cheung *et al.*, 1980; Hoggan *et al.*, 1972; Laughlin *et al.*, 1986; McLaughlin *et al.*, 1988), lack of any associated human disease (Berns *et al.*, 1982), broad host range (Buller *et al.*, 1979; Casto *et al.*, 1967), the ability to infect growth-arrested cells (Wong *et al.*, 1993), and the ability to carry non-viral regulatory sequences without interference from the viral genome (Miller *et al.*, 1993a; Walsh *et al.*, 1992). In addition, there has been no superinfection immunity associated with AAV vectors (Lebkowski *et al.*, 1988; McLaughlin *et al.*, 1988).

AAV is a defective virus with a unique bi-phasic life cycle. It can be propagated either as a lytic virus or maintained as a provirus integrated into the host cell genome (Atchison *et al.*, 1965; Hoggan *et al.*, 1966, 1972) (*Figure 5.1*). In a lytic infection, replication requires co-infection with either adenovirus (Atchison *et al.*, 1965; Hoggan *et al.*, 1966; Melnick *et al.*, 1965), or herpes simplex virus (HSV) (Buller *et al.*, 1981; McPherson *et al.*, 1985); hence the classification of AAV as a 'defective' virus. Vaccinia virus can also provide at least partial helper function (Schlehofer *et al.*, 1986). When AAV infects tissue culture cells in the absence of helper virus, it establishes latency by persisting in the host cell genome as an integrated provirus (Berns *et al.*, 1975; Cheung *et al.*, 1980; Handa *et al.*, 1977; Hoggan *et al.*, 1972). Although AAV physically recombines its DNA into the host cell genome, it can be rescued from the chromosome and re-enter the lytic cycle if these cells are superinfected with helper virus. The lytic phase of the AAV life cycle requires the expression of the adenovirus early gene products (Richardson and Westphal, 1981) E1a (Chang *et al.*, 1989; Richardson and Westphal, 1984), E1b (Richardson and Westphal, 1984; Samulski *et al.*, 1988), E2a (Jay *et al.*, 1979), E4 (Carter *et al.*, 1983; Laughlin *et al.*, 1982; Richardson and Westphal, 1981, 1984), and VA RNA (Janik *et al.*, 1989; West *et al.*, 1987).

Gene Therapy, edited by N.R. Lemoine and D.N. Cooper.
© 1996 BIOS Scientific Publishers Ltd, Oxford.

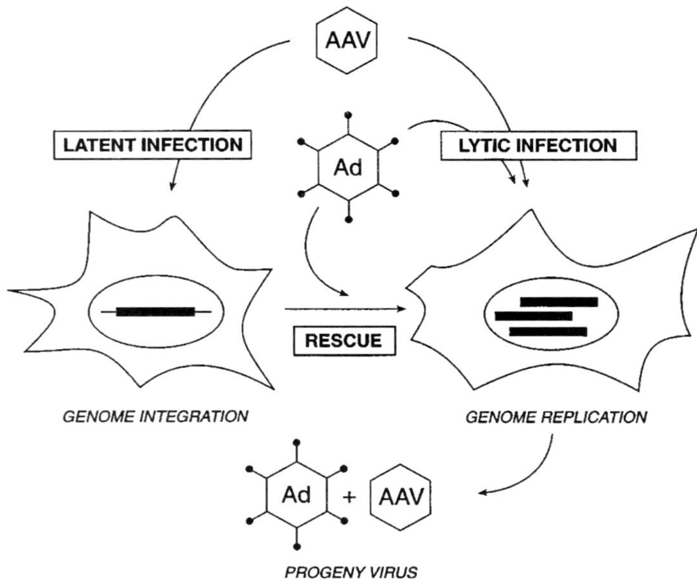

Figure 5.1. The life cycle of AAV has both a latent and lytic component. In the presence of adenovirus helper virus (Ad), AAV proceeds through a lytic infection. In the absence of helper virus, the AAV genome integrates into the host genome, thus establishing latency. The latent virus genome is stable for many generations, although it can be 'rescued' to enter the lytic phase upon subsequent helper virus superinfection.

The AAV genome is encapsidated as a single-stranded DNA molecule of plus or minus polarity. Strands of both polarities are packaged, but in separate virus particles (Berns and Adler, 1972; Berns and Rose, 1970; Mayor *et al.*, 1969; Rose *et al.*, 1969), and both strands are infectious (Samulski *et al.*, 1987). Five serotypes of AAV have been identified, but the most extensively characterized is AAV-2. The non-enveloped virion is icosohedral in shape and one of the smallest that has been described, about 20–24 nm in diameter (Hoggan, 1970; Tsao *et al.*, 1991) with a density of 1.41 g cm^{-3} (de la Maza and Carter, 1980a,b). The relatively high density of AAV particles allows them to be easily separated by CsCl density centrifugation from adenovirus helper virus that has a density of approximately 1.35 g cm^{-3} (de la Maza and Carter, 1980b). In addition, the AAV virion is resistant to a number of physical treatments that inactivate other viruses, such as heat treatment (56°C for 1 h), low pH, detergents and proteases (Bachmann *et al.*, 1979), thereby further permitting the virions to be purified or concentrated.

5.2 AAV structure and genetics

The AAV virion is composed of three structural proteins: VP1, VP2 and VP3 (Johnson *et al.*, 1971, 1977, 1975; Rose *et al.*, 1971). These are called the capsid, or Cap, proteins and have molecular masses of 87, 73 and 61 kDa, respectively. VP3 is the most abundant protein in the virion and comprises about 90% of the total virion protein; VP1 and VP2 each account for about 5% of the total virion protein. All three capsid proteins are *N*-acetylated (Becerra *et al.*, 1985). At present, little is known about the structure of these proteins in the capsid although the crystal structure of a related canine parvovirus has recently been determined (Tsao *et al.*, 1991).

There are at least four non-structural AAV proteins collectively termed the Rep proteins for their role in viral DNA replication. These proteins are often referred to according to their molecular masses (i.e. Rep 78, Rep 68, Rep 52 and Rep 40) (Mendelson *et al.*, 1986; Srivastava *et al.*, 1983; Yang *et al.*, 1994).

The complete nucleotide sequence of AAV-2 consists of 4680 nucleotides (Srivastava *et al.*, 1983). The genome is single-stranded, linear DNA and contains two terminal inverted repeats that are 145 bp long (Gerry *et al.*, 1973; Koczot *et al.*, 1973; Lusby *et al.*, 1980) (*Figure 5.2*). These repeats are thought to form terminal T-shaped hairpin structures at each end of the AAV genome. The internal portion of the AAV genome is divided genetically into two regions that encode the non-structural (Rep) and structural (Cap) viral proteins respectively (see *Figure 5.2*). Three promoters have been identified and named according to their approximate map positions: p5, p19 and p40 (Green *et al.*, 1980a,b,c; Laughlin *et al.*, 1979; Lusby and Berns, 1982). Transcripts initiating at each of these promoters share a common intron and all terminate at the same polyadenylation site at map position 95 (Green and Roeder, 1980b; Laughlin *et al.*, 1979; Srivastava *et al.*, 1983). Both spliced and unspliced transcripts are detectable in infected cells (Green and Roeder, 1980a,b; Laughlin *et al.*, 1979), and at least one alternatively spliced message is also present (Trempe and Carter, 1988a).

The messages encoding the non-structural proteins are transcribed from the p5 and p19 promoters (*Figure 5.2*). The structural mRNAs are transcribed from the p40 promoter (*Figure 5.2*). The amino acid sequence of the major capsid protein VP3 is contained within the two larger and less abundant capsid proteins VP1 and VP2 (Janik *et al.*, 1984). VP2 is synthesized from the same mRNA as VP3 using an upstream ACG start codon. VP1 is synthesized from the alternatively spliced p40 transcript mentioned above. Although infectious virions can be made which lack one or more capsid proteins, very low numbers of infectious particles are produced from these mutants (Hermonat *et al.*, 1984a; Tratschin *et al.*, 1984a).

The two larger Rep proteins, Rep 78 and Rep 68, control DNA replication

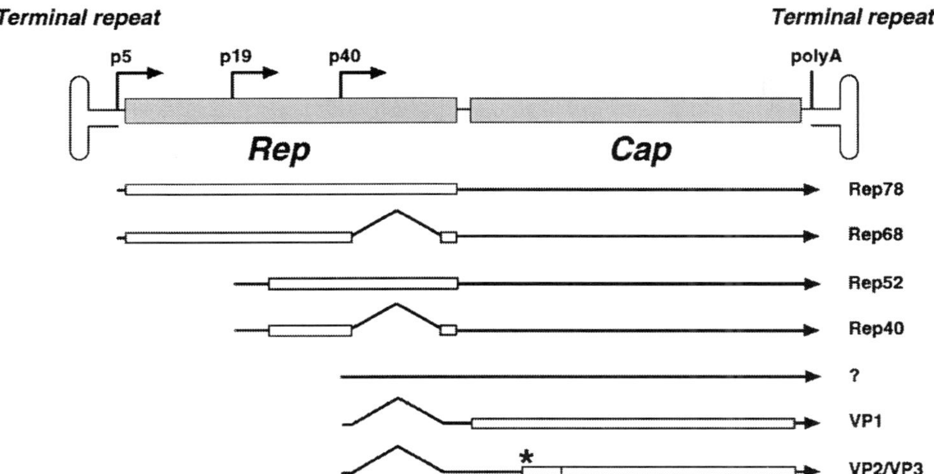

Figure 5.2. The AAV genome. The viral transcripts are shown below the genetic map of AAV. The size of each mRNA and the proteins synthesized from each mRNA are shown on the right. Also shown are the three AAV promoters and the polyadenylation site. VP2 is synthesized from an extended VP3 ORF by using an ACG start codon, indicated by the asterisk.

(Hermonat *et al.*, 1984a; Labow *et al.*, 1986, 1987; Tratschin *et al.*, 1986). The exact role the two smaller Rep proteins play during a productive viral infection is unknown (Owens *et al.*, 1993). Both Rep 68 and Rep 78 have been expressed in non-mammalian systems and purified to homogeneity. Their activities include binding to the terminal hairpin structures (Ashktorab and Srivastava, 1989; Chiorini *et al.*, 1994; Im and Muzyczka, 1989), ATP-dependent DNA helicase activity (Im and Muzyczka, 1990), strand- and sequence-specific DNA endonuclease activity (Im and Muzyczka, 1990), covalent binding to the 5′ end of their DNA substrate (Im and Muzyczka, 1990), and the ability to replicate viral DNA in a cell-free system (Chiorini *et al.*, 1994; Ni *et al.*, 1994). Interestingly, in the absence of helper-virus co-infection, Rep 78 and Rep 68 are able to regulate replication negatively (Berns *et al.*, 1988). In addition, in the absence of helper virus, the Rep proteins can repress both viral and heterologous gene expression (Antoni *et al.*, 1991; Labow *et al.*, 1987; Mendelson *et al.*, 1988; Tratschin *et al.*, 1986; Trempe and Carter, 1988b; West *et al.*, 1987). It has been suggested that this effect is mediated at the translational level for repression of the Cap proteins from the p40 promoter (Trempe and Carter, 1988b). The mechanism for the decreased expression of the Rep proteins from the p5 and p19 promoters is not clear. The idea that the *Rep* gene can autoregulate its own expression is attractive because it may help explain how Rep protein synthesis and DNA replication are turned off when the virus has integrated into the host genome.

5.3 Production of recombinant AAV

Since the initial description of recombinant AAV (rAAV) as a vector for gene transfer, substantial progress has been made towards the ultimate goal of using this vector in human gene therapy. Nevertheless, a number of important biological and technical issues remain to be addressed as rAAV vectors are being considered for human use. One significant technical obstacle is the production of rAAV in sufficient amounts for preclinical studies and, ultimately, human trials. At the present time, it is difficult to estimate how much vector will be needed and how much it will cost to produce individual clinical batches. However, it is anticipated that the amounts of virus needed will substantially exceed the capabilities of current methods of production.

The present method for producing stocks of rAAV uses a two-component plasmid system divided in terms of the *cis* and *trans* components necessary for replication, expression and encapsidation of the recombinant virus. The viral terminal repeats are the only elements required in *cis* and, in the current packaging system, flank the transgene on one of the two plasmids. The second plasmid supplies the necessary *Rep* and *Cap* gene products in *trans*. It is important that the two plasmid DNAs are sufficiently non-homologous to preclude homologous recombination events which could generate wild-type AAV (Samulski *et al.*, 1989). Although there are many variations on this theme, the most widely used methods for generation of rAAV involve co-transfection of the human cell line 293 with these two plasmids (Bartlett and Samulski, 1994; Bartlett *et al.*, 1996b; Samulski *et al.*, 1989). The transfected cells are then infected with adenovirus. Thus, all the components are delivered to the host cell to allow rescue, replication and packaging of the foreign gene into AAV particles. The Repgene products recognize the AAV *cis*-acting terminal repeats on the recombinant plasmid containing the foreign gene, rescue the rAAV genome out of the plasmid, and begin to replicate it. The AAV capsids begin to accumulate, recognize the AAV *cis*-

acting packaging signals located in the AAV terminal repeats, and encapsidate the recombinant viral DNA into an AAV virion. The products of this packaging scheme are then adenovirus helper virus and rAAV particles. Adenovirus can be removed by any of a number of physical separation strategies. In this manner, one can generate high-titre, helper-free stocks of rAAV.

Initially, the amounts of recombinant virus produced by this system were quite low (i.e. 10^4–10^5 transducing units per millilitre). However, recent advances have allowed the production of vector at concentrations above 10^9 transducing units per millilitre. The overall approach has remained the same. The increases in vector titres have come from simply optimizing each step of the procedure. Initially, the feeling was that the relatively low amount of recombinant virus produced by this system was due to the low levels of *Rep* and *Cap* gene products produced from the transfected plasmid templates (Kotin, 1994). In a wild-type lytic infection, the viral genome is amplified several times, which provides a much larger number of templates for Rep and Cap expression than are available following plasmid transfection. Consequently, several attempts aimed at producing higher titre vectors were centred upon strategies to increase the number of AAV genome equivalents available for Rep and Cap production. These included the incorporation of the *Rep* and *Cap* genes into replication-competent vectors, and attempts to establish cell lines with high endogenous copy numbers of these genes (Trempe and Yang, 1993; Vincent *et al.*, 1990). It was thought that the generation of these cell lines would also greatly simplify the production scheme by lessening the reliance on plasmid transfection. However, the construction of these cell lines was hampered by the toxicity of the Rep proteins (Winocour *et al.*, 1992). One cell line (HA25a) (Vincent *et al.*, 1990), which contains integrated copies of both the AAV *Rep* and *Cap* coding regions, was capable of generating low titre recombinant stocks. In another case, a 293 cell line was made in which the *Rep* gene was placed under the inducible control of the metallothionein promoter (Trempe and Yang, 1993). Although the level of inducible Rep expression from this line was shown to complement Rep⁻AAV replication, it is unclear whether any increase in recombinant viral titres will be possible. More recently, Flotte *et al.* (1995) described a packaging system wherein the rAAV vector sequences were integrated stably into 293 cells. To augment recombinant vector yield, this cell line was then transfected with a novel AAV helper plasmid which consisted of the human immunodeficiency virus long terminal repeat (HIV LTR) promoter in place of the endogenous AAV p5 promoter. Recently, yet another cell line has been constructed in which the AAV *Rep* and *Cap* gene segments are provided under control of their endogenous adenovirus-inducible promoters (Reed Clark *et al.*, 1995). The fact that this cell line was constructed in HeLa cells may have been crucial to its ability to be propagated without cytopathic effects. The authors suggest that previous attempts to construct producer lines in 293 cells were hampered by elevated levels of Rep expression due to constitutive adenovirus E1A expression in these cells, which is known to upregulate the AAV p5 and p19 promoters (Chang *et al.*, 1989) through interactions with the cellular transcription factor YY1 (Shi *et al.*, 1991).

Although the AAV packaging cell lines and methodologies realized from this work provided little in terms of increased vector yields, two important phenomena relating to AAV biology were observed: (i) the ability to uncouple expression of the Rep and Cap proteins and still maintain processing of the transcripts from each region and the production of the different Rep and Cap products in the same ratios as is expressed during a wild-type lytic infection; and (ii) the observation that there may be secondary genetic elements that enhance the efficiency of packaging into AAV virions. Identification of these elements and the development of new procedures for generating recombinant viral stocks of higher titre will be imperative for subsequent animal and human studies using AAV as a vector for gene therapy.

5.4 AAV-mediated gene transfer

The mechanism by which AAV vectors transduce cells has not been well defined. In fact, the mode of viral uptake has not been clearly established and no cellular receptor has been identified. Recently, AAV has been shown to associate with an 150kDa cell membrane glycoprotein (Mizukami *et al.*, 1996). However, the exact role that this association plays in infection is not known. In tissue culture, AAV is able to infect nearly all of the established human cell lines so far examined (Laughlin *et al.*, 1986; Lebkowski *et al.*, 1988; McLaughlin *et al.*, 1988; Samulski *et al.*, 1989; Tratschin *et al.*, 1985). However, an erythroid cell line (UT-7/Epo) (Mizukami *et al.*, 1996), as well as some megakaryocytic cell lines (Ponnazhagan *et al.*, 1994; J.S. Bartlett, unpublished observations) may be non-permissive to AAV. These cell lines should prove valuable in further studies aimed at characterizing the cellular receptor for AAV.

5.4.1 Size limitations of AAV vector genomes

Although the number of genes built into AAV vectors and tested *in vitro* remains relatively small, there are few limitations to this step of the procedure. Foremost, the recombinant genomes must be between 50% and 110% of the wild-type AAV size to be efficiently packaged into AAV particles (R.J. Samulski, unpublished) (de la Maza and Carter, 1980b). This means that most rAAV vectors can only accommodate inserts of up to about 4.5 kb in length. This, of course, will not limit most gene therapy strategies. However, there will undoubtedly be a number of genes that are too large to be built into AAV vectors. Even some genes that should fit within these size limitations may be made too large upon inclusion of exogenous transcriptional control elements. Second, a few cases have been found in which a foreign DNA sequence inserted into an AAV vector has inhibited AAV DNA replication. In one instance, the HSV gene for thymidine kinase (*tk*) was shown to reduce replication of a recombinant vector by approximately 50 times (Hermonat and Muzyczka, unpublished). Nonetheless, it should be possible to determine those sequences incompatible with AAV DNA replication and remove them from these constructs. In any event, it seems likely that AAV vectors will be capable of transferring many of the necessary therapeutic genes for the treatment of human diseases.

5.4.2 Transduction of non-dividing cells

Little is known about the relationship between viral infection and the cell cycle. Retroviral vectors based on the murine leukaemia virus require cell division for efficient transduction (Miller *et al.*, 1990). For this reason, there has been intense interest in the ability of rAAV vectors to transduce non-dividing cells. Recently, several reports have begun to shed light on this possibility. Saswati Chatterjee's laboratory has reported that rAAV is able to transduce growth-arrested human fibroblasts or 293 cells at the same, or better, efficiency as actively proliferating cells (Podsakoff *et al.*, 1994; Wong *et al.*, 1993). However, since the cells were allowed to resume growth prior to being scored for transduction, there remained the possibility that the virus simply remained episomal and integrated after the cells entered S phase. A second report seems to confirm this possibility since it was determined that the vector genomes can persist in stationary phase cells but that transduction preferentially occurs in cells after they have entered S phase (Russell *et al.*, 1994). In this case, transduction was assayed without subsequent stimulation and cell division. Although proliferating cells may be preferentially transduced by AAV vectors, proliferation may not be absolutely required for transduction. In fact, *in vivo* experiments have shown that rAAV vectors are able to transduce cell populations that are thought to be largely quiescent at remarkably high efficiencies. These include human bone marrow progenitors (Goodman *et al.*, 1994; Miller *et al.*, 1994; Zhou *et al.*, 1993) in culture, and rat brain *in vivo* (Bartlett *et al.*, 1995, 1996b; Kaplitt *et al.*, 1994; McCown *et al.*, 1996). Although the human bone marrow progenitor (CD34$^+$) cells were maintained in media containing growth factors [interleukin-3 (IL-3), IL-6, and stem-cell factor], analysis of colonies derived from these progenitors should allow the determination of the integration frequency in the primary haematopoeitic cells that gave rise to each colony. Histological analysis of rat brains injected with recombinant virus encoding β-galactosidase has demonstrated long-term expression in many regions of the rat brain (McCown *et al.*, 1996; M.J. During *et al.*, unpublished results). In addition to demonstrating the potential for gene therapy approaches into neuronal cells using AAV, this finding represents the most conclusive demonstration that rAAV vectors can transduce non-dividing cells. It should be noted that AAV transduction measured by gene expression, as described in the majority of reported experiments, does not reflect viral integration since gene expression can take place off episomal or integrated templates.

5.4.3 Integration of AAV vector genomes

One of the most interesting aspects of the AAV life cycle is the ability of the virus DNA to integrate into the host genome in the absence of a helper virus. AAV integration appears to have no effect on cell growth and, in spite of its propensity to integrate into the cellular genome as a rescuable provirus, there is no evidence of AAV functioning as a tumour virus (Handa *et al.*, 1977). The mechanism of viral integration is not known. It appears that the way in which AAV integrates is novel, since there are several aspects that distinguish it from other better characterized viral integration events. (i) No viral gene expression is required for integration to occur (McLaughlin *et al.*, 1988; Samulski *et al.*, 1989), although viral gene products may still be required. Only the AAV terminal repeats appear to be essential for integration (Hermonat and Muzyczka, 1984b; McLaughlin *et al.*, 1988; Samulski *et al.*, 1989; Srivastava *et al.*, 1989; Tratschin *et al.*, 1984b). (ii) Latently infected cells are very stable and capable of maintaining the integrated viral DNA for thousands of passages (Berns *et al.*, 1982; Samulski *et al.*, 1991). Concatemers consisting of 2–4 tandem copies often exist at the integration locus regardless of the initial multiplicity of infection, suggesting that at least a limited amount of viral replication may precede the integration event (Laughlin *et al.*, 1986; McLaughlin *et al.*, 1988). (iii) Integration of the wild-type AAV genome seems to prefer a target sequence located on human chromosome 19q13.3-qter (Kotin *et al.*, 1990, 1991; Samulski *et al.*, 1991). This site-specific integration has been documented in a number of cell types including human T cells, colon, lung, bone marrow stem cells, and monkey kidney cells (Goodman *et al.*, 1994; X. Zhu, X. Xiao and R.J. Samulski, unpublished observations).

rAAV vectors containing only the viral terminal repeat sequences are also capable of integration. While viral gene products are not required for integration *per se*, it is unclear whether viral proteins are required for the targeting of the integration event to chromosome 19. There has been a great deal of debate regarding the role of the Rep proteins in this process. Although latently infected cell lines containing viral DNA in the preferred integration site on chromosome 19 have been produced with *rep⁻/neomycin (neo)⁺* recombinant virus, wild-type virus may have been present in these recombinant virus stocks and Rep proteins could have been supplied in *trans*. Subsequent attempts to target *rep⁻* virus to chromosome 19 using recombinant viral stocks free of contamination with wild-type virus have been less successful as these recombinant viruses appeared to have integrated randomly (Walsh *et al.*, 1992). Analysis of cell lines transfected with *rep⁺/neo⁺* plasmids containing the AAV terminal repeats demonstrated that the *neo* gene had integrated into the favoured AAV DNA integration site on chromosome 19 in a majority of the cells whereas the same constructs without the *Rep* coding region failed to target this region (Shelling and Smith, 1994). The role of the Rep proteins in targeted integration has recently been examined further. Plasmids containing AAV terminal repeat sequences and a marker gene were transfected into cells in the presence and absence of a second plasmid containing the AAV *Rep* coding region. Without selection, the resulting transfectants were examined for the presence of the transferred sequences at the AAV DNA integration site on chromosome 19. Only those cells that received both plasmids had specifically integrated the marker gene into the AAV locus on chromosome 19 (W. Xiao and R.J. Samulski, unpublished observation). These experiments clearly demonstrate the requirement for the AAV *Rep* gene for targeted integration, and suggest that the *Rep* gene products delivered in *trans* in conjunction with the AAV terminal repeats delivered in *cis* are the minimal elements required for targeted integration. The involvement of Rep protein in the integration event is further supported by the recent finding that purified Rep protein may be able to recognize and bind specifically to the AAV DNA integration site on chromosome 19 (Weitzman *et al.*, 1994). The AAV structural proteins do not appear to be required for targeted integration since targeted integration can be achieved upon DNA transfection.

Studies using recombinant plasmid DNAs containing the AAV terminal repeats have attracted considerable interest lately. Philip *et al.* (1994) have reported efficient and sustained gene expression in primary T lymphocytes and primary and cultured tumour cells transfected with AAV plasmid DNA. Although evidence for integration is inconclusive, the AAV terminal repeats may increase the stability of the transfected plasmid DNA and enable longer-lasting gene expression, or promote efficient transfer from the cytoplasm to the nucleus.

5.4.4 Influence of cellular and helper functions on transduction by AAV vectors

A potential barrier to further *in vivo* studies with rAAV vectors has been the low transduction efficiencies of these vectors in some cell types. Recently, it has been shown that an immediate–early event, namely second-strand DNA synthesis, can be rate-limiting in the absence of adenovirus co-infection (Ferrari *et al.*, 1996; Fisher *et al.*, 1996). These studies suggest that the limiting step in our ability to score for gene transduction is not internalization of the virus, but rather the synthesis of a transcriptionally active double-stranded version of the AAV genome. This genomic conversion and subsequent expression of the recombinant reporter or therapeutic gene is greatly facilitated by expression of the adenovirus E4 open reading frame 6 (ORF6) protein (Ferrari *et al.*, 1996; Fisher *et al.*, 1996). The data indicate that expression of the ORF6 protein is both necessary and sufficient to increase the efficiency of gene transduction by a rAAV vector in a way that is dependent upon synthesis of double-stranded viral DNAs. Interestingly, the phenomenon elicited by the adenovirus E4 ORF6 protein can be reproduced to different degrees in rAAV-infected cells by exposure of the cells to heat shock or genotoxic reagents (Ferrari *et al.*, 1996). It is assumed that adenovirus E4 ORF6, genotoxic and physical stresses are acting through a common mechanism and that these effects are linked to the induction of the host cell DNA repair machinery rather than the cell cycle (Ferrari *et al.*, 1996; Fisher *et al.*, 1996). However, the precise mechanism for the increase in AAV transduction has not been defined.

The impact of these findings on the use of AAV vectors for gene therapy is unclear. However, it is apparent that rAAV vector transduction can be improved dramatically in cultured cells by a number of physical and chemical manipulations (Alexander *et al.*, 1994; Ferrari *et al.*, 1996; Fisher *et al.*, 1996; Russell *et al.*, 1995). This suggests that similar reagents could be coupled with current AAV vector strategies to enhance the delivery of therapeutic genes *in vivo*. Unfortunately, the use of these reagents *in vivo* is only beginning to be evaluated and early results do not come close to matching the increases in vector transduction efficiency seen in tissue culture (J.S. Bartlett and R.J. Samulski, unpublished observations). Nonetheless, strategies can be imagined in which rAAV transduction of bone marrow stem cells could be enhanced through the use of hydroxyurea, a reagent that is currently being used in the treatment of sickle cell anaemia (Charache *et al.*, 1995). Similarly, AAV vectors may be especially well suited for cancer gene therapy when combined with cytotoxic agents such as X-rays or chemotherapy that have been shown to enhance transduction of AAV *in vitro*. It is interesting to postulate that the efficiency of gene delivery *in vivo* may be several orders of magnitude higher than we are able to detect based on gene expression. However, it may be that this effect is restricted to specific cell types since some primary cells are transduced very efficiently by AAV vectors, and cannot be improved by adenovirus co-infection. The importance of second-strand synthesis on AAV-mediated gene transduction *in vivo* will have to be determined experimentally for each therapeutic protocol.

5.5 Use of rAAV for *in vivo* gene delivery

rAAV vectors are among the newest gene transfer vehicles. For this reason, many questions remain to be answered regarding the biology of AAV and its applicability to gene therapy. Detailed *in vitro* and *in vivo* studies will be needed to ascertain the true value of this vector. A number of these studies have been reported (see Bartlett *et al.*, 1995). The important areas that need to be addressed by experimentation include:

(i)   the ability of rAAV vectors to carry the transcriptional elements necessary for optimal transgene expression;

(ii) the ability of these control elements to function correctly after vector transduction *in vivo*;

(iii) the persistence of transgene expression *in vivo*;

(iv) the efficiency of vector transduction *in vivo*;

(v) the distribution of viral tropism and tissue distribution in seroconverted individuals.

From the *in vitro* studies presented to date, it appears that several different kinds of transcriptional elements will be active in AAV vectors. These include heterologous high-level viral promoter and enhancer elements (Chatterjee *et al.*, 1992; Hermonat and Muzyczka, 1984b; Lebkowski *et al.*, 1988; Muro-Cacho *et al.*, 1992; Vincent *et al.*, 1990), cellular enhancers linked to viral promoters (Ponnazhagan *et al.*, 1993; Zhou *et al.*, 1993), tissue-specific cellular control regions (Miller *et al.*, 1993a,b, 1994; Walsh *et al.*, 1992; Zhou *et al.*, 1993), and inducible cellular control elements (Walsh *et al.*, 1992). In addition, polymerase III control elements and several snRNA, pol III and pol II transcriptional elements have been used in conjunction with rAAV vectors (Bartlett *et al.*, 1996a; Rossi *et al.*, 1994). In each case, these elements have been shown to function correctly and independently of their location within an AAV vector. Although parallel data *in vivo* has not been reported for each of these elements, such preliminary results in tissue culture are encouraging.

A number of studies have been carried out to determine the potential of rAAV to transduce and express genes in primary haematopoietic progenitor cells (Goodman *et al.*, 1994; LaFace *et al.*, 1988; Miller *et al.*, 1994; Zhou *et al.*, 1993). A high frequency of gene transfer and expression of the β-galactosidase reporter gene was observed in progenitor cells, highly purified by positive immunoselection (Goodman *et al.*, 1994). However, the frequency of transfer of therapeutic genes into these cells has not been as extensively studied. Recently, the gene encoding the human Fanconi anaemia C complementing (FACC) protein and a mutationally marked gene for human γ-globin have been successfully transferred into human haematopoietic progenitor cells (Miller *et al.*, 1994; Walsh *et al.*, 1994). Specific mRNA derived from the rAAV-transduced genes was present at high levels in a majority of colonies and correction of the inherent cellular defects was observed, confirming the ability of rAAV to transduce human haematopoietic progenitors with high frequency. Although these data support the potential of rAAV to infect haematopoietic progenitors, the efficiency of stable integration of the vector genome has not been determined. Interestingly, wild-type AAV has been shown to integrate specifically into the AAV DNA integration site on chromosome 19 at a much lower frequency in these progenitor cells than in established human cell lines (Goodman *et al.*, 1994).

Experiments in tissue culture have demonstrated that rAAV can efficiently transduce a very wide range of cell types. These include several established transformed lines (HeLa, KB, Detroit 6, 293), a normal lymphoblastoid line (NC37), human T-cell lines (CEM, H9), human colon cancer lines (HT29, LIM, CaCo), several leukaemia lines (K562, KG1a, HEL, HL60, U937), an airway epithelial cell line (A549), human lymphoblast cell lines transformed with Epstein–Barr virus (EBV), and a HPV16-immortalized cystic fibrosis tracheal epithelial cell line (CFT-1) (Hermonat and Muzyczka, 1984b; Laughlin *et al.*, 1986; Lebkowski *et al.*, 1988; McLaughlin *et al.*, 1988; Samulski *et al.*, 1989; Tratschin *et al.*, 1985; Walsh *et al.*, 1992, 1994; J.S. Bartlett *et al.*, unpublished observation). Primary cells transduced with recombinant AAV have included human liver hepatocytes, human fibroblasts, human nasal airway epithelial cells, and explanted human glial cells (Wei *et al.*, 1994; J.S. Bartlett and R.J. Samulski, unpublished observation). So far, the versatility of AAV gene transfer in tissue culture is paralleled in primary cells.

Detailed *in vivo* studies have been hampered by the technical difficulties in producing the quantity of recombinant virus needed for these experiments. rAAV containing the β-galactosidase marker gene have been introduced into rat colon and rat brain (Bartlett *et al.*, 1995; McCown *et al.*, 1996). Gene expression was detected in both cases by intense blue staining of the transduced tissue. A portion of the human gene for the cystic fibrosis transmembrane regulator (CFTR) has been delivered intrabronchially to rabbits (Flotte *et al.*, 1993). Transcripts from the introduced gene and CFTR protein have been detected up to 6 months post-transfer. Although it is unclear whether or not sufficient protein is being produced to be of therapeutic value, this is an important experiment because it represents the first reported application of a potentially therapeutic rAAV vector to an animal. Other animal studies that are currently underway include the introduction of rAAV containing the gene for human tyrosine hydroxylase (TH) into rat brain striatum (Kaplitt *et al.*, 1994). Evidence of TH protein has been demonstrated for up to 2 months, suggesting that this vector–therapeutic gene combination may have potential for gene therapy in Parkinson's disease and other neuronal disorders. Importantly, no evidence for toxicity in the animals treated with these vectors was observed. This experiment extends early studies of AAV gene delivery in neuronal cells and demonstrates for the first time long-term expression of a therapeutic gene. These experiments will have important implications for the use of this vector for the treatment of human CNS disease.

References

Alexander IE, Russell DW, Miller AD. (1994) DNA-damaging agents greatly increase the transduction of nondividing cells by adeno-associated virus vectors. *J. Virol.* **68:** 8282–8287.

Antoni BA, Rabson AB, Miller IL, Trempe JP, Chejanovsky N, Cater BJ. (1991) Adeno-associated virus Rep protein inhibits human immunodeficiency virus type I production in human cells. *J. Virol.* **65:** 396–404.

Ashktorab H, Srivastava A. (1989) Identification of nuclear proteins that specifically interact with the adeno-associated virus 2 inverted terminal repeat hairpin DNA. *J. Virol.* **63:** 3034–3039.

Atchison RW, Casto BC, Hammond WM. (1965) Adenovirus-associated defective virus particles. *Science* **149:** 754–756.

Bachmann PA, Hoggan MD, Kurstak E, Melnick JL, Pereira HG, Tattersall P, Vago C. (1979) Parvoviridae: second report. *Intervirology* **11:** 248–254.

Bartlett JS, Samulski RJ. (1994) Methods for the construction and propagation of recombinant adeno-associated virus vectors. In: *Methods in Molecular Biology: Gene Therapy Protocols* (ed. P Robbins). Humana Press, NY.

Bartlett JS, Quattrocchi KB, Samulski RJ. (1995) The development of adeno-associated virus as a vector for cancer gene therapy. In: *The Internet Book of Gene Therapy: Cancer Therapeutics* (eds RE Sobol and KJ Scanlon). Appleton and Lange, Stamford, CT, pp. 27–40.

Bartlett JS, Ramamurthy L, Sethna M, Samulski JS, Marzluff WA. (1996a) Efficient expression of protein coding genes from the murine U1b snRNA promoter. *Proc. Natl Acad. Sci. USA* (in press).

Bartlett JS, Xiao X, Samulski RJ. (1996b) Adeno-asssociated virus vectors for gene transfer. In: *Protocols for Gene Transfer in Neuroscience: Towards Gene Therapy of Neurological Disorders* (eds PR Lowenstein and LW Enquist). John Wiley and Sons, Chichester, UK, pp. 115–127.

Becerra SP, Rose JA, Hardy M, Baroudy BM, Anderson CW. (1985) Direct mapping of adeno-associated virus proteins B and C: a possible ACG initiation codon. *Proc. Natl Acad. Sci. USA* **82:** 7919–7923.

Berns KI, Adler S. (1972) Separation of two types of adeno-associated virus particles containing complementary polynucleotide chains. *J. Virol.* **9:** 394–396.

Berns KI, Rose JA. (1970) Evidence for a single-stranded adeno-associated virus genome: isolation and separation of complementary single strands. *J. Virol.* **5:** 693–699.

Berns KI, Pinkerton TC, Thomas GF, Hoggan MD. (1975) Detection of adeno-associated virus (AAV)-specific nucleotide sequences in DNA isolated from latently infected Detroit 6 cells. *Virology* **68:** 556–560.

Berns KI, Cheung A, Ostrove J, Lewis M. (1982) Adeno-associated virus latent infection. In: *Virus Persistance* (eds BWJ Mahy, AC Minson and GK Darby). Cambridge University Press, Cambridge, UK.

Berns KI, Kotin RM, Labow MA. (1988) Regulation of adeno-associated virus DNA replication. *Biochem. Biophys. Acta* **951**: 425–429.

Buller RM, Straus SE, Rose IA. (1979) Mechanism of host restriction of adenovirus-associated virus replication in African green monkey kidney cells. *J. Gen. Virol.* **43**: 663–672.

Buller RM, Janik JE, Sebring ED, Rose JA. (1981) Herpes simplex virus types 1 and 2 completely help adenovirus-associated virus replication. *J. Virol* **40**: 241–247.

Carter BJ, Marcus-Sekura CJ, Laughlin CA, Ketner G. (1983) Properties of an adenovirus type 2 mutant, Addl807, having a deletion near the right-hand genome terminus: failure to help AAV replication. *Virology* **126**: 505–516.

Casto BC, Armstrong JA, Atchison RW, Hammon WM. (1967) Studies on the relationship between adeno-associated virus type I (AAV-I) and adenoviruses. II. Inhibition of adenovirus plaques by AAV; its nature and specificity. *Virology* **33**: 452–458.

Chang L-S, Shi Y, Shenk T. (1989) Adeno-associated virus p5 promoter contains an adenovirus EIA inducible element and a binding site for the major late transcription factor. *J. Virol.* **63**: 3479–3488.

Charache S, Terrin ML, Moore RD, Dover GJ, Barton FB, Eckert SV, McMahon RP, Bonds DR. (1995) Effect of hydroxyurea on the frequency of painful crises in sickle cell anemia. *New Engl. J. Med.* **332**: 1317–1322.

Chatterjee S, Johnson PR, Wong KJ. (1992) Dual-target inhibition of HIV-1 *in vitro* by means of an adeno-associated virus antisense vector. *Science* **258**: 1485–1488.

Cheung AK, Hoggan MD, Hauswirth WW, Berns KI. (1980) Integration of the adeno-associated virus genome into cellular DNA in latently infected human Detroit 6 cells. *J. Virol.* **33**: 739–748.

Chiorini JA, Weitzman MD, Owens RA, Urcelay E, Safer B, Kotin RM. (1994) Biologically active Rep proteins of adeno-associated virus type 2 produced as fusion proteins in *Escherichia coli*. *J. Virol.* **68**: 797–804.

de la Maza LM, Carter BJ. (1980a) Heavy and light particles of adeno-associated virus. *J. Virol.* **33**: 1129–1137.

de la Maza LM, Carter BJ. (1980b). Molecular structure of adeno-associated virus variant DNA. *J. Biol. Chem.* **255**: 3194–3203.

Ferrari FK, Samulski T, Shenk T, Samulski RJ. (1996) Second-strand synthesis is a rate-limiting step for efficient transduction by recombinant adeno-associated virus vectors. *J. Virol.* **70**: 3227–3234.

Fisher K, Gao G-P, Weitzmann MD, DeMatteo R, Burda JF, Wilson JM. (1996) Transduction with recombinant adeno-associated virus for gene therapy is limited by leading-strand synthesis. *J. Virol.* **70**: 520–532.

Flotte TR, Afione SA, Conrad C, McGrath SA, Solow R, Oka H, Zeitlin PL, Guggino WB, Carter BJ. (1993) Stable *in vivo* expression of the cystic fibrosis transmembrane regulator with an adeno-associated virus vector. *Proc. Natl Acad. Sci. USA* **90**: 10613–10617.

Flotte TR, Barraza-Ortiz X, Solow R, Afione SA, Carter BJ, Guggino WB. (1995) An improved system for packaging recombinant adeno-associated virus vectors capable of *in vivo* transduction. *Gene Ther.* **2**: 29–37.

Gerry HW, Kelly TJJ, Berns KI. (1973) Arrangement of nucleotide sequences in adeno-associated virus DNA. *J. Mol. Biol.* **79**: 207–225.

Goodman S, Xiao X, Donahue RE, Moulton A, Miller J, Walsh C, Young NS, Samulski RJ, Nienhuis AW. (1994) Recombinant adeno-associated virus-mediated gene transfer into hematopoietic progenitor cells. *Blood* **84**: 1492–1500.

Green MR, Roeder RG. (1980a) Definition of a novel promoter for the major adenovirus-associated virus mRNA. *Cell* **1**: 231–242.

Green MR, Roeder RG. (1980b) Transcripts of the adeno-associated virus genome: mapping of the major RNAs. *J. Virol.* **36**: 79–92.

Green MR, Straus SE, Roeder RG. (1980c) Transcripts of the adenovirus-associated virus genome: multiple polyadenylated RNAs including a potential primary transcript. *J. Virol.* **35**: 560–565.

Handa H, Shiroki K, Shimojo H. (1977) Establishment and characterization of KB cell lines latently infected with adeno-associated virus type 1. *Virology* **82**: 84–92.

Hermonat PL, Muzyczka N. (1984b) Use of adeno-associated virus as a mammalian DNA cloning vector: transduction of neomycin resistance into mammalian tissue culture cells. *Proc. Natl Acad. Sci. USA* **81**: 6466–6470.

Hermonat PL, Labow MA, Wright R, Berns KI, Muzyczka N. (1984a) Genetics of adeno-associated virus: isolation and preliminary characterization of adeno-associated virus type 2 mutants. *J. Virol.* **51:** 329–333.

Hoggan MD. (1970) Adeno-associated viruses. *Prog. Med. Virol.* **12:** 211–239.

Hoggan MD, Blacklow NR, Rowe WP. (1966) Studies of small DNA viruses found in various adenovirus preparations: physical, biological, and immunological characteristics. *Proc. Natl Acad. Sci. USA* **55:** 1457–1471.

Hoggan MD, Thomas GF, Thomas FB, Johnson FB. (1972) Continuous carriage of adenovirus associated virus genome in cell culture in the absence of helper adenovirus. In: *Proceedings of the Fourth Lepetite Colloquium*, Cocoyac, Mexico, pp. 243–249.

Im D-S, Muzyczka N. (1989) Factors that bind to the AAV terminal repeats. *J. Virol.* **63:** 3095–3104.

Im D-S, Muzyczka N. (1990) The AAV origin binding protein Rep68 is an ATP-dependent site-specific endonuclease with DNA helicase activity. *Cell* **61:** 447–457.

Janik IE, Huston MM, Rose JA. (1984) Adeno-associated virus proteins: origin of the capsid components. *J. Virol.* **52:** 591–597.

Janik JE, Huston MM, Cho K, Rose JA. (1989) Efficient synthesis of adeno-associated virus structural proteins requires both adenovirus DNA binding protein and VA I RNA. *Virology* **168:** 320–329.

Jay FT, De La Maza LM, Carter BJ. (1979) Parvovirus RNA transcripts containing sequences not present in mature mRNA: A method for isolation of putative mRNA precursor sequences. *Proc. Natl Acad. Sci. USA* **76:** 625–629.

Johnson FB, Ozer HL, Hoggan MD. (1971) Structural proteins of adenovirus-associated virus type 3. *J. Virol.* **8:** 860–863.

Johnson FB, Thomson TA, Taylor PA, Vlazny DA. (1977) Molecular similarities among the adenovirus-associated virus polypeptides and evidence for a precursor protein. *Virology* **82:** 1–13.

Johnson FB, Whitaker CW, Hoggan MD. (1975) Structural polypeptides of adenovirus-associated virus top component. *Virology* **65:** 196–203.

Kaplitt MG, Leone P, Samulski RJ, Xiao X, Pfaff DW, O'Malley KL, During MJ. (1994) Long-term expression and phenotypic correction using adeno-associated virus vectors in the mammalian brain. *Nature Genet.* **8:** 148–154.

Koczot FJ, Carter BJ, Garon CF, Rose JA. (1973) Self-complementarity of terminal sequences within plus or minus strands of adenovirus-associated virus DNA. *Proc. Natl Acad. Sci. USA* **70:** 215–219.

Kotin RM. (1994) Prospects for the use of adeno-associated virus as a vector for human gene therapy. *Hum. Gene Ther.* **5:** 793–801.

Kotin RM, Siniscalco M, Samulski RJ, Zhu X, Hunter L, Laughlin CA, McLaughlin S, Muzyczka N, Rocchi M, Berns KI. (1990) Site-specific integration by adeno-associated virus. *Proc. Natl Acad. Sci. USA* **87:** 2211–2215.

Kotin RM, Menninger JC, Ward DC, Berns KI. (1991) Mapping and direct visualization of a region-specific viral DNA integration site on chromosome 19q 13-qter. *Genomics* **10:** 831–834.

Labow MA, Hermonat PL, Berns KI. (1986) Positive and negative autoregulation of the adeno-associated virus type 2 genome. *J. Virol.* **60:** 251–258.

Labow MA, Graf LH, Berns KI. (1987) Adeno-associated virus gene expression inhibits cellular transformation by heterologous genes. *Mol. Cell. Biol.* **7:** 1320–1325.

LaFace D, Hermonat P, Wakeland E, Peck A. (1988) Gene transfer into hematopoietic progenitor cells mediated by an adeno-associated virus vector. *Virology* **162:** 483–486.

Laughlin CA, Westphal H, Carter BJ. (1979) Spliced adenovirus-associated virus RNA. *Proc. Natl Acad. Sci. USA* **76:** 5567–5571.

Laughlin CA, Jones N, Carter BJ. (1982) Effects of deletions in adenovirus region I genes upon replication of adeno-associated virus. *J. Virol.* **41:** 868–876.

Laughlin CA, Cardellichio CB, Coon HC. (1986) Latent infection of KB cells with adeno-associated virus type 2. *J. Virol.* **60:** 515–524.

Lebkowski JS, McNally MM, Okarma TB, Lerch LB. (1988) Adeno-associated virus: a vector system for efficient introduction of DNA into a variety of mammalian cell types. *Mol. Cell. Biol.* **8:** 3988–3996.

Lusby E, Berns KI. (1982) Mapping of the 5′ termini of two adeno-associated virus 2 RNAs in the left half of the genome. *J. Virol.* **41:** 518–526.

Lusby E, Fife KH, Berns KI. (1980) Nucleotide sequence of the inverted terminal repetition in adeno-associated virus DNA. *J. Virol.* **34:** 402–409.

Mayor HD, Torikai K, Melnick J, Mandel M. (1969) Plus and minus single-stranded DNA separately encapsidated in adeno-associated satellite virions. *Science* **166:** 1280–1282.

McCown TJ, Xiao X, Li J, Breese GR, Samulski RJ. (1996) Differential and persistant expression patterns of CNS gene transfer by an adeno-associated virus (AAV) vector. *Brain Res.* **713**: 99–107.

McLaughlin SK, Collis P, Hermonat PL, Muzyczka N. (1988) Adeno-associated virus general transduction vectors: analysis of proviral structures. *J. Virol.* **62**: 1963–1973.

McPherson RA, Rosenthal LJ, Rose JA. (1985) Human cytomegalovirus completely helps adeno-associated virus replication. *Virology* **147**: 217–222.

Melnick JL, Mayor HD, Smith KO, Rapp F. (1965) Association of 20 millimicron particles with adenoviruses. *J. Bacteriol.* **90**: 271–274.

Mendelson E, Trempe JP, Carter BJ. (1986) Identification of the trans-active rep proteins of adeno-associated virus by antibodies to a synthetic oligopeptide. *J. Virol.* **60**: 823–832.

Mendelson E, Smith MG, Miller IL, Carter BJ. (1988) Effect of a viral rep gene on transformation of cells by an adeno-associated virus vector. *Virology* **166**: 612–615.

Miller DG, Adam MA, Miller AD. (1990) Gene transfer by retrovirus vectors occurs only in cells that are actively replicating at the time of infection. *Mol. Cell. Biol.* **10**: 4239–4242.

Miller JL, Walsh CE, Ney PA, Samulski RJ, Nienhuis AW. (1993a). Single-copy transduction and expression of human gamma-globin in K562 erythroleukemia cells using recombinant adeno-associated virus vectors: The effect of mutations in NF-E2 and GATA-1 binding motifs within the hypersensitivity site 2 enhancer. *Blood* **82**: 1900–1906.

Miller JL, Walsh CE, Samulski RJ, Young NS, Nienhuis AW. (1993b) Transfer and expression of the human γ-globin gene in purified hematopoeitic progenitor cells from Rhesus bone marrow using a recombinant adeno-associated viral (rAAV) vector (Abstract). In: *Fifth Parvovirus Workshop*, 10–14 November 1993, Crystal River, FL.

Miller JL, Donahue RE, Sellers SE, Samulski RJ, Young NS, Nienhuis AW. (1994) Recombinant adeno-associated virus (rAAV) mediated expression of a human γ-globin gene in human progenitor derived erythroid cells. *Proc. Natl Acad. Sci. USA* **91**: 10183–10187.

Mizukami H, Young NS, Brown KE. (1996) Adeno-associated virus type 2 binds to a 150-kilodalton cell membrane glycoprotein. *Virology* **217**: 124–130.

Muro-Cacho C, Samulski RJ, Kaplan D. (1992) Gene transfer in human lymphocytes using a vector based on adeno-associated virus. *J. Immunother.* **11**: 231–237.

Ni T-H, Zhou X, McCarty D, Zolotukhin I, Muzyczka N. (1994) *In vitro* replication of adeno-associated virus DNA. *J. Virol.* **68**: 1128–1138.

Owens RA, Weitzman MD, Kyostio SR, Carter BJ. (1993) Identification of a DNA-binding domain in the amino terminus of adeno-associated virus Rep protein. *J. Virol.* **67**: 997–1005.

Philip R, Brunette E, Kilinski L, Murugesh D, McNally MA, Ucar K, Rosenblatt J, Okarma TB, Lebkowski JS. (1994) Efficient and sustained gene expression in primary T lymphocytes and primary and cultured tumor cells mediated by adeno-associated virus plasmid DNA complexed to cationic liposomes. *Mol. Cell. Biol.* **14**: 2411–2418.

Podsakoff G, Wong KKJ, Chatterjee S. (1994) Efficient gene transfer into nondividing cells by adeno-associated virus-based vectors. *J. Virol.* **68**: 5656–5666.

Ponnazhagan S, Nallari ML, Srivastava A. (1993) Suppression of human alpha-globin gene expression mediated by the recombinant adeno-associated virus 2-based antisense vectors. *J. Exp. Med.* **179**: 733–738.

Ponnazhagan S, Wang XS, Kang LY, Woody MJ, Nallari ML, Munshi NC, Zhou SZ, Srivastava A. (1994) Transduction of human hematopoietic cells by the adeno-associated virus 2 vectors is receptor-mediated. *Blood* **84** (Suppl.): 742a.

Reed Clark K, Voulgaropoulou F, Fraley DM, Johnson PR. (1995) Cell lines for the production of recombinant adeno-associated virus. *Hum. Gene Ther.* **6**: 1329–1341.

Richardson WD, Westphal WD. (1981) A cascade of adenovirus early functions is required for expression of adeno-associated virus. *Cell* **27**: 133–141.

Richardson WD, Westphal WD. (1984) Requirement for either early region la or early region lb adenovirus gene products in the helper effect for adeno-associated virus. *J. Virol.* **51**: 404–410.

Rose JA, Berns KI, Hoggan MD, Koczot FJ. (1969) Evidence for a single-stranded adenovirus-associated virus genome: Formation of a DNA density hybrid on release of viral DNA. *Proc. Natl Acad. Sci. USA* **64**: 863–869.

Rose JA, Maizel JK, Shatkin AJ. (1971) Structural proteins of adenovirus-associated viruses. *J. Virol.* **8**: 766–770.

Rossi JJ, Carbonnelle C, Li S, Chatterjee S, Zaia JA, Larson G, Bertrand E. (1994) Promoter for expression of transduced anti-HIV and SIV ribozymes. In: *Gene Therapy, Keystone Symposia*, Copper Mountain, CO.

Russell DW, Miller AD, Alexander IE. (1994) Adeno-associated virus vectors preferentially transduce cells in S phase. *Proc. Natl Acad. Sci. USA* **91:** 8915–8919.

Russell DW, Miller AD, Alexander IE. (1995) DNA synthesis and topoisomerase inhibitors increase transduction by adeno-associated virus vectors. *Proc. Natl Acad. Sci. USA* **92:** 5719–5723.

Samulski RJ, Shenk T. (1988) Adenovirus E1B 55-Mr, polypeptide facilitates timely cytoplasmic accumulation of adeno-associated virus mRNAs. *J. Virol.* **62:** 206–210.

Samulski RJ, Chang L-S, Shenk T. (1987) A recombinant plasmid from which an infectious adeno-associated virus genome can be excised *in vitro* and its use to study viral replication. *J. Virol.* **61:** 3096–3101.

Samulski RJ, Chang L-S, Shenk T. (1989) Helper-free stocks of recombinant adeno-associated viruses: normal integration does not require viral gene expression. *J. Virol.* **63:** 3822–3828.

Samulski RJ, Zhu X, Xiao X, Brook JD, Housman DE, Epstein N, Hunter LA. (1991) Targeted integration of adeno-associated virus (AAV) into human chromosome 19. *EMBO J.* **10:** 3941–3950.

Schlehofer JR, Ehrbar M, zur Hausen H. (1986) Vaccinia virus, herpes simplex virus, and carcinogens induce DNA amplification in a human cell line and support replication of a helpervirus dependent parvovirus. *Virology* **152:** 110–117.

Shelling A, Smith MG. (1994) Targeted integration of transfected and infected adeno-associated virus vectors containing the neomycin resistance gene. *Gene Ther.* **1:** 165–169.

Shi Y, Seto E, Chang L-S, Shenk T. (1991) Transcription repression by YY1, a human GL1-Krüppel-related protein, and relief of repression by adenovirus E1A protein. *Cell.* **67:** 377–388.

Srivastava A, Lusby EW, Berns KI. (1983) Nucleotide sequence and organization of the adeno-associated virus 2 genome. *J. Virol.* **45:** 555–564.

Srivastava CH, Samulski RJ, Lu L, Larsen SH, Srivastava A. (1989) Construction of a recombinant human parvovirus B19: adeno-associated virus 2 (AAV) DNA inverted terminal repeats are functional in an AAV-B19 hybrid virus. *Proc. Natl Acad. Sci. USA* **86:** 8078–8082.

Tratschin J-D, Miller IL, Carter BJ. (1984a) Genetic analysis of adeno-associated virus: properties of deletion mutants constructed *in vitro* and evidence for an adeno-associated virus replication function. *J. Virol.* **51:** 611–619.

Tratschin J-D, West MHP, Sandbank T, Carter BJ. (1984b) A human parvovirus, adeno-associated virus, as a eukaryotic vector: transient expression and encapsidation of the prokaryotic gene for chloramphenicol acetyltransferase. *Mol. Cell. Biol.* **4:** 2072–2081.

Tratschin J-D, Miller IL, Smith MG, Carter BJ. (1985) Adeno-associated virus vector for high-frequency integration, expression, and rescue of genes in mammalian cells. *Mol. Cell. Biol.* **5:** 3251–3260.

Tratschin J-D, Tal J, Carter BJ. (1986) Negative and positive regulation in *trans* of gene expression from adeno-associated virus vectors in mammalian cells by a viral Rep gene product. *Mol. Cell. Biol.* **6:** 2884–2894.

Trempe JP, Carter BJ. (1988a) Alternate mRNA splicing is required for synthesis of adeno-associated virus VP1 capsid protein. *J. Virol.* **62:** 3356–3363.

Trempe JP, Carter BJ. (1988b) Regulation of adeno-associated virus gene expression in 293 cells: control of mRNA abundance and translation. *J. Virol.* **62:** 68–74.

Trempe JP, Yang Q. (1993) Characterization of a cell line that expresses the AAV replication proteins (Abstract). In: *Fifth Parvovirus Workshop*, 10–14 November 1993, Crystal River, FL.

Tsao J, Chapman MS, Agbandjo M, Keller W, Smith K, Wu H, Luo M, Smith TJ, Rossman MG, Compans RW, Parrish CR. (1991) The three-dimensional structure of canine parvovirus and its functional implications. *Science* **25:** 1456–1464.

Vincent KA, Moore GK, Haigwood NL. (1990) Replication and packaging of HIV envelope genes in a novel adeno-associated virus vector system. *Vaccine* **90:** 353–359.

Walsh CE, Liu JM, Xiao X, Young NS, Nienhuis AW, Samulski RJ. (1992) Regulated high level expression of a human γ-globin gene introduced into erythroid cells by an adeno-associated virus vector. *Proc. Natl Acad. Sci. USA.* **89:** 7257–7261.

Walsh CE, Nienhuis AW, Samulski RJ, Brown MG, Miller JL, Young NS, Liu JM. (1994) Phenotypic correction of Fanconi anemia in human hematopoietic cells with a recombinant adeno-associated virus vector. *J. Clin. Invest.* **94:** 1440–1448.

Wei J-F, Wei F-S, Samulski RJ, Barranger JA. (1994) Expression of the human glucocerebrosidase and arylsulfatase A genes in murine and patient primary fibroblasts transduced by an adeno-associated virus vector. *Gene Ther.* **1:** 261–268.

Weitzman MD, Kyostio SR, Kotin RM, Owens RA. (1994) Adeno-associated virus (AAV) Rep proteins mediate complex formation between AAV DNA and its integration site in human DNA. *Proc. Natl Acad. Sci. USA* **91:** 5808–5812.

West MHP, Trempe JP, Tratschin J-D, Carter BJ. (1987) Gene expression in adeno-associated virus vectors: the effects of chimeric mRNA structure, helper virus, and adenovirus VAI RNA. *Virology* **160**: 38–47.

Winocour E, Puzis L, Etkin S, Koch T, Danovitch B, Mendelson E, Shaulian E, Karby S, Lavi S. (1992) Modulation of the cellular phenotype by integrated adeno-associated virus. *Virology* **190**: 316–329.

Wong KK, Podsakoff G, Lu D, Chatterjee S. (1993) High efficiency gene transfer into growth arrested cells utilizing an adeno-associated virus (AAV)-based vector (Abstract). In: Conference Proceedings, 3–7 December 1993. American Society of Hematology, St Louis, MO.

Yang Q, Chen F, Trempe JP. (1994) Characterization of cell lines that inducibly express the adeno-associated virus Rep proteins. *J. Virol.* **68**: 4847–4856.

Zhou SZ, Broxmeyer HE, Cooper S, Harrington MA, Srivastava A. (1993) Adeno-associated virus 2-mediated gene transfer in murine hematopoietic progenitor cells. *Exp. Hematol.* **21**: 928–933.

6

Liposome delivery systems

Ronald K. Scheule and Seng H. Cheng

6.1 Introduction

Liposomal delivery systems for nucleic acids fall within the larger class of non-viral delivery vehicles. With the construction of vehicles incorporating viral as well as non-viral components, the boundaries between viral and non-viral delivery systems have and will become increasingly blurred. This review will focus on the major categories of non-viral delivery systems and attempt to compare their strengths and weaknesses, with an emphasis on their relative *in vivo* performances.

The major categories of non-viral delivery systems under consideration are depicted schematically in *Figure 6.1*. Not included in this collection is naked plasmid DNA (pDNA), which will also be discussed. Chronologically, liposomal systems designed to encapsulate pDNA were among the first non-viral systems to demonstrate gene delivery. Virosomes, or empty viruses, are currently being used to take advantage of the membrane-fusing activity of enveloped viruses. Liposomes prepared using cationic lipids have also become extremely popular for the delivery of pDNA, largely due to their simplicity and relative effectiveness. In addition, polycationic delivery systems, such as those incorporating poly-L-lysines (pLys), have been developed with the aim of creating well-defined, targetable vehicles. Finally, in an attempt to make these delivery systems more efficient, viral components have been incorporated either in *cis* (i.e. directly conjugated) or in *trans* (i.e. added separately). For example, several variants of adenovirus–polyamine conjugates have been constructed to make use of the entry properties of adenovirus together with the targeting and DNA-compacting capabilities of polycationic structures.

6.2 Naked DNA

Plasmid DNA by itself has transfection activity in some systems. Direct injection of naked pDNA into muscle results in transfection levels that are similar to those obtained by transfecting fibroblasts *in vitro* (Wolff *et al.*, 1990). Although the trans-fecting pDNA does not appear to be integrated (Wolff *et al.*, 1990), expression has been shown to persist for months (Davis *et al.*, 1993; Jiao *et al.*, 1992; Wolff *et al.*, 1990). In addition to skeletal muscle, cardiac muscle can also be transfected by direct injection (Acsadi *et al.*, 1991; Lim *et al.*, 1991; Wang *et al.*, 1991). The susceptibility of muscle to transfection has been hypothesized to be related to the unique architecture

Gene Therapy, edited by N.R. Lemoine and D.N. Cooper.
© 1996 BIOS Scientific Publishers Ltd, Oxford.

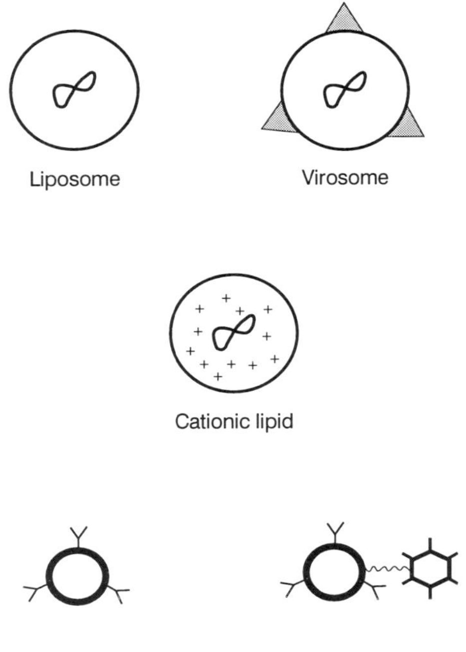

Figure 6.1. Schematic diagrams of DNA delivery vehicles discussed in this chapter. Liposomes and virosomes typically encapsulate DNA within a lipid membrane. Virosomes take advantage of viral fusion proteins, such as influenza haemagglutinin (depicted as triangles) to increase delivery efficiency. Cationic lipid (including lipopolyamine) delivery vehicles most likely adsorb DNA on to their surfaces, although some coating of DNA by lipid may also occur. Polyamine conjugates use polycations, such as poly-L-lysine, to compact the DNA into toroids (depicted as thick rings); targeting moieties (depicted as an antibody) can be incorporated into these structures by attaching them to a polycation and allowing them to interact with the negative charges of the DNA. These polyamine conjugates can be further targeted and their efficiency enhanced by coupling them to adenovirus. This coupling can be covalent or non-covalent, and can be through the adenovirus fibre protein, in which case the normal viral targeting is lost, or through alternative viral capsid proteins, such as hexon, in which case the normal adenovirus targeting is preserved.

of skeletal and cardiac muscle (Ascadi *et al.*, 1991). The transverse tubules of the myofibres, which are in direct contact with the extracellular space, may be responsible for the efficient uptake of pDNA.

Several studies have explored methods to increase the uptake and expression of pDNA by muscle (Danko *et al.*, 1994; Davis *et al.*, 1993; Manthorpe *et al.*, 1993). Preinjection of the muscle with hypertonic sucrose appears to increase the reproducibility and magnitude of expression (Davis *et al.*, 1993). Preinjection with the local anaesthetic, bupivacaine, 5–7 days prior to transfection increased expression to a level even greater than that attained with hypertonic sucrose. Myofibre damage by bupivacaine followed by regeneration may have been responsible for the enhanced uptake and transfection (Danko *et al.*, 1994).

The ability to transfect skeletal muscle has suggested several clinical applications. In principle, the treatment of genetic diseases of skeletal muscle can be approached by delivery of the normal gene directly into muscle (Acsadi *et al.*, 1991). However, to be successful in a disease such as muscular dystrophy, efficient transfection will be required to produce relatively large amounts of transgene product delivered to a large proportion of the body muscle mass. A more immediate application is the expression of heterologous transgenes in muscle for vaccine development (Fuller and Haynes, 1994; Fynan *et al.*, 1993; Lowrie *et al.*, 1994; Webster *et al.*, 1994). For example, gene gun delivery of the cDNA for HIV gp120 in a mouse model was found to generate both humoral and cytotoxic T-lymphocyte responses (Fuller and Haynes, 1994). The simplicity of the approach, the small dose of immunogen expression required, together with the absence of anti-pDNA antibodies, even in primates (Jiao *et al.*, 1992), argues for the clinical usefulness of naked pDNA in this context, assuming that intramuscular injection in primates will elicit immune responses analogous to those seen in other animal models.

6.3 Encapsulation of pDNA

6.3.1 Liposomes

Several approaches to deliver nucleic acids have made use of protocols designed to encapsulate the genetic material inside a liposomal carrier. Early attempts involved encapsulation of viral nucleic acids, and it was shown to be possible to deliver them successfully to cells *in vitro* as alternatives to calcium phosphate or diethylaminoethyl (DEAE) dextran-mediated transfections (Frahley *et al.*, 1980, 1981; Szelei and Duda, 1989; Wilson *et al.*, 1979). It is notable that these early efforts used phosphatidylserine (PS), a lipid bearing a net negative charge, as a major component of the liposomal delivery system. For example, poliovirus RNA was encapsulated into pure PS vesicles using cochleates as the intermediate structures (Wilson *et al.*, 1979), and SV40 DNA was encapsulated into large unilamellar vesicles by reverse phase evaporation using PS and cholesterol (Frahley *et al.*, 1980, 1981). Indeed, in these later studies, PS was found to be the optimal lipid for delivery. The fact that these negatively charged vesicles could successfully deliver nucleic acids to cells implies that a positively charged delivery vehicle is not absolutely essential, as has been previously suggested (Behr *et al.*, 1989; Felgner *et al.*, 1987). Transfection efficiencies resulting from encapsulated nucleic acids in general rivalled those obtained using calcium phosphate.

Lipids with pH sensitivity have also been incorporated into the encapsulating liposome to enhance endosomal disruption and entry of the liposomal contents into the cytoplasm (Wang and Huang, 1987). A recent study (Legendre and Szoka, 1992) directly compared liposomal encapsulation delivery systems, both pH-sensitive and -insensitive, with the cationic lipid, Lipofectin. This study demonstrated the superiority of the cationic lipid over the pH-sensitive lipid delivery method, which in turn was much more efficient at *in vitro* transfection than the non-pH-sensitive lipids. Interestingly, this study again demonstrated that negatively charged delivery vehicles readily associate with cells.

6.3.2 Virosomes

To make liposomal delivery methods more efficient, components of enveloped viruses have been incorporated. The enhancement in transfection efficiencies of these delivery systems over those of liposomes is imparted by the fusion properties of viral envelopes, such as the F protein of Sendai virus (haemagglutinating virus of Japan, or

HVJ). At neutral pH, and with high efficiency, this protein fuses the membrane in which it is embedded to a target membrane. Thus, delivery vehicles bearing the Sendai F protein would be expected to deliver nucleic acid to the cytoplasm by way of the plasma membrane. An early study demonstrated the use of HVJ to deliver either an encapsulated protein or gene to cells in culture (Nakanishi *et al.*, 1985). Liposomes composed of phosphatidylcholine, cholesterol and PS, and encapsulating either fragment A of diphtheria toxin or the thymidine kinase gene of herpes simplex virus (*HSV-tk*) were incubated with UV-light-inactivated HVJ at 4°C to promote binding between the liposomes and HVJ. The bound liposome–HVJ complex was subsequently incubated at 37°C to bring about the fusion of liposomal and viral membranes. These resultant 'virosomes' (see *Figure 6.1*) were then incubated with cells to bring about the fusion of the virosomal membrane with the cell plasma membrane. Gene delivery using such virosomal constructs was found to be relatively efficient, delivering the *tk* gene to approximately 10% of the target mouse L cells *in vitro*. Subsequent improvements on this protocol included the incorporation of gangliosides in the liposome to serve as receptors for HVJ (Kaneda *et al.*, 1987), and the inclusion of a nuclear protein (high mobility group protein-1 or HMG-1) to compact the nucleic acid and increase both its encapsulation efficiency and its ability to translocate to the nucleus (Kaneda *et al.*, 1989a).

Using a three-part system composed of HVJ, red blood cells (RBCs) to encapsulate HMG-1, and liposomes to encapsulate the human insulin gene, the method was reported to be capable of gene delivery *in vivo* as well as *in vitro* (Kaneda *et al.*, 1989b). Thus, in the rat, injection of virosomes prepared from this three-component system resulted in the detection of human insulin mRNA and protein for about 10 days. A rapid decrease in expression observed after day 7 may have been due to an immune response of the host to the human transgene product or to viral proteins of the vector. A similar series of experiments, encapsulating HMG-1 and pDNA in the same liposomes, demonstrated gene delivery of the hepatitis B virus (HBV) surface antigen to rat liver following direct injection. More recently, the HVJ method has also shown promise for the delivery of genes to smooth muscle cells (Morishita *et al.*, 1993a, b, c). In a primary vascular smooth muscle cell (VSMC) model, the HVJ method resulted in approximately 10 times more gene expression than the commercial cationic lipid, Lipofectin (Morishita *et al.*, 1993a). The HVJ method could also transfect endothelium in an organ culture model of rat carotid artery. These results were confirmed and extended in the VSMC model for gene delivery of components of the vascular renin–angiotensin system (Morishita *et al.*, 1993b), where a comparison of HVJ and Lipofectin showed that only the HVJ system was able to effect gene delivery.

The HVJ system has also been used successfully to deliver antisense oligonucleotides *in vivo* to inhibit neointimal hyperplasia in a model of restenosis (Morishita *et al.*, 1993c). The translation initiation sites of proliferating cell nuclear antigen (PCNA) and cyclin-dependent kinase 2 (cdc2), two critical components of cell-cycle progression, were targeted with antisense oligonucleotides. Virosomes in this case did not contain HMG-1, since the aim was to deliver the antisense oligonucleotides to the cytoplasm. Delivery of the HVJ virosomes was accomplished by incubating the virosomes in a ligated carotid artery for 15 min. The combined use of antisense oligonucleotides for both PCNA and cdc2 was found to inhibit intimal hyperplasia in this model for 8 weeks following treatment.

Clearly, the HVJ method represents an efficient virosomal delivery system. It may have optimal use in applications that do not require repeated application of the delivery vehicle, since it is likely to elicit an immune response. For example, it may be useful in treating disorders such as restenosis, in which a single administration is sufficient, or in cancer gene therapy, where the stimulation of an immune response may be beneficial.

6.4 Cationic lipids

Almost 10 years ago, it was found that simply mixing pDNA with liposomes composed of cationic lipids led to the formation of lipid–pDNA complexes that could transfect tissue culture cells (Felgner *et al.*, 1987). In contrast to liposomal and virosomal protocols, which necessitated the encapsulation of pDNA, an intrinsically inefficient step, the complexes between pDNA and cationic lipid appeared to be dominated by simple electrostatic interactions between positively charged lipid and negatively charged DNA. Further, virtually all DNA could be incorporated into complexes. It is now clear that these lipid–pDNA complexes can mediate the transfection of mammalian cells both *in vitro* and *in vivo*, and with efficiencies that exceed those of methods that depend on the encapsulation of pDNA (Legendre and Szoka, 1992). High throughput protocols for optimizing the relevant composition and concentration variables have been developed (Felgner *et al.*, 1993; Loeffler and Behr, 1993).

6.4.1 Structure of the complex

The details of the molecular structure(s) of cationic lipid–pDNA complexes are still uncertain. Electron microscopic evidence has been interpreted as indicating that the cationic lipid 'coats' the DNA, providing a cationic shell that is optimal for interaction with negatively charged cell surfaces (Behr *et al.*, 1989; Felgner *et al.*, 1987; Sternberg *et al.*, 1994). While this hypothesis is logical, recent biophysical data are difficult to reconcile with such a structure (Eastman *et al.*, unpublished results). Zeta potential measurements (see Section 6.4.4) of the surface charge of the complexes indicated a negative net charge at the surface, which is consistent with a lipid particle with DNA adsorbed on to its surface. Even under conditions of a charge excess of cationic lipid, DNA in the complex was found to be accessible to ethidium bromide, and was made more so by an increase in the ionic strength of the medium. These data are most consistent with a model of the complex in which at least some of the DNA is exposed at the surface. Additional studies at the molecular level are needed to define the details of these complexes further, and to assess the heterogeneity that results when DNA is mixed together with lipids of different structure and different ratios of cationic lipid to DNA.

6.4.2 Uses

The ever-increasing variety of cationic lipid structures in use has been the subject of several recent reviews, to which the reader is referred for further information (Behr, 1994; Gao and Huang, 1995; Ledley, 1995). Cationic lipids have been used to transfect mammalian cells *in vitro* and *in vivo* using plasmid DNA. They have also been used to deliver RNA (Lu *et al.*, 1994) and protein (Debs *et al.*, 1990; Farhood *et al.*, 1995). The *in vitro* transfection efficiency of cationic lipid–pDNA complexes generally exceeds (by a factor of at least 10) that which can be obtained using other standard methods, such as calcium phosphate precipitation (Behr *et al.*, 1989; Felgner *et al.*, 1987). Cationic lipids have been used successfully to generate both transient and stable transfectants (Felgner *et al.*, 1987), and have been shown to be capable of transfecting adherent (Behr *et al.*, 1989; Felgner *et al.*, 1987; Gao and Huang, 1991) as well as suspension cells (Loeffler *et al.*, 1993; Ruysschaert *et al.*, 1994), although suspension cells are generally more difficult to transfect (Harrison *et al.*, 1995; Zhou *et al.*, 1991). Finally, cationic lipids have proven useful to deliver RNA (Lu *et al.*, 1994) and pDNA (Gao and Huang, 1993) to the cytoplasm for the purpose of cytoplasmic transcription. As might be expected, expression resulting from the delivery of RNA is transient and precedes that obtained with pDNA transfection. However, the use of RNA eliminates the possibility of integration and oncogenesis.

6.4.3 Advantages and disadvantages

In principle, cationic-lipid-based DNA delivery systems offer a significant advantage over delivery systems that incorporate viral proteins or peptides in that the lipid-based systems should be less immunogenic. The immunogenicity of viral-based systems has been well documented and presents significant obstacles to persistent transgene expression as well as to repeat administrations of the viral vector. Cationic lipid systems devoid of viral components have been found to be free of host immune responses to the vector itself. For example, we have been able to detect neither antibodies nor cytotoxic T lymphocytes directed against the components of cationic lipid–pDNA delivery systems (R.K. Scheule and S.H. Cheng, unpublished results).

The immune advantage of purely cationic lipid systems is offset somewhat by the lower potency of these delivery systems compared with viral systems. For example, we have found that 100–1000 times more DNA is required by cationic liposomes than by adenovirus to produce the same biological end-point (R.K. Scheule and S.H. Cheng, unpublished data). Thus, the same functional end-point can be reached using cationic lipids, but at the expense of using a greater mass of material relative to virus. Although the production of cationic lipid and pDNA is not limiting, this nonetheless means that any toxicity associated with the cationic-lipid-based delivery system becomes an important consideration. This realization has driven the search for new, more efficient cationic lipids, and for strategies and formulations that make existing cationic lipids more potent.

6.4.4 Factors influencing transfection efficiency

In addition to the structure of the cationic lipid itself, several parameters have been recognized that affect the efficacy of cationic-lipid-mediated transfection. Several studies have established that a given cationic lipid has differing abilities to transfect different cell types (Behr *et al.*, 1989; Caplen *et al.*, 1995a; Debs *et al.* 1992; Fasbender *et al.*, 1995; Felgner *et al.*, 1987; Gao and Huang, 1991; Rose *et al.*, 1991; Ruysschaert *et al.*, 1994; Zhou *et al.*, 1991). Several of these studies have also confirmed that cationic-lipid-mediated toxicity is dependent on the cell type being studied, with primary cells generally being both more difficult to transfect and more resistant to lipid-mediated toxicity. *In vitro* transfections are generally carried out in the absence of serum, since its inclusion has been found to decrease transfection efficiencies (Fasbender *et al.*, 1995; Felgner *et al.*, 1993; Zhou *et al.*, 1991). This may be due in part to the presence of nucleases in serum, and the differing abilities of various cationic lipids to protect complexed DNA from degradation. In general, longer incubations ($\geqslant 4$ h) of cationic lipid–pDNA complexes with cells *in vitro* have been found to result in improved transfection efficiencies (Fasbender *et al.*, 1995; Felgner *et al.*, 1994). However, to achieve such a lengthy 'dwell time' of these complexes with the target cells may prove much more problematic *in vivo*.

The cationic–anionic charge ratio of the lipid–pDNA mixture has been hypothesized to be important for efficient transfection, with positively charged complexes being more conducive to uptake and expression of the transgene (Behr *et al.*, 1989; Felgner *et al.*, 1987). However, as noted above (Section 6.3.1), it is well established that cells are capable of taking up negatively charged lipid vesicles (Frahley *et al.*, 1980, 1981; Wilson *et al.*, 1979). It should also be noted that the net charge of the complex is most often calculated based on the stoichiometry of the cationic lipid and pDNA that are mixed together. However, this theoretical argument relies on a knowledge of the charge state(s) of the amine(s) that make up the cationic lipid headgroup, which is usually uncharacterized in this regard. Such calculations also assume that all of the cationic groups of the lipid and anionic phosphates of the DNA are equally available for interaction with each other, which may not be the case. Finally, this argument assumes that the charge neutralization that occurs between the cationic lipid and pDNA determines the surface charge of the complex, which in turn is responsible for the interaction of the complex with the cell and/or biological fluids. A physical parameter that can be measured to reflect the surface charge of the complex is the zeta potential. We have found that although there is a correlation between the net charge balance and the zeta potential, the zeta potential measured is significantly less positive than predicted based on the simple arithmetic of a cationic-anionic charge interaction.

6.4.5 Prediction of transfection efficiency in vivo

As noted above (Section 6.4.4), the transfection of primary cells with a given cationic lipid is in general a less efficient process than is the transfection of immortalized cell lines. Given this fact, plus the noted effects of biological fluids and of different cell lines on transfection efficiency, the usefulness of *in vitro* optimization of cationic lipid-mediated transfection for *in vivo* applications can be questioned. Indeed, a recent comparison of three cationic lipids *in vitro* and *in vivo* concluded that it would have been difficult to predict the most effective lipid *in vivo* based on the *in vitro* rank order (Solodin *et al.*, 1995). Our experience in testing over 150 cationic lipids in the same *in vitro* and *in vivo* assay systems has demonstrated some predictive value of the *in vitro* system. Thus, while not all lipids that transfect well *in vitro* also do so *in vivo*, it was found that those lipids that transfect well *in vivo* were among the best *in vitro* lipids. In other words, a subset of the best *in vitro* lipids was found to retain effectiveness *in vivo*.

6.4.6 Pharmacokinetics and persistence

Recent pharmacokinetic studies (Lew *et al.*, 1995; Mahato *et al.*, 1995) have shown that cationic lipid–pDNA complexes are rapidly cleared from the circulation; DMRIE–DOPE–pDNA complexes have a half-life of less than 5 min in the blood following a tail-vein injection (Lew *et al.*, 1995). Naked pDNA appears to be rapidly taken up by the liver, 60% being cleared by 1.5 min after application (Mahato *et al.*, 1995). Lipid–pDNA complexes are less efficiently cleared than naked pDNA by the liver, with significant amounts being removed by the lung (Mahato *et al.*, 1995).

Whereas the lung appears to be readily transfected by intravenous delivery of cationic lipid–pDNA complexes, the liver is not. Perhaps the complexes removed from the circulation by the liver are scavenged by the Kupffer cells, which may be less susceptible to transfection than hepatocytes. Such a hypothesis is consistent with the size of the complexes, which are generally > 200 nm in diameter, a size generally believed to be too large to penetrate to the hepatocytes.

Several studies have shown that expression following *in vivo* transfection with lipid–pDNA complexes is relatively transient (Brigham *et al.*, 1989, 1993; Canonico *et al.*, 1994; Hazinski *et al.*, 1991; McLachlan *et al.*, 1995; Philip *et al.*, 1993; Stribling *et al.*, 1992; Tsukamoto *et al.*, 1995). Expression generally reaches peak values 2–3 days following transfection and may persist, albeit at much lower levels, for up to 2–3 weeks. Expression from a cytomegalovirus (CMV)-driven pDNA may persist slightly longer than that from a simian virus 40 (SV40) promoter (McLachlan *et al.*, 1995). The transient nature of the expression may be related to the observation that the delivered pDNA is non-integrated, and remains in an episomal form. However, whether this transience is due to actual degradation of the transcriptionally active nuclear pDNA or rather to it being rendered transcriptionally inactive is not known. An exception to these results comes from studies designed to characterize cationic lipid-mediated transfection in a mouse model following intravenous delivery of chloramphenicol acetyltransferase (CAT) plasmids by DOTMA–DOPE (Zhu *et al.*, 1993). Although CAT DNA was found to persist for less than 21 days by Southern analysis, CAT expression was reported to persist for at least 9 weeks by immunostaining; cystic fibrosis transmembrane conductance regulator (CFTR) was still reported to be detectable by immunostaining at least 21 weeks after injection. The incorporation of viral genes that confer on a plasmid an ability to replicate episomally has been shown to extend expression for several months (Thierry *et al.*, 1995). However, the use of viral oncogenes for this purpose in a clinical context might be problematic.

6.4.7 Targeting

Although some targeting of cationic lipid–pDNA complexes may occur naturally, simply by using different cationic lipids and formulations, major increases in cell- and organ-specific uptake of these complexes are likely to be achieved by incorporating targeting moieties into their structure. A few attempts along these lines have been made to date, with testing largely confined to cells in culture (Kichler *et al.*, 1995; Remy *et al.*, 1995; Trubetskoy *et al.*, 1992a, b). Ternary complexes have been made by adding pDNA to cationic liposomes (DC-Chol) and then using the excess negative charges of the DNA to interact with a pLys–antibody conjugate, thereby non-covalently incorporating the antibody into the complex (Trubetskoy *et al.*, 1992a, b). These targeted complexes were shown to increase expression in a cell containing the appropriate cell-surface antigen by a factor of at least 10 compared with analogous complexes containing an irrelevant antibody. Incorporating a thiol-reactive lipid into the cationic lipid–pDNA complex was sufficient to increase transfection efficiency of HepG2 and 3T3 cells by factors of 1000 and 35, respectively (Kichler *et al.*, 1995). However, this strategy only appeared to work at an overall net charge near neutrality, and was hypothesized to result from the covalent reaction of the delivery vehicle with cell-surface groups that could be subsequently endocytosed. Modification of this reactive thiol lipid to form a triantennary galactolipid produced a complex that targeted hepatocytes through the asialoglycoprotein receptor (Remy *et al.*, 1995). These constructs were indeed shown to increase the transfection efficiency of the complex by a factor of 1000 in HepG2 cells. It will be interesting to see if these promising *in vitro* results with targeted cationic lipid–pDNA complexes will be mirrored *in vivo*. Success in these efforts could mean that the overall dose of cationic lipid can be decreased significantly, thereby reducing the toxicity of the delivery system.

6.4.8 Germ-line gene transfer

The transfection of germ-line cells by cationic lipid-mediated gene transfer might be problematic for clinical indications. Gonadal localization following intravenous injection of lipid–pDNA complexes has been evaluated for DMRIE–DOPE complexes in the mouse, and no test DNA was found in the gonads after giving a dose of 50 μg of pDNA (San, 1993). However, using DOTMA–DOPE complexes with 100 μg of a CMV–CAT reporter, CAT activity was detectable in ovarian tissue (Zhu *et al.*, 1993). The extent to which germ cells might have been responsible for this activity was not reported. Finally, and perhaps indirectly related to these studies of germ-line transmission, the intravenous injection of 133 μg of CAT pDNA complexed with DOGS into pregnant mice was found to result in transfection of the progeny *in utero* (Tsukamoto *et al.*, 1995). Although no toxicity to the dams or progeny was noted, this study highlights the need for caution in gene transfer experiments during pregnancy.

6.4.9 Clinical use (trials)

Based on the successful demonstration of cationic lipid-mediated gene transfer *in vivo* (Alton *et al.*, 1993; Brigham *et al.*, 1989; Canonico *et al.*, 1994; Conary *et al.*, 1994; Nabel *et al.*, 1990, 1992) and the lack of significant toxicity at the doses tested (McLachlan *et al.*, 1995; San *et al.*, 1993; Stewart *et al.*, 1992), several clinical trials have been initiated (see Ledley *et al.*, 1995 for a table of trials underway). Results from these trials have confirmed the low toxicity associated with low doses of lipid-based delivery vehicles, and have demonstrated some promising signs of efficacy (Caplen *et al.*, 1995b; Nabel *et al.*, 1993, 1994). DC-Chol was used to deliver a plasmid containing the B7 co-stimulatory molecule as part of an immune system stimulation protocol designed to attack melanoma cells (Nabel *et al.*, 1993). This trial demonstrated the proof of principle that cationic lipid–pDNA complexes, when injected into a tumour, could elicit specific cytotoxic T cells against the tumour antigen in the absence of significant toxicity. An intranasal trial for cystic fibrosis, also using DC-Chol, demonstrated some efficacy in the absence of significant toxicity (Caplen *et al.*, 1995b). Some correction of the genetic defect characteristic of cystic fibrosis was seen. With the use of newer cationic lipids that can achieve higher suspension concentrations of complex without precipitation, it is likely that the efficacies seen in these early clinical trials can be improved.

6.5 Lipopolylysines

Lipopolylysines are delivery vehicles whose structure bridges those of lipid-based delivery vehicles and molecular conjugates (Zhou *et al.*, 1991). Phospholipid (DOPE) was chemically conjugated to the primary amino groups of pLys. An average of two lipid molecules were derivatized to each approximately 16-mer pLys. These lipopolylysine conjugates exhibited about twice the transfection activity of naked DNA on mouse L929 cells. A comparison between lipopolylysines and the cationic lipid Lipofectin showed that the lipopolylysines were approximately three times more efficient. The addition of phospholipids, such as DOPE, to the lipopolylysine, had negative effects on transfection activity, as did serum. Curiously, transfection with lipopolylysines required scraping of the cells during the transfection procedure. In the absence of scraping, Lipofectin was at least 10 times *more* effective than lipopolylysine.

A second study of lipopolylysines as transfection agents investigated variables such as the ratio of lipid:polylysine molecules and the dependence of transfection activity on helper lipids, such as DOPE (Zhou and Huang, 1994). Optimal transfection activity was found at a lipid:polylysine ratio of 2:1. It was also found that the previous dependence on scraping could be overcome by using a much higher lipopolylysine:DOPE ratio. It should be noted that the lipopolylysines synthesized in this study were not based on phospholipids, but used dipalmitoylsuccinylglycerol to impart lipophilicity to the polylysine core. Based on electron microscopic images and the effects of lysosomotropic and fusion-inhibiting agents, transfection was proposed to occur through both endosomal and plasma membrane routes. These reagents hold potential as transfection vehicles because they can be derivatized, for example using antibodies, to add a targeting moiety.

6.6 Polycation molecular conjugates

Molecular conjugates consisting of a polycation such as pLys, which serves to condense or compact DNA together with a targeting moiety, have been constructed as alternative delivery vehicles. *Table 6.1* describes some of these conjugates and provides a sampling of the diversity of structures that have been devised. The majority of these conjugates use pLys as the DNA-binding polycation (Wagner *et al.*, 1990; Wu and Wu, 1988), although other novel ligands have been explored, such as DNA intercalators (Haensler and Szoka, 1993). Several different conjugation chemistries have been used to link the polycation to the targeting moiety, including, for example, heterobifunctional crosslinkers (Wagner *et al.*, 1990; Wu and Wu, 1988) and the biotin–streptavidin system (Gottschalk *et al.*, 1995). A diverse selection of targeting moieties has been explored. Asialorosomucoid (ASOR) has been used to target hepatocytes of the liver (Wu and Wu, 1988; Wu *et al.*, 1989) while transferrin (TFN) has been used as a general targeting molecule to direct DNA to rapidly dividing cells (Cotten *et al.*, 1990; Wagner *et al.*, 1990).

6.6.1 Hepatocyte targeting

Several studies have demonstrated that molecular conjugates can be used to target gene delivery to hepatocytes. These studies have resulted in the expression of both reporter genes as well as functional genes; e.g. low density lipoprotein (LDL) receptor in the liver (Wilson *et al.*, 1992; Wu and Wu, 1988) and HBV antigens in hepatocytes (Liang *et al.*, 1993). ASOR, a galactose-terminal (asialo-) glycoprotein, has been covalently attached to pLys in several studies as a way of targeting conjugates specifically to hepatocytes (Chowdhury *et al.*, 1993; Liang *et al.*, 1993; Wilson *et al.*, 1992; Wu and Wu, 1988; Wu *et al.*, 1989). Alternatively, galactose residues have been coupled to serve the same purpose (Haensler and Szoka, 1993; Perales, 1994). In general, these studies have demonstrated that hepatocytes can be targeted *in vitro* (Liang *et al.*, 1993) as well as *in vivo* (Chowdhury *et al.*, 1993; Wilson *et al.*, 1992; Wu and Wu, 1988; Wu *et al.*, 1989). Gene delivery appears to be specific for the liver (Wu and Wu, 1988). For example,

Table 6.1. Components of molecular conjugates and their targets

| Conjugate | | | |
DNA-binding moiety	Targeting moiety	Target cell/tissue	References
pLys	ASOR	Hepatocytes/liver	Chowdhury *et. al.* (1993); Wilson *et al.* (1992); Wu *et al.* (1988, 1989)
		HuH-7, SK-Hep1/human hepatoma	Liang *et al.* (1993)
pLys	TFN	K562/erythroblasts	Cotten *et al.* (1990); Wagner *et al.* (1990, 1991a); Zenke (1990)
		HBE1, KB, HeLa	Curiel *et al.* (1992b)
pLys	HA-peptides	K562, HeLa, BNL CL.2	Wagner *et al.* (1992b)
pLys	MoAb	Pulmonary endothelium	Trubetskoy *et al.* (1992a)
		Jurkat/human T leukaemia tissue	Thurnher *et al.* (1994)
pLys	SP-B	H441, 3T3, HeLa	Baatz *et al.* (1994)
pLys	Galactose	Liver	Perales *et al.* (1994)
pLys	PFO	sol 8/murine myoblast	Gottschalk *et al.* (1995)
EtD	TFN	K562/erythroblasts	Wagner *et al.* (1991b)
Bisacridine	Galactose	Primary hepatocytes	Haensler and Szoka (1993)

ASOR, asialorosomucoid; EtD, ethidium homodimer; HA, influenza haemagglutinin; MoAb, monoclonal antibody; PFO, perfringolysin O; pLys, poly-L-lysine; SP-B, surfactant-associated protein B; TFN, transferrin.

pharmacokinetic studies showed that while only 17% of naked pDNA was localized in the liver 10 min after intravenous injection, 85% of pDNA complexed to a pLys–ASOR conjugate reached this target organ (Wilson *et al.*, 1992; Wu and Wu, 1988). Around 90% of the conjugate in the liver appeared to be taken up by hepatocytes (Wilson *et al.*, 1992).

Persistence as well as expression of the transgene in the liver appeared to be transient. Although on average there were approximately 1000 copies of the transgene per cell 10 min after administration, this copy number decreased rapidly, such that 1 day after delivery, only one copy per cell remained (Wilson *et al.*, 1992). Maximum transgene expression levels occurred 24 h post-transfection and had decreased to undetectable levels 3–4 days after transfection (Wilson *et al.*, 1992; Wu *et al.*, 1989). Interestingly, induction of hepatocyte replication by partial hepatectomy following administration of the conjugate resulted in prolonged expression of up to several weeks (Chowdhury *et al.*, 1993; Wu *et al.*, 1989). Although initial studies suggest that the transgene may have become integrated as a result of liver regeneration following partial hepatectomy (Wu *et al.*, 1989), more recent evidence indicates that persistence of the transgene may have been due to disruption of the endosomal–lysosomal pathway, leading to an abnormal persistence of the transgene in cytoplasmic vesicles (Chowdhury *et al.*, 1993). Persistent transgene expression in the absence of hepatectomy has been reported using a molecular conjugate composed of galactosylated pLys together with transgenes containing a

PEPCK promoter. Infusion of these conjugates into the caudal vena cava resulted in expression of human factor IX for up to 140 days (Perales, 1994). Although no evidence of integration was apparent, the Southern analysis used may not have been sensitive enough to detect a small amount of integration. With these conjugates, as well as others directed to hepatocytes, it is not clear how the incoming transgene escapes from the lysosomal pathway to result in transfection in the absence of conjugate features that would specifically enhance this process.

6.6.2 Transferrin receptor targeting

Molecular conjugates using TFN as the targeting moiety covalently coupled to pLys (*Table 6.1*) have been used to deliver genes to cells with TFN receptors, such as erythroblasts and erythroleukemic cells (Cotten *et al.*, 1990; Wagner *et al.*, 1990, 1991a; Zenke *et al.*, 1990). As with galactosylated conjugates (Haensler and Szoka, 1993), the use of DNA intercalating agents to compact the pDNA was found to produce conjugates whose transfection activity was inferior to that of the corresponding pLys conjugate. Thus, ethidium homodimers (EtD) conjugated to TFN had less transfection activity than pLys–TFN conjugates, and the activity of the EtD–TFN conjugates could be increased substantially by the addition of pLys, presumably due to the superior DNA-compacting function of pLys (Wagner *et al.*, 1991b).

The pLys–TFN conjugates appeared to enter cells by the receptor-mediated endocytic pathway by way of the TFN receptor (Cotten *et al.*, 1990). Transfection activity in the human erythroleukaemic cell line K562 was dependent on the presence of chloroquine (100 μM), which was proposed to enhance exit from the endosomes (Cotten *et al.*, 1990). Under these conditions, transfection activity was found to be at least 10 times higher than could be obtained using DEAE dextran. Transfection activity was also found to correlate strongly with the ability of the conjugate to compact pDNA into toroids 80–100 nm in diameter estimated to contain approximately 120 TFN molecules.

Since these conjugates also had no apparent means to escape endosomal degradation, additional moieties have been added to increase their transfection activity (Wagner *et al.*, 1991a, 1992b). In the K562 cell model, activity could be enhanced seven times by incorporating histone H4 into the final complex, presumably because of the enhanced nuclear localization imparted by the histone. Peptide analogues of the influenza haemagglutinin (HA) fusion domain were also linked to pLys and incorporated into the complex (Wagner *et al.*, 1992b). The inclusion of these peptides substantially enhanced transfection activity of the pLys–TFN complex in both K562 and HeLa cells as well as in a murine hepatocyte cell line (BNL CL.2), presumably due to the pH-dependent membrane-destabilizing activity of the HA peptides in the endosome (Wagner *et al.*, 1992b).

6.6.3 Epithelial cell targeting

Molecular conjugates have also been evaluated for their ability to transfect epithelial cells. pLys–TNF conjugates have been used to transfect an immortalized respiratory epithelial cell line, HBE1 (Curiel *et al.*, 1992b). The conjugates achieved expression levels that were two to four times that attained using pLys alone, which is minimally above background levels. Again, chloroquine significantly enhanced (by about 10 times) the expression of the reporter gene. The potential of TNF-based conjugates for gene delivery to the epithelial cells of the lung remains to be evaluated. However, the respiratory lining fluid has significant levels of TNF, leading to the possibility that the receptor is present on the apical surface of the epithelium, at least of some cell types. Whether this is true and can be used to target gene delivery vehicles to the lung remains to be seen.

A conjugate formed by covalently attaching surfactant-associated protein B (SP-B) preferentially by its N-terminus to pLys was used to transfect a distal lung airway cell line, H441, derived from a human bronchiolar adenocarcinoma that expresses SP-B (Baatz *et al.*, 1994). Targeting and uptake was presumably a result of a normal recycling mechanism for SP-B in this cell line. Uptake appeared to be endosomal, as transfection activity was enhanced at least 30 times by the co-administration of adenovirus. Although the uptake and role of SP-B have not been fully characterized, it is likely that it is taken up and processed by type II epithelial cells *in vivo*; therefore, conjugates containing SP-B might be a way of targeting these cells from the lumen of the lung. The identification of additional endocytosed markers of airway cells might provide other potential specific targets for gene delivery and treatment of diseases such as cystic fibrosis.

6.6.4 Endothelial cell targeting

Endothelial cells of the lung have been targeted using conjugates composed of two to three pLys molecules covalently attached to a monoclonal antibody (MoAb) against thrombomodulin (Trubetskoy *et al.*, 1992a, b). A direct comparison with a cationic lipid *in vitro* demonstrated that the cationic lipid resulted in approximately 10 times more expression than the pLys–MoAb conjugate. *In vivo*, a comparison between conjugates made with the anti-thrombomodulin antibody and an irrelevant antibody demonstrated a specific (10-fold) accumulation of the thrombomodulin conjugate in the lung; expression was not reported. Clearly, pLys–MoAb conjugates have the potential of organ-specific targeting. However, their usefulness for transgene expression *in vivo* remains to be demonstrated.

6.6.5 Non-specific targeting

Non-specific cell targeting was recently reported using a member of a sulphydryl-activated group of bacterial membrane-active proteins (Gottschalk *et al.*, 1995). Perfringolysin O (PFO), which forms ring-shaped structures 25–30 nm in diameter, has been used to deliver small molecules, proteins, and antisense oligonucleotides to cells by virtue of its affinity for cholesterol-containing mammalian membranes (Ahnert-Hilger *et al.*, 1989; Barry *et al.*, 1993; Bhakdi *et al.*, 1993). PFO was biotinylated and attached to pLys using streptavidin. This linkage resulted in a 10-fold decrease in the haemolytic activity of PFO, but it was nonetheless capable of transfecting 10–20% of the target cells (sol 8 cells, a murine myoblast line). While PFO conjugates may be useful as gene delivery agents in several clinical indications, they may be most useful in situations where the natural cytotoxicity of the bacterial proteins can be used to advantage, such as in the delivery of genes to cancer cells. The potential for an immunologic response to these conjugates may be tempered by the use of small peptides harbouring similar membrane-perturbing activities.

6.7 Viral hybrids

Co-administration of adenovirus (Ad) can significantly enhance the expression resulting from transfection using polylysine molecular conjugates or cationic lipid

(Cristiano *et al.*, 1993a, b; Curiel *et al.*, 1991; Fisher and Wilson, 1994; Harris *et al.*, 1993; Wagner *et al.*, 1992a; Wu *et al.*, 1994). In these studies, Ad has either been added in *trans* (i.e. in a co-delivery mode) or in *cis* (i.e. directly conjugated to the pDNA-containing component of the delivery vehicle). A recent review supplies additional details of these methods (Curiel, 1994).

6.7.1 Virus (Ad) enhancement of polylysine molecular conjugates

The addition of Ad in *trans* together with TFN–pLys conjugates has been shown to result in a 1000-fold increase in the transfection activity of the conjugate in multiple cell types *in vitro* (e.g. HeLa, CFT1, KB, WI-38, MRC-5 cells; Curiel *et al.*, 1991). On HeLa cells, for example, a TFN–pLys conjugate displayed approximately five times the activity of pLys + pDNA or pDNA alone, while the inclusion of Ad in *trans* boosted expression of the TFN–pLys conjugate approximately 1000 times. Ad enhancement of expression was attributed to its ability to aid the escape of the conjugate from the endosomal–lysosomal pathway.

Several methods have been used to couple Ad to pLys and test the ability of the virus in *cis* to enhance conjugate delivery and expression. These include use of antibodies (Curiel *et al.*, 1992a), biotin–streptavidin (Wagner *et al.*, 1992a), and chemical coupling (Cristiano *et al.*, 1993b). Untargeted Ad–pLys conjugates are formed by simply adding pDNA to the conjugate. Targeted conjugates are made by first adding enough of the conjugate to a given amount of pDNA to result in the neutralization of approximately a quarter of the negative charges on the pDNA. A targeting component, such as TFN–pLys, is then added to bind to the Ad–pLys–DNA conjugate and neutralize the remaining negative charge (Wagner, 1992a). In this way, the pDNA acts as a negatively charged 'glue' that serves to bind together both the positively charged Ad–pLys conjugate and the positively charged TFN–pLys targeting component. Other targeting modules, such as ASOR–pLys, have been incorporated into this scheme to target other cells such as primary hepatocytes (Cristiano *et al.*, 1993a, b).

A direct comparison of Ad enhancement of TFN–pLys conjugates in the *cis* and *trans* configurations showed that it was much more effective in *cis* (Cristiano *et al.*, 1993b; Wagner *et al.*, 1992a). For example, Ad in *cis* was 10 times more effective in transfecting hepatocytes, and 1000 times more effective in transfecting K562 cells, an erythroleukaemic cell line expressing TFN receptors. Ad–pLys/DNA/pLys–TFN complexes transduced cells at the level of approximately 60 DNA molecules of DNA added per cell, which represents a significant increase in efficiency over methods such as calcium phosphate, which requires on the order of 10^5 DNA molecules per cell (Sambrook *et al.*, 1989).

At least one study has shown that there is a large difference in the ability of TFN–pLys/DNA/pLys–Ad conjugates to transfect immortalized and primary cells (Harris *et al.*, 1993): a papilloma-virus-immortalized human airway epithelial cell line HBE1 was readily transfected with these conjugates whereas primary airway epithelial cells were not. These differences in transfection efficiency may reflect differences in the TFN-receptor-mediated processes in these two cell models.

The molecular structure of Ad–pLys/DNA/pLys–X complexes is shown schematically in *Figure 6.1*, and has been deduced from electron microscopic images of the complexes (Cristiano *et al.*, 1993b). In these complexes, Ad was coupled to pLys through its penton base, leaving the fibre protein and the normal entry mechanism of the virus intact.

Alternative schemes for coupling Ad in *cis* have been developed with the goal of inhibiting the normal entry mechanism of the virus while retaining its ability to enhance endosomal escape. Coupling ASOR–pLys to Ad through the viral carbohydrates on the Ad fibre protein resulted in an Ad–ASOR–pLys/DNA complex with a greatly decreased ability to transfect asialoglycoprotein receptor (–) cells (HeLa S3, SK Hep1), while retaining its ability to transfect asialoglycoprotein receptor (+) cells (Huh 7) (Wu *et al.*, 1994). Hepatitis B surface antigen expression levels from transfected DNA were 30 times higher with the complex than in the absence of Ad enhancement.

Yet another strategy has been used to create a targeted Ad-based conjugate (Fisher and Wilson, 1994). Starting with a pLys-modified Ad, DNA was added to bind non-covalently. The targeting moiety pLys–ASOR was then added to form the final Ad–pLys/DNA/pLys–ASOR complex. Using electron microscopy, these final complexes appeared as Ad surrounded by a 'cloud' of DNA/pLys–ASOR – that is, a 'coated' Ad. These structures were shown to lack the infectivity characteristic of the wild-type virion but to transfect human hepatoma cell lines (HUH, HepG2) bearing asialoglycoprotein receptors in a specific manner. These structures would appear to have the advantage that the expression-enhancing features of adenovirus may not be compromised by an immune response of the host, since the virus itself would appear to be 'masked' from immune surveillance; this hypothesis awaits experimental testing.

6.7.2 Virus-cationic lipid

Ad has also been used to enhance the activity of cationic-lipid-mediated transfection. In an initial report, Ad infection of several cell lines (COS-7, HeLa, CV-1) 30 min after transfection with Lipofectin resulted in expression levels that were 2–10 times those that could be obtained using Lipofectin alone (Yoshimura *et al.*, 1993). Presumably, the enhancement was a result of the endosomolytic activity of Ad increasing the likelihood of pDNA escape and transit to the nucleus. A more recent study (Raja-Walia *et al.*, 1995) has extended the relevance of this observation to include an *in vivo* model. In this study, a complex was formed between cationic lipid, pDNA and Ad which was then added to either cells *in vitro* or to a rabbit femoral artery that had been dilated and balloon-injured. The structure of the complex was proposed to consist of DNA and Ad non-covalently bound together with cationic lipid. The inclusion of Ad in these complexes increased expression (as luciferase light units) up to 1000 times *in vitro* in a primary endothelial cell model over what could be obtained with cationic lipid and pDNA alone. Expression was also apparent in the femoral artery model as nuclear-localized lacZ activity in both the neointimal smooth muscle cells and the adventitia. Complexes of this kind thus show promise for indications such as restenosis, where transient gene expression may suffice, and where the immunological consequences of Ad administration may not preclude its use therapeutically.

6.8 Summary

Non-viral gene transfer delivery systems have evolved substantially in the last decade. The efficiency with which these systems can deliver polynucleotides to cells has improved dramatically over that of encapsulating liposomal and virosomal systems. The types of delivery systems available today span a wide range of vehicles and complexities, from naked pDNA to viral-assisted molecular conjugates. Each of these systems has its own particular niche in which it performs optimally. However, although all of the non-viral gene transfer vehicles described here are potentially useful clinically, it must be emphasized that at present this promise has yet to be realized. Transfer and validation of the delivery technology from its *in vitro* proving ground to the *in vivo* arena has proven difficult. As a result, many cationic lipids and molecular conjugates have not survived this winnowing process.

Regardless of the difficulties in translating *in vitro* results into *in vivo* practice, it is clear that non-viral systems offer advantages over viral systems in terms of ease of production, lack of potential recombination events and lack of immunogenicity. These characteristics have provided and continue to provide the impetus for further development. Presently, additional direct and quantitative comparative data between the viral and non-viral delivery systems would help to evaluate their relative merits. Indeed, even within the class of non-viral delivery systems, it would be extremely useful to have more standardized assays and consistent internal standards, so that investigators could directly compare the relative performance of their delivery system relative to others in the field. The current disadvantage of relatively low efficiency compared with that of viral vectors is being addressed by many investigators, and may be reduced with the development of more potent non-viral delivery systems. It is to be hoped that these later generation non-viral systems will achieve the goal of patient benefit in the not-too-distant future.

References

Acsadi G, Dickson G, Love DR, Jani A, Walsh FS, Gurusinghe A, Wolff JA, Davies KE. (1991) Human dystrophin expression in *mdx* mice after intramuscular injection of DNA constructs. *Nature* 352: 815–818.

Ahnert-Hilger G, Mach W, Fohr KJ, Gratzl M. (1989) Poration by α-toxin and streptolysin O: An approach to analyze intracellular processes. *Methods Cell Biol.* 31: 63–90.

Alton EWFW, Middleton PG, Caplen NJ, Smith SN, Steel DM, Munkonge FM, Jeffery PK, Geddes DM, Hart SL, Williamson R, Fasold KI, Miller AD, Dickinson P, Stevenson BJ, McLachlan G, Dorin JR, Porteous DJ. (1993) Non-invasive liposome-mediated gene delivery can correct the ion transport defect in cystic fibrosis mutant mice. *Nature Genet.* 5: 135–142.

Baatz JE, Bruno MD, Ciraolo PJ, Glasser SW, Stripp BR, Smyth KL, Korfhagen TR. (1994) Utilization of modified surfactant-associated protein B for delivery of DNA to airway cells in culture. *Proc. Natl Acad. Sci. USA* 91: 2547–2551.

Barry ELR, Gesek FA, Friedman PA. (1993) Introduction of antisense oligonucleotides into cells by permeabilization with streptolysin O. *Biotechniques* 15: 1016–1020.

Behr J-P. (1994) Gene transfer with synthetic cationic amphiphiles: Prospects for gene therapy. *Bioconjugate Chem.* 5: 382–389.

Behr, J-P, Demeneix B, Loeffler J-P, Perez-Mutul J. (1989) Efficient gene transfer into mammalian primary endocrine cells with lipopolyamine-coated DNA. *Proc. Natl Acad. Sci. USA,* 86: 6982–6986.

Bhakdi S, Weller U, Walev I, Martin E, Jonas D, Palmer M. (1993) A guide to the use of pore-forming toxins for controlled permeabilization of cell membranes. *Med. Microbiol. Immunol.* 182: 167–175.

Brigham KL, Meyrick B, Christman B, Magnuson M, King G, Berry LC, Jr. (1989) *In vivo* transfection of murine lungs with a functioning prokaryotic gene using a liposome vehicle. *Am. J. Med. Sci* 298: 278–281.

Brigham KL, Meyrick B, Christman B, Conary JT, King G, Berry LC, Jr, Magnuson MA. (1993) Expression of human growth hormone fusion genes in cultured lung endothelial cells and in the lungs of mice. *Am. J. Respir. Cell Mol. Biol.* 8: 209–213.

Canonico AE, Conary JT, Meyrick BO, Brigham KL. (1994) Aerosol and intravenous transfection of human α1-antitrypsin gene to lungs of rabbits. *Am. J. Respir. Cell Mol. Biol.* 10: 24–29.

Caplen NJ, Kinrade E, Sorgi F, Gao X, Gruenert D, Geddes D, Coutelle C, Huang L, Alton EWFW, Williamson R. (1995a) *In vitro* liposome-mediated DNA transfection of epithelial cell lines using the cationic liposome DC-Chol/DOPE. *Gene Ther.* 2: 603–613.

Caplen NJ, Alton EWFW, Middleton PG, Dorin JR, Stevenson BJ, Gao X, Durham SR, Jeffery PK, Hodson ME, Coutelle C, Huang L, Porteous DJ, Williamson R, Geddes DM. (1995b) Liposome-mediated *CFTR* gene transfer to the nasal epithelium of patients with cystic fibrosis. *Nature Med.* 1: 39–46.

Chowdhury NR, Wu CH, Wu GY, Yerneni PC, Bommineni VR, Chowdhury JR. (1993) Fate of DNA targeted to the liver by asialoglycoprotein receptor-mediated endocytosis *in vivo*. *J. Biol. Chem.* **268:** 11265–11271.

Conary JT, Parker RE, Christman BW, Faulks RD, King GA, Meyrick BO, Brigham KL. (1994) Protection of rabbit lungs from endotoxin injury by *in vivo* hyperexpression of the prostaglandin G/H synthase gene. *J. Clin. Invest.* **93:** 1834–1840.

Cotten M, Langle-Rouault F, Kirlappos H, Wagner E, Mechtler K, Zenke M, Beug H, Birnstiel ML. (1990) Transferrin-polycation-mediated introduction of DNA into human leukemic cells: Stimulation by agents that affect the survival of transfected DNA or modulate transferrin receptor levels. *Proc. Natl Acad. Sci. USA* **87:** 4033–4037.

Cristiano RJ, Smith LC, Kay MA, Brinkley BR, Woo SLC. (1993a) Hepatic gene therapy: Efficient gene delivery and expression in primary hepatocytes utilizing a conjugated adenovirus–DNA complex. *Proc. Natl Acad. Sci. USA* **90:** 11548–11552.

Cristiano RJ, Smith LC, Woo SLC. (1993b) Hepatic gene therapy: Adenovirus enhancement of receptor-mediated gene delivery and expression in primary hepatocytes. *Proc. Natl Acad. Sci. USA* **90:** 2122–2126.

Curiel DT. (1994) High-efficiency gene transfer employing adenovirus–polylysine–DNA complexes. *Nat. Immunol.* **13:** 141–164.

Curiel DT, Agarwal S, Wagner E, Cotten M. (1991) Adenovirus enhancement of transferrin–polylysine-mediated gene delivery. *Proc. Natl Acad. Sci. USA* **88:** 8850–8854.

Curiel DT, Wagner E, Cotten M, Birnstiel ML, Agarwal S, Li C-M, Loechel S, Hu P-C. (1992a) High-efficiency gene transfer mediated by adenovirus coupled to DNA–polylysine complexes. *Hum. Gene Ther.* **3:** 147–154.

Curiel DT, Agarwal S, Romer MU, Wagner E, Cotten M, Birnstiel ML, Boucher RC. (1992b) Gene transfer to respiratory epithelial cells via the receptor-mediated endocytosis pathway. *Am. J. Respir. Cell Mol. Biol.* **6:** 247–252.

Danko I, Fritz JD, Jiao S, Hogan K, Latendresse JS, Wolff JA. (1994) Pharmacological enhancement of *in vivo* foreign gene expression in muscle. *Gene Ther.* **1:** 114–121.

Davis HL, Whalen RG, Demeneix BA. (1993) Direct gene transfer into skeletal muscle *in vivo*: Factors affecting efficiency of transfer and stability of expression. *Hum. Gene Ther.* **4:** 151–159.

Debs RJ, Freedman LP, Edmunds S, Gaensler KL, Duzgunes N, Yamamoto KR. (1990) Regulation of gene expression *in vivo* by liposome-mediated delivery of a purified transcription factor. *J. Biol. Chem.* **265:** 10189–10192.

Debs R, Pian M, Gaensler K, Clements J, Friend DS, Dobbs L. (1992) Prolonged transgene expression in rodent lung cells. *Am. J. Respir. Cell Mol. Biol.* **7:** 406–413.

Farhood H, Gao X, Barsoum J, Huang L. (1995) Codelivery to mammalian cells of a transcriptional factor with *cis*-acting element using cationic liposomes. *Anal. Biochem.* **225:** 889–893.

Fasbender AJ, Zabner J, Welsh MJ. (1995) Optimization of cationic lipid-mediated gene transfer to airway epithelia. *Am. J. Physiol.* **269:** L45–L51.

Felgner PL, Gadek TR, Holm M, Roman R, Chan HW, Wenz M, Northrop JP, Ringold GM, Danielsen M. (1987) Lipofection: A highly efficient, lipid-mediated DNA-transfection procedure. *Proc. Natl Acad. Sci. USA* **84:** 7413–7417.

Felgner J, Bennett F, Felgner PL. (1993) Cationic lipid-mediated delivery of polynucleotides. *Methods* **5:** 67–75.

Felgner JH, Kumar R, Sridhar CN, Wheeler CJ, Tsai YJ, Border R, Ramsey P, Martin M, Felgner PL. (1994) Enhanced gene delivery and mechanism studies with a novel series of cationic lipid formulations. *J. Biol. Chem.* **269:** 2550–2561.

Fisher KJ, Wilson JM. (1994) Biochemical and functional analysis of an adenovirus-based ligand complex for gene transfer. *Biochem. J.* **299:** 49–58.

Frahley R, Subramani S, Berg P, Papahadjopoulos D. (1980) Introduction of liposome-encapsulated SV40 DNA into cells. *J. Biol. Chem.* **255:** 10431–10435.

Frahley R, Straubinger RM, Rule G, Springer EL, Papahadjopoulos D. (1981) Liposome-mediated delivery of deoxyribonucleic acid to cells: enhanced efficiency of delivery related to lipid composition and incubation conditions. *Biochemistry* **20:** 6978–6987.

Fuller DH, Haynes JR. (1994) A qualitative progression in HIV type 1 glycoprotein 120-specific cytotoxic cellular and humoral immune responses in mice receiving a DNA-based glycoprotein 120 vaccine. *AIDS Res. Hum. Retroviruses* **10:** 1433–1441.

Fynan EF, Robinson HL, Webster RG. (1993) Use of DNA encoding influenza hemagglutinin as an avian influenza vaccine. *DNA Cell Biol.* **12:** 785–789.

Gao X, Huang L. (1991) A novel cationic liposome reagent for efficient transfection of mammalian cells. *Biochem. Biophys. Res. Commun.* **179:** 280–285.

Gao X, Huang L. (1993) Cytoplasmic expression of a reporter gene by co-delivery of T7 RNA polymerase and T7 promoter sequence with cationic liposomes. *Nucleic Acids Res.* **21:** 2867–2872.

Gao X, Huang L. (1995) Cationic liposome-mediated gene transfer. *Gene Ther.* **2:** 710–722.

Gottschalk S, Tweten RK, Smith LC, Woo SLC. (1995) Efficient gene delivery and expression in mammalian cells using DNA coupled with perfringolysin O. *Gene Ther.* **2:** 498–503.

Haensler J, Szoka FC, Jr. (1993) Synthesis and characterization of a trigalactosylated bisacridine compound to target DNA to hepatocytes. *Bioconjugate Chem.* **4:** 85–93.

Harris CE, Agarwal S, Hu P-C, Wagner E, Curiel DT. (1993) Receptor-mediated gene transfer to airway epithelial cells in primary culture. *Am. J. Respir. Cell Mol. Biol.* **9:** 441–447.

Harrison GS, Wang Y, Tomczak J, Hogan C, Shpall EJ, Curiel TJ, Felgner PL. (1995) Optimization of gene transfer using cationic lipids in cell lines and primary human CD4+ and CD34+ hematopoietic cells. *Biotechniques* **19:** 816–823.

Hazinski TA, Ladd PA, DeMatteo CA. (1991) Localization and induced expression of fusion genes in the rat lung. *Am. J. Respir. Cell Mol. Biol.* **4:** 206–209.

Jiao S, Williams P, Berg RK, Hodgeman BA, Liu L, Repetto G, Wolff JA. (1992) Direct gene transfer into nonhuman primate myofibers *in vivo*. *Hum. Gene Ther.* **3:** 21–33.

Kaneda Y, Uchida T, Kim J, Ishiura M, Okada Y. (1987) The improved efficient method for introducing macromolecules into cells using HVJ (Sendai virus) liposomes with gangliosides. *Exp. Cell Res.* **173:** 56–69.

Kaneda Y, Iwai K, Uchida T. (1989a) Increased expression of DNA cointroduced with nuclear protein in adult rat liver. *Science* **243:** 375–378.

Kaneda Y, Iwai K, Uchida T. (1989b) Introduction and expression of the human insulin gene in adult rat liver. *J. Biol. Chem.* **264:** 12126–12129.

Kichler A, Remy J-S, Boussif O, Frisch B, Boeckler C, Behr J-P, Schuber F. (1995) Efficient gene delivery with neutral complexes of lipospermine and thiol-reactive phospholipids. *Biochem. Biophys. Res. Commun.* **209:** 444–450.

Ledley FD. (1995) Nonviral gene therapy: The promise of genes as pharmaceutical products. *Hum. Gene Ther.* **6:** 1129–1144.

Legendre J-Y, Szoka FC, Jr. (1992) Delivery of plasmid DNA into mammalian cell lines using pH-sensitive liposomes: Comparison with cationic liposomes. *Pharmaceut. Res.* **9:** 1235–1242.

Lew D, Parker SE, Latimer T, Abai AM, Kuwahara-Rundell A, Doh SG, Yang Z-Y, Laface D, Gromkowski SH, Nabel GJ, Manthorpe M, Norman J. (1995) Cancer gene therapy using plasmid DNA: pharmacokinetic study of DNA following injection in mice. *Hum. Gene Ther.* **6:** 553–564.

Liang TJ, Makdisi WJ, Sun S, Hasegawa K, Zhang Y, Wands JR, Wu CH, Wu GY. (1993) Targeted transfection and expression of hepatitis B viral DNA in human hepatoma cells. *J. Clin. Invest.* **91:** 1241–1246.

Lim CS, Chapman GD, Gammon RS, Muhlestein JB, Bauman RP, Stack RS, Swain JL. (1991) Direct *in vivo* gene transfer into the coronary and peripheral vasculatures of the intact dog. *Circulation* **83:** 2007–2011.

Loeffler J-P, Behr J-P. (1993) Gene transfer into primary and established mammalian cell lines with lipopolyamine-coated DNA. *Methods Enzymol.* **217:** 41–42.

Lowrie DB, Tascon RE, Colston MJ, Silva CL. (1994) Towards a DNA vaccine against tuberculosis. *Vaccine* **12:** 1537–1540.

Lu D, Benjamin R, Kim M, Conry RM, Curiel DT. (1994) Optimization of methods to achieve mRNA-mediated transfection of tumor cells *in vitro* and *in vivo* employing cationic liposome vectors. *Cancer Gene Ther.* **1:** 245–252.

Mahato RI, Kawabata K, Takakura Y, Hashida M. (1995) *In vivo* disposition characteristics of plasmid DNA complexed with cationic liposomes. *J. Drug Target.* **3:** 149–157.

Manthorpe M, Cornefert-Jensen F, Hartikka J, Felgner J, Rundell A, Margalith M, Dwarki V. (1993) Gene therapy by intramuscular injection of plasmid DNA: Studies on firefly luciferase gene expression in mice. *Hum. Gene Ther.* **4:** 419–431.

McLachlan G, Davidson DJ, Stevenson BJ, Dickinson P, Davidson-Smith H, Dorin JR, Porteous DJ. (1995) Evaluation *in vitro* and *in vivo* of cationic liposome-expression construct complexes for cystic fibrosis gene therapy. *Gene Ther.* **2:** 614–622.

Morishita R, Gibbons GH, Kaneda Y, Ogihara T, Dzau VJ. (1993a) Novel *in vitro* gene transfer method for study of local modulators in vascular smooth muscle cells. *Hypertension* **21**: 894–899.

Morishita R, Gibbons GH, Kaneda Y, Ogihara T, Dzau VJ. (1993b) Novel and effective gene transfer technique for study of vascular renin angiotensin system. *J. Clin. Invest.* **91**: 2580–2585.

Morishita R, Gibbons GH, Ellison KE, Nakajima M, Zhang L, Kaneda Y, Ogihara T, Dzau VJ. (1993c) Single intraluminal delivery of antisense cdc2 kinase and proliferating-cell nuclear antigen oligonucleotides results in chronic inhibition of neointimal hyperplasia. *Proc. Natl Acad. Sci. USA* **90**: 8474–8478.

Nabel EG, Plautz G, Nabel GJ. (1990) Site-specific gene expression *in vivo* by direct gene transfer into the arterial wall. *Science* **249**: 1285–1288.

Nabel EG, Gordon D, Yang Z-Y, Xu L, San H, Plautz GE, Wu B-Y, Gao X, Huang L, Nabel GJ. (1992) Gene transfer *in vivo* with DNA–liposome complexes: Lack of autoimmunity and gonadal localization. *Hum. Gene Ther.* **3**: 649–656.

Nabel GJ, Nabel EG, Yang Z-Y, Fox BA, Plautz GE, Gao X, Huang L, Shu S, Gordon D, Chang AE. (1993) Direct gene transfer with DNA-liposome complexes in melanoma: Expression, biologic activity, and lack of toxicity in humans. *Proc. Natl Acad. Sci. USA* **90**: 11307–11311.

Nabel EG, Yang Z, Muller D, Chang AE, Gao X, Huang L, Cho KJ, Nabel GJ. (1994) Safety and toxicity of catheter gene delivery to the pulmonary vasculature in a patient with metastatic melanoma. *Hum. Gene Ther.* **5**: 1089–1094.

Nakanishi M, Uchida T, Sugawa H, Ishiura M, Okada Y. (1985) Efficient introduction of contents of liposomes into cells using HVJ (Sendai virus). *Exp. Cell Res.* **159**: 399–409.

Perales JC, Ferkol T, Beegen H, Ratoff OD, Hanson RW. (1994) Gene transfer *in vivo*: sustained expression and regulation of genes introduced into the liver by receptor-targeted uptake. *Proc. Natl Acad. Sci. USA* **91**: 4086–4090.

Philip R, Liggitt D, Philip M, Dazin P, Debs R. (1993) *In vivo* gene delivery: efficient transfection of T lymphocytes in adult mice. *J. Biol. Chem* **268**: 16087–16090.

Raja-Walia R, Webber J, Naftilan J, Chapman GD, Naftilan AJ. (1995) Enhancement of liposome-mediated gene transfer into vascular tissue by replication deficient adenovirus. *Gene Ther.* **2**: 521–530.

Remy J-S, Kichler A, Mordvinov V, Schuber F, Behr J-P. (1995) Targeted gene transfer into hepatoma cells with lipopolyamine-condensed DNA particles presenting galactose ligands: A stage toward artificial viruses. *Proc. Natl Acad. Sci. USA* **92**: 1744–1748.

Rose JK, Buonocore L, Whitt MA. (1991) A new cationic liposome reagent mediating nearly quantitative transfection of animal cells. *Biotechniques* **10**: 520–525.

Ruysschaert J-M, El Ouahabi A, Willeaume V, Huez G, Fuks R, Vandenbranden M, Di Stefano P. (1994) A novel cationic ampiphile for transfection of mammalian cells. *Biochem. Biophys. Res. Commun.* **203**: 1622–1628.

Sambrook J, Fritsch EF, Maniatis T. (1989) Expression of cloned genes in cultured mammalian cells. In: *Molecular Cloning: A Laboratory Manual, Vol. 3.* Cold Spring Harbor Laboratory Press, Cold Spring Harbor, NY, pp. 16.39–16.40.

San H, Yang Z-Y, Pompili VJ, Jaffe ML, Plautz GE, Xu L, Felgner JH, Wheeler CJ, Felgner PL, Gao X, Huang L, Gordon D, Nabel GJ, Nabel EG. (1993) Safety and short-term toxicity of a novel cationic lipid formulation for human gene therapy. *Hum. Gene Ther.* **4**: 781–788.

Solodin I, Brown CS, Bruno MS, Chow C-Y, Jang E-H, Debs RJ, Heath TD. (1995) A novel series of amphiphilic imidazolinium compounds for *in vitro* and *in vivo* gene delivery. *Biochemistry* **34**: 13537–13544.

Sternberg B, Sorgi FL, Huang L. (1994) New structures in complex formation between DNA and cationic liposomes visualized by freeze-fracture electron microscopy. *FEBS Lett* **356**: 361–366.

Stewart MJ, Plautz GE, Buono LD, Yang ZY, Xu L, Gao X, Huang L, Nabel EG, Nabel GJ. (1992) Gene transfer *in vivo* with DNA–liposome complexes: Safety and acute toxicity in mice. *Hum. Gene Ther.* **3**: 267–275.

Stribling R, Brunette E, Liggitt D, Gaensler K, Debs R. (1992) Aerosol gene delivery *in vivo*. *Proc. Natl Acad. Sci. USA* **89**: 11277–11281.

Szelei J, Duda E. (1989) Entrapment of high-molecular-mass DNA molecules in liposomes for the genetic transformation of animal cells. *Biochem. J.* **259**: 549–553.

Thierry AR, Lunardi-Iskandar Y, Bryant J, Rabinovich P, Gallo RC, Mahan LC. (1995) Systemic gene therapy: Biodistribution and long-term expression of a transgene in mice. *Proc. Natl Acad. Sci. USA* **92**: 9742–9746.

Thurnher M, Wagner E, Clausen H, Mechtler K, Rusconi S, Dinter A, Birnstiel ML, Berger EG, Cotten M. (1994) Carbohydrate receptor-mediated gene transfer to human T leukaemic cells. *Glycobiology* **4:** 429–435.

Trubetskoy VS, Torchilin VP, Kennel S, Huang L. (1992a) Cationic liposomes enhance targeted delivery and expression of exogenous DNA mediated by N-terminal modified poly(L-lysine)–antibody conjugate in mouse lung endothelial cells. *Biochim. Biophys. Acta* **1131:** 311–313.

Trubetskoy VS, Torchilin VP, Kennel SJ, Huang L. (1992b) Use of N-terminal modified poly(L-lysine)–antibody conjugate as a carrier for targeted gene delivery in mouse lung endothelial cells. *Bioconjugate Chem* **3:** 323–327.

Tsukamoto M, Ochiya T, Yoshida S, Sugimura T, Terada M. (1995) Gene transfer and expression in progeny after intravenous DNA injection into pregnant mice. *Nature Genet* **9:** 243–248.

Wagner E, Zenke M, Cotten M, Beug H, Birnstiel ML. (1990) Transferrin–polycation conjugates as carriers for DNA uptake into cells. *Proc. Natl Acad. Sci. USA* **87:** 3410–3414.

Wagner E, Cotten M, Foisner R, Birnstiel ML. (1991a) Transferrin–polycation–DNA complexes: The effect of polycations on the structure of the complex and DNA delivery to cells. *Proc. Natl Acad. Sci. USA* **88:** 4255–4259.

Wagner E, Cotten M, Mechtler K, Kirlappos H, Birnstiel ML. (1991b) DNA-binding transferrin conjugates as functional gene-delivery agents: Synthesis by linkage of polylysine or ethidium homodimer to the transferrin carbohydrate moiety. *Bioconjugate Chem.* **2:** 226–231.

Wagner E, Zatloukal K, Cotten M, Kirlappos H, Mechtler K, Curiel DT, Birnstiel ML. (1992a) Coupling of adenovirus to transferrin–polylysine/DNA complexes greatly enhances receptor-mediated gene delivery and expression of transfected genes. *Proc. Natl Acad. Sci. USA* **89:** 6099–6103.

Wagner E, Plank C, Zatloukal K, Cotten M, Birnstiel ML. (1992b) Influenza virus hemagglutinin HA-2 N-terminal fusogenic peptides augment gene transfer by transferrin–polylysine–DNA complexes: Toward a synthetic virus-like gene-transfer vehicle. *Proc. Natl Acad. Sci. USA* **89:** 7934–7938.

Wang C-Y, Huang L. (1987) pH-sensitive immunoliposomes mediate target-cell-specific delivery and controlled expression of a foreign gene in mouse. *Proc. Natl Acad. Sci. USA* **84:** 7851–7855.

Wang J, Jiao SS, Wolff JA, Knechtle SJ. (1991) Gene transfer and expression into rat cardiac transplants. *Transplantation* **53:** 703–705.

Webster RG, Fynan EF, Santoro JC, Robinson H. (1994) Protection of ferrets against influenza challenge with a DNA vaccine to the hemagglutinin. *Vaccine* **12:** 1495–1498.

Wilson JM, Grossman M, Wu CH, Chowdhury NR, Wu GY, Chowdhury JR. (1992) Hepatocyte-directed gene transfer *in vivo* leads to transient improvement of hypercholesterolemia in low density lipoprotein receptor-deficient rabbits. *J. Biol. Chem.* **267:** 963–967.

Wilson T, Papahadjopoulos D, Taber R. (1979) The introduction of poliovirus RNA into cells via lipid vesicles (liposomes). *Cell* **17:** 77–84.

Wolff JA, Malone RW, Williams P, Chong W, Acsadi G, Jani A, Felgner PL. (1990) Direct gene transfer into mouse muscle *in vivo*. *Science* **247:** 1465–1468.

Wu CH, Wilson JM, Wu GY. (1989) Targeting genes: Delivery and persistent expression of a foreign gene driven by mammalian regulatory elements *in vivo*. *J. Biol. Chem.* **264:** 16985–16987.

Wu GY, Wu CH. (1988) Receptor-mediated gene delivery and expression *in vivo*. *J. Biol. Chem.* **263:** 14621–14624.

Wu GY, Zhan P, Sze LL, Rosenberg AR, Wu CH. (1994) Incorporation of adenovirus into a ligand-based carrier system results in retention of original receptor specificity and enhances targeted gene expression. *J. Biol. Chem.* **269:** 11542–11546.

Yoshimura K, Rosenfeld MA, Nakamura H, Scherer EM, Pavirani A, Lecocq J-P, Crystal RG. (1992) Expression of the human cystic fibrosis transmembrane conductance regulator gene in the mouse lung after *in vivo* intratracheal plasmid-mediated gene transfer. *Nucleic Acids Res.* **20:** 3233–3240.

Zenke M, Steinlein P, Wanger E, Cotten M, Beug H, Birnstiel ML. (1990) Receptor-mediated endocytosis of transferrin–polycation conjugates: An efficient way to introduce DNA into hematopoietic cells. *Proc. Natl Acad. Sci. USA* **87:** 3655–3659.

Zhou X, Huang L. (1994) DNA transfection mediated by cationic liposomes containing lipopolylysine: Characterization and mechanism of action. *Biochim. Biophys. Acta* **1189:** 195–203.

Zhou X, Klibanov AL, Huang L. (1991) Lipophilic polylysines mediate efficient DNA transfection in mammalian cells. *Biochim. Biophys. Acta* **1065:** 8–14.

Zhu N, Liggitt D, Liu Y, Debs R. (1993) Systemic gene expression after intravenous DNA delivery into adult mice. *Science* **261:** 209–211.

7

Development of mammalian artificial chromosome vectors: prospects for somatic gene transfer

Zoia Larin

7.1 Introduction

The technical problems inherent in the safe and efficient delivery of genes intact to somatic cells where they are expressed appropriately are a major focus of gene therapy research. Existing vectors and transfer systems have been useful, but the real and potential problems of poor gene expression, insertional mutagenesis and generation of an antigenic response have demonstrated the major limitations of these methods. Mammalian artificial chromosomes (MACs) are being proposed as novel vectors for gene transfer and, potentially, they may be useful for somatic gene therapy (Huxley, 1994). In contrast to existing gene transfer vectors, MACs will remain extrachromosomal and segregate as a normal chromosome. This will ensure that genes will be expressed in a natural chromosome context and be maintained at the correct level throughout the lifetime of the cell. Also, MACs should not cause any mutagenic or oncogenic effects as a result of insertion into a host chromosome. As potential agents for clinical somatic gene therapy, MAC vectors should be non-cytotoxic and provide a safe alternative system to other vectors.

The development of a functional MAC requires knowledge of the structure and function of natural chromosomes, and the processes that affect their behaviour during the cell cycle. This involves determining the DNA sequences that comprise the essential chromosomal elements, the *trans*-acting factors involved in their function, and establishing the minimum size for mitotic stability. On the basis of this understanding, systems can be established *in vivo* for studying chromosome function. Linear yeast artificial chromosomes (YACs) were developed according to such criteria. The essential chromosomal elements (telomeres, centromeres and replication origins) in yeast were defined and also the *trans*-acting factors associated with them. MACs, however, are proving more difficult to construct. We are beginning to understand some of

Gene Therapy, edited by N.R. Lemoine and D.N. Cooper.
© 1996 BIOS Scientific Publishers Ltd, Oxford.

the processes that affect chromosome function in mammalian cells such as DNA replication and segregation, but not all of the structural chromosomal components have been defined. The structure and function of human telomeres is known since they comprise only a few kilobases of repetitive DNA. By contrast, the sequence requirements of human centromeres are still unresolved since they are much larger and may consist of several sequence elements. There has been some progress in identifying human replication origins, but their structure is also complex. This chapter will outline several aspects of eukaryotic chromosome structure and function as a basis for understanding the requirements for constructing MACs.

7.2 Properties of a functional chromosome in the eukaryotic cell cycle

The basic function of a chromosome in an eukaryotic cell is to propagate a faithful copy of its DNA content throughout the cell cycle and from one generation to the next. This process involves the precise initiation and control of a series of events including replication and segregation of the chromosome throughout mitosis and meiosis. Replication of the chromosome occurs from specific origins which are located at many sites along its length, and is initiated and completed during the synthetic or S phase of the cell cycle. The process has been studied extensively in yeast, but our knowledge is less advanced in humans, although it appears that many of the mechanisms are similar. Control of replication is important in order to ensure that S phase occurs once per cell cycle. Proper segregation of chromosomes during metaphase requires attachment of the chromosome to the spindle at a specific region of the chromatin, the centromere, which in yeast is composed of defined DNA sequence elements. A proteinaceous unit, the kinetochore, forms at the centromere and directs binding of specific centromere proteins. Lastly, telomeres are the sequences which protect the ends of chromosomes, and maintain their linearity. Replication of the ends of chromosomes is directed by an enzyme, telomerase, and involves a special mechanism to ensure that the ends do not shorten with each replication cycle.

The sequence of events that occurs in the mitotic cell cycle is controlled by different protein kinases which phosphorylate other protein complexes to activate different parts of the cell cycle (Kearsey *et al.*, 1996; Nigg, 1995). The kinases, termed cyclin-dependent kinases (cdks), are regulated by other proteins, the cyclins, which activate or inactivate them at specific parts of the cell cycle. This mechanism ensures that the cell cycle is completed in adequate time and that DNA is replicated once per cell cycle. Prior to and during replication, the chromatin structure changes to allow accessibility of different protein complexes for initiation of unwinding DNA (via helicases) and replication. Replication units are then activated in an ordered sequence which is dependent on the nature of chromatin (Wolffe, 1995). Normally, DNA is folded into chromatin by core (two molecules of H2A, H2B, H3 and H4) and linker histones which interact with themselves and also recruit other chromosomal proteins to form specialized structures. The repeat element of chromatin is the nucleosome which is composed of 146 bp of DNA wrapped around an octamer of core histones. At the start of replication, the structure of chromatin changes to allow for duplication of the DNA and histones, and afterwards the chromatin structure is restored with the formation of new histones. Disruption of the cell cycle at specific stages can lead to DNA damage, genetic instability, tumorigenesis, cell death and apoptosis. Errors include disruption of replication or repair, chromosomal rearrangements (translocations, amplification or deletions), spindle abnormalities resulting in aneuploidy and loss of telomere sequences resulting in senescence (Hartwell and Kastan, 1994).

7.3 Linear yeast artificial chromosomes

The development of YACs was based on work by Murray and Szostak (1983), who observed that faithful replication and correct segregation of linear yeast plasmids during mitosis requires centromere (CEN), autonomously replicating (ARS) and telomeric sequences, and a minimum size for mitotic stability. Understanding the structure and function of these sequences has provided a framework for determining the equivalent human components for the assembly of MACs.

7.3.1 Telomeres

Telomeres are short (few kb) G-rich repeat sequence elements which define the ends of a linear chromosome, and a chromosome end without a telomere is progressively lost (Zakian, 1995). They were first described in the ciliated protozoans, *Tetrahymena* and *Oxytricha*, and the sequence similarity extends from protozoa, fungi and plants to mammals. Replication of the ends of the chromosome requires the enzyme, telomerase, a protein–RNA complex. The genes encoding the RNA component have been cloned in ciliates, yeast, mice and humans. Telomerase elongates the leading G-rich strand of the telomere in a 5′ to 3′ direction. The enzyme carries an RNA template for synthesizing the DNA sequences, and completion of the lagging strand is carried out by a DNA polymerase. Adding on sequences to the chromosome ends is a processive process, with the incorporation of many repeats at one time before the enzyme dissociates; however, in most organisms the restoration of the ends does not result in a defined number of repeats, which accounts for the heterogeneity in length of each chromosome end. In yeast, the chromatin structure at the ends of the chromosome, the telosome, is distinct from the adjacent DNA which is formed into nucleosomes, and consists of transcriptionally inactive chromatin. It also appears that a similar telosome-like structure is present at human telomeres. A structural protein, Rap1p, identified in *Saccharomyces cerevisiae*, binds along the length of the telomere and is probably important for telomere stability and regulating position effects in this specialized area of chromatin. In addition, Rap1p has other roles as a transcriptional transactivator or repressor at other DNA binding sites.

7.3.2 Autonomously replicating sequences

Yeast ARS elements allow plasmids, in the presence of a selectable marker, to be replicated and maintained as extrachromosomal molecules. They were first identified in *S. cerevisiae* as short sequence elements (100–200 bp) producing a high frequency of transformation, and containing an 11-bp AT-rich autonomous consensus sequence, ACS (Donavan and Diffley, 1996). A complex of proteins, the origin recognition complex (ORC), binds to the ACS. The ORC consists of six proteins and, although required for initiation of replication, it remains bound to the ACS throughout the cell cycle. The ORC also appears to have additional roles such as involvement in silencing at mating type loci (Ehrenofer-Murray *et al.*, 1995). Pre-replication complexes (Pre-RCs) in *S. cerevisiae* consist of ORC and additional proteins such as CDC6. They are present in S phase only, and appear to control the initiation of replication. These complexes themselves are regulated by cdks which in turn are inactivated by cyclins until S phase begins again. This process limits replication to once per cell cycle. Less is known about *Schizosaccharomyces pombe* origins, but recently Dubey *et al.* (1996) reported that they contain a similar type of structure to *S. cerevisiae*. The origins are much larger, spanning 30–55 kb, and may bind a larger complex of proteins for initiation of replication. Also, Pre-RCs and cdks have been identified, which also play a major role in ensuring replication occurs in S phase only.

7.3.3 Centromeres

The CEN sequences confer stability on ARS-containing DNA, and allow linear yeast plasmids to segregate normally in the cell cycle. The structure of the region has been studied extensively in *S. cerevisiae* and *S. pombe* (Pluta *et al.*, 1995). In *S. cerevisiae*, the elements span about 130 bp and are composed of three sequence elements CDEI, CDEII and CDEIII. A 25 bp element in CDEIII is essential for chromosome segregation and binds a protein complex called CBF3. The 8 bp CDEI element also binds a protein called Cpf1. CDEII lies between CDEI and III and it appears that the length of this element (varying between 78 and 86 bp) is important for the formation of the chromatin structure. The *S. cerevisiae* CEN DNA also appears to bind a single microtubule.

By contrast, *S. pombe* centromeres span several kilobases (40–100 kb) and several sequence elements may be involved in regulating kinetochore formation and spindle assembly. The organization of the centromere is chromosome specific, as found in human chromosome centromeres. The region is composed of a series of direct and inverted repeats, and all centromeres contain a central core of 4–7 kb which confers a special chromatin structure and contains several protein-binding sites. The complex kinetochore region also binds bundles of two to four microtubules. The functional sequences include the core sequences and part of the K-type repeat element (spanning 2.1 kb), which are separated by a few kilobases at the natural centromere. The same sequences assembled in a yeast construct form a functional minichromosome (Baum *et al.*, 1994). The stability of the minichromosome depends on the spatial organization of these elements and whether they can fold into stable chromatin (Steiner and Clarke, 1994). Recently, the minimal sequence requirements for the *Drosophila* centromere were defined in a functional minichromosome. The core element consists of 220 kb of complex DNA known as *Bora Bora*, but for full centromere function this region must be flanked either 5′ or 3′ by another 200 kb of simple sequence DNA (Murphy and Karpen, 1995). The implication is that the region forms a special chromatin structure essential for kinetochore formation and protein binding. This organization suggests parallels with the *S. pombe* centromere and provides pointers for human centromere structure.

Normal eukaryotic chromosomes are linear, and the mitotic stability of linear artificial chromosomes in yeast seems to depend on size, increasing with length (Murray *et al.*, 1986). Linear plasmids of 10–16 kb are less stable than circular plasmids of the same size, but linear artificial chromosomes ranging from 15 to 137 kb increase in mitotic stability with size. However, mitotic stability does not appear to be affected by the distance between the centromeric and telomeric sequences. The identification of these elements led Burke *et al.* (1987) to construct the original pYAC vectors which contain both bacterial and yeast sequences for propagation as a circular or linear molecule in each host system. The construction of YAC libraries has contributed significantly to the long-range physical, genetic and functional mapping of complex genomes, and proved to be invaluable in localizing and defining new genes.

7.4 Components required for human artificial chromosomes

7.4.1 Human replication origins

While it may not be necessary to identify specific replication origins for incorporation in a MAC, it is important to understand the nature of these elements in human cells. Different regions of the chromosome replicate at different times during S phase, depending on their position in the chromatin. For example, centromeric heterochromatin tends to replicate late in S phase. The precise *cis*-acting sequences responsible for initiation sites is controversial. Origins have been located both 3′ and 5′ to genes, and some reports describe short replication origins whereas others describe broad initiation zones. Also, some origins appear to be regulated by local sequences and others by sequences quite distant from the initiation site. Recently, two groups have identified human-specific origins (Donavan and Diffley, 1996). One spans 474 bp and lies 3′ of the human lamin B2 gene, and the other lies within 2 kb 5′ of the β-globin gene. Interestingly, the locus control region (LCR) upstream from the β-globin locus appears to contain sequences essential for the replication origin to be activated, which suggests that another initiator-binding site is present and that elements controlling transcription or chromatin structure are involved. It has been difficult to determine how many proteins initiate replication and the nature of these proteins, but it is probable that there may be human homologues of the *S. cerevisiae* ORC, since two ORC-related proteins have been identified in *Drosophila* (Gavin *et al.*, 1995; Gossen *et al.*, 1995).

Although human replication origins do not allow the construction of small autonomously replicating plasmids in human cells equivalent to yeast ARS plasmids, large molecules introduced into mouse cells can replicate autonomously (Featherstone and Huxley, 1993), so the incomplete understanding of human replication origins should not hinder the construction of large MACs.

7.4.2 Human telomeres

Human telomeres are composed of a tandem array of the sequence TTAGGG maintained by telomerase. Telomerase also occasionally heals chromosome ends if they are broken or fragmented (Melek and Shippen, 1996) and its activity is important for maintenance of telomere length; low activity is an indicator of cell senescence. In germ cells, telomerase is active and length is maintained. In somatic cells, telomeres progressively shorten since telomerase is less active but, in some immortalized cells which have escaped cell crisis, telomerase is active again and telomere length is stabilized (Wright and Shay, 1995). Recently, a structural protein that binds to double-stranded repeat arrays in human cells has been identified, known as the telomere repeat factor (TRF), and is required for telomere function (Chong *et al.*, 1995).

Human telomeres were cloned by a functional assay in yeast. A functional human telomere was identified by its ability to confer stability on a YAC containing one telomere (Cooke, 1995). Cloned human telomeres have been shown to function in mammalian cells and seed a new telomere, either after direct introduction of a telomere-containing construct by electroporation, or as a YAC following yeast–mammalian cell fusion. These results indicate that the telomere requirements for MAC construction can readily be met.

7.4.3 Human centromeres

Human centromeres span several megabases and contain several different satellite DNAs and other repeat sequence elements. Low copy repeat sequences are found interspersed between the satellites, but single copy sequences have yet to be detected in these regions. Although many of the repeat families are present at several centromeres, the sequence organization of each region varies. Five major repeat families have been identified: 170, 5, 68, 42 and 48 bp, classified here simply according to the repeat length [see Tyler-Smith and Willard (1993) for the corresponding satellite].

The 170 bp repeat corresponds to the α satellite or alphoid DNA and is the only repeat found at all centromeres. The repeat itself is organized into larger units which then form major arrays spanning between 200 kb and several megabases, but there are other smaller arrays located in the pericentric regions of most chromosomes. To date, it is the only sequence which has been shown to exhibit any centromere function when introduced as exogeneous DNA into monkey and human cells (Haaf *et al.*, 1992; Larin *et al.*, 1994). Several proteins are associated with alphoid DNA: CENP-B (80 kDa) recognizes a 17-bp DNA binding site in a subset of alphoid monomers at all human centromeres, except on the Y chromosome. This sequence is conserved in *Mus musculus*, but CENP-B recognizes a reduced consensus sequence (nine out of 17 bp) in *Mus caroli* and African green monkey. An alphoid-derived 9 bp repeat sequence located at the junction of α satellite DNA and the 5 bp sequence (conserved on chromosomes 13, 14 and 21) binds a novel protein pJα (Gaff *et al.*, 1994). Also, HMG-I/Y (10 kDa) is associated with alphoid DNA, and is involved in altering or bending DNA. The 42-bp repeat is the only other sequence which has shown some DNA–protein binding activity.

Other human centromeric proteins have been identified, localized to the centromere and shown to be involved in different functions (for review, see Pluta *et al.*, 1995). CENP-A (17 kDa) is a histone-related protein with homology to histone H3, and shares structural similarities with the yeast protein Cse4p; it has a diverged amino-terminal domain and a conserved carboxyl-terminal domain. In yeast, Cse4p is required for correct chromosome segregation, and perhaps CENP-A has a similar role. CENP-C (140 kDa) is involved primarily in assembly of the kinetochore plate, but is also a DNA-binding protein (Sugimoto *et al.*, 1994) and associated with a nucleolar transcription factor UBF or NOR90 in interphase cells. CENP-E (312 kDa), and CENP-F (400 kDa) are associated with the outer kinetochore plate and are involved in chromosome movement and stabilization of the spindle. A further group of proteins, the INCENPs (135–150 kDa) are associated with the pairing domain of the kinetochore plate, and may be involved in sister chromatid pairing.

Since the centromere is large and composed of highly repetitive DNA, it has proven impossible to analyse and clone the entire region in a single YAC, but more detailed physical maps are now emerging across some centromeric regions. The structure of the Y chromosome centromere region is one of the most complete, possibly because it is one of the smallest. It comprises a large alphoid array spanning 200–1600 kb (in different males) flanked on the short (p) and long (q) arms by four major repeat sequence families, 68, 48 and 5 bp and YαII, and several other novel low copy repeats (Tyler-Smith and Willard, 1993). Also, a detailed structural analysis of several rearranged human Y chromosomes has defined a centromeric interval of approximately 500 kb, including 200 kb of alphoid DNA and 300 kb of short arm (Yp) flanking DNA (Tyler-Smith *et al.*, 1993). A functional assay using cloned α satellite DNA in a YAC identified these sequences as exhibiting partial centromere function when integrated into human chromosomes (Larin *et al.*, 1994). Thus, alphoid DNA is the best candidate for a human centromere sequence, but it is unclear whether the centromere consists only of alphoid DNA.

7.5 Strategies for constructing MACs

Three different approaches to MAC construction currently are being tested. One strategy involves breaking natural chromosomes at the human centromere by telomere-associated chromosome fragmentation (TACF) to form a truncated chromosome with a new telomere (Barnett *et al.*, 1993; Farr *et al.*, 1991). Another approach is to assemble the essential human components in yeast by recombining YACs to produce a single construct, and then re-introduce the artificial chromosomes into mammalian cells (Monaco and Larin, 1994). A third approach is to develop human artificial episomal chromosomes based on a vector which can replicate independently in mammalian cells using the replication origin from Epstein–Barr virus (EBV) (Sun *et al.*, 1994), and use them to clone other chromosomal components. A major limitation to all these strategies is the lack of understanding of the sequence requirements of a human centromere, and the development of reliable assays to test centromere function in mammalian cells.

7.5.1 TACF

TACF using cloned human telomeric DNA has been used successfully to break the human X or Y chromosome, and produce minichromosomes containing the natural q or p arm by random integration and non-homologous recombination. The efficiency of non-homologous recombination was increased by employing a positive–negative selection mechanism to select for the correct clones (Bollag *et al.*, 1989). TACF has generated minichromosomes of about 5–7 Mb by a second round of breakage of the human X and Y chromosomes (Farr *et al.*, 1995; Heller *et al.*, 1996). The minichromosomes remain extrachromosomal and appear to segregate normally. Further breakage of the minichromosomes by targeting sequences close to the centromeric regions would then determine the minimal centromere sequence and size requirements for autonomous chromosome function (*Figure 7.1*).

The major difficulty with this approach is that the efficiency of targeted recombination in human somatic cells is low when comparing homologous with non-homologous recombination ($\sim 10^{-4}$ to 10^{-5}) (Brown *et al.*, 1994; Itzhaki *et al.*, 1992). An alternative strategy may be to target sequences to a specific chromosome in other recombination-proficient cell lines, and then transfer the modified chromosome back to human cells (Brown *et al.*, 1996). Recent reports have indicated that human chromosome 11 was targeted efficiently in an avian leukosis virus (ALV)-induced chicken pre-B cell line (DT40) (Dieken *et al.*, 1996). Initially, a tagged human chromosome 11 was transferred to DT40 cells by microcell fusion. Then the β-globin gene locus was altered by targeted homologous recombination in the DT40–human microcell hybrid with resulting efficiencies of 4–15%. Following transfer of chromosome 11 back to murine erythroleukaemia (MEL) cells by microcell fusion, an appropriate level of

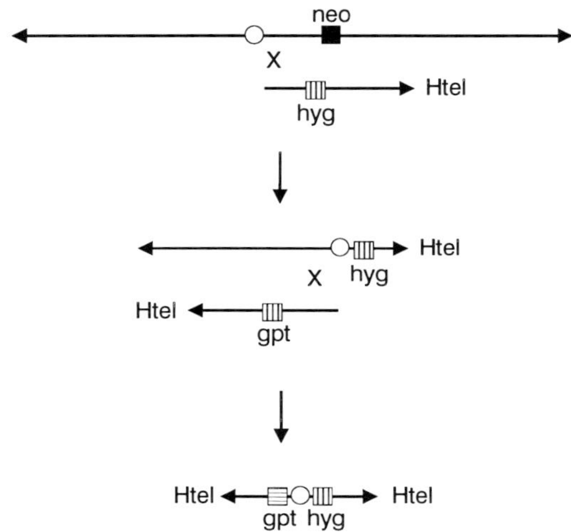

Figure 7.1. Schematic diagram depicting telomere-associated chromosome fragmentation (TACF) of an existing chromosome in a human cell to identify a functional centromere, and define the minimal chromosomal elements of a stable human minichromosome. Selection for the correct recombinant in mammalian cells is by resistance to the mammalian markers, neomycin (neo), hygromycin (hyg) and guanine phosphoribosyltransferase (gpt).

β-globin expression was demonstrated, which was not affected by transfer of the chromosome through the chicken cells.

Other systems are also available which can efficiently target sequences into a defined location following introduction of a specific site by homologous recombination. These include the site-specific recombinase systems, the FLP–*FRT* system from *S. cerevisiae* (O'Gorman *et al.*, 1991), and Cre–*lox*P from bacteriophage P1 (Sauer and Henderson, 1988). These may be useful strategies for incorporating known genes into minichromosomes.

7.5.2 Construction of MACs in yeast

The second approach is to generate candidate MAC vectors in yeast by recombining YACs containing human telomeres, putative centromere sequences and replication origins, and to transfer the constructs intact into mammalian cells. The advantages of the recombination approach are that it will define the precise sequence requirements for chromosome function, and allow extensive manipulations of artificial chromosomes in yeast, including incorporating genes into the MAC vectors. In addition, MACs can be shuttled between yeast and mammalian cells. The disadvantages are that human telomeres are modified with yeast telomeric DNA and, although they are still functional in mammalian cells, the frequency is low compared with unmodified telomeres (Taylor *et al.*, 1994).

Recently, candidate MACs containing human Y chromosome centromeric sequences and human telomeres were assembled in a single large construct in yeast and introduced into cultured mouse cells. The resulting analyses indicated that they produced circular extrachromosomal elements which did not segregate or form active centromeres (Taylor *et al.*, 1996). There may be several reasons for this. Firstly, circularization of the constructs indicates that the telomeric sequences were not functioning properly, either

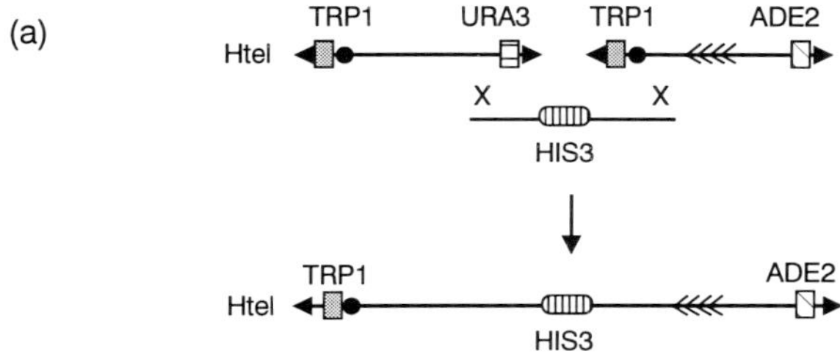

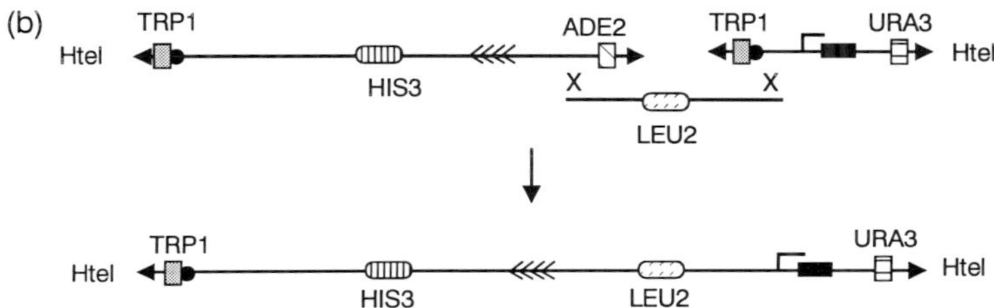

Figure 7.2. (a) Mitotic recombination in yeast of non-overlapping human centromeric YACs and human telomeres (Htel) with plasmids containing homologous sequences to the vector arms. Recombinants are selected for the auxotrophic marker *his3*. α satellite sequences from the centromeric region are depicted as four arrows in tandem. (b) Recombination of a MAC construct with a YAC carrying a known gene. Recombinants are selected for the auxotrophic marker *leu2*.

because of the presence of additional yeast telomeric sequences, or because the telomeres were lost or degraded when introduced into the mouse cells. Secondly, the constructs did not contain all of the sequences from the centromeric interval which may be needed for an active centromere. Thirdly, human-specific factors not present in mouse cells may be required for full centromeric activity, or other epigenetic effects such as chromatin structure or *de novo* DNA methylation may be playing a role.

Ideally, a construct containing the entire centromeric region is required for testing chromosome function in human cells, but it has not been possible to clone this in a single construct, since large tracts of repetitive DNA are unstable in YACs (Neil *et al.*, 1989). Reconstruction of the centromeric region by recombination of YACs is one possibility. However, instability of α satellite-containing YACs and adjacent sequences has been observed following meiotic recombination (Taylor, 1996). An alternative procedure is mitotic recombination of non-overlapping YACs that can link up to four YACs in a single stable construct (Larin *et al.*, 1996). It is envisaged that centromeric YACs with human telomeres and YACs carrying genes could then be recombined stably in a single MAC construct (*Figure 7.2*).

7.5.3 Human artificial episomal chromosomes

Cloning large DNA as virus-based episomes in human cells has been demonstrated by Sun *et al.* (1994). The construction of human artificial episomal chromosomes (HAECs) is based on a cloning vector containing the EBV replication origin (*ori*P), which is activated by its viral protein counterpart Epstein–Barr nuclear antigen-1

(EBNA-1). These vectors can stably maintain up to 350 kb of human genomic DNA as a circular minichromosome, and are potentially useful as gene transfer vectors in somatic cells if expression is maintained (Vos *et al.*, 1995). Sun *et al.* (1994) also suggest the possibility of cloning a functional centromere in a HAEC, and developing an artificial chromosome by adding human telomeres, presumably following removal of the EBV *ori*P or EBNA-1 (*Figure 7.3*).

7.6 Prospects of MACs for somatic gene transfer

The development of MACs for human somatic gene therapy is premature, since the construction of vectors is still at a preliminary stage. Once a stable and functional MAC vector is developed, then methods to deliver large DNA molecules expressing the appropriate gene into both cultured cells (*in vitro*) and somatic cells in mice (*in vivo*) will be required. The major problems to be encountered will be the development of suitable transfer systems for large DNA, and identifying methods to ensure long-term gene expression. It will also be important to consider systems which can stably transfer a large DNA molecule into the host cell nucleus so that it can form and fold into proper chromatin structure via the assembly of DNA and histones.

Currently, candidate MACs in yeast or the TACF minichromosomes are being shuttled to different cultured cells either by fusing yeast and mammalian cells (Featherstone and Huxley, 1993) or by forming microcells (Brown *et al.*, 1994). Since these methods are not suitable for somatic cell transfer *in vivo*, more direct methods to introduce DNA into different cell types are required. It will also be essential to identify efficient and non-toxic methods which can target cells and deliver DNA intact to the nucleus. Several approaches are being investigated, including transfer by liposomes (Behr *et al.*, 1989), virally mediated delivery systems (Curiel *et al.*, 1992) and cell surface receptor-mediated delivery (Cristiano *et al.*, 1993).

Lipofection involving a complex of cationic liposomes and DNA is an efficient means of transfecting plasmid DNA for expression of genes in cultured cells. Liposomes effectively mediate condensation of DNA prior to cell uptake and, after binding to the negatively charged cell surface, can transfect cells reasonably efficiently via endocytosis. Gao and Huang (1995) review in detail the cationic liposomes currently available and their mode of action. Large DNA molecules such as YACs have been transferred relatively intact by lipofection to mammalian cultured cells, including human cells, and mouse embryonic stem (ES) cells for generation of transgenic mice (Larin, 1995). The disadvantage of this method is that it involves isolating large DNA by pulsed-field gel electrophoresis in sufficient quantities for transfection, and will therefore require methods to increase the concentration of MAC DNA. There are several examples of successful gene transfer to mouse somatic cells by liposomes either by topical application or systemic administration, with maintenance of long-term gene expression and no immunogenic response (Gao and Huang, 1995; Zhu *et al.*, 1993). As a result, liposomes are now approved for use in human gene therapy experiments. For example, a clinical trial involving cystic fibrosis patients was approved to investigate the effect of delivery of the cystic fibrosis transmembrane regulator (*CFTR*) gene–liposome complex via the nasal epithelium and as an aerosol to the lung (Caplen *et al.*, 1995). Already, some restoration of chloride channel activity is apparent (20%).

Curiel *et al.* (1992) developed a strategy for introducing DNA by viral-mediated delivery into cells by capitalizing on the efficiency of adenovirus entry. They con-

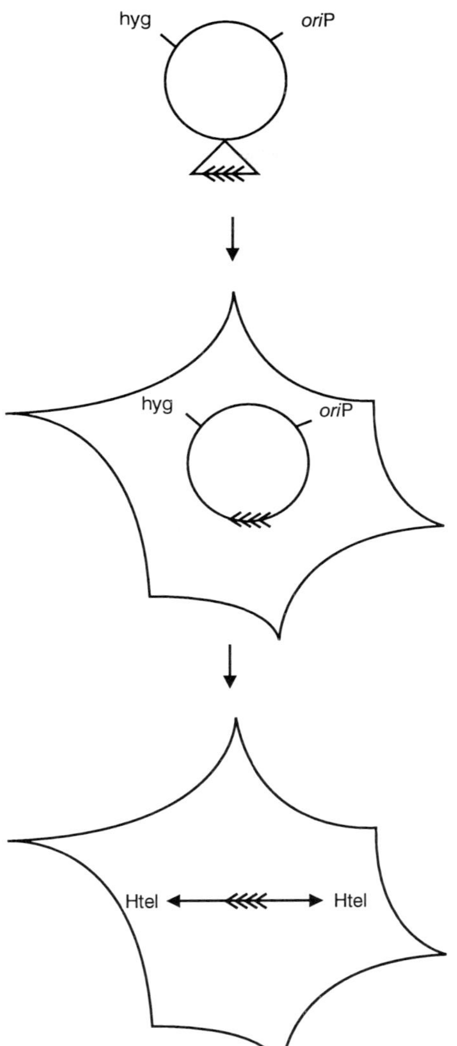

Figure 7.3. Schematic diagram depicting the cloning of a functional human centromere (four arrows in tandem) in a HAEC vector, which is maintained as a stable episome in a human cell. Addition of human telomeres (Htel) and removal of *ori*P generates a linear minichromosome.

structed DNA–polylysine complexes which bind to the exterior of the virion using a monoclonal antibody to an adenovirus epitope covalently linked to polylysine. The DNA–polylysine–adenovirus complex is then taken up into the cell by adenoviral receptors. Because DNA is linked to the outside of the virion, there is no restriction on size of foreign DNA that is transferred, so this may be a feasible method to transfer MAC DNA. Also, the method is relatively safe since only the entry mechanisms and not the viral genome are required.

Somatic therapy should ideally introduce the gene into all tissues. However, some diseases may only require specific targeting to the defective tissue. Methods to target DNA to cell surface receptors have been developed, in association with liposomes or replication-deficient adenovirus for enhanced transfer. For example, high efficiency transfer to hepatocytes has been demonstrated using the receptor-mediated uptake of asialoorosomucoid–polylysine–DNA complexes co-internalized with replication-deficient adenovirus (Cristiano *et al.*, 1993). Also, there has been success in targeting to transferrin receptors which are more ubiquitous cell surface receptors.

A major obstacle with most methods is to avoid lysosomal degradation of DNA once across the cell membrane and to enhance delivery into the nucleus (Schofield and Caskey, 1995). Following endocytosis, DNA is usually degraded within the acidified endosome, and so better methods to protect the DNA or to transfer DNA directly to the nucleus will need to be developed. Possible methods include the use of substances to reduce endosomal acidification such as chloroquine, coating DNA with nuclear protein HMG-I, which is efficient at targeting the nucleus, using viruses to avoid the endosome such as Sendai virus (haemagglutinating virus of Japan, HVJ), and lastly developing synthetic peptides which can disrupt endosomes (e.g. haemagglutin, an influenza virus membrane protein).

Mouse models of a particular human disease will be invaluable for determining the feasibility of delivering MACs to somatic cells in mice. The possibility of transferring MACs through the germ line in mice would determine whether the MAC could be transferred successfully to every cell, and would show the expected phenotype. This would be a distinct advantage for somatic therapy. For example, germ-line transmission could be tested by transferring the MAC to ES cells, and positive cells used to produce chimeric mice. The opportunity for determining the fate of the transgene and the meiotic stability of the MAC within the germ line could also be established.

7.7 Conclusions

Progress on the development of MACs is ongoing, and the prospect of a usable vector for somatic gene transfer is still in the early stages. Currently, the major limitation to the development of a MAC is identifying the human centromere. Several groups are actively engaged in defining the minimal centromere requirements by fragmenting chromosomes *in vivo* or recreating an active centromere *in vitro*. Once we know the sequence requirements of the centromere, this will be a major breakthrough for MAC construction; once defined, MACs will be of enormous benefit for studying chromosome function and behaviour during mitosis and meiosis.

The idea of using MACs for gene therapy is an exciting one, since it could offer a long-term solution for the correction of many diseases. As an extrachromosomal gene transfer vector, MACs would provide significant advantages over current vectors for maintenance of gene expression without any adverse effects. Several major problems will need to be overcome for adapting MACs as gene transfer vectors to somatic cells. Efforts would need to concentrate on developing high efficiency MAC transfer to different cell types. Targeting to tissues composed of non-dividing cells will require the development of a sophisticated and effective delivery system for introducing the MAC into every cell. Delivery of MACs to dividing cells or stem cells which later become differentiated cells, would be more ideal.

Acknowledgements

I thank Chris Tyler-Smith for comments on the manuscript. Z.L. is supported by the Wellcome Trust.

References

Barnett M, Buckle VJ, Evans EP, Porter ACG, Rout D, Smith AG, Brown WRA. (1993) Telomere directed fragmentation of mammalian chromosomes. *Nucleic Acids Res.* **21**: 27–36.

Baum M, Ngan VK, Clarke L. (1994) The centromeric K type repeat and the central core are together sufficient to establish a functional *Schizosaccharomyces pombe* centromere. *Mol. Biol. Cell* **5**: 747–761.

Behr J-P, Demeneix B, Loeffler J-P, Perez-Mutul J. (1989) Efficient gene transfer in mammalian primary endocrine cells with lipopolyamine-coated DNA. *Proc. Natl Acad. Sci. USA* **86**: 6982–6986.

Bollag RJ, Waldman AS, Liskay RM. (1989) Homologous recombination in mammalian cells. *Annu. Rev. Genet.* **23**: 199–225.

Brown KE, Barnett MA, Burgtorf C, Shaw P, Buckle VJ, Brown WRA. (1994) Dissecting the centromere of the human Y chromosome with cloned telomeric DNA. *Hum. Mol. Genet.* **3**: 1227–1237.

Brown W, Heller R, Loupart M-L, Shen M-H, Chand A. (1996) Mammalian artificial chromosomes. *Curr. Opin. Genet. Dev.* **6**: 281–288.

Burke DT, Carle GF, Olson MV. (1987) Cloning of large segments of exogenous DNA into yeast by means of artificial chromosome vectors. *Science* **236**: 806–812.

Caplen NJ, Kinrade E, Sori F, Gruenert D, Geddes D, Coutelle C, Huang L, Alton EWFW, Williamson R. (1995) *In vitro* liposome-mediated DNA transfection of epithelial cell lines using the cationic liposome DC-Chol/DOPE. *Gene Ther.* **2**: 603–613.

Chong L, van Stensal B, Broccoli D, Erdjument-Bromage H, Hanish J, Tempst P, de Lange T. (1995) A human telomeric protein. *Science* **270**: 1663–1671.

Cooke H. (1995) Non-programmed and engineered chromosome breakage. In: *Telomeres* (eds EH Blackburn, CW Greider). Cold Spring Harbor Laboratory Press, Cold Spring Harbor NY, pp. 219–245.

Cristiano RJ, Smith LC, Woo SLC. (1993) Hepatic gene therapy: adenovirus enhancement of receptor-mediated gene delivery and expression in primary hepatocytes. *Proc. Natl Acad. Sci. USA* **90**: 2122–2126.

Curiel DT, Wagner E, Cotten M, Birnstiel ML, Agarwal S, Li C-M, Loechel S, Hu P-C. (1992) High-efficiency gene transfer mediated by adenovirus coupled to DNA–polylysine complexes. *Hum. Gene Ther.* **3**: 147–154.

Dieken ES, Epner EM, Fiering S, Fournier REK, Groudine MG. (1996) Efficient modification of human chromosomal alleles using recombination-proficient chicken/human microcell hybrids. *Nature Genet.* **12**: 174–182.

Donavan S, Diffley J. (1996) Replication origins in eukaryotes. *Curr. Opin. Genet. Devel.* **6**: 203–207.

Dubey DD, Kim S-M, Todorov IT, Huberman JA. (1996) Large, complex modular structure of a fission yeast DNA replication origin. *Curr. Biol.* **6**: 467–473.

Ehrenofer-Murray AE, Gossen M, Pak DTS, Botchan MR, Rine J. (1995) Separation of origin recognition complex functions by cross species complementation. *Science* **270**: 1671–1674.

Farr C, Fantes J, Goodfellow P, Cooke H. (1991) Functional reintroduction of human telomeres into mammalian cells. *Proc. Natl Acad. Sci. USA* **88**: 7006–7010.

Farr CJ, Bayne RAL, Kipling D, Mills W, Critcher R, Cooke HJ. (1995) Generation of human X-derived minichromosome using telomere-associated chromosome fragmentation. *EMBO J.* **14**: 5444–5454.

Featherstone T, Huxley C. (1993) Extrachromosomal maintenance and amplification of yeast artificial chromosome DNA in mouse cells. *Genomics* **17**: 267–278.

Gaff C, du Sart D, Kalitsis P, Iannello R, Nagy A, Choo KHA. (1994) A novel nuclear protein binds centromeric alpha satellite DNA. *Hum. Mol. Genet.* **3**: 711–716.

Gao X, Huang L. (1995) Cationic liposome-mediated gene transfer. *Gene Ther.* **2**: 710–722.

Gavin KA, Hidaka M, Stillman B. (1995) Conserved initiator proteins in eukaryotes. *Science* **270**: 667–670.

Gossen M, Pak DTS, Hansen SK, Acharya JK, Botchan MR. (1995) A *Drosophila* homolog of the yeast origin recognition complex. *Science* **270**: 1674–1677.

Haaf MW, Warburton PE, Willard HF. (1992) Integration of human alpha satellite DNA into simian chromosomes: centromere protein binding and disruption of normal chromosome segregation. *Cell* **70**: 681–689.

Hartwell LH, Kastan MB. (1994) Cell cycle control and cancer. *Science* **266**: 1821–1828.

Heller R, Brown KE, Burgtorf C, Brown WRA. (1996) Minichromosome derivatives of the human Y chromosome by telomere directed chromosome breakage. *Proc. Natl Acad. Sci. USA* **93**: 7125–7130.

Huxley C. (1994) Mammalian artificial chromosomes: a new tool for gene therapy. *Gene Ther.* **1**: 7–12.

Itzhaki JE, Barnett MA, MacCarthy AB, Buckle VJ, Brown WRA, Porter ACG. (1992) Targeted breakage of a human chromosome mediated by cloned telomeric DNA. *Nature Genet.* **2**: 283–287.

Kearsey SE, Labib K, Macorano D. (1996) Cell cycle control of eukaryotic DNA replication. *Curr. Opin. Genet. Devel.* **6**: 208–214.

Larin Z. (1995) Functional analysis of genomes using YACs. In: *Pulsed Field Gel Electrophoresis: A Practical Approach* (ed. AP Monaco). IRL Press at Oxford, University Press, Oxford, pp. 139–158.

Larin Z, Fricker MD, Tyler-Smith C. (1994) *De novo* formation of several features of a centromere following introduction of a Y alphoid YAC into mammalian cells. *Hum. Mol. Genet.* **3**: 689–695.

Larin Z, Taylor SS, Tyler-Smith C. (1996) A method for linking yeast artificial chromosomes. *Nucl. Acids Res.* (in press).

Melek M, Shippen DE. (1996) Chromosome healing: spontaneous and programmed *de novo* telomere formation by telomerase. *BioEssays* **18**: 301–308.

Monaco AP, Larin Z. (1994) YACs, BACs, PACs, and MACs: artificial chromosomes as research tools. *Trends Biotechnol.* **12**: 280–286.

Murphy TD, Karpen GH. (1995) Localization of centromere function in a *Drosophila* minichromosome. *Cell* **82**: 599–609.

Murray AW, Szostak JW. (1983) Construction of artificial chromosomes in yeast. *Nature* **305**: 189–193.

Murray AW, Schultes NP, Szostak JW. (1986) Chromosome length controls mitotic segregation in yeast. *Cell* **45**: 529–536.

Neil DL, Villasante A, Fisher RB, Vetrie D, Cox B, Tyler-Smith C. (1989) Structural instability of human tandemly repeated DNA sequences cloned in yeast artificial chromosome vectors. *Nucleic Acids Res.* **18**: 1421–1428.

Nigg EA. (1995) Cyclin-dependent protein kinases: key regulators of the eukaryotic cell cycle. *BioEssays* **17**: 471–479.

O'Gorman S, Fox DT, Wahl F. (1991) Recombinase-mediated gene activation and site-specific integration in mammalian cells. *Science* **251**: 1351–1355.

Pluta AF, Mackay AM, Ainsztein AM, Goldberg IG, Earnshaw WC. (1995) The centromere: hub of chromosomal activities. *Science* **270**: 1591–1594.

Sauer B, Henderson N. (1988) Site-specific DNA recombination in mammalian cells by the Cre recombinase of bacteriophage P1. *Proc. Natl Acad. Sci. USA* **85**: 5166–5170.

Schofield JP, Caskey CT. (1995) Non-viral approaches to gene therapy. In: *Gene Therapy* (eds AML Lever, P Goodfellow). Churchill Livingstone, London, pp. 56–72.

Steiner NC, Clarke L. (1994) A novel epigenetic effect can alter centromere function in fission yeast. *Cell* **79**: 865–874.

Sugimoto K, Yata H, Muro Y, Himeno M. (1994) Human centromere protein C (CENPC) is a DNA binding protein which possesses a novel DNA binding motif. *J. Biochem.* **116**: 877–881.

Sun T-Q, Fenstermacher DA, Vos J-MH. (1994) Human artificial episomal chromosomes for cloning large DNA fragments in human cells. *Nature Genet.* **8**: 33–41.

Taylor SS. (1996) Manipulation of YACs to construct an artificial chromosome. D. Phil Thesis, Oxford University.

Taylor SS, Larin Z, Tyler-Smith C. (1994) The addition of functional human telomeres to YACs. *Nucleic Acids Res.* **3**: 1383–1387.

Taylor SS, Larin Z, Tyler-Smith C. (1996) Analysis of extrachromosomal structures containing human centromeric alphoid satellite DNA sequences in mouse cells. *Chromosoma* **105**: 70–81.

Tyler-Smith C, Oakey R, Larin Z, Fisher RB, Crocker M, Affara NA, Ferguson-Smith MA, Muenke M, Zuffardi O, Jobling MA. (1993) Localization of the DNA sequences required for human centromere function through an analysis of rearranged Y chromosomes. *Nature Genet.* **5**: 368–375.

Tyler-Smith C, Willard H. (1993) Mammalian chromosome structure. *Curr. Opin. Genet. Dev.* **3**: 390–397.

Vos J-MH, Livanos E, Banerjee S. (1995) Therapeutic gene delivery in human B-lymphoblastoid cells by engineered non-transforming infectious Epstein–Barr virus. *Nature Med.* **1**: 1303–1307.

Wolffe A. (1995) Chromatin and nuclear assembly. In: *Chromatin. Structure and Function* (ed. A Wolffe). Academic Press, London, pp. 105–147.

Wright WE, Shay JW. (1995) Time, telomeres, and tumours: is cellular senescence more than an anticancer mechanism? *Trends Cell Biol.* **5**: 293–297.

Zakian VA. (1995) Telomeres: beginning to understand the end. *Science* **270**: 1601–1606.

Zhu N, Liggitt D, Liu Y, Debs R. (1993) Systemic gene expression after intravenous DNA delivery into adult mice. *Science* **261**: 209–211.

8

Infectious herpes vectors for gene therapy

Jean-Michel H. Vos, Eva-Maria Westphal and Subrata Banerjee

8.1 Overview

Human herpes viruses represent large, double-stranded DNA, enveloped viruses with genomes ranging in size from 125 kb to 229 kb. Eight different types of human herpes simplex virus (HSV) have been identified (reviewed by Roizman, 1990; Vos, 1995), the latest one being the HSV-like sequences isolated from Kaposi sarcoma tissue (Chang *et al.*, 1994). Herpes viruses are ubiquitous, and few humans escape being infected by them (Strauss, 1990). One feature shared by all herpes viruses is the capacity to induce life-long latent infection in their natural hosts (Stevens, 1989). With the exception of vari-cella zoster virus (VZV), most herpes virus infections are transmitted asymptomatically. *In vivo*, specific mechanisms must be engaged to prevent completion of the virus replicative cycle and to foil immune surveillance of virally infected cells. The life-long persistence of herpes viruses in the adult human population world-wide suggests the possibility for long-term genetically based treatments of both inherited and acquired human diseases. The target cells for latency vary with the herpes virus, but appears gen-erally restricted to a limited subset of tissues; for example, HSV-1 is neurotropic, whereas Epstein–Barr virus (EBV) is a lymphotropic virus with latency occurring either in neurons or in B lymphocytes. On the other hand, herpes virion production can occur either in a variety of human cell types, as is found in the case of HSV-1, or can be very restricted, as in the case of EBV. Species tropism can be extremely wide, as in HSV-1, or can be very limited, as in the infectious cycle of EBV in some primates. The narrow tro-pism of herpes viral latency indicates its potential application to organ-specific *in vivo* targeting; for example, HSV-1 to neurons and EBV to B lymphocytes.

Two alternative genetic engineering strategies have been developed for the produc-tion of infectious herpes viruses: (i) the generation of 'helper-virus-free' recombinant viruses by homologous insertion of the foreign gene into the viral genome; and (ii) the 'helper-dependent' mini-viruses by construction of plasmids carrying the mini-mal *cis* elements required for replication and packaging into infectious virions. Because herpes viral latency appears generally associated with episomal maintenance, the long-term episomal persistence avoids the potentially damaging random integra-tion of the vector into cellular genes or other functional regions. Hence, the proper-ties of these viruses appear to make them suitable for transducing large genes and other chromosomal regions of functional interest. More generally, this family of enveloped double-stranded DNA viruses offers the potential for the development and

Gene Therapy, edited by N.R. Lemoine and D.N. Cooper.
© 1996 BIOS Scientific Publishers Ltd, Oxford.

delivery of human artificial episomal chromosomes (HAECs). However, the complex biology of these large viruses and the poorly understood pathology associated with herpes viral infections, such as the cytotoxicity of HSV-1 and the oncogenicity of EBV, underscores the necessity for careful work in order to establish safe and efficient herpes viral vectors for human gene therapy.

8.2 Molecular biology of herpes viruses

A number of studies have been carried out to test the feasibility of engineering human herpes viruses as infectious vectors for gene delivery and expression *in vitro* (i.e. cultured cells) and *in vivo* (i.e. whole animals). Since most work has concentrated on developing infectious vectors derived from the lymphotropic EBV and the neurotropic HSV-1, we shall limit our review to these two prototype herpes viruses. The long-term potential of other human herpes viruses for gene therapy has been recently summarized elsewhere (Vos, 1995).

8.2.1 Epstein–Barr virus

Epstein–Barr virus (EBV) is a 172 kb, double-stranded DNA, γ herpes virus (*Figure 8.1*). More than 90% of the human population world-wide has been estimated to be infected with EBV. The virus is believed to be transmitted by saliva and infects nasopharyngeal cells and B lymphocytes. Initially, it was suggested that oropharyngeal epithelial cells are the target of primary EBV infection and also the site of EBV persistence. Virus infection of B lymphocytes was considered to be a secondary event. However, several recent studies point to the B-lymphoid compartment as the main site of EBV latency and possibly of primary EBV infection (Chen *et al.*, 1995; Tierney *et al.*, 1994). EBV infects human B cells by binding specifically to the type 2 complement receptor (CD21/CR2), followed by receptor-mediated endocytosis (Cooper *et al.*, 1988). However, for EBV infection of epithelial cells, a CD21-mediated mechanism is not yet well established, although there are reports of CD21 mRNAs in certain EBV-infected smooth muscle tumours (McClain *et al.*, 1995). There have been conflicting reports regarding the association of EBV with cervical carcinoma (Hilton *et al.*, 1993; Landers *et al.*, 1993) and breast epithelial cell tumours (Labrecque *et al.*, 1995). Alternative mechanisms including cell fusions between virus-carrying B cells and epithelial cells or immunoglobulin (Ig) A-mediated infection have been suggested (Sixbey and Yao, 1992).

Two types of cellular infection are possible (*Figure 8.1*). In a lytic infection, viral DNA, RNA and protein synthesis begin, followed by the assembly of viral proteins and lysis of the host cell. Alternatively, the more common latent, non-lytic infection can occur, in which the viral DNA is incorporated into the host genome and is maintained through subsequent cell divisions. EBV has profound effects on B-lymphocyte growth characteristics *in vitro:* (i) the virus is a potent, T-cell independent polyclonal activator of B-cell proliferation; and (ii) EBV can immortalize normal human B cells so that they will proliferate in culture indefinitely. The resulting long-term B lymphoblastoid cell lines (LCLs) are latently infected with the virus and express various virally encoded antigens termed Epstein–Barr nuclear antigens (EBNAs).

About 100 different genes have been identified in the large 172 kb genome of EBV, of which only 10 are expressed in LCLs. The mechanism of action of these viral genes and their role in B-cell immortalization are still largely unknown. Several of them appear to co-operate in the initiation and maintenance of B-cell immortalization and

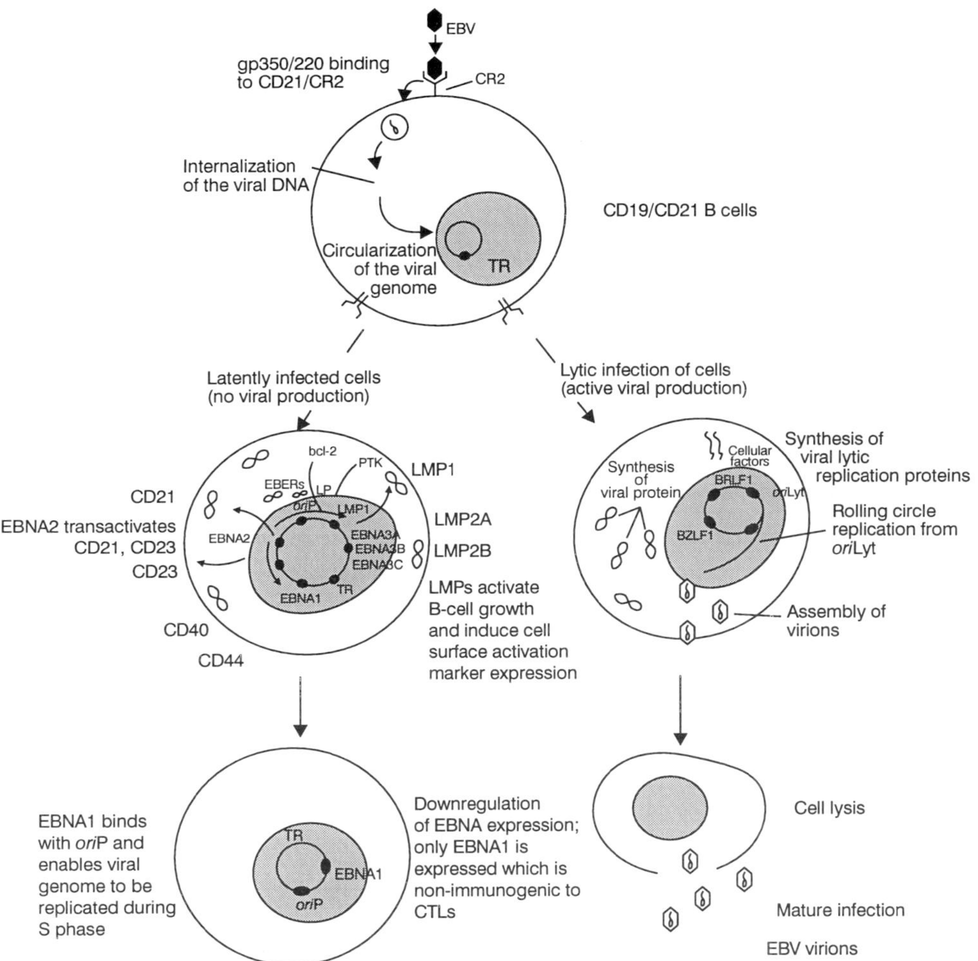

Figure 8.1. Life cycle of EBV. EBV infects oropharyngeal epithelial cells, where it may replicate and shed progeny that may then infect localized B cells. Alternatively, the B cells themselves are infected in a primary infection. Infection of B cells is mediated by the CD21/CR2 receptors and the viral protein gp350/220. The virus replicates in the infected cells and induces B-cell proliferation. The infected B cells expressing virally encoded EBNAs and LMPs are then targeted by the host's primary immune response. At this time a small fraction of the expanded EBV-carrying cells switch on different viral promoters to downregulate expression of EBNAs, while expressing only the immunogenic EBNA1. Latent replication of the virus is maintained in these B-cells via the interaction of *ori*P and EBNA1. However, with immune supression induced by other viruses or drugs, the EBV-infected B cells start to proliferate again; a number of viral genes such as *BZLF1*, *BRLF1* and *BMLF1* are activated. EBV replicates in an uncontrolled fashion in this state which potentially leads to lymphoproliferative disorders. Furthermore, high viral titres and impaired immune surveillance in this state may allow the dissemination of EBV to distant epithelial cells. The infected epithelial cells subsequently may form tumours.

have regulatory activities affecting the expression of viral and cellular genes including the anti-apoptotic gene *bcl-2* (Henderson *et al.*, 1991), B-cell activation markers and adhesion molecules. EBV normally infects resting cells and is maintained in an episomal latent fashion in the host B cells. The various latent states of EBV have been broadly classified as latency I (EBNA1 expression only), latency II (EBNA1, LMP1 and LMP2 expression) and latency III (EBNA1, 2A, 2B, 3A, 3B, 3C, LP, LMP1, LMP2 expression) and are controlled by a complex machinery involving the use of alternative promoters (Rowe *et al.*, 1992).

In latently infected B cells, multiple copies of the viral genome are maintained predominantly as episomes that are replicated once per cell cycle. Latency replication proceeds from *ori*P which has multiple binding sites for the viral EBNA1 gene product. *EBNA2* is one of the first genes to be expressed during EBV infection and is a transcriptional activator for a large number of cellular and viral genes required for progression through the cell cycle. The gene is also essential for cellular immortalization and has recently been implicated in the transcriptional suppression of the immunoglobulin *M* gene and hence may have an antiproliferative effect in Burkitt lymphoma (Jochner *et al.*, 1996). EBNA-LP acts in co-operation with EBNA2 to bring about the induction of cyclin D2 in primary B cells and plays an essential role in cellular immortalization (Sinclair *et al.*, 1994). The exact role of the gene product is unknown but it has been reported to bind both Rb and p53 proteins *in vitro* (Szekely *et al.*, 1993). The three genes *EBNA3A*, *3B* and *3C* are adjacent in the viral genome and may perform related functions. EBNA3A may only be essential in the early stages of the immortalization process (Kempkes *et al.*, 1995) and EBNA3C has been implicated in regulating expression of genes such as *CD21* and *LMP1* (Wang *et al.*, 1990). Of the six essential EBV-immortalizing genes, *LMP1* is the only one reported to transform rodent fibroblasts in culture rendering them tumorigenic in mice (Wang *et al.*, 1985). *LMP1* is one of the few EBV genes to be expressed in both phases of the virus life cycle and has also recently been implicated in the immunogenicity of the virus (Rowe, 1995). As well as the nuclear antigens (EBNAs) and the membrane proteins (LMPs), the small non-polyadenylated RNAs EBERs have been found to be expressed in latently infected cells. The exact role of EBERs is still unclear but they have been implicated in cellular growth via interaction with the interferon-inducible protein kinase PKR (Sharpe *et al.*, 1993).

The switch from latent to lytic infection is mediated by host cellular factors and the expression of the viral regulatory proteins BZLF1, BRLF1 and BMLF1 (Kieff and Liebowitz, 1990). The concerted action of these proteins leads to activation of the complete cascade of early and late EBV gene expression such as the EBV DNA polymerase. Lytic DNA replication proceeds from a separate origin, *ori*Lyt, and results in a several-hundred-fold amplification of the genome via concatemeric intermediates. The concatemers are then cleaved to generate virion DNAs which are packaged into viral capsids. The tandemly arranged terminal repeats (TR) of EBV virion DNA are believed to play a critical role in this cleavage and packaging process as well as in the circularization event of the viral genome in latently infected cells.

8.2.2 Herpes simplex virus 1

Herpes simplex virus types 1 and 2 (HSV-1 and HSV-2) are 152 kb, double-stranded DNA, α herpes viruses (*Figure 8.2*). HSV-1 and HSV-2 are remarkable in their ability to infect essentially all cell types in many species. They either undergo lytic infection,

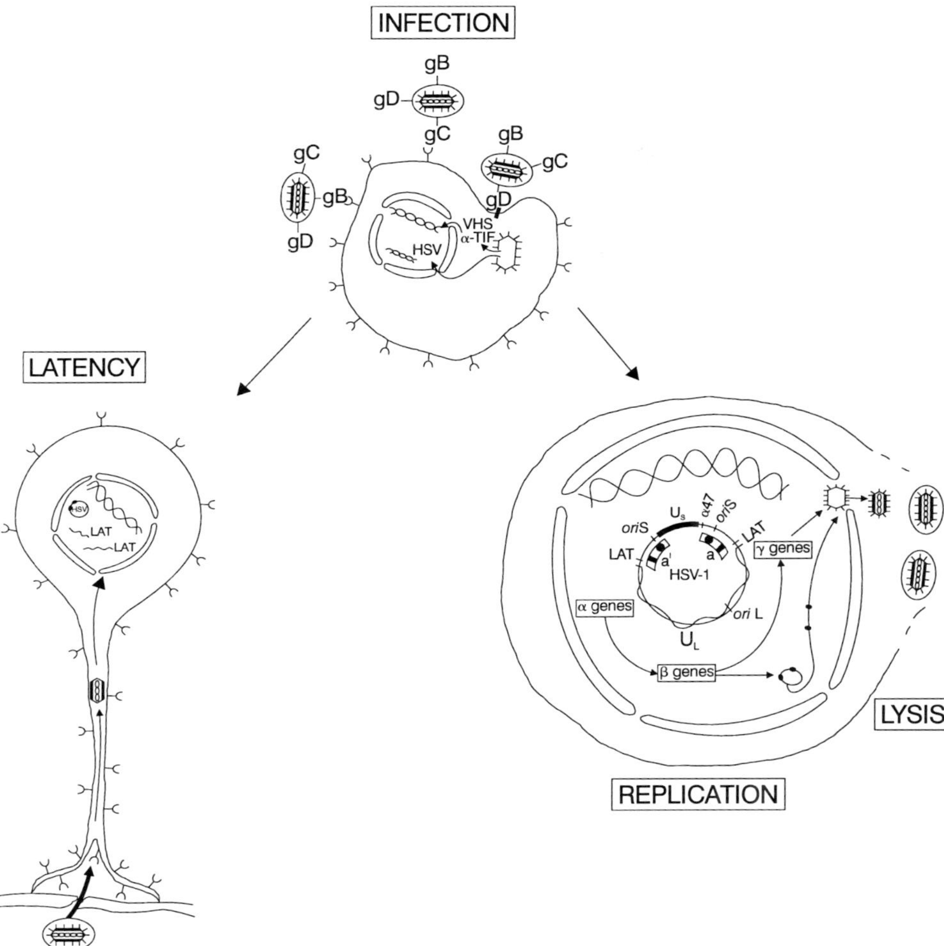

Figure 8.2. Life cycle of HSV-1.

Entry / Infection	Latency	Replication and Lysis
Virus envelope glycoproteins (gB, gC, gD) interact with heparan sulphate molecules in the membrane of any cell type. VHS, virion host-shut-off protein; α-TIF, transactivator of HSV and other genes; Y = heparan sulphate.	Episomal maintenance of HSV in the nucleus of sensory neurons; the only active part of HSV, latency associated transcripts (LAT), are shown.	DNA is replicated in a rolling circle mechanism, cleaved and packaged into the assembled virions which are transported to the cell surface where lysis of the host cell occurs. U_S, unique short region; U_L, unique long region; a, a', a sequence (cleavage and packaging signal); *ori*S, *ori*L, origins of replication; *α47* encodes ICP47 (immune escape). β genes: metabolism (*tk*), DNA replication (helicase primase complex); γ genes: replication, cleavage and packaging of DNA respectively, virion proteins.

destroying the host cell, or establish life-long persistence with the possibility of reactivation. The wide host range is due to the binding of glycoproteins (gB, gC) to heparan sulphate molecules found ubiquitously in all cell membranes (WuDunn and Spear, 1989). Uptake of the virus particles involves glycoprotein gD and fibroblast growth factor (FGF) receptor (Kaner *et al.*, 1990). The nucleocapsids contain linear HSV DNA, which becomes circularized immediately after release into the nucleus, and viral proteins including the virion host shut-off protein (unspecific RNase) and VP16/α-*trans*-inducing factor (α-TIF). A complex between (α-TIF) and cellular *oct*-1 activates transcription of the five immediate–early (IE) or α genes of HSV (*α0, α4, α22, α27, α47*) (Stern *et al.*, 1989). *α0* and *α4* positively regulate the expression of other HSV genes. *α4* represses itself and *α0*. A defect in *α4* results in an inability to replicate. Four of the immediate–early proteins induce a group of approximately 14 β genes which are involved in nucleic acid metabolism (thymidine kinase; tk) and in the rolling circle mechanism of DNA replication. Three origins of replication have been identified: each of the two repeat regions of the HSV genome contains one copy of *ori*S (*Figure 8.2*); a third origin of replication (*ori*L) has been located in the unique long region of HSV. Activation of the approximately 41 late or γ genes of HSV depends on these previous steps of the viral life cycle. Late proteins are involved in DNA cleavage and packaging (signalled by the 'a' sequence in the repeats) and in assembly of the virion particles. The viral glycoproteins and ICP34.5 (which is necessary for HSV replication in central and peripheral neurons) are also among the late proteins. Towards the end of the lytic cycle, HSV virions are transported through the endoplasmic reticulum to the cell surface.

This lytic infection can occur in virtually any cell type, whereas latent infection is predominantly found in ganglia of sensory neurons. Recently, blood cells of all types, bone marrow cells (Cantin *et al.*, 1994), regions of the CNS (brain stem, olfactory bulbs, cerebrum and cerebellum; Drummond *et al.*, 1994) and stromal cells of the cornea (Perng *et al.*, 1994) have been implicated as potential additional sites of latency. Only one region of the 152 kb HSV genome appears to be active during latent infection. It overlaps the *α0* gene but is oriented in the antisense direction and encodes a set of non-polyadenylated RNAs, the latency-associated transcripts (LATs). These transcripts are also produced during lytic infection and are found in the cytoplasm associated with ribosomes (Nicosia *et al.*, 1994). During latency, splicing occurs and LATs are restricted to the nuclei of the cells (Wagner *et al.*, 1988). LATs are negatively regulated by ICP0 and ICP4 and are not necessary for establishing latency (Farrell *et al.*, 1994; Rivera-Gonzalez *et al.*, 1994) since even in the absence of the LAT region, HSV can undergo latent infection. However, LAT promoters seem to be indispensable for reactivation of HSV (Bloom *et al.*, 1994; Fareed and Spivack, 1994). No LAT-specific protein has been identified, yet. The role of LATs may be to keep the *α0* region open for immediate entry into lytic replication after a reactivation signal (Latchman, 1994).

8.3 Engineering therapeutic herpes viruses

As mentioned earlier, two complementary strategies have been developed for the construction of infectious herpes viral vector (reviewed by Vos, 1995). In one, the entire virus is manipulated to insert the gene of interest into the viral genome, resulting in a recombinant 'helper virus-free' approach. The other, the 'helper-dependent' mini-virus approach, derives from the identification of the minimal viral *cis* elements required for the replication and packaging into infectious particles. *Table 8.1* summarizes recent developments based on these original strategies.

Table 8.1. Strategies for engineering infectious human herpes viral vectors

Virus	Strategy	Viral vector size (kb)	Insert packaging range[a]	Helper-dependent[b]	Latency-proficient[c]	Reference
EBV	Recombinant	150–170	Small	–	+	Lee *et al.* (1992); Wang *et al.* (1991)
	Hemi-EBV[d]	65–75	Medium	+	+	Kempkes *et al.* (1995); Robertson and Kieff (1995)
	Mini-EBV	15–25	Small	+	+	Hammerschmidt and Sugden (1989); Sun and Vos (1992)
	Mini-EBV/BAC[e]	20–30	Large	+	+	Banerjee and Vos (1996)[f]
HSV-1	Recombinant	120–150	Medium	–	+	Roizman and Jenkins (1985)
	Amplicon	10–15	Small	+	–	Stow *et al.* (1986); Vlazny *et al.* (1982)
	Mini-HSV-1	15–20	Small	+	Not tested	Wang and Vos (1996)
	Mini-HSV-1/BAC	20–25	Large	+	Not tested	Westphal and Vos (1996)[d]

[a] Small, ≤ 15 kb; medium, ≤ 50 kb; large, ≤ 150 kb.
[b] Helper-dependent for production of infectious particles.
[c] Latency-proficient for long-term maintenance after infection.
[d] Hemi-EBV, approx. half-size EBV vector.
[e] BAC, bacterial artificial chromosome.
[f] Unpublished data.

8.3.1 Therapeutic mini-herpes viruses

A simple and powerful strategy for the insertional expression of foreign genes into herpes viruses has resulted from the identification of the minimal viral elements required for their replication and packaging as infectious virions (reviewed by Vos, 1995). As with the packaging systems for retroviruses (reviewed by Morgan, 1995) and adeno-associated viruses (reviewed by Samulski, 1995), these virus-based vectors are defective in viral production and therefore require a helper virus to provide the missing viral proteins in *trans*. Two analogous systems have been developed for lymphotropic EBV and neurotropic HSV-1 viruses.

Mini-EBV. Ever since it was shown that a vector carrying the latent replication origin *ori*P and the viral nuclear antigen gene *EBNA1* was sufficient for plasmid replication in LCLs (Yates *et al.*, 1984), investigators have tried to develop EBV-based vectors for gene transfer (*Table 8.2*). Indeed, a number of genes have been cloned using plasmids carrying the EBV elements *ori*P and *EBNA1* (Legerski and Peterson, 1992; Strathdee *et al.*, 1992). In our laboratory, a novel EBV-based vector was shown to be capable of carrying inserts in the range of 300 kb (Sun *et al.*, 1994) when transfected into mammalian cells. Such a large cloning capacity of an EBV-based vector led us to develop an infectious B-lymphotropic mini-EBV for use in gene therapy. This EBV-based vector was built with minimal EBV sequence containing both latent and lytic origins of replication, *ori*P and *ori*Lyt, the TRs, and a drug resistance marker for selection. Infectious virions were generated and were shown not only to express a reporter gene such as *lac*Z in the infected B-lymphoblastoid cells (Sun and Vos, 1992) but also to correct phenotypically the defects in LCLs established from a Fanconi anaemia patient (Banerjee *et al.*, 1995) and a Lesch-Nyhan patient (Sun *et al.*, 1996). It was demonstrated that this infectious mini-EBV was able to carry inserts in the range of 140–160 kb and was also episomal in the infected B cells. The mini-EBV approach is dependent on the cell line HH-514 for providing lytic replication and packaging functions and hence on co-formation of non-transforming helper virus. Gains in the efficiency of selectively cloning large viral DNA in prokaryotes such as BAC/PAC vectors will ultimately translate into a packaging cell line defective in producing helper virus but supporting lytic replication of mini-EBV.

Mini-HSV-1. Experiments using HSV amplicon vectors (*Table 8.1*) have been performed to show the functionality of transgenes as well as to analyse conditions affecting promoter activity and duration of gene expression. Amplicon vectors with glucocorticoid-inducible expression of human growth hormone have been described (Mester *et al.*, 1995). With respect to potential target conditions, sequences encoding tyrosine hydroxylase (Parkinson's disease; Geller *et al.*, 1995), nerve growth factor (Alzheimer's disease; Geschwind *et al.*, 1994), brain-derived neurotrophic factor (maturation and function of auditory neurons; Geschwind *et al.*, 1996) and the growth-associated phosphoprotein B-50/GAP43 (Verhaagen *et al.*, 1994) have been cloned into HSV amplicon vectors under the control of the HSV IE4/5 or the cytomegalovirus (CMV) promoter. Expression and functionality of the transgenes were detected for up to 10 days post-infection. Lowenstein *et al.* (1994) showed that after co-infection of glial cells with helper virus and *lac*Z-containing amplicon particles, the intracellular localization of β-galactosidase was altered from cytoplasmatic to nuclear. In the same experiment, the subcellular localization of the plasma membrane-targeted tissue

Table 8.2. Therapeutic gene delivery with EBV in human disease cells

Hereditary syndrome	Target cells	Tranduced gene	Vector type	Maintenance mode	Helper virus	Reference
Fanconi's anaemia group C	B lymphocyte	Fanconi's anaemia group C (FACC cDNA)	Mini-EBV	Multimeric episomal	Non-transforming	Banerjee *et al.* (1995)
Insulin-dependent diabete mellitus	B lymphocyte	Transporter-associated antigen processing (Tap1/Tap2 cDNA)	Mini-EBV	Multimeric episomal	Transforming	Wang F. *et al.* (1995)
Lesch-Nyhan	B lymphocyte	Hypoxanthine phosphoribosyl transferase (HPRT cDNA)	Mini-EBV	Multimeric episomal	Non-transforming	Sun *et al.* (1996)

inhibitor of metalloproteinases (TIMP)-Thy1 was not altered during co-infection. A study of long-term promoter activity in cultured sensory neurons revealed that β-galactosidase expression driven by IE1, IE3 and IE4/5 promoters was detectable for up to 10 weeks after gene transfer, whereas the same promoters in the context of the whole HSV genome are known to be silent during latency (Smith *et al.*, 1995). Amplicon-based vectors contain only replication and packaging signals and so should therefore allow the cloning of large DNA fragments. Indeed, multiple copies of fragments up to 15 kb seem to be stably propagated, adding up to 150 kb, rather than larger pieces at a lower copy number (Vos, 1995). Earlier studies with HSV-1 mini-viruses have indicated the instability of large inserts cloned into HSV-1 vectors (Bear *et al.*, 1984; Kwong and Frenkel, 1984). The mechanism of potential instability of larger DNA fragments has not yet been determined, but very long stretches of human DNA are more likely to contain repeats which could favour the occurrence of homologous recombination during the amplification of HSV DNA in the lytic life cycle (Vos, 1995).

8.3.2 Therapeutic recombinant herpes viruses

Because of the large size of the herpes virus genomes, the simple and elegant molecular cloning techniques so instrumental to the development of small to medium-size viral vectors such as retroviruses (reviewed by Morgan, 1995), parvoviruses (reviewed by Samulski, 1995) and adenoviruses (reviewed by Stratford-Perricaudet and Perricaudet, 1995) could not be used. However, the pioneering works of Roizman and colleagues with HSV-1 (reviewed by Vos, 1995) and Moss, Paoletti and colleagues with vaccinia virus (reviewed by Cox *et al.*, 1995) have demonstrated that large human viral DNAs can be engineered following strategies based on the occurrence of homologous recombination in herpes viral and poxviral genomes during their passage in cultured cells. Provided that suitable selection schemes are available to identify and purify the viral recombinants, such as the complementation of potentially replication-defective viruses, virtually any region of the herpes viral genome can be targeted for homologous DNA insertion and/or deletion (reviewed by Vos, 1995). Similar strategies have been derived for the engineering of infectious recombinants of the neurotropic HSV-1 and HSV-2 viruses, and of the lymphotropic EBV for transducing marker genes as well as more recently therapeutic rodent and human genes (*Table 8.1*).

EBV recombinants. As an alternative to the mini-EBV approach (Section 8.3.1), a recombinant strategy is being developed to generate EBV-based gene therapy vectors relying on the large insert cloning capacity of EBV (*Table 8.1*). This was initially based on the observation that targeted gene disruption in EBV is feasible (Lee *et al.*, 1992; Marchini *et al.*, 1992). In this method, the 172 kb viral genome was selectively deleted to build a B-lymphocyte transforming vector. This approach has demonstrated that non-coding exons and introns of *EBNA* transcripts and most of the other lytic genes between *EBNA1* and *LMP1* can be deleted to yield a transforming 64 kb vector which can carry inserts in the range of 80–100 kb (Kempkes *et al.*, 1995; Robertson and Kieff, 1995). So far, no therapeutic gene delivery has yet been reported using recombinant EBV.

HSV-1 recombinants. Since not all of the approximately 70 genes of HSV are necessary for growth in cell culture, the non-essential ones could be replaced by foreign genes in recombinant vectors (*Table 8.1*). Up to 15 kb of HSV DNA have been replaced by different genes (reviewed in Vos, 1995). In a series of experiments using recombinant HSV vectors, investigators have been able to demonstrate that non-dividing cells other than neurons could be rapidly and efficiently infected followed by release of the gene product (e.g. human growth hormone) into the medium (of transient human liver cell cultures) (Fong *et al.*, 1995). The gene product was also shown to be functional, as exemplified by the case of murine interferon α_1, which prevented HSV-1 replication and vesicular stomatitis virus (VSV) superinfection in mouse L cells (Mester *et al.*, 1995), and by the case of the catalytic subunit of the cAMP-dependent protein kinase A, which supported survival of rat sympathetic neurons in the absence of nerve growth factor (Buckmaster and Tolkovsky, 1994). Wang, M.J. *et al.* (1995) cloned mouse nerve growth factor cDNA, under the control of a LAT promoter modified to include a Rous sarcoma virus (RSV) enhancer element, into an *α4/VP16*-deficient HSV vector and obtained growth factor expression for at least 7 days after infection and differentiation of the infected mouse PC12 cells. The double mutant showed less cytotoxicity and the RSV enhancer contributed to prolonged gene expression. Roemer *et al.* (1995) presented evidence that the context of the HSV genome might reduce the neuron specificity of the rat enolase promoter. Since only LATs remain active during latent infection in neurons, the LAT promoters were extensively studied. Removal of the ICP4 binding site resulted in increased mRNA expression of a *tk* gene driven by a mutated LAT promoter (Rivera-Gonzalez *et al.*, 1994). Although the LAT promoter region is also active during lytic infection, there appear to be differences in the localization of sequences directing chloramphenicol acetyltransferase gene expression in fibroblast and neuroblastoma cell lines (Morrow and Rixon, 1994).

8.4 Therapeutic gene delivery in animals

Only a limited number of studies have been carried out to test the feasibility of human herpes viruses as viral vectors for gene delivery into whole organisms. Although most of the *in vivo* work has been concentrated on developing HSV-1 as a neurotropic virus, we also outline potential *in vivo* strategies for testing EBV-mediated gene therapy (*Table 8.3*).

8.4.1 EBV

No therapeutic gene delivery experiments have as yet been reported using engineered infectious EBV. Although EBV (or EBV-transformed cells) injected into several species of New World primates can lead to productive and persistent infection (Frank *et al.*, 1976; Ishida and Yamamoto, 1987), the lack of therapeutic data with this lymphotropic herpes virus may be due to its intrinsic inability to infect species other than primates. However, experiments on rodents with EBV have been conducted in two different ways. In one, lymphoid lines or epithelial cells containing EBV were tested for oncogenicity by transplantation. There are reports that epithelial cells containing latent EBV form tumours when transplanted into nude mice. Alternatively, with the availability of a human lymphoid, severe combined immunodeficiency (SCID) mouse haematopoietic model, a number of experiments have been performed to evaluate the role of EBV in post-transplantation lymphoproliferative diseases (Mosier *et al.*, 1992). Experiments with SCID mice are generally performed in two ways: either the mouse is grafted with human bone marrow cells resulting in the Hu-SCID haematopoietic model; or CD19/CD21 B cells collected from human blood are injected intraperitoneally. There have been conflicting reports regarding peripheral blood from EBV-seropositive patients causing tumours in Hu-SCID mice but most LCLs transformed by EBV do form tumours in these systems (Mosier *et al.*, 1992).

Table 8.3. Therapeutic gene delivery with HSV-1 in model diseases

Model disease	Target organ	Correcting gene	Vector type	Experimental time frame[a]	Reference
Haemophillia B	Mouse liver	Canine factor IX	Recombinant	Long-term	Miyanohara et al. (1992)
Neuron loss (hypoxaemia, hypoglycaemia, seizures)	Rat hippocampus	Rat brain glucose transporter	Amplicon	Short-term	Ho et al. (1993)
Parkinson's disease	Rat striatum (CNS)	Human tyrosine hydroxylase	Amplicon	Long-term	During et al. (1994)
Cancer	Rat brain tumour (gliosarcoma)	HSV-1-tk	Recombinant	Long-term	Boviatsis et al. (1994)
Neuron loss (kainic acid-induced seizure)	Rat hippocampus (dentate, CA3 cell field)	Rat brain glucose transporter	Amplicon	Short-term	Lawrence et al. (1995)

[a] Short, a few days; medium, 2–4 weeks; long, ≥ 4 weeks.

8.4.2 HSV-1

Because of the biology of HSV-1, HSV-based vectors may be suitable tools for gene delivery into non-dividing cells as a potential therapy for inherited diseases as well as in targeting the rapidly dividing cells of malignant tumours, especially in the brain (*Table 8.3*). Therapy of disease requires an ability to reach the defective cells and to maintain long-term expression of the corrective gene(s). That delivery of HSV particles is not always dependent on invasive methods was elegantly demonstrated by Davar *et al.* (1994) who inoculated recombinant HSV vectors on to the cornea and snout of mice and detected HSV-derived material in the central trigeminal ganglia of the animals. These experiments were based on the fact that HSV particles are transported from the periphery of sensory neurons into their ganglia by fast axonal tansport.

Miyanohara *et al.* (1992) injected recombinant HSV particles carrying a canine factor IX cDNA into the portal vein of mice and were able to detect factor IX expression in the liver for up to two months. Ho *et al.* (1993) directed HSV amplicons containing the rat brain glucose transporter cDNA into the rat hippocampus by stereotactic injection but observed expression for only a few days. Lawrence *et al.* (1995) also obtained only short-term expression. However, their amplicons containing the rat glucose transporter cDNA under the control of the hCMV or the $\alpha 4$ HSV promoter enhanced the glucose uptake into the hippocampus cells *in vitro* and *in vivo* and were able to reduce neuron loss even after the onset of kainic-acid-induced seizures in rats. By contrast, During *et al.* (1994) observed behavioural recovery in Parkinsonian rats for up to 1 year after stereotactic administration of tyrosine hydroxylase cDNA cloned into the amplicon vector pHSV.

For stable long-term expression, the choice of the correct promoter seems to be very important. The neuron-specific enolase promoter conferred gene activity for only 2 weeks (Davar *et al.*, 1994), and the promoters of the genes for rabbit β-globin and human α_2-globin required the presence of HSV ICP4 for activity (Smiley and Duncan, 1992). Lokensgard *et al.* (1994) tested a series of recombinant HSV vectors with different viral and cellular promoters driving a marker gene. The addition of transcription factor binding sites from the upstream LAT region to the metallothionein promoter or the Moloney murine leukemia virus long terminal repeat (LTR) resulted in increased activity during acute infection of the murine dorsal root ganglia, whereas after establishing latency, only LAT–LTR remained active. A more thorough analysis of the LAT promoter region revealed a second element (LAP2; Goins *et al.*, 1994) which resembled in its G+C content the promoters of housekeeping genes and was able to permit long-term expression for up to 300 days of a LAP2–lacZ construct cloned into gC of an HSV recombinant virus applied to the mouse cornea.

For short-term expression in HSV vectors, as, for example, in cancer treatment, promoter choice seems to be less critical. Towards a potential use of HSV in the therapy of human brain tumours, Boviatsis *et al.* (1994) and Kramm *et al.* (1996) tested a recombinant HSV vector (hrR3) which contained a *lacZ* marker gene and an intact HSV-*tk* gene for simultaneous ganciclovir application. Because of ribonucleotide reductase deficiency, the vector does not appear to replicate in non-dividing neurons but replicates preferentially in dividing cells where the enzyme can be provided by the host cell. Co-treatment of rats bearing brain tumours with hrR3 and ganciclovir resulted in long-term survival of 48% of the animals whereas among the rats treated with hrR3 alone only 20% survived longer than 5 months (Boviatsis *et al.*, 1994). hrR3 was also able to target disseminated tumour cells in the brain of adult rats. In the study performed by Kramm *et al.* (1996), no ganciclovir was given during inoculation with hrR3 and significant cytotoxicity was observed in the brain of the animals as leptomeningeal inflammation.

Although considerable progress has been made in designing HSV vectors to overcome safety concerns and to reduce toxicity, much still remains to be done. Expanding the cloning capacity of amplicon vectors and extending long-term gene expression may serve as examples. The long-term approach might be useful for non-dividing cells since not every host cell receives HSV molecules after cell division. The short-term lytic approach to kill tumour cells appears promising, as shown by the experiments of Boviatsis *et al.* (1994).

8.5 Safety issues

8.5.1 Potential pathogenesis of recombinant and helper virus

EBV. There is a wide spectrum of *sequelae* to infection by EBV. Most people are infected during childhood and do not experience any symptoms; viral replication is apparently controlled by humoral and T-cell-mediated immune responses. In previously uninfected young adults, upon EBV infection a disease termed infectious mononucleosis typically develops (Liebowitz and Kieff, 1993). This disease is characterized by sore throat, fever, and general lymphadenopathy. Large, morphologically atypical T cells are abundant in the peripheral blood of infectious mononucleosis patients. These cells activate cytotoxic CD8$^+$ T lymphocytes (CTLs) with specificity for EBV-encoded antigens. Previously infected, healthy individuals harbour the virus for the rest of their lives in latently infected B cells. It is estimated that one of every million B cells in a previously EBV-infected individual is latently infected (Miyashita *et al.*, 1995). However, in immunocompromised individuals, there is a strong correlation for viral infection and certain kinds of malignancy.

HSV-1. Despite the advantages of HSV as a basis for gene therapy vectors, such as potentially large cloning capacity, broad host range, high infectivity and targeting of non-dividing cells (neurons), there are serious safety problems to overcome. The most severe complication is a potentially lethal encephalitis caused by widespread lytic CNS infection. Long-lasting immune response and inflammation have been observed around the injection site in the brain of laboratory animals (Wood *et al.*, 1994). Toxic reactions were attributed to the presence of HSV α genes in recombinant vectors (Johnson *et al.*, 1992) and to *VP16* expression (Johnson and Friedmann, 1994) but were also observed after administration of lysates prepared from helper cell lines that did not contain any virus (Ho *et al.*, 1995). Therefore, the purification of infectious particles deserves careful attention since the toxicity in this case appears to be caused by components of the helper cell, not of the virus. Recombinant viruses carrying double mutants in *α4* and *VP16* were completely replication-defective and much less cytotoxic than single mutants in *α4* but also showed reduced expression of the transgene (Johnson and Friedmann, 1994). Finally, recombinant vectors might have the potential to recombine with latent HSV in the host organism, thereby creating a fully replication-competent HSV vector which might be spread during virus reactivation and start of a lytic replication cycle.

Amplicon vectors do not contain α genes but are currently delivered as a mixture with helper virus. If replication-competent, the latter could start a lytic infection in the host. Also it has been shown that amplicon and helper virus can infect the very same host cell (Lowenstein *et al.*, 1994), thereby providing the potential for recombination. In addition, integration of retroviruses into the HSV genome has been described with the potential that the recombination product acquires the host range of HSV (Isford *et al.*, 1994).

These safety problems can be overcome by using amplicon vectors in combination with replication-defective helper virus mutants carrying deletions of *α4*, for instance. They need to be propagated in cell lines stably transfected with *α4* to provide its function in *trans* (Paterson and Everett, 1990). A disadvantage of this system still to be overcome is that the helper:amplicon ratio varies from one preparation to the next so that until now it has been difficult to obtain reproducible titres of the amplicon (Efstathiou and Minson, 1995).

8.5.2 Immune escape

EBV. There is compelling evidence that T cell-mediated immunity is required for the control of EBV infections and, in particular, for the killing of EBV infected B cells. (i) Individuals with deficiencies in T cell-mediated immunity often have uncontrolled, widely disseminated, and perhaps lethal, acute EBV infection. (ii) EBV-infected B cells isolated from infectious mononucleosis patients can be propagated *in vitro* indefinitely, but only if the patient's T cells are completely removed or inactivated by drugs such as cyclosporin A (Tosato *et al.*, 1979). In fact, immortalization of normal peripheral blood B cells by *in vitro* infection with EBV is possible only if the donor's T cells are removed or inactivated. (iii) CTLs specific for EBV-encoded antigens are present in both acutely infected and in completely recovered infectious mononucleosis patients. Cloned CTL lines have been established *in vitro* that specifically lyse EBV-infected B cells, and these CTLs most often recognize peptide fragments of EBNA proteins in association with class 1 major histocompatibility complex (MHC) molecules (Khanna *et al.*, 1992). It is possible that EBV-specific T cells are required *in vivo* to limit the proliferation of infected B cells as well as to kill potentially immortalized clones of latently infected B cells. A loss of normal T cell-mediated immunity [e.g. in malaria-infected populations in Africa, in immunocompromised AIDS patients, and in allograft recipients receiving immunosuppressive drugs] allows latently infected B cells to progress towards malignant transformation (Miller, 1990).

Recently, the viral nuclear antigen EBNA1 expressed during latency was shown to play a pivotal role in evading the host immune surveillance (Levitskaya *et al.*, 1995; *Figure 8.3*). It was shown that the presence of a glycine–alanine repeat within EBNA1 prevents the presentation by MHC class I proteins of antigenic peptides derived from non-repetitive regions of the protein. The mechanism of this unique EBNA1-mediated effect is still to be established. However, based on these findings, one may speculate as to the co-existence of EBV in human cells. Infection of B cells by EBV activates production of B-cell growth factors and proliferation of EBV-infected B cells. Primary immune response results in rejection of virally infected cells. A small fraction of the expanded population of EBV-carrying cells may downregulate EBNA2 and EBNA3 production and switch to EBNA1 instead. Only the non-immunogenic EBNA1 protein would be expressed, thereby safeguarding the maintenance of viral episomes.

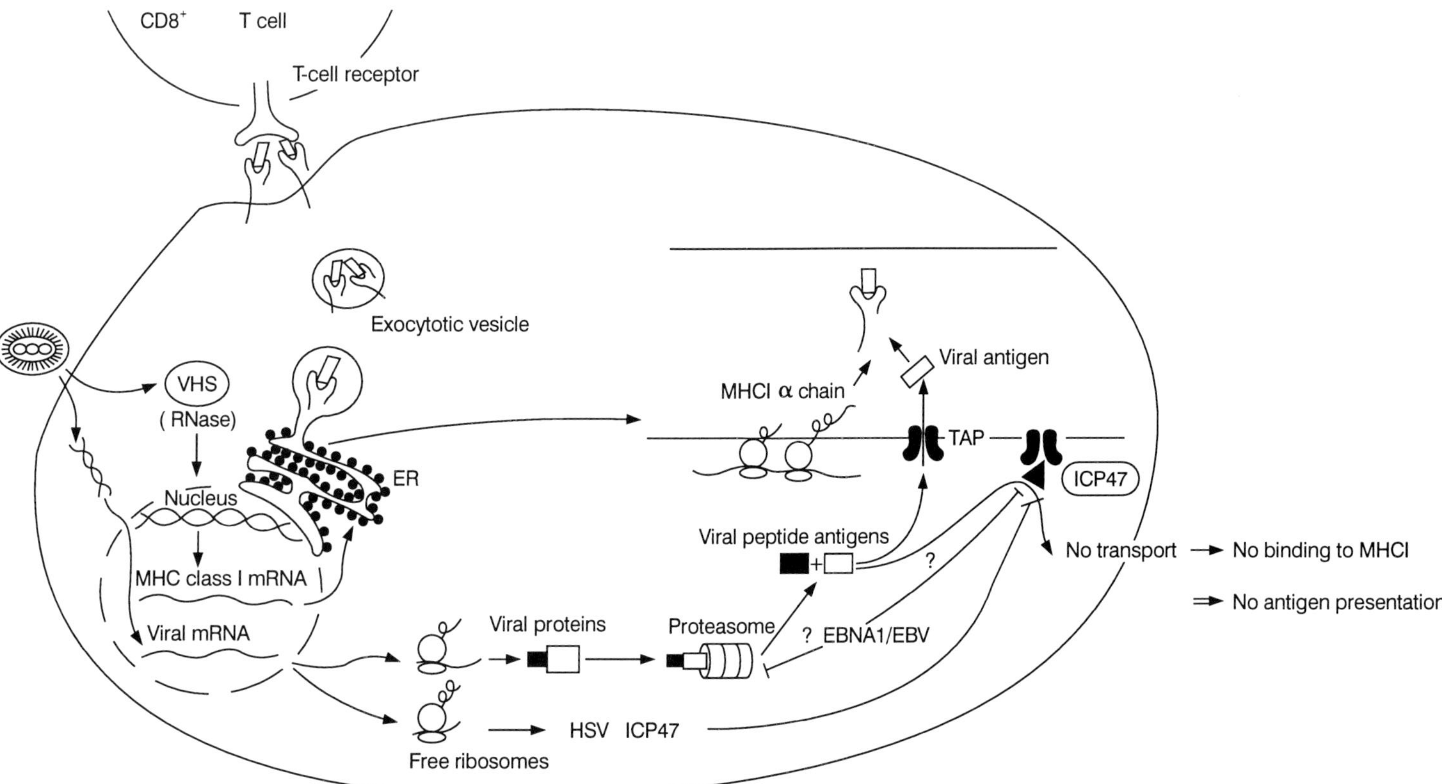

Figure 8.3. Schematic strategy for the immune escape by EBV and HSV-1. The HSV protein ICP47 binds to the transporter TAP, thereby preventing antigen transport into the endoplasmic reticulum (ER) and the binding of these antigenic peptides to MHC I α chains: no antigens are therefore presented. The virion host shut-off protein (VHS) causes non-specific RNA degradation leading to a net excess of viral RNA. The glycine–alanine repeat within EBNA1 of the EBV genome prevents the presentation by MHC class I proteins of antigenic peptide. Redrawn and updated from Lewin (1995).

HSV-1. Normally, cells expressing foreign proteins are eliminated by CTLs. This process involves processing of those proteins into peptides, their transport into the endoplasmic reticulum of the cells where they bind to MHC class I molecules, and presentation of these complexes on the outer cell membrane to CTLs. Viruses have developed various strategies to counteract immune surveillance. In the case of HSV, several strategies are possible (*Figure 8.3*). The virion host shut-off protein is present very early in infection. Its non-specific RNase activity degrades all kinds of RNAs including virus-specific ones and MHC class I mRNAs. Because of their higher transcription rate, HSV-specific RNAs are less affected. One obvious way to escape immune surveillance is not to express viral proteins. This happens during HSV latency where only LAT RNAs are found. The most important way has been discovered recently (*Figure 8.3*; Früh *et al.*, 1995; Hill *et al.*, 1995): ICP47, the product of an immediate– early gene, binds to the transporter associated with antigen presentation (TAP), thereby preventing the transport of viral antigens into the endoplasmic reticulum. Consequently, no binding of HSV peptides to MHC class I proteins and no antigen presentation to CTLs can occur. Including α47 in a gene therapy vector may therefore bypass the problem of unwanted elimination of cells expressing therapeutic proteins.

8.6 Future directions

8.6.1 Potential diseases for herpes viral-based gene therapy

The development of herpes viruses as potential vectors for gene therapy is at an early stage. With the exception of CMV, the human herpes viral family can be divided into two broad subfamilies: the neurotropic HSV-1, HSV-2 and VZV; and the lymphotropic EBV, HSV-6, HSV-7 and HSV-8. Of the known eight human herpes viruses, only HSV-1 has received some attention in experiments aiming at gene therapy (reviewed by Breakefield and DeLuca, 1991; Geller, 1995; Vos, 1995). Hence, the following discussion is by necessity rather speculative. Nonetheless, it may help to define potential direction for the future of herpes viral-based applications in gene therapy.

EBV as a prototype lymphotropic therapeutic vector for blood diseases/factors and genetic-based vaccination. Although human lymphotropic herpes viruses have not yet attracted much interest from the community of gene therapists, the following discussion illustrates their potential for development as therapeutic genetic vectors. The paradigm of human lymphotropic herpes viruses is represented by EBV, whose tropism is restricted to B lymphocytes (*Figure 8.4*) and some other cell types, including those of epithelial origin (Miller, 1990). The B lymphotropism of EBV suggests that it may be particularly well adapted for the treatment of diseases involving circulating and/or diffusible gene products (*Table 8.4*). Thus, inherited recessive monogenic disorders of serum proteins such as blood clotting factors (e.g. haemophilia), hormones such as insulin (diabetes) or enzymes such as glucocerebrosidase (Gaucher disease), α_1-antitrypsin (inherited emphysema) and β-glucuronidase (Sly syndrome) may be suitable candidates for EBV-based gene therapy. In addition, the selective EBV tropism for human B lymphocytes may allow the development of strategies for genetic-based vaccination. Hence, EBV may also be useful for treatment of acquired diseases such as cancer and infectious diseases. However, several potential difficulties will have to be overcome before EBV can be safely and effectively used in human clinical trials. Human B lymphocytes, the primary target for EBV-based gene therapy, are mostly short-lived

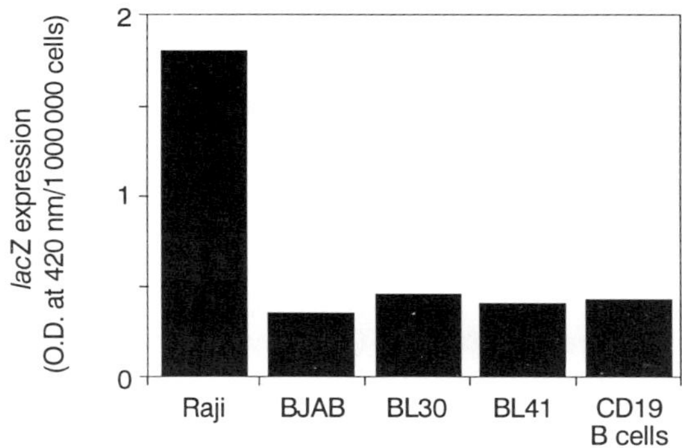

Figure 8.4. Functional gene delivery into primary human B cells by an infectious mini-EBV. Primary human CD19-positive B lymphocytes collected from peripheral blood lymphocytes (PBLs) of normal donors were incubated with aliquots of mini-EBV stocks transducing a constitutively expressed reporter *lacZ* gene. As control, the infectivity levels were compared with an EBV-positive cell line (Raji) and three EBV-negative Burkitt's lymphoma B-cell lines (BJAB, BL30 and BL41). Cells were analysed for *lacZ* expression 48 h post-infection using the enzymatic assay as described previously (Banerjee *et al.*, 1995; Sun and Vos, 1992; Sun *et al.*, 1996).

cells. In addition, as discussed above, EBV has been associated with specific types of human cancer (i.e. Burkitt's lymphoma and nasopharyngeal carcinoma; Epstein and Achong, 1986; Klein, 1989); such long-term potential oncogenicity of EBV, particularly in immuno-compromised individuals, will have to be overcome before this lymphotropic herpes virus can be employed for therapeutic purposes. On the other hand, other human lymphotropic herpes viruses (HSV-6, HSV-7 and possibly HSV-8) and non-human primate lymphotropic herpes viruses (saimiri, papio, *Ateles* strain 810, *aotus* type 2) may become prime vectors to target specific diseases, especially if characterized by a wider tropism and lack of oncogenicity in comparison with EBV.

HSV-1 as a prototype neurotropic therapeutic vector for neurological deficiencies and cancer. Most of the current work with the neurotropic subfamily has focused on HSV-1 as a genetic vehicle to treat neurodegenerative diseases affecting the peripheral and central nervous systems. The organizational complexity and difficulty in accessing most brain cells make viruses such as HSV-1 very attractive to direct 'therapeutic' genes inside this fundamental organ (Friedmann, 1989; Mulligan, 1993; Vos, 1995). The capacity of HSV-1 to infect quiescent cells such as those of the brain or liver represents an advantage over viruses requiring active cell growth for infection. Hence, conditions such as Sly syndrome, Lesch-Nyhan syndrome, Huntington's disease, Alzheimer's disease and Parkinson's disease may be prime candidates for HSV-1 vectors expressing specific human genes. Indeed, the successful HSV-1-mediated transduction of various genes in the rodent CNS (*Table 8.3*) supports the long-term development of genetic-based treatment of Lesch-Nyhan syndrome, Sly syndrome and Alzheimer's disease. However, such enthusiasm should be tempered by several complicating factors, such as disorders affecting neurons distributed diffusely in the CNS, multi-factorial/-genic CNS diseases, the lack of knowledge of normal CNS function and possible pathogenic side effects of HSV-1 infection on CNS function. In addition, as discussed above in the context of the long-term potential oncogenicity of EBV, the short-term cytopathic effects of HSV-1, particu-

Table 8.4. Human diseases as potential targets for gene therapy mediated by infectious herpes virus vectors

Virus	Disease (examples)	Target cells	Latency[a]	Vector[b]
EBV	Blood disorders (haemophilia, insulin-deficiency)	Circulating B lymphocytes	Yes	Mini-EBV and/or recombinant
	Genetic-based vaccination (infectious agents, cancer metastasis)	Circulating B lymphocytes	No	Mini-EBV
HSV-1	Neurological disorders (Alzheimer's, Parkinson's)	Neurons of brain (CNS)	Yes	Recombinant
	Cancer (brain tumour)	Tumour mass	No	Recombinant and/or mini-HSV-1

[a] Long-term maintenance expected for beneficial outcome.
[b] Type of infectious vector predicted to be most effective.

larly in neurologically compromised individuals, will have to be overcome, before this neurotropic herpes virus can be adopted for therapeutic purposes.

On the other hand, the wide tissue tropism of HSV-1 also suggests its potential use for treating non-CNS disorders of inherited and acquired origins (Wang and Vos, 1996), as illustrated by the successful transfer and expression of genes into the liver via either direct injection or through the portal vein (Miyanohara *et al.*, 1992). Since the liver plays a central role in human metabolism and in the expression of many genetic diseases, such studies stress the potential clinical usefulness of the combination of liver/HSV-1 infection for gene therapy of a variety of inherited and acquired disorders, such as familial hypercholesterolaemia, inherited emphysema, phenylketonuria and hyperammonaemia. One possible limitation is the apparent inability of HSV-1 to establish latency in liver and other non-CNS organs which may limit its applicability to short-term and/or repetitive treatments.

In addition, HSV-1 may also become a useful vector for the treatment of acquired disorders such as cancer, as illustrated by the successful use of a *tk*-deleted HSV-1 mutant attenuated for neurovirulence as an oncolytic 'suicide' genetic vector against experimentally induced rodent gliomas (Martuza *et al.*, 1991). More generally, other neurotropic herpes viruses, including non-human ones such as the cercopithicine primate and the bovine herpes viruses (Roizman, 1990), will probably be added to the list of vectors being evaluated for gene therapy, especially for disorders affecting non-growing cell tissues (Vos, 1995).

8.6.2 Other potential herpes viral vectors for gene therapy

Several studies have reported the successful engineering of infectious recombinants of herpes viral strains from various non-human species. Most of the work has concentrated on the primate herpes virus saimiri (HVS or saimirine herpes virus, also called squirrel monkey virus). In early experiments, Desrosiers *et al.* (1985) were able to insert a constitutively expressed gene for bovine growth hormone in HVS following the homologous gene targeting strategy (reviewed by Roizman and Jenkins, 1985). New World primates experimentally infected with this HVS construct produced circulating bovine growth hormone for up to 9 weeks post infection, whereas stable expression in infected tissue cultures was observed for up to 6 months (the longest time tested). In a series of *in vitro* studies by Fleckenstein and colleagues, infectious HVS recombinants expressing dominant resistance markers for stable selection in mammalian cells were generated which could persistently infect established human cell lines derived from various tissues, including T and B lymphocytes, fibroblasts, epithelial, myeloid and erythroid cells (Alt *et al.*, 1991; Grassmann and Fleckenstein, 1989; Simmer *et al.*, 1991). The potential of recombinant HVS as a general viral gene transfer vector has been documented with the development of a strategy to identify candidate oncogenes by the conversion of the original non-transforming HVS into a transforming virus, as shown with the X region of the human T-cell leukaemia retrovirus type I (Grassmann *et al.*, 1989). Such apparent lack of transforming ability of HVS in human cells combined with its wide tissue cell tropism confers several distinct qualities upon this simian herpes virus as a possible vector for future human gene therapy. Through an elegant, *in vitro*, site-specific recombination experiment with the pseudorabies virus (PRV, suid herpes virus 1), Enquist and colleagues demonstrated the feasibility of engineering this swine pathogen for gene expression purposes (Sauer *et al.*, 1987). However, no follow-up study has yet appeared on this interesting herpes viral system.

8.6.3 Delivery of mammalian artificial chromosomes

Engineering therapeutic human artificial chromosomes will require the development of strategies to deliver large, functional, self-replicating, extrachromosomal DNA into target cells. Members of the human herpes viral family are among the largest known double-stranded episomal DNA viruses. In contrast to bulky recombinant viral vectors, small mini-herpes viral vectors are characterized by a large cloning potential. Hence, herpes viruses such as EBV and HSV-1 may become paradigm vectors for delivering large inserts (Sun *et al.*, 1996; Vos, 1995). *Figure 8.5* illustrates the two potential strategies for using mini-herpes viral vectors to package non-viral DNA into infectious virions. The chosen strategy depends on the size of the non-viral cloned DNA to be packaged. Mini-herpes viruses carrying small inserts are packaged as linear multimeric forms (Sun and Vos, 1992; Vos 1995). The number of multimers in virions is dictated by the relative size of the monomeric vector and total packageable DNA following the equation:

$$\text{Number of multimeric vectors per virion} = \frac{\text{(Herpes viral genome size)}}{\text{(Mini-herpes vector size)}}$$

For example, a mini-EBV of 20 kb is expected to be packaged as 8- and 9-mers in virions normally carrying the 172 kb EBV genome. The measurement of the upper and lower limits of sizes for packaging into herpes virions can be estimated experimentally. Mini-herpes viruses derived from HSV-1 and EBV are packaged as linear multimeric forms. Using pulsed-field gel electrophoresis and Southern blotting of vector DNA extracted from the EBV virions produced from human lymphoblastoid helper

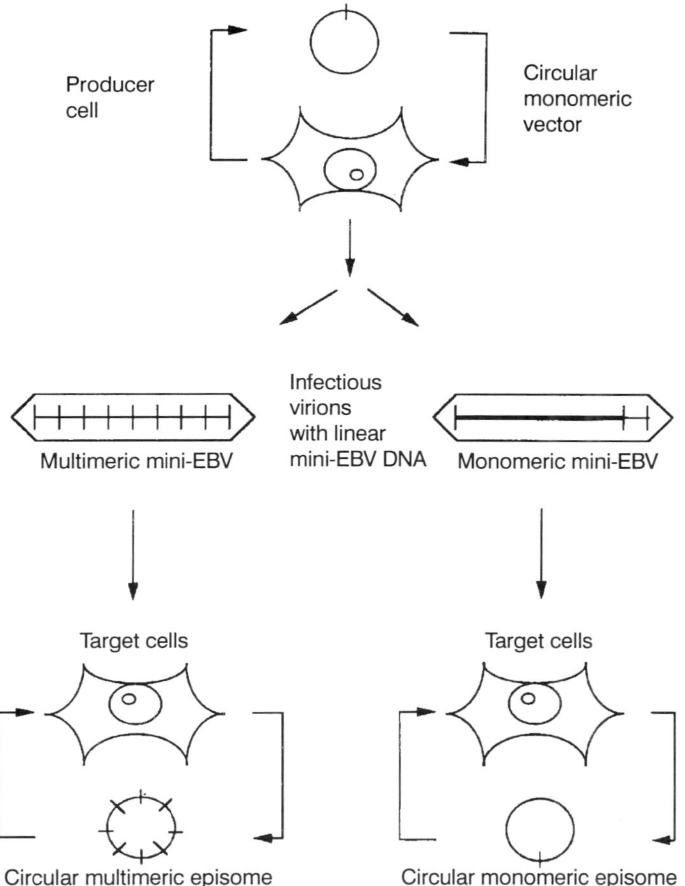

Figure 8.5. Schematic life cycle of infectious engineered mini-EBV. During the lytic phase, the mini-EBV DNA replicates in the presence of helper virus via a rolling circle mechanism to generate head-to-tail concatemeric molecules for small inserts and a single large molecule for large inserts, which are then cleaved and packaged into infectious EBV virions. After infection of susceptible human cells, the incoming mini-EBV linear DNA recircularizes and episomal replication occurs. The left and right panels illustrate the packaging of small multimeric and large monomeric inserts into EBV virions, respectively.

cells latently infected with EBV, Sun and Vos (1992) measured a packaging size range of 150–200 kb for the mini-EBV. Such a simple assay to determine the minimal and maximal size of DNA packaged into EBV virions should be applicable to other herpes viruses, provided that they replicate via a rolling circle mechanism.

In theory, the largest herpes mini-viral vector that is expected to be packaged into virions will be in a monomeric form. Such a vector would correspond roughly to the size of the whole viral genome (e.g. 150 kb for HSV-1 and 172 kb for EBV). Hence, the cloning size range of these viruses is potentially 10–20 times that of current viral systems used in gene therapy clinical trials (Sun *et al.*, 1996). With the available 10–20 kb mini-HSV-1 and mini-EBV mini-viral vectors, up to approximately 140–150 kb of foreign DNA could be inserted, which would represent an unusually large cloning capacity for an infectious human viral vector. Although no cloned DNA fragment larger than 15 kb has yet been propagated into infectious EBV or HSV-1, very large inserts have been stably propagated on EBV-based episomal vectors. For example, up to 330 kb of random human DNA segments have been stably cloned in an *ori*P-based episome in human cells (Sun *et al.*, 1994). The human DNA inserts in individual clones were derived from different regions of the human genome and appeared to represent faithful copies of the source DNA. Hence, such HAEC systems allowed the stable propagation on a latent mini-EBV of DNA inserts in the size range of the EBV genome. This strategy may set the stage for the future delivery of human artificial chromosomes by infectious herpes viruses. Such systems may also allow the transfer of large therapeutic human genes, such as those encoding factor VIII, cystic fibrosis transmembrane conductance regulator and dystrophin – either as cDNA or as genomic clones including regulatory elements. Alternatively, the large packaging capacity of herpes viruses may allow the co-delivery of multiple genes, which may be particularly useful in developing gene therapy protocols for common polygenic diseases such as cancer (Sun *et al.*, 1996).

In summary, bearing in mind that herpes viruses from human and non-human origin appear to share basic characteristics such as large genomic size and episomal latency, one could envisage a series of herpes viral systems for the transduction of large, therapeutic, self-replicating, functional extrachromosomal elements (i.e. human artificial chromosomes) in various human cell types. In view of the wide spectrum of tropism displayed by the members of the large mammalian herpes viral family, such a possibility could offer a general strategy for performing successful gene therapy on a variety of human diseases. Hence, with the help of powerful ways of engineering various herpes viruses, one may predict the rapid development of this large and pleiotropic family of DNA viruses as 'giants' of viral-based gene therapy.

Acknowledgements

The authors wish to thank members of the laboratory for helpful suggestions, in particular Z. Kelleher and T.-Q. Sun for many useful discussions and K. Neal for excellent secretarial assistance. We are grateful to many colleagues for communicating unpublished data, in particular X. Breakefield, P. Johnson, S. Kenney, E. Kieff, J. Pagano, N. Raab-Traub and B. Sugden. Research from the principal investigator's laboratory was supported by grants from the Life and Health Insurance Medical Research Fund, the Elsa U. Pardee Foundation, the Department of Energy (Human Genome Initiative #DEFG0591ER611350), and the National Institutes of Health (1-PO1-HL51818–01). S.B. was recipient of a Fanconi's Anemia Research Foundation postdoctoral fellowship. J.-M.H.V. was supported by a Basil O'Connor Starter Scholar Research Award from the March of Dimes Birth Defects Foundation (F494–0171) and an American Cancer Society Junior Faculty Award (JFRA-330).

References

Alt M, Fleckenstein B, Grassman R. (1991) A pair of selectable herpesvirus vectors for simultaneous gene expression in human lymphoid cells. *Gene* **102**: 265–269.

Banerjee S, Livanos E, Vos J-MH. (1995) Therapeutic gene delivery in human B-lymphocytes with engineered Epstein-Barr virus. *Nature Med.* **1**: 1303–1308.

Bear SE, Colberg-Poley AM, Court DL, Carter BJ, Enquist LW. (1984) Analysis of two potential shuttle vectors containing herpes simplex virus defective DNA. *J. Mol. Appl. Genet.* **2**: 471–484.

Bloom DC, Devi-Rao GB, Hill JM, Stevens JG, Wagner EK. (1994) Molecular analysis of herpes simplex virus type I during epinephrine-induced reactivation of latently infected rabbits *in vivo*. *J. Virol.* **68**: 1283–1292.

Boviatsis EJ, Park JS, Sena-Esteves M, Kramm CM, Chase M, Efird JT, Wei MX, Breakefield XO, Chiocca EA. (1994) Long-term survival of rats harboring brain neoplasms treated with ganciclovir and a herpes simplex virus vector that retains an intact thymidine kinase gene. *Cancer Res.* **54**: 5745–5751.

Breakefield XO, DeLuca NA. (1991) Herpes simplex virus for gene delivery to neurons. *New Biol.* **3**: 203–218.

Buckmaster EA, Tolkovsky AM. (1994) Expression of the cyclic AMP-dependent protein kinase A (PKA) catalytic subunit from a herpes simplex virus vector extends the survival of rat sympathetic neurons in the absence of NGF. *Eur. J. Neurosci.* **6**: 1316–1327.

Cantin E, Chen J, Gaidulis L, Valo Z, McLaughlin-Taylor, E. (1994) Detection of herpes simplex virus DNA in human blood and bone marrow cells. *J. Med. Virol.* **42**: 279–286.

Chang Y, Cesarman E, Pessin MS, Lee F, Culpepper J, Knowles DM, Moore PS. (1994) Identification of herpesvirus-like DNA sequences in AIDS associated Kaposi's sarcoma. *Science* **266**: 1865–1869.

Chen F, Zou JZ, di Rienzo L. (1995) A subpopulation of latently infected normal B-cells resembles Burkitt lymphoma (BL) in expressing EBNA1 but not EBNA2 or LMP1. *J. Virol.* **69**: 3752–3756.

Cooper NR, Moore MD, Nemerow GR. (1988) Immunobiology of CR2, the B-lymphocyte receptor for Epstein-Barr Virus and the C3d complement fragment. *Annu. Rev. Immunol.* **6**: 85–113.

Cox WI, Gettig RR, Paoletti E. (1995) Poxviruses as genetic vectors. In: *Viruses in Human Gene Therapy* (ed. J-MH Vos). Carolina Academic Press and Chapman & Hall, Durham, NC and London, UK, pp. 141–178.

Davar G, Kramer MF, Garber D, Roca AL, Andersen JK, Bebrin W, Coen DM, Kosz-Vnenchak M, Knipe DM, Breakefield XO. (1994) Comparative efficacy of gene delivery to mouse sensory neurons using herpes virus vectors. *J. Comp. Neurol.* **339**: 3–11.

Desrosiers RC, Kamine J, Bakker A, Silva D, Woychick RP, Sakai DD, Rottman FM. (1985) Synthesis of bovine growth hormone in primates by using a herpesvirus vector. *Mol. Cell. Biol.* **5**: 2796–2803.

Drummond CW, Eglin RP, Esiri MM. (1994) Herpes simplex virus encephalitis in a mouse model: PCR evidence for CNS latency following acute infection. *J. Neurol. Sci.* **172**: 159–163.

During MJ, Naegele JR, O'Malley KL, Geller AI. (1994) Long-term behavioral recovery in parkinsonian rats by an HSV vector expressing tyrosine hydroxylase. *Science* **266**: 1399–1403.

Efstathiou S, Minson AC. (1995) Herpes virus-based vectors. *Br. Med. Bull.* **51**: 45–55.

Epstein MA, Achong BG. (1986) *The Epstein-Barr Virus: Recent Advances.* John Wiley and Sons, NY.

Fareed MU, Spivak JG. (1994) Two open reading frames (ORF1 and ORF2) within the 2.0 kb latency-associated transcript of herpes simplex virus type 1 are not essential for reactivation from latency. *J. Virol.* **68**: 8071–8081.

Farrell MJ, Margolis TP, Gomes WA, Feldman LT. (1994) Effect of the transcription start region of the herpes simplex virus type I latency associated transcript promoter on expression of productively infected neurons *in vivo*. *J. Virol.* **68**: 5337–5343.

Fong Y, Federoff HJ, Brownlee M, Blumberg D, Blumgart LH, Brennan MF. (1995) Rapid and efficient gene transfer in human hepatocytes by viral vectors. *Hepatology* **22**: 723–729.

Frank A, Andiman W, Miller G. (1976) Epstein-Barr virus and non-human primates: natural and experimental infection. *Adv. Cancer. Res.* **23**: 171–210.

Friedmann T. (1989) Progress toward human gene therapy. *Science* **244**: 1275–1281.

Früh K, Ahn K, Djaballah H, Sempé P, van Endert PM, Tampé R, Peterson PA, Yang Y. (1995) A viral inhibitor of peptide transporters for antigen presentation. *Nature* **375**: 415–418.

Geller AI, During JJ, Oh YJ, Freese A, O'Malley K. (1995) An HSV-1 vector expressing tyrosine hydroxylase causes production and release of L-dopa from cultured rat striatal cells. *J. Neurochem.* **64**: 487–496.

Geschwind MD, Kessler JA, Geller AI, Federoff HJ. (1994) Transfer of the nerve growth factor gene into cell lines and cultured neurons using a defective herpes simplex virus vector. *Brain Res. Mol. Brain Res.* **24**: 327–335.

Geschwind MD, Hartnick CJ, Liu W, Amat J, Van de Water TR, Federoff HJ. (1996) Defective HSV-1 vector expressing BDNF in auditory ganglia elicits neurite outgrowth: model for treatment of neuron loss following cochlear degeneration. *Hum. Gene Ther.* **7**: 173–182.

Goins WF, Sternberg LR, Croen KD, Krause PR, Hendricks RL, Fink DJ, Straus SE, Levine M, Glorioso JC. (1994) A novel latency-active promoter is contained within the herpes simplex virus type 1 UL flanking repeats. *J. Virol.* **68**: 2239–2252.

Grassmann R, Fleckenstein B. (1989) Selectable recombinant herpesvirus saimiri is capable of persisting in a human T-cell line. *J. Virol.* **63**: 1818–1821.

Grassmann R, Dengler C, Muller-Fleckenstein I, Fleckenstein B, McGuire K, Dokhelar M-C, Sodroski JG, Haseltine WA. (1989) Transformation to continuous growth of primary human T-cell leukemia virus type I X-region genes transduced by a herpesvirus saimiri vector. *Proc. Natl Acad. Sci. USA* **86**: 3351–3355.

Hammerschmidt W and Sugden B. (1989) Genetic analysis of immortalizing functions of Epstein-Barr virus in human B-lymphocytes. *Nature* **340**: 393–397.

Henderson S, Rowe M, Gregory C, Groom-Carter D, Wang F, Longnecker R, Kieff E, Rickinson A. (1991) Induction of bc12 expression by Epstein-Barr virus Latent membrane protein 1 protects infected B cells from programmed cell death. *Cell* **65**: 1107–1115.

Hill A, Jugovic P, York I, Russ G, Bennink J, Yewdell J, Ploegh H, Johnson D. (1995) Herpes simplex virus turns off the TAP to evade host immunity. *Nature* **375**: 411–415.

Hilton DA, Brown LJR, Pringle JH, Nandha H. (1993) Absence of EBV in carcinoma of cervix. *Cancer* **72**: 1946–1948.

Ho DY, Mocarski ES, Sapelsky RM. (1993) Altering central nervous system physiology with a defective herpes simplex virus vector expressing the glucose transporter gene. *Proc. Natl Acad. Sci. USA* **90**: 3655–3659.

Ho DY, Fink SL, Lawrence MS, Meier TJ, Saydam TC, Dash R, Sapolsky RM. (1995) Herpes simplex virus vector system: analysis of its *in vivo* and *in vitro* cytopathic effects. *J. Neurosci. Methods* **57**: 205–215.

Isford RJ, Witter R, Kung HJ. (1994) Retrovirus insertion into herpesviruses. *Trends Microbiol.* **2**: 174–177.

Ishida T, Yamamoto K. (1987) Survey of non human primates for antibodies reactive with Epstein-Barr virus (EBV) antigens and susceptibility of their lymphocytes for immortalization with EBV. *J. Med. Primatol.* **16**: 359–371.

Jochner N, Eick D, Zimber-Strobl U, Pawlita M, Bornkamm GW, Kempkes B. (1996) Epstein-Barr virus nuclear antigen 2 is a transcriptional suppressor of the immunoglobin m gene: implication for the expression of the translocated c-*myc* gene in Burkitt's lymphoma cells. *EMBO J.* **15**: 375–382.

Johnson PA, Friedmann T. (1994) Replication-defective recombinant herpes simplex virus vectors. In: *Methods in Cell Biology* (ed. MG Roth). Academic Press, New York, pp 211–230.

Johnson PA, Miyanohara A, Levine F, Cahill T, Friedmann T. (1992) Cytotoxicity of a replication-defective mutant of herpes simplex virus type 1. *J. Virol.* **66**: 2952–2965.

Kaner RJ, Baird A, Manusukhani A, Basilico C, Summers BD, Florkiewicz RZ, Hajjar DP. (1990) Fibroblast growth factor receptor is a portal of cellular entry for herpes simplex virus type I. *Science* **248**: 1410–1413.

Kempkes B, Pich D, Zeidler R, Sugden B, Hammerschmidt W. (1995) Immortalization of human B-Lymphocytes by a plasmid containing 71 Kilobase pair of Epstein-Barr virus DNA. *J. Virol.* **69**: 231–238.

Khanna R, Borrows S, Kurilla M. (1992) Localization of Epstein-Barr virus cytotoxic T cell epitopes using recombinant vaccinia: implication for vaccine development. *J. Exp. Med.* **176**: 169–176.

Kieff E, Liebowitz D. (1990) Epstein-Barr virus and its replication. In: *Virology*, 2nd Edn (eds BN Fields, DK Knipe). Raven Press, NY, pp. 1889–1920.

Klein G. (1989) Viral latency and transformation: the strategy of Epstein-Barr virus. *Cell* **58**: 5–8.

Klein G. (1994) Epstein-Barr virus strategy in normal and neoplastic cells. *Cell* **77**: 791–793.

Kramm CM, Rainov NG, Sena-Esteves M, Chase M, Pechan PA, Chiocca EA, Breakefield XO. (1996) Herpes vector-mediated delivery of marker genes to disseminated central nervous system tumors. *Hum. Gene Ther.* **7**: 291–300.

Kwong AD, Frenkel N. (1984) Herpes simplex virus amplicon: effect of size on replication of constructed defective genomes containing eukaryotic DNA sequences. *J. Virol.* **51**: 595–603.

Labrecque LG, Barnes DM, Fentiman IS, Griffin BE. (1995) Epstein-Barr virus in epithelial cell tumors: a breast cancer study. *Cancer Res.* **55**: 39–45.

Landers RJ, O'Leary JJ, Crowley M. (1993) Epstein Barr virus in normal, pre-malignant and malignant lesions of the uterine cervix. *J. Clin. Pathol.* **46**: 931–935.

Latchman DS. (1994) Herpes simplex virus vectors for gene therapy. In: *Molecular Biotechnology* (ed. JM Walker). Human Press Inc., Totowa, NJ, pp. 179–195.

Lawrence MS, Ho DY, Dash R, Sapolsky RM. (1995) Herpes simplex virus vectors overexpressing the glucose transporter gene protect against seizure-induced neuron loss. *Proc. Natl Acad. Sci. USA* **92**: 7247–7251.

Lee MA, Kim OJ, Yates JL. (1992) Targeted gene disruption in Epstein-Barr virus. *Virology* **189**: 253–265.

Legerski R, Peterson C. (1992) Expression cloning of a human DNA repair gene involved in Xeroderma Pigmentosum group C. *Nature* **360**: 610–614.

Levitskaya J, Coram M, Levitsky V, Imreh S, Steigerwald-Mullen PM, Klein G, Kurilla MG, Masucci MG. (1995) Inhibition of antigen processing by the internal repeat region of the Epstein-Barr virus nuclear antigen-1. *Nature* **375**: 685–688.

Lewin DI. (1995) Herpes and EBV survive by antigenic stealth. *J. Natl. Inst. Health Res.* **7**: 49–53.

Liebowitz D, Kieff E. (1993) Epstein Barr virus. In: *The Human Herpesviruses* (eds B Roizman, RJ Whitley, C Lopez). Raven Press, NY, pp. 107–172.

Lokensgard JR, Bloom DC, Dobson AT, Feldman LT. (1994) Long-term promoter activity during herpes simplex virus latency. *J. Virol.* **68**: 7148–7158.

Lowenstein PR, Fournel S, Bain D, Tomasec P, Clissold P, Castro MG, Epstein AL. (1994) Herpes simplex virus 1 (HSV-1) helper co-infection affects the distribution of an amplicon encoded protein in glia. *Neuroreport* **5**: 1625–1630.

Marchini A, Longnecker R, Kieff E. (1992) Epstein-Barr virus (EBV)-negative B-lymphoma cell lines for clonal isolation and replication of EBV recombinants. *J. Virol.* **66**: 4972–4981.

Martuza RL, Malick A, Markert JM, Ruffner KL, Coen DM. (1991) Experimental therapy of human glioma by means of a genetically engineered virus mutant. *Science* **252**: 854–856.

McClain KL, Leach CT, Jenson HB, Joshi VV, Pollock BH, Parmley RT, DiCarlo FJ, Chadwick EG, Murphy SB. (1995) Association of Epstein-Barr virus with leiomyosarcoma in children with AIDS. *New Engl. J. Med.* **332**: 12–18.

Mester JC, Pitha PM, Glorioso JC. (1995) Antiviral activity of herpes simplex virus vectors expressing murine alpha 1-interferon. *Gene Ther.* **2**: 187–196.

Miller G. (1990) Epstein-Barr virus: biology, pathogenesis and medical aspects. In: *Virology*, 2nd Edn (eds BN Fields, DK Knipe). Raven Press, NY, pp. 1921–1958.

Miyanohara A, Johnson PA, Elam RL, Dai Y, Witztum JL, Verma IM. (1992) Direct gene transfer to the liver with herpes simplex virus type 1 vectors: transient production of physiologically relevant levels of circulating factor IX. *New Biol.* **4**: 238–246.

Miyashita EM, Yang B, Lam KMC, Crawford DH, Thorley-Lawson DA. (1995) A novel form of Epstein-Barr virus latency in normal B-cells *in vivo*. *Cell* **80**: 593–601.

Morgan RA. (1995) Retroviral vectors in human gene therapy. In: *Viruses in Human Gene Therapy* (ed J-MH Vos). Carolina Academic Press and Chapman & Hall, Durham, NC, and London, UK, pp. 77–108.

Morrow JA, Rixon FJ. (1994) Analysis of sequences important for herpes simplex virus type 1 latency-associated transcript promoter activity during lytic infection of tissue culture cells. *J. Gen. Virol.* **75**: 309–316.

Mosier DE, Picchio GR, Kirven MB, Garnier JL, Torbett BE, Baird SM, Kobayashi R, Kipps TJ. (1992) EBV-induced human B-cell lymphomas in hu-PBL-SCIID mice. *AIDS Res. Hum. Retroviruses* **8**: 735–740.

Mulligan RC. (1993) The basic science of gene therapy. *Science* **260**: 926–932.

Nicosia M, Zabolotny JM, Lirette RP, Fraser NW. (1994) The HSV-1 2kb latency-associated transcript is found in the cytoplasm comigrating with ribosomal subunits during productive infection. *Virology* **204**: 717–728.

Paterson T, Everett RD. (1990) A prominent serine-rich region in Vmw175, the major regulatory protein of herpes simplex virus type 1 is not essential for virus growth in tissue culture. *J. Gen. Virol.* **71**: 1775–1783.

Perng GC, Zwaagstra JC, Ghiasi H, Kaiwar R, Brown DJ, Nesburn AB, Wechsler SL. (1994) Similarities in regulation of the HSV-1 LAT promoter in corneal and neuronal cells. *Invest. Ophthalmol. Visual Sci.* **35**: 2981–2989.

Rivera-Gonzalez R, Imbalzano AN, Deluca NA. (1994) The role of ICP4 repressor activity in temporal expression of the IE-3 and latency-associated transcript promoters during HSV-1 infection. *Virology* **202**: 550–564.

Robertson E, Kieff E. (1995) Reducing the complexity of the transforming Epstein-Barr virus genome to 64 kilobase pairs. *J. Virol.* **69**: 983–993.

Roemer K, Johnson PA, Friedmann T. (1995) Transduction of foreign regulatory sequences by a replication-defective herpes simplex virus type 1: the rat neuron-specific enolase promoter. *Virus Res.* **35:** 81–89.

Roizman B. (1990) Herpesviridae: a brief introduction. In: *Virology,* 2nd Edn (eds BN Fields, DK Knipe). Raven Press, NY, pp. 1787–1793.

Roizman B, Jenkins FJ. (1985) Genetic engineering of novel genomes of large DNA viruses. *Science* **229:** 1208–1214.

Rowe M. (1995) The EBV latent membrane protein – (LMP1): a tale of two functions. *Epstein Barr Virus Report* **2:** 99–102.

Rowe M, Lear AL, Croom-Carter D, Davies AH, Rickinson AB. (1992) Three pathways of Epstein-Barr virus gene activation from EBNA1 positive latency in B-lymphocytes. *J. Virol.* **66:** 122–131.

Samulski RJ. (1995) Adeno-associated viral vectors. In: *Viruses in Human Gene Therapy* (ed. J-MH Vos). Carolina Academic Press and Chapman & Hall, Durham, NC, and London, UK, pp. 53–76.

Sauer B, Whealy M, Robbins A, Enquist L. (1987) Site-specific insertion of DNA into a pseudorabies virus vector. *Proc. Natl Acad. Sci. USA* **84:** 9108–9112.

Sharpe TV, Schwemmle M, Jeffrey I, Laing K, Mellor H, Proud CG, Hilse K, Clemens MJ. (1993) Comparative analysis of the regulation of the interferon inducible protein kinase PKR by Epstein-Barr virus RNAs EBER1 and EBER2 and adenovirus VA1 RNA. *Nucleic. Acids Res.* **21:** 4483–4490.

Simmer B, Alt M, Buckreus I, Berthold S, Fleckenstein B, Platzer E, Grassman R. (1991) Persistence of selectable herpesvirus saimiri in various human hematopoietic and epithelial lines. *J. Gen. Virol.* **72:** 1953–1958.

Sinclair AJ, Palmero I, Peters G, Farell PJ. (1994) EBNA-2 and EBNA-LP cooperate to cause G0 to G1 transition during immortalization of resting human B-lymphocytes by Epstein-Barr virus. *EMBO J.* **14:** 3321–3328.

Sixbey JW, Yao QY. (1992) Immunoglobin A-induced shift of EBV tissue tropism. *Science* **255:** 1578–1580.

Smiley JR, Duncan J. (1992) The herpes simplex virus type 1 immediate–early polypeptide ICP4 is required for expression of globin genes located in the viral genome. *Virology* **190:** 538–541.

Smith RL, Geller AI, Escudero KW, Wilcox CL. (1995) Long-term expression in sensory neurons in tissue culture from herpes simplex virus type 1 (HSV-1) promoters in an HSV-1-derived vector. *J. Virol.* **69:** 4593–4599.

Stern S, Tanaka M, Herr W. (1989) The oct-1 homeodomain directs formation of a multiprotein–DNA complex with the HSV transactivator VP16. *Nature* **341:** 624–630.

Stevens JG. (1989) Human herpesviruses: a consideration of the latent state. *Microbiol. Rev.* **53:** 318–332.

Stow ND, Murray MD, Stow EC. (1986) *Cis*-acting signals involved in the replication and packaging of herpes simplex virus type-1 DNA. *Cancer Cells* **4:** 497–507.

Stratford-Perricaudet LD, Perricaudet M. (1995) Adenovirus-mediated *in vivo* gene therapy. In: *Viruses in Human Gene Therapy* (ed. J-MH Vos). Carolina Academic Press and Chapman & Hall, Durham NC, and London, UK, pp. 1–32.

Strathdee CA, Gavish H, Shannon WR, Buchwald M. (1992) Cloning of cDNA for Fanconi's anemia by functional complementation. *Nature* **356:** 763–767.

Strauss SE. (1990) Introduction to herpesviridae. In: *Principles and Practice of Infectious Diseases.* (eds GL Mandell, RG Douglas Jr, JE Bennett). Churchill Livingstone, NY, pp. 1139–1144.

Sun T-Q, Vos J-MH. (1992) Packaging of 200 kb engineered DNA as infectious Epstein-Barr virus. *Int. J. Genome Res.* **1:** 45–57.

Sun T-Q, Fenstermacher D, Vos J-MH. (1994) Human artificial episomal chromosomes for cloning large DNA in human cells. *Nature Genet.* **8:** 33–41.

Sun T-Q, Livanos E, Vos J-MH. (1996) Engineering a mini-herpesvirus as a general strategy to transduce up to 180 kb of functional self-replicating human mini-chromosomes. *Gene Ther.* (in press).

Szekely L, Selivanova G, Magnusson KP, Klein G, Wiman KG. (1993) EBNA-5, an Epstein-Barr virus encoded nuclear antigen, binds to the retinoblastoma and p53 proteins. *Proc. Natl Acad. Sci USA* **90:** 5455–5459.

Tierney RJ, Steven N, Young LS, Rickinson AB. (1994) Epstein-Barr virus latency in blood mononuclear cells: analysis of viral gene transcription during primary infection and in carrier state. *J. Virol.* **68:** 7374–7385.

Tosato G, Magrath I, Koski I, Dooley N, Blaese M. (1979) Activation of suppressor T cells during Epstein-Barr virus induced infectious mononucleosis. *New Engl. J. Med.* **301:** 1133–1137.

Verhaagen J, Hermens WTJMC, Oestreicher AB, Gispen WH, Rabkin SD, Pfaff DW, Kaplitt MG. (1994) Expression of the growth-associated protein B-50/GAP 43 via a defective herpes simplex virus vector results in profound morphological changes in non-neuronal cells. *Brain Res. Mol. Brain Res.* **26:** 26–36.

Vlazny DA, Kwong A, Frenkel N. (1982) Site-specific cleavage/packaging of herpes simplex virus DNA and the selective maturation of nucleocapsids containing full-length viral DNA. *Proc. Natl Acad. Sci. USA* **79:** 1423–1427.

Vos J-MH. (1995) Herpesviruses as genetic vectors. In: *Viruses in Human Gene Therapy* (ed. J-MH Vos). Carolina Academic Press and Chapman & Hall, Durham, NC, and London, UK, pp. 109–140.

Wagner EK, Flanagan WM, Davi-Rao GB, Zhang YF, Hill JM, Anderson P, Stevens JG. (1988) The herpes simplex virus latency-associated transcript is spliced during the latent phase of infection. *J. Virol.* **62:** 4577–4585.

Wagner R, Radman M. (1995) Mismatch repair and human disease. In: *DNA Repair Mechanisms: Impact on Human Diseases and Cancer* (ed. J-MH Vos). RG Landes and Springer-Verlag, Austin, TX, and Heidelberg, Germany, pp. 151–159.

Wang B, Liebowitz D, Kieff E. (1985) An EBV membrane protein expressed in immortalized lymphocytes transforms established rodent cells. *Cell* **43:** 831–840.

Wang F, Gregory C, Sample C, Rowe M, Liebowitz D, Murray R, Rickinson A, Kieff E. (1990) Epstein-Barr virus latent membrane protein (LMP1) and nuclear protein 2 and 3C are effectors of phenotypic changes in B-lymphocytes: EBNA2 and LMP1 cooperatively induce CD23. *J. Virol.* **64:** 2309–2318.

Wang F, Marchini A, Kieff E. (1991) Epstein Barr virus recombinants: use of positive selection markers to rescue mutants in EBV negative B lymphoma cells. *J. Virol.* **65:** 1701–1709.

Wang F, Li X, Annis B, Faustman DL. (1995) Tap-1 and Tap-2 gene therapy selectively restores conformationally dependent HLA class I expression in type-I diabetic cells. *Hum. Gene Ther.* **6:** 1005–1017.

Wang MJ, Friedmann T, Johnson PA. (1995) Differentiation of PC12 cells by infection with an HSV-1 vector expressing nerve growth factor. *Gene Ther.* **2:** 323–335.

Wang S, Vos J-M. (1996) A hybrid herpes infectious vector based on Epstein-Barr virus and herpes simplex 1 for efficient gene transfer into human cells *in vitro* and *in vivo*. *J. Virol.* (in press).

Wood MJA, Byrnes AP, Pfaff DW, Rabkin SD, Charlton HM. (1994) Inflammatory effects of gene transfer into the CNS with defective HSV-1 vector. *Gene Ther.* **1:** 283–291.

WuDunn D, Spear PG. (1989) Initial interaction of herpes simplex virus with cells is binding to heparan sulfate. *J. Virol.* **63:** 52–58.

Yates JL, Warren N, Reisman D, Sugden BA. (1984) A *cis*-acting element from the Epstein-Barr viral genome that permits stable replication of recombinant plasmids in latently infected cells. *Proc. Natl Acad. Sci. USA* **81:** 3806–3810.

9

Role of animal models in gene therapy

Julia R. Dorin and David J. Porteous

9.1 Why use animals?

The development of functional and safe somatic gene therapy protocols is key to the success of this approach to treat inherited disease. *In vitro* work is an essential primary step in the development of novel strategies and for efficacy assessment. Ultimately, however, whole-organ or tissue effects need to be addressed and this requirement demands *in vivo* studies. The majority of basic metabolism is shared between mammals and despite some species differences in development and physiology, comparative study can be extremely valuable in dissecting disease pathogenesis and the potential for intervention. Preclinical safety testing and checks for absence of toxicity or germ-line transmission can be carried out using normal animals while gene delivery can be evaluated by introducing reporter genes into normal animals. However, an animal model with a specific gene deficiency, and particularly one which mimics the cardinal features of the disease, is invaluable both for monitoring successful gene delivery and determining whether the therapy protocol can safely reverse the disease phenotype.

9.2 What animal models are available?

Several distinct types of animal model exist that mimic human genetic diseases (*Table 9.1*). An important distinction must be made between phenocopies (animals that display a disease state that is of unknown or other genetic basis or is environmentally induced) and animals mutant in the homologue of a human disease gene. The potential value of phenocopies is elegantly illustrated by the demonstration that sheep grazing on copper-deficient soil in Australia have coats resembling the 'kinky' hair of infants with Menkes disease (Hamer, 1993). Menkes disease patients display characteristic brittle, depigmented hair as well as neurological degeneration and mental retardation together with connective tissue and vascular defects. The phenotypic effect of copper deficiency in the sheep indicated that the Menkes disease gene may involve copper metabolism. This reasoning was borne out in 1993 when the strongest similarity match of the cloned gene was shown to be to a bacterial Cu^{2+} export protein (Chelley *et al.*, 1993; Mercer *et al.*, 1993; Vulpe *et al.*, 1993).

The relevance of phenocopies to gene therapy is in their role in addressing questions of gene delivery (in a compromised physiological environment) and safety.

Gene Therapy, edited by N.R. Lemoine and D.N. Cooper.
© 1996 BIOS Scientific Publishers Ltd, Oxford.

Table 9.1. Animal models of disease

Mutation in the animal	Human disease	Inheritance pattern	Animal mutant
Spontaneous and/or randomly induced by chemical or radiation	Duchenne muscular dystrophy	X-linked	*mdx* mouse
	GCPS	Autosomal dominant	Extra toe (*Xt*)
	Albinism	Autosomal recessive	Albino (*c*)
	Diabetes mellitus type I	Multifactorial	Non-obese diabetic mouse (*NOD*)
Classical transgenic	Sickle cell anaemia	Autosomal recessive	Human Hbs transgene
ES cell gene targeting	Cystic fibrosis	Autosomal recessive	*cftr*^{m1HGU}, *cftr*^{m1UNC}, *cftr*^{m1Cam}, *cftr*^{m1Bay}, *cftr*^{m1HSC}, *cftr*$^{\Delta F508}$, *cftr*^{G551D}
	Gaucher's syndrome	Autosomal recessive	Glucocerebrosidase (–/–)

Their obvious limitation is in assessing gene complementation. This can only be approached using a mutant genocopy; that is, where there is loss of function of the disease gene homologue. The choice of which species to use as a model of a human disease would vary depending on which organ was most affected. For example, cystic fibrosis (CF) affects mainly the lungs, and the sheep or pig would be the optimal choice for respiratory tree similarity. Primate models, while being ethically much less acceptable, may be important models for diseases affecting sophisticated brain function. Animal choice is not without restriction, however, and for several reasons, the majority of animal models that now exist are murine. From a practical viewpoint, mice are easy to house and handle, have a quick breeding time, large litters and are cheap to maintain. From a scientific point of view, the power of the detailed mouse genetic map is immense. Recombinant inbreds, outbred and specific mutant or marker strains are all commercially available and extremely well characterized. Comparative genetic mapping reveals a striking degree of conservation of gene and linkage groups between human and mouse.

Functional complementation in somatic cell hybrids and a high degree of sequence similarity supports the assumption that mouse and human share a high degree of basic metabolic function. The commonality of protein function between these two species has been reinforced by studies of genes such as that encoding the cystic fibrosis transmembrane conductance regulator (*CFTR*). It is apparent that those amino acid residues which are affected by clinical mutation are strikingly conserved evolutionarily (Dorin *et al.*, 1994a).

The main reason for the number of murine animal models is a technical one. Only in this species can the loss of a particular pre-determined gene function be specifically achieved. Providing the gene responsible is available, it is now possible to generate homozygous mutant mice by gene targeting into pluripotent embryonal stem (ES) cells. These 'knockout' animals can only be created in the mouse because only murine ES cells can be grown successfully in tissue culture. Several laboratories are making progress in culturing totipotent rat ES cells but at the time of writing, this has not been achieved (Mullins and Mullins, 1996). The advantage of the rat would lie in the detailed physiological data, particularly on brain development, that already exist and the fact that its slightly larger size would make some experimental manipulations significantly easier.

So, what animal models exist and how have they been produced?

9.3 Spontaneous mutants and mutants induced by random mutagenesis

Some animal models for human disease have arisen by spontaneous mutation and have been recognized by researchers. Many of these naturally occurring mutants affect dominant or visible phenotypes such as coat or eye colour as these are obviously easy to detect. One example of this is the *albino* mouse where the lack of a functional *tyrosinase* gene results in the absence of visible pigment. This pigment is also absent in tyrosinase-negative individuals with albinism. The *albino* phenotype has been elegantly rescued by direct pronuclear injection of a 250 kb tyrosinase genomic transgene in the form of a recombinant yeast artificial chromosome (YAC) (Schedl *et al.*, 1993). While this is not yet a feasible strategy in humans, the fact that the YAC transgene showed position-independent, copy-number-dependent expression is important. This is particularly pertinent in dominant disorders where dosage will be an important factor.

Random mutagenesis schemes have produced many mutant animals and again the dominant or visual phenotypes are easiest to pick up. Exposure of male mice to chemicals such as ethylnitrosourea (ENU), viral or radiation-induced mutation, and screening the resultant offspring for mutant phenotypes is the most common approach. Recessive mutations are revealed by sib crossing. One useful disease-associated mouse model derived by ENU mutagenesis is the *Min* mouse which is defective in the mouse homologue of the *APC* (*adenomatous polyposis coli*) gene; it has a dominant phenotype and develops multiple intestinal tumours. Many spontaneous mutants have also been derived using large-scale mutagenesis screens. One example of a human disease model which has arisen both by spontaneous and chemical mutagenesis is Greig's cephalopolysyndactyly syndrome (GCPS). This is a dominant disorder in humans which presents with craniofacial defects and polydactyly. The dominant mouse mutation, extra toes (*Xt*), results in similar dysmorphology to GCPS and is due to mutation in the murine homologue of the human disease gene *Gli3* (Hui and Joyner, 1993). It is difficult to imagine somatic gene therapy strategies for disorders which already have phenotypic consequence at birth but mouse models such as *Xt* could provide a valuable test for *in utero* intervention.

In view of the significant experimental advantages of the rat for physiological and behavioural studies and the present failure to develop ES cell technology in the rat, it is worth considering the possibility of applying random mutagenesis schemes. With the rapidly improving genetic map of the species (Jacob *et al.*, 1995), as long as there is a readily recognizable phenotype in the rat for the function of interest, it should be possible to apply a random mutagenesis, genetic linkage and positional cloning strategy.

Mucopolysaccharidosis (MPS) type VII (Sly syndrome) is a lysosomal storage disorder which is due to a lack of the enzyme β-glucuronidase. Feline, canine and murine animal models have been identified for MPS (Birkenmeier *et al.*, 1989; Haskins *et al.*, 1984; Jezyk *et al.*, 1977). An alternative treatment regime to enzyme replacement is continuous delivery of the correctly processed lysosomal enzyme from an autologous implant of genetically modified cells. This strategy has been successfully proven using the Sly mouse model. Mutant mouse skin fibroblasts were grown in culture and transfected with a retroviral vector carrying cells with the human β-glucuronidase cDNA (Moullier *et al.*, 1993). These cells expressed enzyme with the correct modifications to allow efficient receptor-mediated cellular uptake. These cells were then embedded into collagen lattices and implanted intraperitoneally into host Sly mutant mice. The vascularized neo-organs survived over 5 months and continuously secreted the human β-glucoronidase enzyme. This resulted in the disappearance of lysosomal storage lesions in the liver and spleen. The mouse experiments are extremely encouraging but also highlight some obvious limitations of gene therapy strategies for this type of disease. The first is that the macroscopic changes associated with the MPS VII mice (skeletal muscle abnormalities, disturbed gait, deafness) were not reversed. These phenotypic features are due to storage lesions in the articular cartilage and in the brain. Some reversal of developmental defects is unlikely to be possible, since therapy would need to be initiated neonatally when bone remodelling can occur. The second consideration is that whereas circulating factors in adults may not cross the blood–brain barrier, this may be possible neonatally so that brain development is likely to be less impaired. These questions can only be adequately addressed in a relevant animal model such as this. Adenoviral-mediated gene transfer of the gene for human β-glucuronidase into the eyes of the Sly mutant mice has been achieved by intravitreal or subretinal injection. By 3 weeks after injection, it was impossible to detect the storage vacuoles normally present as a consequence of the β-glucuronidase deficiency (Li and Davidson, 1995).

Another example of a naturally occurring mutant animal is the Watanabe heritable hyperlipidaemic rabbit. This rabbit model displays dramatically raised plasma cholesterol levels as a result of the loss of four amino acids from the ligand-binding domain of the low density lipoprotein receptor (*LDLR*) gene (Yamamoto *et al.*, 1986). Mutation in this gene in humans is the molecular defect responsible for familial hypercholesterolaemia (FH). These rabbits are being used to determine the potential of various gene transfer methods for the treatment of hyperlipidaemias. The retroviral transfection of hepatocytes *ex vivo* and subsequent repopulation of the liver has shown some improvement in the cholesterol levels in the rabbit and these encouraging results have led to the first human somatic gene therapy trial for FH, using a similar protocol (Grossman *et al.*, 1995). Other strategies appear to be able to reduce significantly the cholesterol levels in this model, including *LDLR* gene delivery using an adenoviral vector (Li *et al.*, 1995).

9.4 Transgenic animals

When dominant disorders are the result of gain-of-function mutations in one copy of the gene, then classical transgenesis can be considered to create a relevant animal model (Gordon and Ruddle, 1981). Classical transgenesis is the introduction of a DNA sequence into the pronucleus of a fertilized oocyte where the transgene integrates into a random chromosomal site (Jaenisch, 1988). The DNA-injected eggs are then reimplanted into pseudopregnant foster mothers and the offspring analysed for the presence of the foreign DNA in the genome. Expression of the introduced gene is largely under the control of the promoter included in the recombinant construct. However, the chromosomal site will also effect expression, although genomic transgene constructs which contain some *cis*-acting signals will help to reduce this.

Although the majority of transgenic experiments are carried out in mice, it is possible to produce rat, goat, rabbit, sheep and zebrafish transgenics (Mullins and Mullins, 1996). An animal model for sickle-cell anaemia has been created by injection into mouse oocytes of a human β-globin transgene carrying the sickle-cell HbS mutation (Greaves *et al.*, 1990). The resulting transgenic mice display typical red cell sickling *in vitro* under conditions of low oxygen tension.

It should be noted that classical transgenesis introduces exogenous DNA into all tissues of the developing animal, unlike gene therapy in which it is presently only ethically acceptable to introduce DNA into somatic cells. The use of transgenic animals is thus a stringent toxicity test of a potential somatic gene therapy gene. The therapeutic gene will be present in all tissues and, if under the control of a promoter which allows expression in all tissues, this should allow assessment of the potential toxic effect of inappropriate or high-level expression in the whole animal. Different founder animals will have different numbers of transgenes integrated at different chromosomal sites which will modulate transgene expression.

9.5 Antisense models

The majority of inherited diseases in humans arise as a consequence of a loss rather than a gain of function. One method for achieving specific inhibition of gene expression is using antisense oligonucleotides or recombinant constructs (reviewed by Pollock and Gäken, 1995). Antisense oligonucleotides can be modified chemically to improve their stability and potency (Wagner, 1994). However, their success in achieving total gene inhibition *in vivo* such as is necessary in an animal model is still limited. Antisense recombinant constructs to myelin basic protein have been injected into mouse oocytes and this has induced a phenotype in mice similar to the *shiverer* mutant. This notwithstanding, the strategy of choice when considering making a mouse model for a recessive disease is to use gene targeting in ES cells (Katsuki *et al.*, 1988).

9.6 ES cell transgenics

Providing that the gene responsible has been cloned, it is possible to generate homozygous mutant mice ('knockouts') by gene targeting in pluripotent ES cells (Capecchi, 1989). ES cells are derived from the inner cell mass of 3.5-day-old, pre-implantation embryos and have the remarkable capacity to be propagated, manipulated and mutated in culture while still retaining their ability to contribute to all tissues of a developing mouse when reintroduced into host blastocysts. Gene targeting directs chromosomal integration to a specific site by virtue of DNA homology. Normally, the targeting of exogenous DNA sequence will result in gene disruption but mutations corresponding precisely to the clinical situation can also be introduced. The production of homozygous mutant mice is achieved by mating heterozygous sibs. There are now many models of human genetic disease generated by ES cell gene targeting. As soon as the disease gene is cloned, the next logical step is to create a mutant mouse.

The first human disease to be modelled in the mouse using ES cell gene targeting was Gaucher's disease (Tybulewicz *et al.*, 1992). This is a hereditary deficiency of glucocerebrosidase which results in the accumulation of cerebrosidases predominantly in macrophages. The phenotype of the null mutant animals appeared to be similar to the phenotype of the rarer type-2 form of the disease which is more severe. The mutant mice died early after birth from a phenotype consistent with a nervous system dysfunction similar to the neurological disease observed in these patients. Since the macrophage appears to be a major affected cell type and is derived from bone-marrow stem cells, this disease must be a good candidate for gene therapy. However, any gene therapy treatment protocol must be safer, more effective and cheaper than the protein replacement treatment currently available.

Clearly, gene disruptions are more relevant to recessive diseases or X-linked diseases where loss of function is predicted to have an effect. However, understanding the normal function of the protein involved in a dominant disease is important in order to access points of entry to disease treatment. Huntington's disease is an autosomal dominant neurological disease associated with CAG repeat expansion in a widely expressed gene that causes selective, progressive neuronal death. Mice homozygously mutated in the homologue of the gene for Huntington's disease (*HD*) die before embryonic day 8.5 (Nasir *et al.*, 1995). Heterozygous mice display increased motor activity and cognitive deficits as well as apparent neuronal loss in the subthalamic nucleus. It is most likely that the clinical phenotype of HD is attributable to the gain of novel function(s). However, this mouse model implies that therapeutic strategies aimed at decreasing the expression of the *HD* gene may not ameliorate the adverse clinical phenotype without incurring novel adverse phenotypic consequences.

9.7 Cancer models

Animal models for cancer fall into two distinct categories. The first includes animals where tumours have been induced by environmental stimuli. The second is where a genetically inherited predisposition results in tumour development. An example of a spontaneously arising cancer model is the *Min* mouse discussed in Section 9.3. This has since been reproduced by targeted mutagenesis of the murine *Apc* gene in ES cells (Fodde *et al.*, 1994).

Several tumour suppressor genes have been disrupted or modified in mice. However, the results have been imperfect in terms of producing accurate models of human cancer. Mutation in the gene for p53 (*TP53*) is involved in many inherited human cancers. Homozygous p53 'null' mice created by gene targeting are viable (Donehower *et al.*, 1992) despite an abundance of experimental data indicating that p53 plays a pivotal role in in cell division and differentiation. The p53 'knockout' mice develop a wide range of neoplasms within the first 6 months of life. This is not dissimilar to the features which typify patients with Li-Fraumeni syndrome. Such individuals inherit one normal and one mutant (can be a null allele) *TP53* gene. The heterozygous mice develop spontaneous tumours at a modest rate. Thus, both heterozygous and homozygous p53 mutant mice will be valuable for testing potential carcinogens and cancer therapies.

Another tumour suppressor gene that has been knocked out in mice is the Wilms' tumour protein *wt1* (Kreidberg *et al.*, 1993). The homozygous mutant mice are embryonic lethal at day 11 and display kidney, gonadal, lung and heart defects, which supports the involvement of the Wilms' tumour gene in genitourinary development. However, unlike the situation in humans, heterozygotes do not develop tumours.

Homozygous mutants of the retinoblastoma (*Rb-1*) gene in mice are embryonic lethal at 12–15 days old and appear to have neural and haematopoietic defects (Clarke *et al.*, 1992; Jacks *et al.*, 1992; Lee *et al.*, 1992). However, heterozygotes do develop tumours. Unexpectedly, these tumours are localized to the pituitary, a situation which contrasts with the retinal tumours observed in humans. In fact, the differences between cancer aetiology in humans and mice does not pose too much of a disadvantage for the use of these and other cancer-prone mice for testing gene therapy protocols. As with any cancer therapy, the problem is delivering the toxin or cell tag, antisense oligonucleotides or therapeutic gene to 100% of the cancer cells. In this regard, the mice available provide a stringent test system to assess the success of these approaches, with tumour regression as a relevant end-point.

9.8 Cystic fibrosis murine models – a demonstration of how relevant an animal model can be

Animal models have proved crucial in assessing the feasibility of gene therapy strategies for cystic fibrosis (CF). Detailed below are the major advances in which experimentation in CF mutant mice has contributed to the design of human gene therapy trials.

Of the inherited gene disorders, CF is one of the most common, affecting over 50 000 individuals world-wide. The disease is characterized by defective ion transport in epithelial cells and results in abnormal mucus accumulation in the exocrine glands. The major affected organ is the lungs, with 95% of morbidity and mortality being due to recurrent respiratory tract infections. Several features make CF a good target for somatic gene therapy. (i) The disease is recessive, which means that introduction of wild-type copies of the *CFTR* gene to the relevant cells should restore the normal phenotype. (ii) The lung disease is progressive, with newborns having essentially normal lung histology and morphology; this implies that early intervention could prevent onset of the disease entirely. (iii) The target organ is accessible and non-germline gene delivery to the lung should be possible. Despite these encouraging points, several pertinent questions remained unanswered. The first was whether over-expression in respiratory cells would be deleterious. This question was addressed by Whitsett *et al.* (1992) who expressed human *CFTR* in transgenic mice under the control of a respiratory-tract-specific (surfactant protein C) promoter with no evidence of toxicity. Once the murine CF gene (*Cftr*) gene was cloned, ES cell gene targeting could be used to produce an animal model. As mentioned in Section 9.2, mice are the only animals in which this targeted mutagenesis scheme can be achieved at present. A sheep or pig model may have closer respiratory tract similarities, and ferrets have an abundance of submucosal glands (the predominant site of *CFTR* gene expression in the lung). However, the probability that a genetic mouse model for CF would mimic the clinical phenotype was given weight by the demonstration of Smith *et al.* (1992) that the ion transport properties of the respiratory epithelium in wild-type mice are similar to those of normal human subjects. In addition, when the murine *Cftr* gene was isolated, the high degree of amino acid similarity (78%) between the human and mouse proteins, particularly within exons 10 and 11 (Tata *et al.*, 1991; Yorifuji *et al.*, 1991), was striking.

Initially four *cftr* mouse models were produced using *Cftr* gene disruption in ES cells (reviewed by Dorin *et al.*, 1994a). The gene-targeting schemes in these models were slightly different, with three groups targeting exon 10 of the *Cftr* gene and one group targeting exon 3. Two phenotypes were produced. One phenotype displayed by the mutant mice from three of the four groups led to a high level (80%) of perinatal death from gut blockage reminiscent of meconium ileus in humans. This is the failure to pass the first stool due to gut obstruction and is present in 10–15% of CF neonates. The other phenotype was characteristic of the mutant mice produced by insertional gene targeting into exon 10 ($cftr^{m1HGU}$) and, in these mice, only 7% of mice died from gut blockage. The reason for this phenotypic difference was that the $cftr^{m1HGU}$ mice produced a low level (~10%) of wild-type *Cftr* mRNA due to aberrant splicing and exon skipping (Dorin *et al.*, 1994b), whereas the other gene disruptions were complete and true 'nulls'. The different phenotypes of the mice were complementary, as they both displayed the electrophysiological abnormality in the chloride channel characteristic of CF. CFTR has been shown to be a cAMP-mediated chloride channel which is apically localized in epithelial cells. CF patients can be distinguished easily from normal individuals on the basis of their raised baseline potential differences (which reflects increased sodium transport) and their reduced electrophysiological response to cAMP agonists (reflecting decreased chloride transport) (Alton *et al.*, 1990; Knowles *et al.*, 1981). The *cftr* 'null' mutant mice have no CFTR and no cAMP-mediated chloride response in the gut so that there is an 80% death rate by 30 days due to gut blockage. However, the $cftr^{m1HGU}$ mutants display 10% residual *Cftr* mRNA and 30% residual cAMP-mediated chloride response in the gastrointestinal tract, resulting in only a very low incidence of intestinal blockage. This implies that a low level of CFTR is sufficient to rescue the chloride transport defect partially and to have a significant effect on phenotype (i.e. survival). We have also shown that compound heterozygotes of one 'null' allele and one $cftr^{m1HGU}$ allele express only 5% of wild-type levels of CFTR, which is still enough to rescue the gut blockage phenotype. This point is fundamental when considering gene therapy approaches because low-level rescue will be the most realistic outcome. These mouse experiments demonstrate that partial correction may be enough to achieve clinical benefit (Dorin *et al.*, 1996).

The need for efficient delivery to the lung epithelium led researchers to consider a viral approach for somatic gene therapy. The lung epithelium has a slow cell turnover time and the surface-lining cells are a non-dividing, terminally differentiated population. Adenoviral vectors appeared to be the perfect solution to this problem as adenovirus is naturally tropic for the lung and can be engineered to accommodate the human *CFTR* cDNA. Rosenfeld *et al.* (1992) used cotton rats to demonstrate that the lung epithelium could be efficiently infected with potentially therapeutic levels of *CFTR* recombinant adenovirus. On the basis of this and additional studies in primates, Zabner *et al.* (1993) initiated the first gene therapy trial in CF patients in the USA. The trial of three patients displayed an encouraging but transient correcting effect on the characteristic CF bioelectric defect. Since then, several other groups in the USA and France have completed adenoviral trials for CF. These trials have reported variable success, but transfection efficiency does not seem to be high. Using the *cftr* 'null' mutant mouse, Grubb *et al.* (1994) achieved efficient adenoviral transfection of basal-like cells but not of the columnar epithelial cells that line the respiratory tract. Repeated high doses of *CFTR* adenovirus to the mice only partially corrected the CF defect in chloride ion transport *in vivo*, and did not correct the sodium transport defect at all. This poor transfection efficiency was mirrored in a clinical trial of 12 patients who received increasing doses via the nose (Knowles *et al.*, 1995). At low doses there were no toxic effects but neither was there evidence for successful gene transfer. At high doses, expression of the new gene could be detected but only in an estimated 1% of cells. At the highest doses, two out of three patients developed mucosal inflammation and there was no evidence that the chloride conductance defect had been corrected. This concern over an inflammatory reaction in patients' lungs already compromised by infection is a serious one and represents an important limitation for the use of adenoviruses, particularly if high viral titres are necessary for efficient delivery.

An alternative strategy is to use DNA–cationic liposome complexes. Liposomes are hypoimmunogenic, chemically defined and easily mass-produced to pharmacological standard. Their disadvantages are that at present their delivery is indiscriminate and less efficient than viral infection. Two related experiments in the *cftr* mutant mice have demonstrated the feasibility of this approach for CF gene therapy. Hyde *et al.* (1993) reported direct tracheal instillation of *CFTR* cDNA complexed with DOTMA–DOPE (Lipofectin™) to young 'null' *cftr* homozygous mutants. Evidence was presented that four of six treated animals showed complete correction of the electrophysiological defect in tracheal explants. The other two animals showed no evidence of correction, which the authors put down to failure of delivery. Alton *et al.* (1993) reported a larger study in the *cftr*[mIHGU], long-term survival, mice in which the DNA was delivered as an aerosol by nebulization. The human *CFTR* cDNA was complexed to DC-Chol–DOPE and the treated mice showed evidence of partial to complete rescue of the electrophysiological defect in both the nose and trachea. The lack of toxicity of this protocol in these mutant mice and correction of the basic chloride transport defect led directly to a human 'nose-only' trial with a DC-Chol–DOPE liposome–human *CFTR* complex. Results from this trial were very similar to the mouse trial, giving partial correction of the electrophysiological defect with no detectable toxicity (Caplen *et al.*, 1995).

9.9 Refinement of the *cftr* mouse model

The *cftr* mutant mice have proved to be very useful in assessing gene therapy protocols, but one obvious limitation of the model is the lack of lung disease. The absolute 'null' animals may not survive long enough to develop lung disease and the *cftr*[mIHGU] mutant did not develop lung disease when maintained in a conventional animal house. However, when these mice were exposed repeatedly to aerosolized CF-associated lung pathogens, the mutant mice developed pathogen-specific lung disease consistent with CF (Davidson *et al.*, 1995). The fact that these mice can be induced to develop lung disease dramatically increases the value of the model because it can be used to assess efficiency of gene delivery in pathological tissue and to determine whether somatic gene delivery can halt the development of lung disease.

Although some CF patients have nonsense mutations and produce no CFTR protein, the majority of patients (70% of cases) carry the ΔF508 mutation which is a 3 bp deletion resulting in the loss of a phenylalanine amino acid residue from the first nucleotide-binding fold of the protein. This results in incorrect localization of the mutant protein; that is, it does not reach the apical membrane. Thus, the majority of gene therapy trials will take place on a background of mutant protein. In order to produce this precise mouse model, three groups have targeted the gene in ES cells and gone on to produce homozygous *ΔF508* mutant mice (Colledge *et al.*, 1995; van Doorninck *et al.*, 1995; Zeiher *et al.*, 1995).

Introduction of another precise CF clinical mutation has demonstrated that genotype–phenotype correlations established in humans can be verified in the mouse on a controlled genetic background. Delaney *et al.* (1996) used ES gene targeting to introduce precisely the G551D mutation into the mouse gene. A key aspect of this missense mutation (results in aspartic acid instead of glycine at position 551 in the protein) is the threefold reduction in the incidence of neonatal intestinal blockage in CF patients carrying the G551D mutation compared with patients homozygons for the most common *CFTR* mutation, ΔF508. The G551D homozygous mutant mice, under the same animal house conditions as the 'null' *cftr* mutant animals, displayed a significantly increased survival rate due to a reduction in gut blockage. Where species differences do exist, this can also be informative. For example, none of the *cftr* mutant mice display the pancreatic abnormalities associated with 85% of CF patients. This difference invalidates the assumption that meconium ileus is associated with pancreatic sufficiency.

9.10 Future prospects

It is clear that animal models, and in particular mouse models, can contribute directly to a better understanding of disease pathogenesis and assessment of therapeutic interventions by pharmacological approaches or gene therapy. The variable phenotype of many essentially monogenic gene disorders implies that other genes can affect disease penetrance. The ability to assess gene allele variants for their contribution to an unlinked disease phenotype in the mouse is beginning to reveal the activity of modifier genes. For example, the *Mom* (modifier of *Min*) (Dietrich *et al.*, 1993) gene modifies the number of intestinal neoplasms present in the *Min* mouse. In addition, Rozmahel *et al.* (1996) created *cftr*-deficient mice and used interstrain crossing to demonstrate that the severity of intestinal disease is genetically determined. A genome scan showed that the major modifier locus maps near the centromere of mouse chromosome 7. As these interactive genes are characterized, their potential use in gene therapy must be considered.

Systemic gene delivery of independent chromosomal entities to, or correction of the mutation by *ex vivo* targeting in, self-renewing stem cells must be the ultimate goal of gene therapy. Patient trials will always be the only true test of therapeutic efficacy but the technical difficulty of devising the perfect approach to somatic gene therapy serves to emphasize the value of a relevant animal model to speed safe and effective progress towards this goal.

References

Alton EWFW, Currie D, Logan-Sinclair R, Warner JO, Hodson ME and Geddes DM. (1990) Nasal potential difference: a clinical diagnostic test for cystic fibrosis. *Eur. Respir. J.* **3**: 922–926.

Alton EWFW, Middleton PG, Caplen NJ, Smith SN, Steel DM, Munkonge FM, Jeffery PK, Geddes DM, Hart SL, Williamson R, Fasold KI, Miller AD, Dickinson P, Stevenson BJ, McLachlan G, Dorin JR, Porteous DJ. (1993) Non-invasive liposome-mediated gene delivery can correct the ion transport defect in cystic fibrosis mutant mice. *Nature Genet.* **5**: 135–142.

Birkenmeier EH, Davisson MT, Beamer WG, Ganschow RE, Vogler CA, Gwynn B, Lyford KA, Maltais LM, Wawrzyniak CJ. (1989) Characterization of a mouse with β-glucuronidase deficiency. *J. Clin. Invest.* **83**: 1258–1266.

Capecchi M. (1989) The new mouse genetics: altering the genome by gene targeting. *Trends Genet.* **5**: 70–76.

Caplen NJ, Alton EWFW, Middleton PG, Dorin JR, Stevenson BJ, Gao X, Durham SR, Jeffery PK, Hodson ME, Coutelle C, Huang L, Porteous DJ, Williamson R, Geddes DM. (1995) Liposome-mediated *CFTR* gene transfer to the nasal epithelium of patients with cystic fibrosis. *Nature Med.* 1: 39–46.

Chelly J, Zeynep T, Tonnesen T, Petterson A, Ishikawabrush Y, Tommerup N, Horn N, Monaco AP. (1993) Isolation of a candidate gene for Menkes disease that encodes a potential heavy metal binding protein. *Nature Genet.* 3: 14–19.

Clarke AR, Mandaag ER, van Roon M, van der Lugt NM, Valk ML, Berns A, te Riele H. (1992) Requirement for a functional *Rb-1* gene in murine development. *Nature* 359: 328–330.

Colledge WH, Abella BS, Southern KW, Ratcliff R, Jiang C, Cheng SH, MacVinish LJ, Anderson JR, Cuthbert AW, Evans MJ. (1995) Generation and characterization of a ΔF508 cystic fibrosis mouse model. *Nature Genet.* 10: 445–450.

Davidson DJ, Dorin JR, McLachlan G, Ranaldi V, Lamb D, Doherty C, Govan J, Porteous DJ. (1995) Lung disease in the cystic fibrosis mouse exposed to bacterial pathogens. *Nature Genet.* 9: 351–357.

Delaney, SJ, Alton EWFW, Smith SN, Lunn DP, Farley R, Lovelock PK, Thomson SA, Hume DA, Lamb D, Porteous DJ, Dorin JR, Wainwright BJ. (1996) Cystic fibrosis mice carrying the missense mutation G551D replicate human genotype–phenotype correlations. *EMBO J.* 15: 955–963.

Dietrich WF, Lander ES, Smith JS, Moser AR, Gould KA, Luongo C, Borenstein N, Dove W. (1993) Genetic identification of *Mom-1*, a major modifier locus affecting min-induced intestinal neoplasia in the mouse. *Cell* 75: 631–639.

Donehower L, Harvey M, Slagle BL, McArthur MJ, Montgomery CA, Butel JS, Bradley A. (1992) Mice deficient for P53 are developmentally normal but susceptible to spontaneous tumours. *Nature* 356: 215–221.

Dorin JR, Alton EWFW, Porteous DJ. (1994a) Mouse models for cystic fibrosis. In: *Cystic Fibrosis Current Topics* (eds JA Dodge, DJH Brock and JH Widdicombe). John Wiley & Sons, Chichester, pp. 3–31.

Dorin JR, Stevenson BJ, Fleming S, Alton EWFW, Dickinson P, Porteous DJ. (1994b) Long-term survival of the exon 10 insertional cystic fibrosis mutant mouse is a consequence of low level residual wild-type *Cftr* gene expression. *Mammal. Genome* 5: 465–472.

Dorin JR, Farley R, Webb S, Smith SN, Farini E, Delaney SJ, Wainwright BJ, Alton EWFW, Porteous DJ. (1996) A demonstration using mouse models that successful gene therapy for cystic fibrosis requires only partial gene correction *Gene Ther.* (in press).

Fodde R, Edelmann W, Yang K, van Leeuwen C, Carlson C, Renault B, Breukel C, Alt E, Lipkin M, Khan PM, Kucherlapati R. (1994) A targeted chain-termination mutation in the mouse *Apc* gene results in multiple intestinal tumours. *Proc. Natl Acad. Sci. USA* 91: 8969–8973.

Gordon JW, Ruddle FH. (1981) Integration and stable germline transmission of genes injected into the mouse pronucleus. *Science* 214: 1244–1246.

Greaves DR, Fraser P, Vidal MA, Hedges MJ, Ropers D, Luzzatto LGF. (1990) A transgenic mouse model of sickle cell disorder. *Nature* 343: 183–185.

Grossman M, Rader DJ, Muller DWM, Kolansky DM, Kozarsky K, Clark BJ, Stein EA, Lupien PJ, Brewer HB, Raper SE, Wilson JM. (1995) A pilot study of *ex vivo* gene therapy for homozygous familial hypercholesterolaemia. *Nature Med.* 1: 1148–1154.

Grubb BR, Pickles RJ, Ye H, Yankaskas JR, Vick RN, Engelhardt JF, Wilson JM, Johnson LG, Boucher RC. (1994) Inefficient gene transfer by adenovirus vector to cystic fibrosis airway epithelia of mice and humans. *Nature* 371: 802–806.

Hamer DH. (1993) 'Kinky hair' disease sheds light on copper metabolism. *Nature Genet.* 3: 3–4.

Haskins ME, Desnick RJ, DiFerrante N, Jezyk PF, Patterson DF. (1984) β-Glucuronidase deficiency in a dog: a model of mucopolysaccharidosis VII. *Pediatr. Res.* 18: 980–984.

Hui CC, Joyner AL. (1993) A mouse model of Greig cephalopolysyndactyly syndrome: the extra-toes[J] mutation contains an intragenic deletion of the *Gli3* gene. *Nature Genet.* 3: 241–246.

Hyde SC, Gill D, Higgins CF, Trezise AEO, MacVinish LJ, Cuthbert AW, Ratcliff R, Evans MJ, Colledge WH. (1993) Correction of the ion transport defect in cystic fibrosis transgenic mice by gene therapy. *Nature* 362: 250–255.

Jacks T, Fazeli A, Schmitt EM, Bronson RT, Goodell MA, Weinberg RA. (1992) Effects of an *Rb* mutation in the mouse. *Nature* 359: 295–300.

Jacob HJ, Brown DM, Bunker RK, Daly MJ, Dzau VJ, Goodman A, Koike G, Kren V, Kurtz T, Lernmark A, Levan G, Mao YP, Pettersson A, Pravenec M, Simon JS, Szpirer C, Szpirer J, Trolliet MR, Winer ES, Lander ES. (1995) A genetic linkage map of the laboratory rat, *Rattus norvegicus*. *Nature Genet.* **9**: 63–69.

Jaenisch R. (1988) Transgenic animals. *Science* **240**: 1468–1474.

Jezyk PF, Haskins ME, Patterson DF. (1977) Mucopolysaccharidosis in a cat with aryl-sulfatase B deficiency: a model of Maroteaux-Lamy syndrome. *Science* **198**: 834–836.

Katsuki M, Sato M, Kimura M, Yokoyama M, Kobayashi K, Nomura T. (1988) Conversion of normal behaviour to *shiverer* by myelin basic protein antisense cDNA in transgenic mice. *Science* **241**: 593–595.

Knowles MR, Gatzy J, Boucher RC. (1981) Increased bioelectric potential difference across respiratory epithelia in cystic fibrosis. *New Engl. J. Med.* **305**: 1489–1495.

Knowles MR, Hohneker KW, Zhaoqing Z, Olsen JC, Noah TL, Hu PC, Leigh MW, Engelhardt JF, Edwards LJ, Jones KR, Grossman M, Wilson JM, Johnson LG, Boucher RC. (1995) A controlled study of adenoviral-vector-mediated gene transfer in the nasal epithelium of patients with cystic fibrosis. *New Engl. J. Med.* **333**: 823–831.

Kreidberg JA, Sariola H, Loring JM, Maeda M, Pelletier J, Housman D, Jaenisch R. (1993) WT-1 is required for early kidney development. *Cell* **74**: 679–691.

Lee EYHP, Chang C, Hu N, Wang YC, Lai CC, Herrup K, Lee WH, Bradley A. (1992) Mice deficient for Rb are nonviable and show defects in neurogenesis and haematopoiesis. *Nature* **359**: 288–330.

Li J, Fang B, Eisensmith R, Li XHC, Nasonkin I, Linlee YC, Mims MP, Hughes A, Montgomery CD, Roberts JD, Parker TS, Levine DM, Woo SLC. (1995) *In vivo* gene therapy for hyperlipidemia: Phenotypic correction in Watanabe rabbits by hepatic delivery of the rabbit LDL receptor gene. *J. Clin. Invest.* **95**: 768–773.

Li T, Davidson BL. (1995) Phenotype correction in retinal pigment epithelium in murine mucopolysaccharidosis VII by adenovirus-mediated gene transfer. *Proc. Natl Acad. Sci. USA* **92**: 7700–7704.

Mercer J, Livingston J, Hall B, Paynter JA, Begy C, Chandrasekharappa S, Lockhart P, Grimes A, Bhave M, Siemieniak D, Glover TW. (1993) Isolation of a partial candidate gene for Menkes disease by positional cloning. *Nature Genet.* **3**: 20–25.

Moullier P, Bohl D, Heard J-M, Danos O. (1993) Correction of lysosomal storage in the liver and spleen of MPS VII mice by implantation of genetically modified skin fibroblasts. *Nature Genet.* **4**: 154–159.

Mullins LJ, Mullins JJ. (1996) Transgenesis in the rat and larger animals. *J. Clin. Invest.* **97**: 1557–1560.

Nasir J, Floresco SB, O'Kusky JR, Diewert VM, Richman JM, Zeisler J, Borowski A, Marth JD, Philips AG, Hayden MR. (1995) Targeted disruption of the Huntington's Disease gene results in embryonic lethality and behavioural and morphological changes in heterozygotes. *Cell* **81**: 811–823.

Pollock D, Gäken J. (1995) Antisense oligonucleotides: a survey of recent literature, possible mechanisms of action and therapeutic progress. In: *Functional Analysis of the Human Genome* (eds F Farzaneh and DN Cooper). BIOS Scientific Publishers, Oxford, pp. 241–265.

Rosenfeld MA, Yoshimura K, Trapnell BC, Yoneyama K, Rosenthal ER, Dalemans W, Fukayama M, Bargon J, Stier LE, Stratford-Perricaudet L. (1992) *In vivo* transfer of the human cystic fibrosis transmembrane conductance regulator gene to the airway epithelium. *Cell* **68**: 143–155.

Rozmahel R, Wilschanski M, Matin A, Plyte S, Oliver M, Auerbach W, Moore A, Forstner J, Durie P, Nadeau J, Bear C, Tsui LC. (1996) Modulation of disease severity in cystic fibrosis transmembrane conductance regulator deficient mice by a secondary genetic factor. *Nature Genet.* **12**: 280–287.

Schedl A, Montolui L, Kelsey G, Schutz G. (1993) A yeast artificial chromosome covering the tyrosinase gene confers copy number-dependent expression in transgenic mice. *Nature* **362**: 258–261.

Smith SN, Alton EWFW, Geddes DM. (1992) Ion transport characteristics of the murine trachea and caecum. *Clin. Sci.* **82**: 667–672.

Tata F, Stanier P, Wicking C, Halford S, Kruyer H, Lench NJS, Hansen PJ. (1991) Cloning the mouse homolog of the human cystic fibrosis transmembrane conductance regulator gene. *Genomics* **10**: 301–307.

Tybulewicz VLJ, Tremblay ML, LaMarca ME, Willemsen R, Stubblefield BK, Winfield S, Zablocka B, Sidrans A. (1992) Animal model of Gaucher's disease from targeted disruption of the mouse glucocerebrosidase gene. *Nature* **357**: 407–410.

van Doorninck JH, French PJ, Verbeek E, Peters RH, Morreau H, Bijman J, Scholte BJ. (1995) A mouse model for the cystic fibrosis ΔF508 mutation. *EMBO J.* **14**: 4403–4411.

Vulpe C, Levinson B, Whitney S, Packman S, Gitschier J. (1993) Isolation of a candidate gene for Menkes disease and evidence that it encodes a copper-transporting ATPase. *Nature Genet.* **3**: 7–13.

Wagner RW. (1994) Gene inhibition using antisense oligodeoxynucleotides. *Nature* **372**: 333–335.

Whitsett JA, Dey C, Stripp BR, Wikenheiser KA, Clark JC, Wert SE, Gregory RJ, Smith AE, Cohn JA, Wilson JM. (1992) Human cystic fibrosis transmembrane conductance regulator directed to respiratory epithelial cells of transgenic mice. *Nature Genet.* **2**: 13–20.

Yamamoto TR, Bishop RW, Brown MS, Goldstein JL, Russell DW. (1986) Deletion in cysteine-rich region of LDL receptor impedes transport to cell surface in WHHL rabbit. *Science* **232**: 1230–1237.

Yorifuji T, Lemna WK, Ballard CF, Rosenbloom CL, Rozmahel RP, Tsui N. (1991) Molecular cloning and sequence analysis of the murine cDNA for the cystic fibrosis transmembrane conductance regulator. *Genomics* **10**: 547–550.

Zabner J, Couture LA, Gregory RJ, Granham SM, Smith AE, Welsh MJ. (1993) Adenovirus-mediated gene transfer transiently corrects the chloride transport defect in nasal epithelia of patients with cystic fibrosis. *Cell* **75**: 207–216.

Zeiher BG, Eichwald E, Zabner J, Smith JJ, Puga AP, McCray PBJ, Capecchi MR, Welsh MJ, Thomas KR. (1995) A mouse model for the ΔF508 allele of cystic fibrosis. *J. Clin. Invest.* **96**: 2051–2064.

10

Gene targeting as an approach to gene therapy

Andrew Porter

10.1 Introduction

In the majority of gene therapy strategies, it is important for therapeutic DNA to persist in recipient cells. This can be achieved by stable integration of the DNA into the genome. The only known way to integrate DNA into a defined part of the genome of a living mammalian cell is by gene targeting, a technique that makes use of the cells' natural ability to carry out homologous recombination. A key attraction of gene targeting in the context of gene therapy is that it can be used to repair a genetic defect; this contrasts with all other other approaches in which the affected gene remains unchanged (*Figure 10.1*). The most obvious potential of gene targeting for gene therapy is therefore for the repair of inherited genetic mutations such as those causing cystic fibrosis and the haemoglobinopathies. It is also possible that the precision afforded by gene targeting could be useful in approaches requiring the efficient expression of genes other than disease genes themselves.

Gene targeting already can be used to repair genetic defects in cell culture, but the process is inefficient and much remains to be done if it is to be used in a therapeutic context. This chapter reviews some of the capabilities of gene targeting and their relevance for gene therapy, and considers the current limitations in applying gene targeting to gene therapy and how they might be overcome.

10.2 Homologous recombination and gene targeting

Most, if not all, organisms have the ability to exchange or transfer information between homologous stretches of DNA (Kucherlapati and Smith, 1988). Cells can use this process to generate genetic diversity [for example, during meiosis (Moens, 1994)] or to repair chromosomal breaks (Shinohara and Ogawa, 1995). The molecular partners in these natural processes may be homologous chromosomes or sister chromatids (interchromosomal homologous recombination) or homologous sequences on the same chromosome (intrachromosomal homologous recombination). However, one of the partners may also be a piece of DNA, homologous to a particular chromosomal locus of interest, that has been introduced artificially into the cell. Experimental homologous recombination of this kind has become known as gene targeting (Capecchi, 1989; Sedivy and Joyner, 1992; Vega, 1995) and is unique in its ability to make defined alterations to the genome of living cells. (The term 'gene targeting' is

Gene Therapy, edited by N.R. Lemoine and D.N. Cooper.
© 1996 BIOS Scientific Publishers Ltd, Oxford.

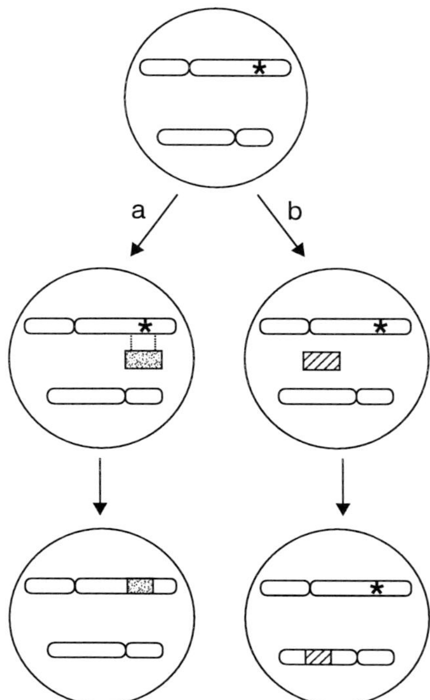

Figure 10.1. Schematic illustration of gene therapy via targeted repair (a) or via other approaches (b). Nuclei are shown containing two different chromosomes, one of which carries a disease mutation (*). Transfection with DNA designed for homologous recombination with the target locus (stippled box) repairs the mutation. Other approaches involve therapeutic DNA (striped) that integrates at a random locus.

sometimes used for the important but quite separate problem of DNA delivery to specific cell types. This subject, for which the term 'cell targeting' might be more appropriate, is dealt with in Chapter 2.)

The process of homologous recombination can be represented simply (see *Figure 10.2*). It is convenient to distinguish two types of targeting event: replacement and insertion. In replacement events, a stretch of genomic DNA is effectively replaced by a homologous stretch from the transfected DNA (replacement construct); sequences in the replacement construct that lie outside the region of homology are lost. In insertion events, the entire piece of transfected DNA (insertion construct) integrates at the region of homology to generate a duplicated region of homology. Insertion constructs are usually linearized within the region of homology. The left and right ends of homology are joined by non-homologous vector DNA that separates the duplicated sequences following integration. Significantly, the duplicated regions of homology can undergo a second, now intrachromosomal, recombination event, resulting in excision of the targeting construct. If, as is likely, the crossover positions for insertion and excision differ, the net effect of insertion plus excision is analogous to replacement.

Although useful, diagrams such as those in *Figure 10.2* do not reflect the complex biochemistry and genetics of homologous recombination which are far from being fully understood. Several mechanisms based on genetic studies of meiotic recombination in fungi have been proposed (Holliday, 1964; Lin *et al.*, 1984; Meselson and Radding,

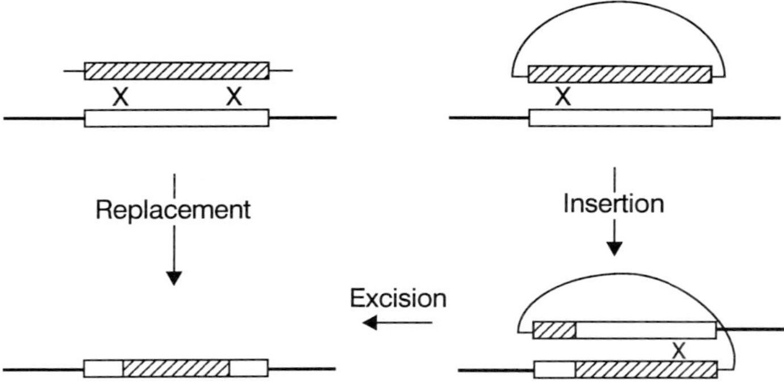

Figure 10.2. Simple representation of gene targeting involving replacement or insertion. Non-homologous chromosomal and vector DNA are shown as thick and thin lines respectively. Regions of homology originating in the target locus and targeting constructs are shown as white and striped boxes respectively. The Xs represent positions of information exchange/transfer during homologous recombination.

1975; Szostak *et al.*, 1983). These mechanisms, and molecular genetic studies of the extent to which they apply to gene targeting, are reviewed elsewhere (Bollag *et al.*, 1989; Waldman, 1995) and will not be discussed further here. The biochemistry of homologous recombination is best characterized in *Escherichia coli* for which various recombination steps can be carried out in the test tube using purified proteins (Dunderdale and West, 1994; West, 1995). Recent progress in the identification of mammalian homologues of recombination genes from simpler organisms (Shinohara and Ogawa, 1995) has raised hopes that a clearer understanding of the mechanism of gene targeting in mammalian cells may soon lead to more efficient targeting protocols.

A further simplification of the diagrams in *Figure 10.2* is that they do not show the genes conferring resistance to drugs which, for reasons described in Section 10.3, are usually included in targeting constructs.

10.3 Dealing with illegitimate recombination and low targeting frequencies

Gene targeting in mammalian cells is an inefficient process. This reflects the variety of fates a targeting construct can have not only *en route* to, but also after reaching, the nucleus (*Figure 10.3*). The great majority of gene targeting experiments to date have delivered DNA by electroporation (Chu *et al.*, 1987) or calcium phosphate precipitation (Graham and Van der Eb, 1973). In these methods, the journey to the nucleus is rather ill defined and inefficient, with many opportunities for the DNA to be lost or degraded. Stable transfection efficiencies are therefore usually quite low, typically in the range 10^{-2}–10^{-5}. Some targeting experiments have used nuclear microinjection as a means of delivering DNA (Thomas *et al.*, 1986; Zimmer and Gruss, 1989). Although this method can achieve transient and stable transfection efficiencies of 50–100% and 20%, respectively (Capecchi, 1980), the number of cells that can be transfected in this way (e.g. up to 1000 h^{-1}) is severely limiting. Nucleic acids can also be delivered with high efficiency by recombinant viral vectors, but such vectors can place limitations on the size and design of the nucleic acid and have generally been avoided for gene targeting experiments.

Once in the nucleus, the targeting construct may still be lost or degraded but, most

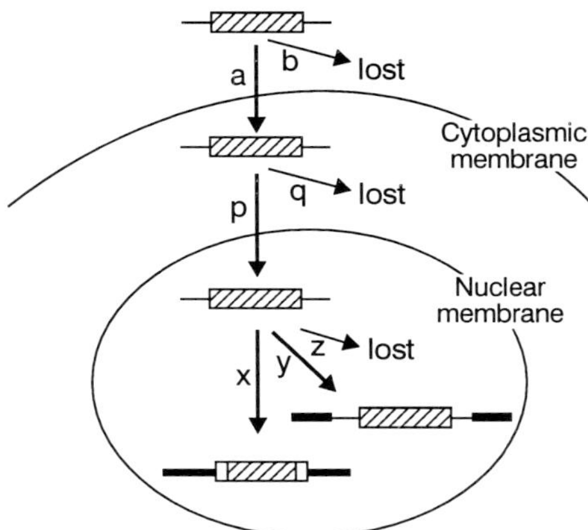

Figure 10.3. Desirable and undesirable paths of a targeting construct. For gene targeting to occur, the targeting construct (striped bar) must enter the cell (a), enter the nucleus (p) and undergo homologous recombination with its target locus (x). At each stage, there are competing fates for the DNA: it may fail to enter the cell (b) or, once in the cell, may fail to reach the nucleus (q). If it does reach the nucleus, it may undergo illegitimate recombination (y) or fail to integrate altogether (z). If the letters represent the fractions of cells following these various pathways then $a + b = 1, p + q = 1, x + y + z = 1$, absolute targeting frequency $(T) = apx$, stable transfection frequency $(S) = apy$ (for $y >> x$) and the ratio of targeted to random integration $(R) = x/y = T/S$. Values for R can range from 10^{-5} to 1, but are more typically in the range of 10^{-3}– 10^{-1}. Values for S vary with cell type and transfection method, but for electroporation of human fibroblasts are typically of the order of 10^{-4}. Typical values for T in human fibroblasts are therefore in the range 10^{-5}–10^{-7}.

significantly, it may also integrate into essentially random sites in the genome by a process of non-homologous (illegitimate) recombination (Roth and Wilson, 1989). Illegitimate recombination is a particular problem because it is typically 10- to 1000-fold more efficient than homologous recombination and because special effort (see below) is required to distinguish homologous from non-homologous recombinants. The combined problems of low transfection efficiencies and low ratios of homologous to illegitimate recombination means that the proportion of transfected cells which become targeted (the absolute targeting frequency), is very low, typically of the order of 10^{-5}–10^{-7} (see legend to *Figure 10.3*).

The challenge of gene targeting is therefore not unlike that of finding a needle in a haystack: not only are the needles (targeted cells) very rare, but the haystack (all other cells surviving transfection) is very large. One way to deal with this is to use selection methods to reduce the size of the haystack. It is on this basis that targeting constructs (or indeed any constructs used to generate stable transfectants) usually include a gene (e.g. *neo*) conferring resistance to a drug (e.g. G418). Such positively selectable genes are positioned within the region of homology in replacement constructs or form part of the vector sequences of insertion constructs. By selecting for drug-resistant clones, one can eliminate all cells that have failed to take up the DNA and incorporate it into their genomes.

Simple positive selection of this kind can reduce the size of the haystack considerably, but we are still left with the residual haystack of non-homologous recombinants. There are two ways to select against non-homologous recombinants. The first, positive–negative selection (Mansour *et al.*, 1988), applies only to targeting with replacement constructs and involves the additional use of a gene [e.g. the thymidine kinase (*tk*) gene of herpes simplex virus] which will kill cells in the presence of a drug (e.g. ganciclovir). If such a negatively selectable gene is included in the targeting construct outside the region of homology it will be lost during a replacement event but will (usually) be retained during illegitimate recombination. Drugs can therefore be used selectively to eliminate non-targeted clones. This approach is widely used, but the degree of enrichment achieved can vary unpredictably from as high as 10 000-fold (Thomas and Capecchi, 1990) to as little as twofold (Mortensen *et al.*, 1992; Tybulewicz *et al.*, 1991; Zijlstra *et al.*, 1989). A second way to eliminate illegitimate recombinants selectively applies only to targets that are genes transcribed in the cells to be transfected. In this approach, expression of a positively selectable marker in the targeting construct is made dependent upon integration by homologous recombination. This can be achieved by fusing a promoterless marker gene to a promoterless region of homology with the target gene in such a way that, following a targeting event, the marker gene is expressed as part of a transcript driven by the target gene promoter. Depending on the nature of the marker and target genes and how they are fused, this promoterless approach can be implemented efficiently in a variety of ways (Charron *et al.*, 1990; Dorin *et al.*, 1989; Itzhaki and Porter, 1991; Jasin and Berg, 1988; Jasin *et al.*, 1990; Mountford *et al.*, 1994; Sedivy and Sharp, 1989).

Selection methods are never perfect, and screening methods are also used, either on their own or, more often, following a preliminary selection step, to identify rare homologous recombinants amongst an excess of illegitimate recombinants. Such screens usually employ Southern blots or polymerase chain reaction (PCR) to identify clones, or pools of clones, that contain a particular DNA structure characteristic of a targeted clone. Depending on the number of clones involved, screening can require quite a lot of manipulation since cultures must be duplicated, one culture for DNA analysis, the other for further growth following positive identification. Even if extensive screening is not required, Southern analysis is usually the best way of confirming that the desired targeting event has taken place. Since a variety of rare targeting-related events [e.g. 'pick-up' events (Adair *et al.*, 1989; Itzhaki and Porter, 1991) and 'one-sided' events (Berinstein *et al.*, 1992)], can survive selection and primary screening procedures, a fairly extensive Southern analysis is advisable.

The screening and selection procedures described above are designed to deal with existing low targeting efficiencies but not to improve them. An alternative way to address the needle in a haystack problem is to increase the number of needles per haystack, that is to improve the targeting efficiency. This can be achieved by improving the delivery system or by improving the efficiency of homologous recombination. Some promising ways of doing this are discussed at the end of this chapter. At present, however, most gene targeting experiments make do with relatively inefficient DNA delivery (by electroporation) and there is little consensus on how best to improve the efficiency of homologous recombination. It has been demonstrated in several mammalian systems that increasing the length of homology improves efficiency (Deng and Capecchi, 1992; Shulman *et al.*, 1990; Smith and Kalogerakis, 1990; Thomas and Capecchi, 1987). The relationship appears to be logarithmic, at least in the range 1–15 kb, and lengths are commonly of the order of 10 kb. The use of isogenic DNA (DNA that is isolated from the cell/strain to be targeted and therefore 100% identical to the target) has also been shown in at least three mammalian systems to improve efficiencies by 5- to 50-fold

(Deng and Capecchi, 1992; te Riele *et al.*, 1992; van Deursen and Wieringa, 1992). Linearization of the targeting construct is routine, although the optimal site for linearization and other aspects of target construct topology are controversial (see below).

10.4 Cell type as a variable in gene targeting

Mouse embryonic stem (ES) cells are by far the most frequently used cells for gene targeting. By modifying target genes in these totipotent cells, it is possible to generate strains of mice with specific genetic alterations. Hundreds of such experiments have been reported (Brandon *et al.*, 1995a–c) and this approach probably represents the most important technical advance in modern genetics. Of specific relevance to gene therapy is the fact that targeted mice that model known human genetic diseases can be generated (Moore and Melton, 1995; Porteous and Dorin, 1993; Smithies, 1993; see Chapter 9). Such animals will be useful in assessing novel therapies and may indeed be useful in developing experimental protocols for gene therapy by gene targeting. The derivation and targeted correction of ES cells from such animals is not envisaged: this would be an example of germ-line gene therapy which, in humans, presently is considered to be unethical, inappropriate or unnecessary (but see Latchman, 1994), and in mice would simply be a reversal of the process by which the strains were created, and hardly informative. However, animals models should be useful for developing protocols involving targeted correction in the appropriate somatic cells (see Section 10.7).

Although the great majority of gene targeting experiments have involved mouse ES cells, gene targeting has also been carried out in somatic cells. Indeed, all the early mammalian gene targeting experiments were carried out in cell lines derived from mouse, human or hamster somatic tissues. With one exception (Smithies *et al.*, 1985), these experiments involved special target genes, such as stably transfected mutant *neo* (Smith and Berg, 1984; Song *et al.*, 1987) or *tk* (Lin *et al.*, 1985) genes, or *aprt* (Adair *et al.*, 1989) and immunoglobulin (Baker and Shulman, 1988), genes for which targeting events resulted in a directly selectable phenotype. Such systems are still very useful for assessing the key parameters in gene targeting.

More recently, gene targeting has been achieved with non-selectable target genes in several somatic cell types including mouse pre-B cell (Charron *et al.*, 1990), chicken B cell (Buerstedde and Takeda, 1991), mouse fibroblast (Accili and Taylor, 1991), human fibrosarcoma (Itzhaki and Porter, 1991; Porter and Itzhaki, 1993), human hepatoma (Farese *et al.*, 1992) human T cell (Manjunath *et al.*, 1993), human colon carcinoma (Shirasawa *et al.*, 1993) and rat fibroblast (Finney and Bishop, 1993) cell lines as well as primary mouse myoblasts (Arbones *et al.*, 1994). With the exception of chicken B cells, which provide a clear and curious example of a cell line with a consistently high targeting efficiency (Bezzubova and Buerstedde, 1994; Buerstedde and Takeda, 1991), targeting efficiencies in somatic cells are unexceptional. There is no reason yet to suppose that somatic cells, as opposed to ES cells, are particularly resistant or prone to targeting. Some extremely efficient targeting experiments have been reported in ES cells (e.g. te Riele *et al.*, 1992) but there also are some reported (and probably many more unreported) examples of very low targeting efficiencies (e.g. Thomas and Capecchi, 1990). These differences probably reflect differences in several key variables such as target locus, isogenicity, construct topology, etc., combinations of which have been explored more thoroughly in ES cells than in somatic cells. In cases where such variables have been held constant whilst comparing targeting in ES and somatic cells, one study

(Arbones *et al.*, 1994) found the absolute targeting efficiency in somatic (myoblast) cells to be eightfold greater than in ES cells, while another (Charron *et al.*, 1990) found no major difference. In the former study, the higher absolute targeting efficiency was a reflection of a much greater transfection efficiency in myoblast cells (120 times that of ES cells) rather than of a higher ratio of homologous to illegitimate recombination, which was actually lower in the myoblasts (15-fold less than in ES cells).

10.5 Choice of genetic modification

In contradiction to its name, gene targeting is not limited to the modification of genes: in principle, any region of the genome can be modified, although it is possible that certain regions (e.g. heterochromatin) are refractory to targeting. Indeed, gene targeting has been used to modify regulatory sequences quite remote from genes (Fiering *et al.*, 1993, 1995; Kim *et al.*, 1992). Nor are the types of modification restricted: any kind of mutation can now been made by gene targeting, from a single base change (Hasty *et al.*, 1991a) to chromosomal engineering involving megabase pairs of DNA (see below).

While the potential for using gene targeting for gene repair has always been apparent, a more immediate use has been to mutate a target locus of interest in order to learn more about its function. Targeted disruptions (or 'knockouts') are the most frequently chosen modifications. Knockouts are most often achieved with replacement constructs, in which a positively selectable drug resistance gene disrupts the target locus, but insertion constructs can also be used. Such crude disruptions can be very informative, but it is often desirable to introduce more subtle, even single base change, mutations. Subtle mutations can be introduced for reconstructing known human disease mutations (Colledge *et al.*, 1995; Yang *et al.*, Zeiher *et al.*, 1995) in mice or for studying the function of specific residues in a target locus (Askew *et al.*, 1993) and may or may not be accompanied by the introduction of selectable markers. Most importantly, in the context of gene therapy, the various methods for introducing subtle mutations are equally applicable for the repair of mutations and examples of this are shown in *Figure 10.4*.

A single replacement event can be used to correct a genetic defect (*Figure 10.4a*). This approach has been used to correct the mutation in the β-globin gene responsible for sickle-cell disease (Shesely *et al.*, 1991). A potential problem with this approach is that a region close to the target gene retains the positively selectable marker gene. It is possible that this foreign DNA could affect expression of the target gene, or of some other nearby gene, with potentially damaging consequences. To avoid this, a second replacement event can be used to remove the drug resistance gene. There are several ways in which this two-step approach can be planned to allow for selection of the second step; the approach shown in *Figure 10.4b* requires a negatively selectable gene (*tk*) to be included in the first targeting construct so that the second targeting event can be selected (Askew *et al.*, 1993). A rather different way of removing the selectable marker genes is to flank them with *loxP* sequences in the original targeting construct (Baubonis and Sauer, 1993). *LoxP* is the 36 bp recognition sequence for the bacteriophage site-specific recombinase, Cre. Following targeting, Cre (which can be introduced into the cell as a protein or be expressed from a transfected plasmid) catalyses intrachromosomal site-specific recombination between the *loxP* sequences, resulting in excision of the marker genes (*Figure 10.4e* and *f*). This leaves behind a single copy of *loxP* which is much less likely to cause problems.

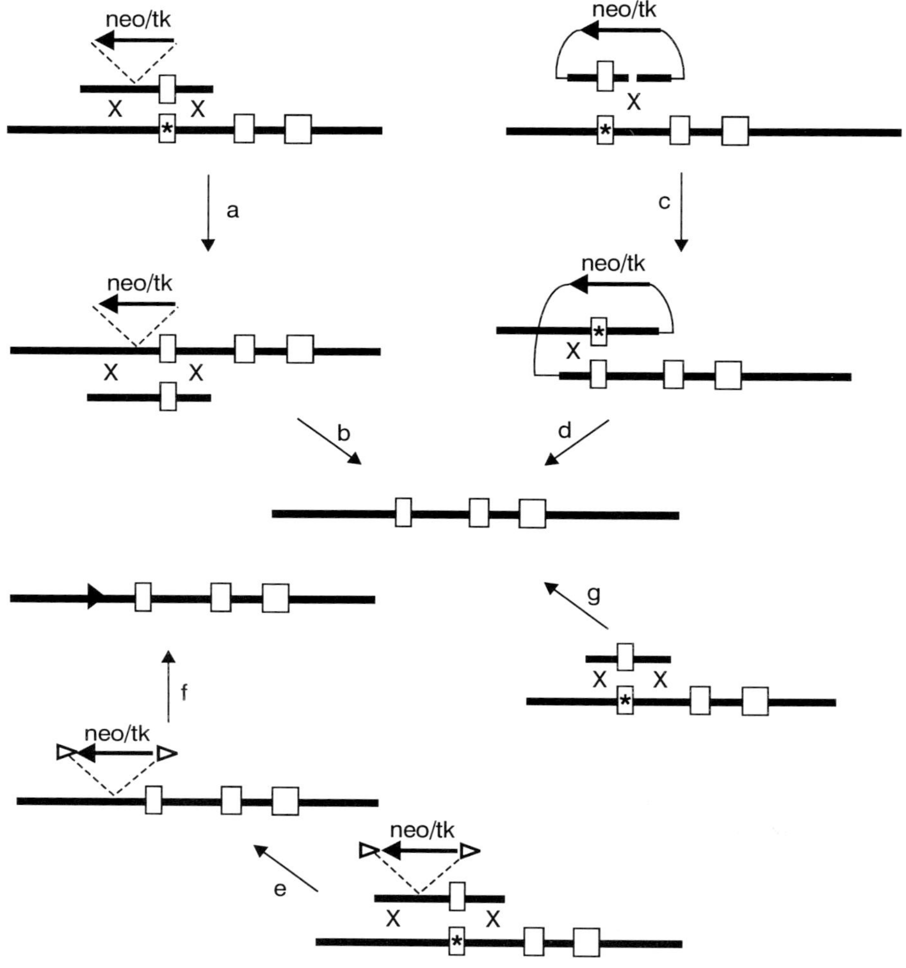

Figure 10.4. Theoretical examples of targeted repair. Targeted repair by replacement (a), double replacement (a and b), insertion (c), insertion/excision (c and d), replacement and Cre-mediated deletion (e and f) and unselected replacement (g) is shown. The target is an imaginary three-exon target gene bearing a mutation (*) in exon 1. *LoxP* sites are represented by open arrowheads. Selected single-step events (a or c) might be sufficient in some situations, but only the selected two-step events (a and b or c and d) or unselected replacement (g) would result in perfect repair. Selection for *neo* and against *tk* would enrich for first and second step events respectively. For further details see text.

A single insertion event can also be designed to repair a genetic lesion as shown in *Figure 10.4c*. This basic approach has been used to correct a mutated *hprt* gene (encoding hypoxanthine guanine phosphoribosyl transferase) in mouse ES cells (Doetschman *et al.*, 1987). As with the single replacement step, gene repair is accompanied by incorporation of unwanted and potentially problematic sequences close to the target gene. In this case, however, the unwanted sequences can, in theory, be removed by spontaneous intrachromosomal homologous recombination (*Figure 10.4d*) requiring neither a second targeting construct nor use of the Cre–*lox* system.

It is theoretically possible to achieve complete repair in a single targeting step by the use of a replacement construct consisting simply of an uninterrupted stretch of homology with the target locus (*Figure 10.4g*). Targeted clones would be identified by a screen involving PCR or Southern analysis. As we have seen, such an approach will only be practical if the DNA is delivered extremely efficiently to avoid the screen being swamped with excessive numbers of untransfected cells. This basic approach was used to introduce a subtle mutation into the *Hox1.1* gene of ES cells which had been microinjected with the targeting construct (Zimmer and Gruss, 1989). The fact that selection-free targeting of this kind has not been adopted widely probably reflects the expense and effort involved in nuclear microinjection which may become limiting in the context of gene therapy (see below).

It is possible to imagine targeted alterations that have therapeutic value but that do not involve the repair of the disease gene itself. For instance, upregulation of the fetal globin gene by natural mutation of its promoter region is known to compensate for mutations in the β-globin gene (Berry *et al.*, 1992). Using the same methods discussed for the repair of disease mutations, gene targeting could be used to introduce such a compensatory mutation. In this way, a single targeting construct may be used for therapy of a variety of different β-haemoglobinopathies, whereas several different constructs would probably be required to repair all of the known β-globin gene mutations.

A relatively recent development is the use of gene targeting for chromosome engineering. Thus, by incorporating into targeting constructs substrates for telomerase or a site-specific recombinase such as Cre, it is has been possible to truncate (Itzhaki *et al.*, 1992), translocate (Smith *et al.*, 1995) or delete (Ramirez Solis *et al.*, 1995) large (involving megabase pairs of DNA) but defined chromosomal regions. This may one day be of therapeutic value (either in the construction and/or manipulation of mammalian artificial chromosomes (Huxley, 1994) or for the repair of gross chromosomal abnormalities), but will not be considered further here.

10.6 Advantages of gene targeting for gene therapy

The attraction of using gene targeting for gene therapy is based mostly on its ability to return a defective gene to its normal state. This is clearly of most relevance to inherited genetic disorders where the underlying mutation has been characterized. In these cases, gene targeting would have several advantages (discussed below and elsewhere; Vega, 1991) over other approaches in which attempts are made to introduce an extra copy of the gene. It is less clear at present how gene targeting may be applied in gene therapy for acquired genetic diseases (cancer, viral infections), where therapy is often based on the unnatural gene expression. It is possible, however, that the precision afforded by gene targeting could be useful in achieving high levels of expression that will often be required for effective therapy.

10.6.1 Stable and appropriate expression

For gene therapy of some diseases (e.g. acquired genetic disease such as cancer or infections), only transient gene expression may be necessary and in others (e.g. haemophilia), gene expression at a fraction of normal levels may be therapeutic. In most situations, however, sustained gene expression and/or expression at optimal levels will be preferable and, in many cases, (e.g. the haemoglobinopathies) essential. A mutated disease gene when returned to its normal state by gene targeting is, by definition, guaranteed to be expressed appropriately, that is stably, at the optimal level and in response to the normal signals. This cannot be said for all other approaches to gene therapy which attempt to express an additional copy of the affected gene, often stripped of its introns and attached to a heterologous promoter, from a chromosomal or extrachromosomal site that bears no relation to its normal chromosomal environment. Bereft of the normal *cis*-acting control sequences, the new gene will probably be expressed sub-optimally. Moreover, genes inserted randomly into the genome will be subject to silencing by chromosomal position effects (Clark *et al.*, 1994), whereas those that remain part of non-replicating extrachromosomal vectors will gradually be lost with successive cell divisions. It may be possible to build into these approaches the necessary *cis*-acting elements such as locus control regions (Plavec *et al.*, 1993), that ensure appropriate expression and/or maintenance, but this will be a demanding task and different for each gene of interest. Gene targeting would, therefore, be the surest way of achieving stable and appropriate gene expression.

10.6.2 Overcoming dominant disease mutations

A large number of inherited genetic disorders are caused by dominant mutations (e.g. Bonne *et al.*, 1995; Chance *et al.*, 1994, Van Adelsberg and Frank, 1995). By definition, a gene with a dominant mutation will not be complemented by its normal counterpart. It is possible that expressing unnaturally large amounts of gene product from an extra copy of the gene may overcome the dominance of the faulty product, but such overexpression may be difficult to achieve and could have deleterious consequences. A much surer solution would be to repair the offending mutation by a gene targeting approach.

10.6.3 No restrictions on size of target gene

Many gene therapy approaches are limited in the amount of DNA they can deliver. For example, retroviral, adenoviral or adeno-associated viral vectors have upper limits of approximately 7, 7.5 and 5 kb respectively. Yet the coding DNA for many disease genes approaches or exceeds these limits (e.g. Koenig *et al.*, 1987; The European Polycystic Kidney Disease Consortium, 1994). If *cis*-acting sequences are to be included in the recombinant viral genome, this size restriction becomes even more of a problem. Gene targeting does not rely on viral delivery of DNA and in any case requires only a portion of the target gene spanning the disease mutation. The size of the target gene is therefore irrelevant.

10.6.4 Safety advantages

DNA is a mutagen by virtue of its ability to integrate into the genome. Thus, any approach that can result in the essentially random integration of DNA into the genome runs the risk of activating an oncogene or inactivating a tumour suppressor gene. Such risks may be low but could be significant, particularly in protocols where transient expression leads to a requirement for repeated introduction of DNA into a patient's cells. Targeted correction clearly avoids these risks altogether. Further risks are associated with approaches that rely on the use of viral vectors. Although such vectors are engineered to be replication deficient, replicating viruses might be generated by recombination with helper viruses during viral production, or even with endogenous latent viruses in the recipient cells. Gene targeting does not rely on the use of viral vectors, so that such concerns need not apply.

10.7 The challenges of gene targeting for therapy

We have seen how existing gene targeting methods can be used to repair genes in cultured cells and how such repair would be better than other approaches towards gene therapy. Now we must consider how these methods might be used therapeutically to modify specific somatic cells of a patient or animal model; it is here that we face new challenges.

For gene therapy to be of long-term benefit, it is important to modify the somatic stem cells that give rise to an appropriate tissue in the patient. Modification of more differentiated cells will usually be only transiently therapeutic because of the natural cell turnover in most tissues. Unfortunately, putative somatic stem cells have been described for only a few tissues including blood (Uchida *et al.*, 1993; Whetton and Dexter, 1993), muscle (Morgan *et al.*, 1993) and epithelium (Jones *et al.*, 1995). Somatic stem cells are very rare, and difficult to assay unequivocally as well as to maintain and expand in culture without losing their pluripotency. Furthermore, under normal conditions, stem cells are likely to be quiescent. This may explain the difficulties experienced in transducing haematopoietic stem cells (HSCs) with recombinant retroviruses, and may also have implications for gene targeting of stem cells since there is evidence that homologous recombination requires cycling cells (Wong and Capecchi, 1987).

These problems become especially challenging when one is considering the possibilities of gene targeting in somatic stem cells. This can be illustrated by considering the gene targeting of HSCs, which have been estimated to be present in bone marrow at a frequency of approximately 5×10^{-5} (Sutherland *et al.*, 1990). For a standard targeting procedure, with a targeting efficiency of say 10^{-6}, one would therefore need to transfect 2×10^{10} bone marrow cells in order to obtain a single targeted stem cell. This figure is similar to the number of bone marrow cells typically harvested from a 70 kg person for transplantation. Although, by definition, a single stem cell can reconstitute the haematopoietic system, in practice only a fraction (e.g. $10^{-1}–10^{-4}$) of infused stem cells will succeed in lodging productively in the bone marrow, presenting an overall shortfall in efficiency.

There are clearly two ways to tackle this problem: to increase the targeting efficiency or to develop a method for obtaining stem cells in greater numbers. Two extreme approaches can be envisaged, each based on a dramatic improvement in one of these areas. The first depends on a breakthrough that allows a patient's somatic stem cells to be purified and expanded in culture. This would allow the stem cells to be transfected with a targeting construct, and for rare targeted clones to be identified, expanded and returned to the patient (as illustrated for mice in *Figure 10.5j–n*). Such an *ex vivo* approach would be analogous to gene targeting in ES cells (*Figure 10.5a–i*), except that the modified stem cells would contribute to a single tissue rather than to all tissues. At the other extreme is an *in vivo* approach which does not require an ability to grow and expand stem cells in culture. In this approach, the targeting construct would be administered directly to the patient's tissue. This approach would certainly require major developments not only in the efficiency of gene targeting but also in the ability to target delivery to the appropriate stem cells (i.e. cell targeting).

A third possibility is that a combination of relatively modest improvements in both stem cell culture and in cell and gene targeting would allow a successful combination of *in vivo* and *ex vivo* approaches. Thus, stem cells might be purified, maintained *ex vivo* with little or no expansion and, following transfection, returned to the patient. Such an approach would not be unlike those currently used in gene marking (Brenner

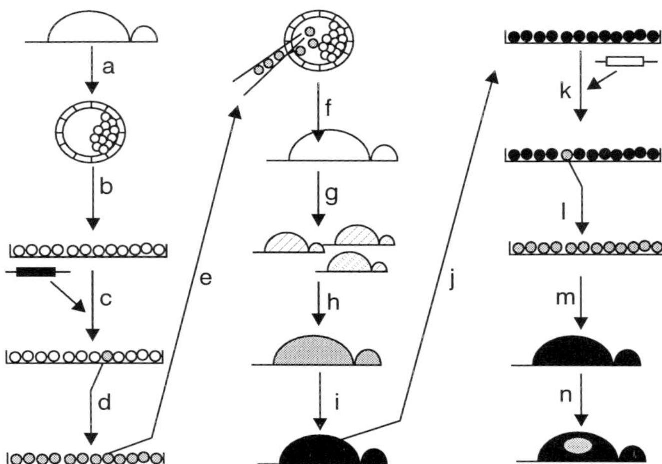

Figure 10.5. Schematic representation of the use of gene targeting for the creation (a–i, current technology) and somatic cell gene therapy (j–n, prospective technology) of genetic disease in mice. Removal of blastocyst (a), establishment of ES cell line (b), transfection of ES cells with a targeting construct carrying disease mutation (c), isolation and expansion of a targeted ES cell clone heterozygous for disease mutation (d), injection of targeted ES cells into a new blastocyst (e), development of a blastocyst in a foster mother (f), generation of chimeric heterozygous offspring (g), breeding of germ-line chimeras to true heterozygotes (h), breeding of mice homozygous for disease mutation (i), isolation and expansion of somatic stem cells of the relevant tissue (j), transfection with a targeting construct designed for disease gene repair (k), isolation and expansion of a targeted clone heterozygous for a disease mutation (l), introduction of targeted somatic stem cells into appropriate tissue (m), *in vivo* regeneration of genetically repaired tissue (n).

et al., 1993) and gene therapy protocols involving retroviral transduction of HSCs (Brenner *et al.*, 1995; Hoogerbrugge *et al.*, 1995; Salvetti *et al.*, 1995). In these protocols, enrichment of stem cells is carried out *in vitro* following bone marrow harvesting by isolation of cells positive for stem cell surface markers (e.g. CD34 for human cells). These are then transduced with recombinant retrovirus before returning them to the patient. In one gene marking study (Brenner *et al.*, 1993), there was evidence that as many as 15% of clonogenic progenitor cells carried the transgene after autologous bone marrow transplantation. Long-term (>2 years) maintenance of the transgene in parts of the haematopoietic system of some patients has been demonstrated. The proportion of true stem cells in the enriched populations and the efficiency with which they are transduced are difficult to judge, but it is unlikely that such procedures can tolerate much reduction in the efficiency of gene transfer. If gene targeting is to be combined with these protocols, its absolute efficiency, with respect to the stem cell population, must approach the efficiency with which stem cells are transduced by recombinant retroviruses.

The design and construction of the targeting construct would also require some special considerations. To maximize the frequency of gene targeting, it would be desirable to use DNA isogenic to the target locus. This requires cloning of the target locus from the patient by making and screening a genomic mini-library or by cloning a high-fidelity PCR product. The appropriate genetic change (e.g. base change) would then need to be introduced (unless the disease concerned was autosomal dominant and the normal allele happened to be isolated) by standard procedures. These steps alone, though technically straightforward, would be time consuming.

The possible use of selectable marker genes would then require consideration. If methods for *ex vivo* growth of somatic stem cells are available, selectable marker genes could be chosen and used as normal (*Figure 10.4*). On the other hand, restrictions on the amount of expansion *ex vivo* might mean that only the single-step procedures (*Figure 10.4a* and *c*) are practical. In the case of the insertion construct, however, the option of selecting *in vivo* with ganciclovir for the second step (excision, *Figure 10.4d*) would still be open. If *ex vivo* expansion is not possible, positive selection *in vivo* for targeted, or at least stably transfected, clones would be an advantage. In a few situations (Blaese *et al.*, 1995; Vega *et al.*, 1994), targeted repair itself may confer a useful selective advantage *in vivo*. In most cases, however, a positively selectable marker that could be used *in vivo*, as *neo* is used *ex vivo*, would be useful. The multidrug resistance gene might be considered in this context (Sorrentino *et al.*, 1992). Alternatively, if the frequency of gene targeting is sufficiently high to yield enough targeted cells to re-populate the affected tissue, it may be possible to omit positive selection completely, in which case single-step replacement (*Figure 10.4e*) would achieve complete repair.

10.8 Future developments

10.8.1 Improving culture of somatic stem cells

The ability to grow and genetically manipulate any one of a number of putative somatic stem cells in culture would be a major medical advance. It is likely that HSCs will continue to be subjected to the most intense efforts in this regard. This reflects the many diseases, both inherited and acquired, that would respond to reconstitution of the haematopoietic system, and the particular accessibility of the haematopoietic system to manipulation. The potential for genetic intervention only adds to the incentive for improved understanding of HSCs. Continued searches for optimal tissue sources (Pettengell *et al.*, 1994) and combinations of growth factors for both *in vivo* mobilization and *ex vivo* maintenance of both mouse and human stem cells are likely to yield important advances (Hoffman *et al.*, 1993; Whetton and Dexter, 1993). Such procedures need to be optimized to ensure that they do not compromise the pluripotency and self-renewing capability of stem cells. Improved assays for stem cell properties would be very helpful. The best assay, reconstitution after lethal irradiation (Smith *et al.*, 1991; Uchida *et al.*, 1994), is clearly unsuitable for human studies. Reconstitution of irradiated, immunodeficient mice with human cells is a promising assay for human stem cells (Chen *et al.*, 1994).

In the absence of such culture conditions, a useful experimental approach involves conditional immortalization of stem cells. Thus, muscle precursor cells with stem cell-like properties can be isolated from an oncomouse, which conditionally expresses the immortalizing T antigen of simian virus 40. These can be grown and genetically manipulated in culture in conditions that allow T antigen expression, and returned to the muscle of an animal where, in the absence of T antigen expression, they contribute to the formation of mature muscle (Morgan *et al.*, 1994). Such a system might be adapted to allow the isolation and characterization of other cells with the properties of somatic stem cells.

Another experimental approach to the isolation of stem cells, which has so far been used only for HSCs, is the differentiation of ES cells in culture (Palacios *et al.*, 1995). While somewhat removed from the clinic, this approach offers a tractable system for analysis of stem cell formation and differentiation in culture.

10.8.2 Improving delivery of targeting constructs

Inefficient DNA delivery is one of the two key factors responsible for low absolute targeting efficiencies (see *Figure 10.3*). It would be a major advantage if targeting constructs could be delivered with the efficiency that viruses deliver their genomes. Some preliminary experiments involving retrovirus (Ellis and Bernstein, 1989) and adenovirus (Fujita *et al.*, 1995; Mitani *et al.*, 1995; Wang and Taylor, 1993) have been conducted to explore this possibility, and the results were encouraging. More work in this area is desirable. An exciting possibility with such approaches is that they may eventually be combined with 'pseudo-typing' studies in which the genes encoding viral coat proteins are engineered so as to alter the species or tissue specificity of the virus (Chu *et al.*, 1994; Michael and Curiel, 1994). In this way, it may be possible to develop an infective particle designed for both cell and gene targeting.

The use of cationic lipids as a means of delivery in gene targeting experiments has not been reported widely despite much interest in their use for gene delivery. As with recombinant viruses, opportunities exist for modifying such lipids in order to control their efficiency and specificity of delivery (Schofield and Caskey, 1995). Cationic lipids may, therefore, provide an alternative to pseudo-typed viruses as a route for the construction of vectors capable of both cell and gene targeting and would seem to warrant more attention in the context of gene targeting.

10.8.3 Improving efficiency of homologous recombination

The reason that homologous recombination is so inefficient in mammalian cells is unclear. Since homologous recombination is part of a natural repair mechanism, it might be induced by DNA-damaging agents and there is some evidence that these can be used to promote gene targeting (Kardinal *et al.*, 1995). The effects seem to be modest, however, and are likely to be accompanied by unwanted mutagenesis.

The nature of the target gene is likely to be an important variable in determining targeting frequencies. Because transcription through the target locus can cause a modest increase in efficiency (Thyagarajan *et al.*, 1995), where possible, transcriptional induction of the target would seem advisable. Mini- and microsatellites appear to promote extrachromosomal homologous recombination (Wahls *et al.*, 1990a,b), and it is possible that these, or perhaps other sequences, will promote gene targeting if they are positioned appropriately with respect to the target locus. If so, it may be feasible in future to choose the regions of homology used in targeting constructs so as to optimize targeting efficiencies.

The variable over which the experimenter has most control is indeed the nature of the targeting construct itself. The importance of the length and fidelity of the homology are well established and have already been mentioned. It is generally accepted that linearization of constructs promotes targeting but the optimal site for linearization and the relative efficiencies of insertion or replacement constructs are unclear. In some systems, replacement and insertion vectors targeted with equal efficiency (Deng and Capecchi, 1992; Nairn *et al.*, 1993; Ward *et al.*, 1993), but in others insertion vectors were more efficient (Dickinson *et al.*, 1993; Hasty *et al.*, 1991b). These differences may be related to whether or not isogenic DNA was used (Deng and Capecchi, 1992), but it also appears that the nature of the target locus is crucial in determining whether or not insertion constructs are more efficient than replacement constructs (Hasty *et al.*, 1994). For insertion vectors, a linearization site positioned centrally in the region of homology seems to promote targeting events (Dickinson *et al.*, 1993; Hasty *et al.*, 1992). Nevertheless, although they reduce the frequency of targeting, terminal heterologies can be compatible with targeting; the heterologies are removed during targeting or, if they are short, they can be incorporated into the target locus (Hasty *et al.*, 1992; Kumar and Simons, 1993). Future experiments may identify further features of the targeting construct that affect efficiency. For example, the methylation status of the targeting DNA may be important and, to date, efforts have not been made to match this with the target locus.

It seems, perhaps surprisingly, that the rate-limiting step in gene targeting is not the search for homology between target and transfected DNA. This conclusion is based on studies where the concentration of target (Zheng and Wilson, 1990) or transfecting sequences (Thomas *et al.*, 1986) was varied without effect. These experiments are not entirely conclusive since they can also be interpreted in terms of competing reactions. Nevertheless, the same conclusion has been reached following kinetic analyses of an *in vitro* homology search system (Yancey Wrona and Camerini Otero, 1995). The steps following recognition are therefore likely to be rate limiting, in which case isolation and over-expression of the genes for the proteins involved could promote homologous recombination. Some candidate mammalian genes have already been identified on the basis of homology to recombination genes from bacteria and yeast (Shinohara and Ogawa, 1995; Shinohara *et al.*, 1993). A potential danger of this approach, as with any approach that attempts to produce a 'global' increase in homologous recombination, is that it could enhance the frequency of intra- and interchromosomal homologous recombination, and so destabilize the genome. It may, therefore, be preferable to form a complex between the purified protein products of these genes and the targeting DNA prior to transfection. In this way, homologous recombination would be promoted specifically at the target locus. Other proteins, including histones, might usefully be complexed with the targeting DNA before transfection in order to make it more closely related to a natural chromatin substrate.

Another possible explanation for low targeting frequencies is that illegitimate recombination and homologous recombination represent competing pathways so that the high efficiency of illegitimate recombination not only masks gene targeting events but reduces their frequency. If this is the case, any method that specifically impairs illegitimate recombination would not only help to remove the background of illegitimate recombinants but would also increase the absolute frequency of gene targeting. The observation that illegitimate and homologous recombination show differential sensitivity to the ADP ribosylase inhibitor 3-methoxybenzamide (Waldman and Waldman, 1990) therefore suggests that this compound might promote targeting events. Perhaps other components of the non-homologous recombination system might be similarly inhibited in future.

It has been shown recently that linearization of the chromosomal target DNA with rare-cutting endonucleases can increase the frequency of gene targeting by up to three orders of magnitude (Choulika *et al.*, 1995; Rouet, 1994; Smih *et al.*, 1995). In its present form, this procedure will not be useful for promoting single-step targeting since it requires the prior introduction of a recognition sequence for the endonuclease into the target sequence. This situation might change if target-specific cleavage can be achieved in cells through the use of triplex-forming oligonucleotides (see Section 10.8.4). Endonuclease-induced target cleavage could be useful, however, in promoting the second step of two-step strategies (*Figure 10.4b* and *d*) if the endonuclease recognition sequence is incorporated into the targeting construct used in step 1.

10.8.4 Targeting with oligonucleotides

The possibility of using oligonucleotides as therapeutic agents is very attractive. Oligonucleotides can be produced with much greater speed and flexibility than plasmid DNA or viral particles and are not associated with any risk of biological contamination. Unfortunately, as has already been discussed, homologous recombination depends critically on homology length, and this presents a major obstacle to the use of oligonucleotides for gene targeting. Nevertheless, single-stranded oligonucleotides have been used for repairing an extrachromosomal *neo* gene by homologous recombination in human cells (Campbell *et al.*, 1989). This work has yet to be extended to include gene targeting, and it may be that targeting frequencies are undetectably low. It is possible that single- or double-stranded oligonucleotides complexed with purified recombination proteins will undergo targeting at detectable frequencies.

Even if oligonucleotides cannot be induced to participate in gene targeting themselves, the ability of certain sequences to form triplexes with homologous target sequences within cells might be useful in promoting targeting events. Oligonucleotides can be used in the test tube to promote target-specific cleavage or mutation, either by linkage to agents such as EDTA.Fe (Moser and Dervan, 1987) or psoralen (Havre *et al.*, 1993) or by protecting the target sequence from methylation and therefore promoting subsequent endonucleolytic cleavage (Strobel *et al.*, 1991). Based on work described in Section 10.8.3, it seems likely that these methods could be used to promote gene targeting if they can be made to work in living cells. Unmodified triplex-forming oligonucleotides can form triplexes with target sequences in living mammalian cells (Ing *et al.*, 1993; Postel *et al.*, 1991) where recently they have been shown to promote mutagenesis via host-mediated repair mechanisms (Wang *et al.*, 1996). Further work is likely to extend the range of oligonucleotide sequences that can form triplexes and to elucidate and exploit triplex interactions with host repair and recombination systems.

10.9 Concluding remarks

Gene targeting methods for making specific modifications to the mammalian genome are well established, and their potential advantages for gene therapy are easy to appreciate. The major limitations in using gene targeting for gene therapy are our limited ability to isolate and grow somatic stem cells, the inefficiency of suitable DNA delivery methods and the inefficiency of mammalian homologous recombination itself. In each of these areas, there is a need for greater understanding of basic parameters involved so that they can be manipulated to allow effective targeted gene therapy.

Acknowledgements

I am grateful to Maggie Dallman, John Porter and Rafael Yáñez for helpful discussions.

References

Accili D, Taylor SI. (1991) Targeted inactivation of the insulin receptor gene in mouse 3T3-L1 fibroblasts via homologous recombination. *Proc. Natl Acad. Sci. USA* **88**: 4708–4712.

Adair GM, Nairn RS, Wilson JH, Seidman MM, Brotherman KA, MacKinnon C, Scheerer JB. (1989) Targeted homologous recombination at the endogenous adenine phosphoribosyltransferase locus in Chinese hamster cells. *Proc. Natl Acad. Sci. USA* **86**: 4574–4578.

Arbones ML, Austin HA, Capon DJ, Greenburg G. (1994) Gene targeting in normal somatic cells: inactivation of the interferon-gamma receptor in myoblasts. *Nature Genet.* **6**: 90–97.

Askew GR, Doetschman T, Lingrel JB. (1993) Site-directed point mutations in embryonic stem cells: a gene-targeting tag-and-exchange strategy. *Mol. Cell. Biol.* **13**: 4115–4124.

Baker MD, Shulman MJ. (1988) Homologous recombination between transferred and chromosomal immunoglobulin kappa genes. *Mol. Cell. Biol.* **8**: 4041–4047.

Baubonis W, Sauer B. (1993) Genomic targeting with purified Cre recombinase. *Nucleic Acids Res.* **21**: 2025–2029.

Berinstein N, Pennell N, Ottaway CA, Shulman MJ. (1992) Gene replacement with one-sided homologous recombination. *Mol. Cell. Biol.* **12**: 360–367.

Berry M, Grosveld F, Dillon N. (1992) A single point mutation is the cause of the Greek form of hereditary persistence of fetal haemoglobin. *Nature* **358**: 499–502.

Bezzubova OY, Buerstedde JM. (1994) Gene conversion in the chicken immunoglobulin locus: a paradigm of homologous recombination in higher eukaryotes. *Experientia* **50**: 270–276.

Blaese RM, Culver KW, Miller AD, Carter CS, Fleisher T, Clerici M, Shearer G, Chang L, Chiang Y, Tolstoshev P, Greenblatt JJ, Rosenberg SA, Klein H, Berger M, Mullen CA, Ramsey WJ, Muul L, Morgan RA, Anderson WF. (1995) T lymphocyte-directed gene therapy for ADA-SCID: initial trial results after 4 years. *Science* **270**: 475–480.

Bollag RJ, Waldman AS, Liskay RM. (1989) Homologous recombination in mammalian cells. *Annu. Rev. Genet.* **23**: 199–225.

Bonne G, Carrier L, Bercovici J, Cruaud C, Richard P, Hainque B, Gautel M, Labeit S, James M, Beckmann J, Weissenbach J, Vosberg H-P, Fiszman M, Komajda M, Schwartz K. (1995) Cardiac myosin binding protein-C gene splice acceptor site mutation is associated with familial hypertrophic cardiomyopathy. *Nature Genet.* **11**: 438–440.

Brandon EP, Idzerda RL, McKnight GS. (1995a) Knockouts. Targeting the mouse genome: a compendium of knockouts (Part I). *Curr. Biol.* **5**: 625–634.

Brandon EP, Idzerda RL, McKnight GS. (1995b) Targeting the mouse genome: a compendium of knockouts (Part II). *Curr. Biol.* **5**: 758–765.

Brandon EP, Idzerda RL, McKnight GS. (1995c) Targeting the mouse genome: a compendium of knockouts (Part III). *Curr. Biol.* **5**: 873–881.

Brenner MK, Rill DR, Holladay MS, Heslop HE, Moen RC, Buschle M, Krance RA, Santana VM, Anderson WF, Ihle JN. (1993) Gene marking to determine whether autologous marrow infusion restores long-term haemopoiesis in cancer patients. *Lancet* **342**: 1134–1137.

Brenner MK, Cunningham JM, Sorrentino BP, Heslop HE. (1995) Gene transfer into human hemopoietic progenitor cells. *Br. Med. Bull.* **51**: 167–191.

Buerstedde JM, Takeda S. (1991) Increased ratio of targeted to random integration after transfection of chicken B cell lines. *Cell* **67**: 179–188.

Campbell CR, Keown W, Lowe L, Kirschling D, Kucherlapati R. (1989) Homologous recombination involving small single-stranded oligonucleotides in human cells. *New Biol.* **1**: 223–227.

Capecchi MR. (1980) High efficiency transformation by direct microinjection of DNA into cultured mammalian cells. *Cell* **22**: 479–488.

Capecchi MR. (1989) Altering the genome by homologous recombination. *Science* **244**: 1288–1292.

Chance PF, Abbas N, Lensch MW, Pentao L, Roa BB, Patel PI, Lupski JR. (1994) Two autosomal dominant neuropathies result from reciprocal DNA duplication/deletion of a region on chromosome 17. *Hum. Mol. Genet.* **3**: 223–228.

Charron J, Malynn BA, Robertson EJ, Goff SP, Alt FW. (1990) High-frequency disruption of the N-*myc* gene in embryonic stem and pre-B cell lines by homologous recombination. *Mol. Cell. Biol.* **10**: 1799–1804.

Chen BP, Galy A, Kyoizumi S, Namikawa R, Scarborough J, Webb S, Ford B, Cen DZ, Chen SC. (1994) Engraftment of human hematopoietic precursor cells with secondary transfer potential in SCID-hu mice. *Blood* **84**: 2497–2505.

Choulika A, Perrin A, Dujon B, Nicolas JF. (1995) Induction of homologous recombination in mammalian chromosomes by using the I–SceI system of *Saccharomyces cerevisiae. Mol. Cell. Biol.* **15:** 1968–1973.

Chu G, Hayakawa H, Berg P. (1987) Electroporation for the efficient transfection of mammalian cells with DNA. *Nucleic Acids Res.* **15:** 1311–1326.

Chu TH, Martinez I, Sheay WC, Dornburg R. (1994) Cell targeting with retroviral vector particles containing antibody–envelope fusion proteins. *Gene Ther.* **1:** 292–299.

Clark AJ, Bissinger P, Bullock DW, Damak S, Wallace R, Whitelaw CB, Yull F. (1994) Chromosomal position effects and the modulation of transgene expression. *Reprod. Fertil. Devel.* **6:** 589–598.

Colledge WH, Abella BS, Southern KW, Ratcliff R, Jiang C, Cheng SH, MacVinish LJ, Anderson JR, Cuthbert AW, Evans MJ. (1995) Generation and characterization of a delta F508 cystic fibrosis mouse model. *Nature Genet.* **10:** 445–452.

Deng C, Capecchi MR. (1992) Reexamination of gene targeting frequency as a function of the extent of homology between the targeting vector and the target locus. *Mol. Cell. Biol.* **12:** 3365–3371.

Dickinson P, Kimber WL, Kilanowski FM, Stevenson BJ, Porteous DJ, Dorin JR. (1993) High frequency gene targeting using insertional vectors. *Hum. Mol. Genet.* **2:** 1299–1302.

Doetschman T, Gregg RG, Maeda N, Hooper ML, Melton DW, Thompson S, Smithies O. (1987) Targeted correction of a mutant HPRT gene in mouse embryonic stem cells. *Nature* **330:** 576–578.

Dorin JR, Inglis JD, Porteous DJ. (1989) Selection for precise chromosomal targeting of a dominant marker by homologous recombination. *Science* **243:** 1357–1360.

Dunderdale HJ, West SC. (1994) Recombination genes and proteins. *Curr. Opin. Genet. Devel.* **4:** 221–228.

Ellis J, Bernstein A. (1989) Gene targeting with retroviral vectors: recombination by gene conversion into regions of nonhomology. *Mol. Cell. Biol.* **9:** 1621–1627.

Farese RV Jr, Flynn LM, Young SG. (1992) Modification of the apolipoprotein B gene in HepG2 cells by gene targeting. *J. Clin. Invest.* **90:** 256–261.

Fiering S, Kim CG, Epner EM, Groudine M. (1993) An 'in–out' strategy using gene targeting and FLP recombinase for the functional dissection of complex DNA regulatory elements: analysis of the beta-globin locus control region. *Proc. Natl Acad. Sci. USA* **90:** 8469–8473.

Fiering S, Epner E, Robinson K, Zhuang Y, Telling A, Hu M, Martin DI, Enver T, Ley TJ, Groudine M. (1995) Targeted deletion of 5'HS2 of the murine beta-globin LCR reveals that it is not essential for proper regulation of the beta-globin locus. *Genes. Devel.* **9:** 2203–2213.

Finney RE, Bishop JM. (1993) Predisposition to neoplastic transformation caused by gene replacement of H-*ras*1. *Science* **260:** 1524–1527.

Fujita A, Sakagami K, Kanegae Y, Saito I, Kobayashi I. (1995) Gene targeting with a replication-defective adenovirus vector. *J. Virol.* **69:** 6180–6190.

Graham FL, Van der Eb AJ. (1973) A new technique for the assay of infectivity of human adenovirus 5 DNA. *Virology* **52:** 456–467.

Hasty P, Ramirez Solis R, Krumlauf R, Bradley A. (1991a) Introduction of a subtle mutation into the *Hox-2.6* locus in embryonic stem cells. *Nature* **350:** 243–246.

Hasty P, Rivera Perez J, Chang C, Bradley A. (1991b) Target frequency and integration pattern for insertion and replacement vectors in embryonic stem cells. *Mol. Cell. Biol.* **11:** 4509–4517.

Hasty P, Rivera Perez J, Bradley A. (1992) The role and fate of DNA ends for homologous recombination in embryonic stem cells. *Mol. Cell. Biol.* **12:** 2464–2474.

Hasty P, Crist M, Grompe M, Bradley A. (1994) Efficiency of insertion versus replacement vector targeting varies at different chromosomal loci. *Mol. Cell. Biol.* **14:** 8385–8390.

Havre PA, Gunther EJ, Gasparro FP, Glazer PM. (1993) Targeted mutagenesis of DNA using triple helix-forming oligonucleotides linked to psoralen. *Proc. Natl Acad. Sci. USA* **90:** 7879–7883.

Hoffman R, Tong J, Brandt J, Travcoff C, Bruno E, McGuire BW, Gordon MS, McNiece I, Srour EF. (1993) The *in vitro* and *in vivo* effects of stem cell factor on human hematopoiesis. *Stem Cells Dayt.* **2:** 76–82.

Holliday RA. (1964) A mechanism for gene conversion in fungi. *Genet. Res. Camb.* **5:** 282.

Hoogerbrugge PM, von Beusechem VW, Kaptein LC, Einerhand MP, Valerio D. (1995) Gene therapy for adenosine deaminase deficiency. *Br. Med. Bull.* **51:** 72–81.

Huxley C. (1994) Mammalian artificial chromosomes: a new tool for gene therapy. *Gene Ther.* **1:** 7–12.

Ing NH, Beekman JM, Kessler DJ, Murphy M, Jayaraman K, Zendegui JG, Hogan ME, O'Malley BW, Tsai MJ. (1993) *In vivo* transcription of a progesterone-responsive gene is specifically inhibited by a triplex-forming oligonucleotide. *Nucleic Acids Res.* **21:** 2789–2796.

Itzhaki JE, Porter AC. (1991) Targeted disruption of a human interferon-inducible gene detected by secretion of human growth hormone. *Nucleic Acids Res.* 19: 3835–3842.

Itzhaki JE, Barnett MA, MacCarthy AB, Buckle VJ, Brown WR, Porter AC. (1992) Targeted breakage of a human chromosome mediated by cloned human telomeric DNA. *Nature Genet.* 2: 283–287.

Jasin M, Berg P. (1988) Homologous integration in mammalian cells without target gene selection. *Genes Devel.* 2: 1353–1363.

Jasin M, Elledge SJ, Davis RW, Berg P. (1990) Gene targeting at the human CD4 locus by epitope addition. *Genes Devel.* 4: 157–166.

Jones PH, Harper S, Watt FM. (1995) Stem cell patterning and fate in human epidermis. *Cell* 80: 83–93.

Kardinal C, Hooijberg E, Lang P, Zeidler R, Mocikat R. (1995) Integration vectors for antibody chimerization by homologous recombination in hybridoma cells. *Eur. J. Immunol.* 25: 792–797.

Kim CG, Epner EM, Forrester WC, Groudine M. (1992) Inactivation of the human beta-globin gene by targeted insertion into the beta-globin locus control region. *Genes Devel.* 6: 928–938.

Koenig M, Hoffman EP, Bertelson CJ, Monaco AP, Feener C, Kunkel LM. (1987) Complete cloning of the Duchenne muscular dystrophy (DMD) cDNA and preliminary genomic organization of the DMD gene in normal and affected individuals. *Cell* 50: 509–517.

Kucherlapati R, Smith GR. (1988) *Genetic Recombination.* American Society for Microbiology, Washington, DC.

Kumar S, Simons JP. (1993) The effects of terminal heterologies on gene targeting by insertion vectors in embryonic stem cells. *Nucleic Acids Res.* 21: 1541–1548.

Latchman DS. (1994) Germline gene therapy? *Gene Ther.* 1: 277–279.

Lin FL, Sperle K, Sternberg N. (1984) Model for homologous recombination during transfer of DNA into mouse L cells: role for DNA ends in the recombination process. *Mol. Cell. Biol.* 4: 1020–1034.

Lin FL, Sperle K, Sternberg N. (1985) Recombination in mouse L cells between DNA introduced into cells and homologous chromosomal sequences. *Proc. Natl Acad. Sci. USA* 82: 1391–1395.

Manjunath N, Johnson RS, Staunton DE, Pasqualini R, Ardman B. (1993) Targeted disruption of CD43 gene enhances T lymphocyte adhesion. *J. Immunol.* 151: 1528–1534.

Mansour SL, Thomas KR, Capecchi MR. (1988) Disruption of the proto-oncogene *int-2* in mouse embryo-derived stem cells: a general strategy for targeting mutations to non-selectable genes. *Nature* 336: 348–352.

Meselson MS, Radding CM. (1975) A general model for genetic recombination. *Proc. Natl Acad. Sci. USA* 72: 358–361.

Michael SI, Curiel DT. (1994) Strategies to achieve targeted gene delivery via the receptor-mediated endocytosis pathway. *Gene Ther.* 1: 223–232.

Mitani K, Wakamiya M, Hasty P, Graham FL, Bradley A, Caskey CT. (1995) Gene targeting in mouse embryonic stem cells with an adenoviral vector. *Somat. Cell Mol. Genet.* 21: 221–231.

Moens PB. (1994) Molecular perspectives of chromosome pairing at meiosis. *BioEssays* 16: 101–106.

Moore RC, Melton DW. (1995) Models of human disease through gene targeting. *Biochem Soc. Trans.* 23: 398–403.

Morgan JE, Pagel CN, Sherratt T, Partridge TA. (1993) Long-term persistence and migration of myogenic cells injected into pre-irradiated muscles of *mdx* mice. *J. Neurol. Sci.* 115: 191–200.

Morgan JE, Beauchamp JR, Pagel CN, Peckham M, Ataliotis P, Jat PS, Noble MD, Farmer K, Partridge TA. (1994) Myogenic cell lines derived from transgenic mice carrying a thermolabile T antigen: a model system for the derivation of tissue-specific and mutation-specific cell lines. *Devel. Biol.* 162: 486–498.

Mortensen RM, Conner DA, Chao S, Geisterfer Lowrance AA, Seidman JG. (1992) Production of homozygous mutant ES cells with a single targeting construct. *Mol. Cell. Biol.* 12: 2391–2395.

Moser HE, Dervan PB. (1987) Sequence-specific cleavage of double helical DNA by triple helix formation. *Science* 238: 645–650.

Mountford P, Zevnik B, Duwel A, Nichols J, Li M, Dani C, Robertson M, Chambers I, Smith A. (1994) Dicistronic targeting constructs: reporters and modifiers of mammalian gene expression. *Proc. Natl Acad. Sci USA* 91: 4303–4307.

Nairn RS, Adair GM, Porter T, Pennington SL, Smith DG, Wilson JH, Seidman MM. (1993) Targeting vector configuration and method of gene transfer influence targeted correction of the APRT gene in Chinese hamster ovary cells. *Somat. Cell Mol. Genet.* 19: 363–375.

Palacios R, Golunski E, Samaridis J. (1995) *In vitro* generation of hematopoietic stem cells from an embryonic stem cell line. *Proc. Natl Acad. Sci. USA* 92: 7530–7534.

Pettengell R, Luft T, Henschler R, Hows JM, Dexter TM, Ryder D, Testa NG. (1994) Direct comparison by limiting dilution analysis of long-term culture-initiating cells in human bone marrow, umbilical cord blood, and blood stem cells. *Blood* **84**: 3653–3659.

Plavec I, Papayannopoulou T, Maury C, Meyer F. (1993) A human beta-globin gene fused to the human beta-globin locus control region is expressed at high levels in erythroid cells of mice engrafted with retrovirus-transduced hematopoietic stem cells. *Blood* **81**: 1384–1392.

Porteous DJ, Dorin JR. (1993) How relevant are mouse models for human diseases to somatic gene therapy? *Trends Biotechnol.* **11**: 173–181.

Porter AC, Itzhaki JE. (1993) Gene targeting in human somatic cells. Complete inactivation of an interferon-inducible gene. *Eur. J. Biochem.* **218**: 273–281.

Postel EH, Flint SJ, Kessler DJ, Hogan ME. (1991) Evidence that a triplex-forming oligodeoxyribonucleotide binds to the c-*myc* promoter in HeLa cells, thereby reducing c-*myc* mRNA levels. *Proc. Natl Acad. Sci. USA* **88**: 8227–8231.

Ramirez Solis R, Liu P, Bradley A. (1995) Chromosome engineering in mice. *Nature* **378**: 720–724.

Roth D, Wilson J. (1989) Illegitimate recombination in mammalian cells. *Genetic Recombination* (eds R Kucherlapati, G Smith). American Society for Microbiology, Washington, DC, pp. 621–653.

Rouet P, Smih F, Jasin M. (1994) Introduction of double-strand breaks into the genome of mouse cells by expression of a rare-cutting endonuclease. *Mol. Cell. Biol.* **14**: 8096–8106.

Salvetti A, Heard JM, Danos O. (1995) Gene therapy of lysosomal storage disorders. *Br. Med. Bull.* **51**: 106–122.

Schofield JP, Caskey CT. (1995) Non-viral approaches to gene therapy. *Br. Med. Bull.* **51**: 56–71.

Sedivy JM, Joyner AL. (1992) *Gene Targeting.* Freeman, New York.

Sedivy JM, Sharp PA. (1989) Positive genetic selection for gene disruption in mammalian cells by homologous recombination. *Proc. Natl Acad. Sci. USA* **86**: 227–231.

Shesely EG, Kim HS, Shehee WR, Papayannopoulou T, Smithies O, Popovich BW. (1991) Correction of a human beta S-globin gene by gene targeting. *Proc. Natl Acad. Sci. USA* **88**: 4294–4298.

Shinohara A, Ogawa T. (1995) Homologous recombination and the roles of double-strand breaks. *Trends Biochem. Sci.* **20**: 387–391.

Shinohara A, Ogawa H, Matsuda Y, Ushio N, Ikeo K, Ogawa T. (1993) Cloning of human, mouse and fission yeast recombination genes homologous to *RAD51* and *recA*. *Nature Genet.* **4**: 239–243.

Shirasawa S, Furuse M, Yokoyama N, Sasazuki T. (1993) Altered growth of human colon cancer cell lines disrupted at activated *Ki-ras*. *Science* **260**: 85–88.

Shulman MJ, Nissen L, Collins C. (1990) Homologous recombination in hybridoma cells: dependence on time and fragment length. *Mol. Cell. Biol.* **10**: 4466–4472.

Smih F, Rouet P, Romanienko PJ, Jasin M. (1995) Double-strand breaks at the target locus stimulate gene targeting in embryonic stem cells. *Nucleic Acids Res.* **23**: 5012–5019.

Smith AJ, Berg P. (1984) Homologous recombination between defective *neo* genes in mouse 3T6 cells. *Cold Spring Harbor Symp. Quant. Biol.* **49**: 171–181.

Smith AJ, Kalogerakis B. (1990) Replacement recombinant events targeted at immunoglobulin heavy chain DNA sequences in mouse myeloma cells. *J. Mol. Biol.* **213**: 415–435.

Smith AJ, De Sousa MA, Kwabi Addo B, Heppell Parton A, Impey H, Rabbitts P. (1995) A site-directed chromosomal translocation induced in embryonic stem cells by Cre–*loxP* recombination. *Nature Genet.* **9**: 376–385.

Smith LG, Weissman IL, Heimfeld S. (1991) Clonal analysis of hematopoietic stem-cell differentiation *in vivo*. *Proc. Natl Acad. Sci. USA* **88**: 2788–2792.

Smithies O. (1993) Animal models of human genetic diseases. *Trends Genet.* **9**: 112–116.

Smithies O, Gregg RG, Boggs SS, Koralewski MA, Kucherlapati RS. (1985) Insertion of DNA sequences into the human chromosomal beta-globin locus by homologous recombination. *Nature* **317**: 230–234.

Song KY, Schwartz F, Maeda N, Smithies O, Kucherlapati R. (1987) Accurate modification of a chromosomal plasmid by homologous recombination in human cells. *Proc. Natl Acad. Sci. USA* **84**: 6820–6824.

Sorrentino BP, Brandt SJ, Bodine D, Gottesman M, Pastan I, Cline A, Nienhuis AW. (1992) Selection of drug-resistant bone marrow cells *in vivo* after retroviral transfer of human *MDR1*. *Science* **257**: 99–103.

Strobel SA, Doucette Stamm LA, Riba L, Housman DE, Dervan PB. (1991) Site-specific cleavage of human chromosome 4 mediated by triple-helix formation. *Science* **254**: 1639–1642.

Sutherland HJ, Landsdorp PM, Henkelman DH, Eaves AC, Eaves CJ. (1990) Functional characterization of individual human hematopoietic stem cells cultured at limiting dilution on supportive marrow stromal layers. *Proc. Natl Acad. Sci. USA* **87**: 3584–3588.

Szostak JW, Orr Weaver TL, Rothstein RJ, Stahl FW. (1983) The double-strand-break repair model for recombination. *Cell* **33**: 25–35.

te Riele H, Maandag ER, Berns A. (1992) Highly efficient gene targeting in embryonic stem cells through homologous recombination with isogenic DNA constructs. *Proc. Natl Acad. Sci. USA* **89**: 5128–5132.

The European Polycystic Kidney Disease Consortium. (1994) The polycystic kidney disease 1 gene encodes a 14 kb transcript and lies within a duplicated region on chromosome 16. *Cell* **77**: 881–894.

Thomas KR, Capecchi MR. (1987) Site-directed mutagenesis by gene targeting in mouse embryo-derived stem cells. *Cell* **51**: 503–512.

Thomas KR, Capecchi MR. (1990) Targeted disruption of the murine *int-1* proto-oncogene resulting in severe abnormalities in midbrain and cerebellar development. *Nature* **346**: 847–850.

Thomas KR, Folger KR, Capecchi MR. (1986) High frequency targeting of genes to specific sites in the mammalian genome. *Cell* **44**: 419–428.

Thyagarajan B, Johnson BL, Campbell C. (1995) The effect of target site transcription on gene targeting in human cells *in vitro*. *Nucleic Acids Res.* **23**: 2784–2790.

Tybulewicz VL, Crawford CE, Jackson PK, Bronson RT, Mulligan RC. (1991) Neonatal lethality and lymphopenia in mice with a homozygous disruption of the c-*abl* proto-oncogene. *Cell* **65**: 1153–1163.

Uchida N, Fleming WH, Alpern EJ, Weissman IL. (1993) Heterogeneity of hematopoietic stem cells. *Curr. Opin. Immunol.* **5**: 177–184.

Uchida N, Aguila HL, Fleming WH, Jerabek L, Weissman IL. (1994) Rapid and sustained hematopoietic recovery in lethally irradiated mice transplanted with purified Thy-1.1^{lo} Lin$^-$Sca-1$^+$ hematopoietic stem cells. *Blood* **83**: 3758–3779.

Van Adelsberg JS, Frank D. (1995) The *PKD1* gene produces a developmentally regulated protein in mesenchyme and vasculature. *Nature Med.* **1**: 359–364.

van Deursen J, Wieringa B. (1992) Targeting of the creatine kinase M gene in embryonic stem cells using isogenic and nonisogenic vectors. *Nucleic Acids Res.* **20**: 3815–3820.

Vega MA. (1991) Prospects for homologous recombination in human gene therapy. *Hum. Genet.* **87**: 245–253.

Vega MA. (ed.) (1995) *Gene Targeting.* CRC Press, Boca Raton, FL.

Vega MA, Goosens M, Besmond C. (1994) A powerful method for *in vitro* selection of normal versus cystic fibrosis airway epithelial cells. *Gene. Ther.* **1**: 59–63.

Wahls WP, Wallace LJ, Moore PD. (1990a) Hypervariable minisatellite DNA is a hotspot for homologous recombination in human cells. *Cell* **60**: 95–103.

Wahls WP, Wallace LJ, Moore PD. (1990b) The Z-DNA motif d(TG)^{30} promotes reception of information during gene conversion events while stimulating homologous recombination in human cells in culture. *Mol. Cell. Biol.* **10**: 785–793.

Waldman AS. (1995) Molecular mechanisms of homologous recombination. In: *Gene Targeting* (ed. M Vega). CRC Press, Boca Raton, FL, pp. 45–64.

Waldman BC, Waldman AS. (1990) Illegitimate and homologous recombination in mammalian cells: differential sensitivity to an inhibitor of poly(ADP-ribosylation). *Nucleic Acids Res.* **18**: 5981–5988.

Wang G, Seidman MM, Glazer PM. (1996) Mutagenesis in mammalian cells induced by triple helix formation and transcription-couple repair. *Science* **271**: 802–805.

Wang Q, Taylor MW. (1993) Correction of a deletion mutant by gene targeting with an adenovirus vector. *Mol. Cell. Biol.* **13**: 918–927.

Ward MA, Abramow Newerly W, Roder JC. (1993) Effect of vector topology on homologous recombination at the CHO *aprt* locus. *Somat. Cell Mol. Genet.* **19**: 257–264.

West SC. (1995) Formation, translocation and resolution of Holliday junctions during homologous genetic recombination. *Philos. Trans. R. Soc. Lond. B. Biol. Sci.* **347**: 21–25.

Whetton AD, Dexter TM. (1993) Influence of growth factors and substrates on differentiation of haemopoietic stem cells. *Curr. Opin. Cell. Biol.* **5**: 1044–1049.

Wong EA, Capecchi MR. (1987) Homologous recombination between coinjected DNA sequences peaks in early to mid-S phase. *Mol. Cell. Biol.* **7**: 2294–2295.

Yancey Wrona JE, Camerini Otero RD. (1995) The search for DNA homology does not limit stable homologous pairing promoted by RecA protein. *Curr. Biol.* **5**: 1149–1158.

Yang B, Kirby S, Lewis J, Detloff PJ, Maeda N, Smithies O. (1995) A mouse model for β^0-thalassemia. *Proc. Natl Acad. Sci. USA* **92:** 11608–11612.

Zeiher BG, Eichwald E, Zabner J, Smith JJ, Puga AP, McCray PB Jr, Capecchi MR, Welsh MJ, Thomas KR. (1995) A mouse model for the delta F508 allele of cystic fibrosis. *J. Clin. Invest.* **96:** 2051–2064.

Zheng H, Wilson JH. (1990) Gene targeting in normal and amplified cell lines. *Nature* **344:** 170–173.

Zijlstra M, Li E, Sajjadi F, Subramani S, Jaenisch R. (1989) Germ-line transmission of a disrupted beta 2-microglobulin gene produced by homologous recombination in embryonic stem cells. *Nature* **342:** 435–438.

Zimmer A, Gruss P. (1989) Production of chimaeric mice containing embryonic stem (ES) cells carrying a homoeobox *Hox 1.1* allele mutated by homologous recombination. *Nature* **338:** 150–153.

11

Cystic fibrosis

N.J. Caplen and E.W.F.W. Alton

11.1 Introduction

Cystic fibrosis (CF) is the most common lethal inherited disease in the Caucasian population. Approximately 1 in 20 people are heterozygous for the abnormal gene. It is inherited as an autosomal recessive character and approximately 1 in 2000 live births in Northern Europe and North America are children with CF. Recent studies have identified the underlying molecular defect in CF. With the identification of the CF gene, gene therapy has become a potential novel form of treatment.

11.1.1 Clinical features of CF

CF is usually diagnosed during the first 5 years of life. Approximately 10% of cases are diagnosed at birth as the infants suffer intestinal blockage; the remainder usually present with an increased susceptibility to respiratory infection. The principal pathology centres around the respiratory and intestinal tracts. In both cases, these hollow, epithelial-lined organs become filled with thickened, tenacious secretions. The intestinal symptoms are generally milder and are of two types. The first results from obstruction of the small intestine (meconium ileus); this is usually observed peri-natally and can generally be treated without the need for surgery. The second is mal-absorption of gut contents resulting from the blockage and atrophy of the pancrea-tic ducts and hence reduced secretion of the pancreatic enzymes needed for absorp-tion. Pancreatic insufficiency eventually occurs in 80–90% of patients, and can be con-trolled with pancreatic and other dietary supplements.

The principal clinical problem of CF is lung damage and respiratory failure as a result of bacterial colonization and recurrent chest infections. Early studies in children with CF dying from non-pulmonary causes showed obstruction of submucosal glands, found throughout the upper respiratory tract, with mucus. Clinically, however, pathol-ogy is usually associated with the bronchiolar region; involvement of the alveoli is uncommon. The main early bacterial infections are due to *Staphylococcus aureus* and *Haemophilus influenzae*, with *Pseudomonas aeruginosa* and *Burkholderia cepacia* subse-quently becoming the principal pathogens. Colonization with these bacteria con-tributes significantly to the cycles of inflammation which lead to bronchiectasis and respiratory failure in CF patients. Other clinical features of CF include the develop-ment of liver disease in approximately 10–15% of CF individuals as a result of block-age of the intrahepatic bile ducts, and azoospermia as a result of absence or obstruction of the vas deferens which renders CF males sterile (Boat *et al.*, 1989). Some 50 years ago the mean life expectancy for an individual with CF was 1 year; this has now risen in the

Gene Therapy, edited by N.R. Lemoine and D.N. Cooper.
© 1996 BIOS Scientific Publishers Ltd, Oxford.

UK to approximately 30 years. This increase is mainly due to improved antibiotics, the use of intensive physiotherapy and (in rare cases) heart and lung transplantation. However, CF remains a disease associated with significant morbidity and mortality and a huge social and financial cost; hence, the urgent need for a better understanding of the pathogenesis of CF and the development of new therapies.

11.1.2 The cystic fibrosis transmembrane conductance regulator (CFTR)

One of the earliest observations in young children with CF was the presence of abnormal levels of chloride in sweat secretions (di Sant' Agensi *et al.*, 1953). The presence of elevated chloride in sweat is still a diagnostic test for CF and is indicative of the underlying cellular defect. The epithelial cells lining the secretory ducts in CF individuals exhibit absent or significantly reduced cAMP-mediated chloride transport across the apical membrane of those cells. The protein affected, the cystic fibrosis transmembrane conductance regulator (CFTR) is a cAMP-regulated chloride channel (Welsh *et al.*, 1992). The CFTR protein consists of two transmembrane regions, two nucleotide-binding domains and a regulatory region containing a number of potential phosphorylation sites (Riordan *et al.*, 1989).

Expression of CFTR is low in the lung, with respect to both mRNA (Trapnell *et al.*, 1991) and protein levels (Crawford *et al.*, 1991; Kartner *et al.*, 1992). The distribution of CFTR within the lung matches the pathologic distribution, with the highest level of expression in submucosal glands (Engelhardt *et al.*, 1992a, 1994a), which predominate in the proximal airways. Additionally, expression of CFTR in a subpopulation of both ciliated and non-ciliated cells is seen more peripherally, although the precise cell types have not yet been identified (Engelhardt *et al.*, 1994a). CFTR protein expression has also been localized within intestinal crypts, the exocrine ducts of the pancreas, the intra-hepatic bile ducts of the liver and in the seminiferous epithelium of the testis (Cohn *et al.*, 1994).

11.1.3 Mutations in the CFTR gene

The *CFTR* gene lies on the long arm of chromosome 7 and was cloned in 1989 (Riordan *et al.*, 1989; Rommens *et al.*, 1989). The gene contains 27 exons, covers approximately 250 kb and encodes a mRNA of approximately 6.5 kb. The most common mutation associated with CF is a 3 bp deletion in exon 10 of the *CFTR* gene (ΔF508) (Kerem *et al.*, 1989); this mutation accounts for approximately 70% of CF chromosomes and removes a phenylalanine residue within the first nucleotide binding fold of CFTR. The mutation is temperature sensitive; at 37°C the abnormal protein is defectively processed, never reaching the cell surface (Cheng *et al.*, 1990). However, at lower temperatures the protein can be translocated to the surface (Denning *et al.*, 1992) where it can conduct chloride normally.

Over 400 CF mutations have now been identified (Tsui 1995), including missense, nonsense, frameshift and splice-junction mutations, as well as single amino acid deletions such as ΔF508. These mutations can be grouped depending on their effect on CFTR function. Most CF mutations, including the ΔF508 mutation, produce mutant protein that becomes arrested within the cell, failing to reach the apical membrane, whereas others produce channels that fail to remain open for as long as normal channels in response to stimulation by cAMP. Yet another class of mutations express channels unable to conduct chloride at the normal rate. Finally, there are those mutations which result in a 'null phenotype' as no full length CFTR is produced at all (Welsh and Smith, 1993). There is a fairly strong correlation between particular CF genotypes and pancreatic disease (Kristidis *et al.*, 1992) but there appears to be little relationship between pulmonary phenotype and genotype (CF Genotype–Phenotype Consortium, 1993), suggesting the influence of other genetic or environmental factors (Rozmahel *et al.*, 1996).

11.1.4 CFTR and the pathogenesis of CF airway disease

The relationship between alterations in the function of CFTR and the pathology of CF is only poorly understood and the matter of a great deal of speculation. One theory suggests that the opening of the CFTR channel to allow chloride secretion from the cell on to the mucosal surface enables water to follow by osmosis. This would provide a means by which the cell surface can be hydrated. In turn, this is likely to be important in the airways in the process of mucociliary clearance (MCC). MCC ensures the removal of inhaled particles and bacteria from the airways by the synchronized beating of cilia on the mucosal surface of the epithelial cells. These cilia beat in a thin fluid layer which maximizes the efficiency of this process. Chloride secretion is thus likely to provide an important contribution to the normal maintenance of this layer. Another factor is that sodium absorption from the airway is abnormal in CF patients, being increased two to three times (Boucher *et al.*, 1986). The link between the chloride defect and the secondary sodium abnormality is presently unclear. However, since water will again follow sodium movement, this second abnormality will also tend to dehydrate the airway surface liquid. The likely net result of these abnormalities is that water movement is reduced, leading to a suboptimal periciliary layer in the airways and impaired MCC. This will reduce clearance of bacteria, leading to the repeated infections which predispose the CF individual to lung damage. More direct links between the abnormal CFTR protein and increased bacterial adherence or reduced bacterial removal are also being described.

11.1.5 New treatments for CF

New treatments for the basic defect or symptoms of CF are being developed, although results to date are limited. One potential therapeutic approach is to administer the sodium-channel blocker amiloride to the airways of these patients, so inhibiting sodium hyperabsorption (Graham *et al.*, 1993; Knowles *et al.*, 1990). A second possible pharmacological approach is to try to bypass the defective CFTR chloride channel and upregulate other types of chloride channel present in the apical membrane of airway epithelial cells. In particular, ATP and UTP have been suggested to be of potential interest (Knowles *et al.*, 1991). Both these agents increase intracellular levels of calcium and open a chloride channel distinct from CFTR in airway epithelium. CFTR protein can be purified (Bear *et al.*, 1992) and two groups have attempted to deliver it *in vitro* and *in vivo* (Ramjeesingh *et al.*, 1995; Scheule *et al.*, 1995). However, the most obvious, and we suggest the most elegant, way of treating CF would be to introduce a new, normal copy of the *CFTR* gene into the respiratory tract of CF individuals.

11.2 Gene transfer systems applicable to CF

The principal gene delivery system for gene therapy at present is retroviral-mediated gene transfer. However, retroviruses were considered not to be applicable to CF as they require cell proliferation for proviral integration and gene expression and, until recently, it was not known that the terminally differentiated surface epithelium and the collecting ducts of submucosal glands can divide (Leigh *et al.*, 1995). Hence, most groups have concentrated on the development of alternative delivery systems for gene transfer to the airway. Only the salient points of these systems, relevant to this discussion, will be summarized here. Further details are provided in other chapters within this publication.

11.2.1 Adenovirus

Four features of adenovirus (Ad) biology make this type of virus an ideal candidate as a vector for gene transfer to the airway:

  (i)   humans are a permissive host;
 (ii)   they have a natural tropism for the lung;
(iii)   they have a well-developed system for escape from the endosomal pathway into which they are taken up by cells;
(iv)   they can infect non-dividing cells.

Adenoviral vectors have been developed from adenoviruses of the subgenus C, principally Ad2 and Ad5. The first generation of recombinant adenoviruses included deletions of sequences spanning the regions *E1a* and *E1b*; second-generation vectors also include a defective *E2a* gene.

11.2.2 Adeno-associated virus

Adeno-associated virus (AAV) is a single-stranded DNA virus which has the potential to integrate into the host genome, often in the same location on chromosome 19. Helper viruses such as adenovirus or herpes virus are required for productive viral infection and thus AAV is naturally defective for replication. The maximum packaging capacity of AAV vectors is approximately 4.5–5 kb, which will just allow the insertion and packaging of the *CFTR* cDNA (Flotte *et al.*, 1992).

11.2.3 DNA–liposomes

Cationic liposome-mediated DNA transfer makes use of a charge interaction between DNA (usually in the form of a plasmid), which is negatively charged, and a mixture of lipids (usually cationic and neutral) to form a complex of lipid-coated DNA. The exact nature of this interaction is unknown and may well be different for different cationic lipids. The mechanism of cell uptake and trafficking of these complexes through the cell is also poorly understood. Several different cationic liposomes have been used to transfer the *CFTR* cDNA including DOTMA–DOPE (Life Technologies, Gaithersburg, Maryland, USA), DOTAP (Boehringer Mannheim, Mannheim, Germany), DMRIE (Vical Inc., San Diego, California, USA) and DC-Chol–DOPE (Gao and Huang, 1991).

11.2.4 CFTR expression cassettes

All of the gene transfer studies described here use full-length cDNA versions of *CFTR* of approximately 4.7 kb which excludes the 5′ and 3′ untranslated regions of the endogenous transcript. Several forms of the *CFTR* cDNA have been generated, primarily because of the necessity to include nucleotide changes within the gene to ensure stable propagation in prokaryotic systems (Drumm *et al.*, 1990; Gregory *et al.*,

1990). In both preclinical and clinical studies, gene expression is usually under the control of ubiquitous eukaryotic viral promoters [e.g. CFTR expression driven by the cytomegalovirus (CMV), simian virus 40 (SV40) or Rous sarcoma virus (RSV) promoters] or mammalian housekeeping gene promoters (e.g. the promoters of the genes for β-actin or phosphoglycerate kinase). These promoters have the advantage of ensuring relatively high levels of expression in the widest range of cell types. However, in the long term it may be advantageous to use endogenous genetic elements to control expression more precisely.

11.3 Gene transfer studies I: heterologous expression of CFTR in model systems

11.3.1 Assays of CFTR gene expression

There are several ways in which *CFTR* gene expression can be demonstrated. Reverse transcriptase–polymerase chain reaction (RT–PCR) and *in situ* hybridization can be used to detect exogenous *CFTR*-derived transcripts. Various antibodies have been raised against CFTR and can be used to assess the gross level of CFTR protein produced by immunoprecipitation or used to localize CFTR expression by immunohistochemistry. *In vitro* electrophysiological measurements can be undertaken using a variety of techniques including patch-clamping, epifluorescence microscopy, radiolabelled halide efflux and Ussing chambers. Each technique provides different and complementary assessments of ion movement. Advantages of *in vitro* studies include measurements under steady-state conditions, studies on individual cells or cell types, and greater certainty of the component of ion transport that may be altered by gene transfer.

11.3.2 In vitro: *cell lines and primary cells*

The first reports of *in vitro* correction of the CF chloride channel defect came in 1990. Drumm and co-workers (1990) used retroviral-mediated transfer of the *CFTR* cDNA to correct the chloride transport defect exhibited by a CF pancreatic carcinoma cell line. The presence of normal *CFTR* mRNA was demonstrated and cAMP-mediated chloride movement induced, as shown by patch-clamp and radiolabelled efflux studies. This was followed by a second report in which vaccinia virus was used to transfect a CF airway epithelial cell line (Rich *et al.*, 1990). Again, chloride movement was restored, as shown by epifluoresence and patch-clamping, following transfection with normal but not mutant *CFTR* cDNA. Stable retrovirally mediated correction of a differentiated CF tracheal epithelial cell line has been observed for up to 6 months (Olsen *et al.*, 1992).

Subsequently, many groups have repeated these findings in culture with several different gene transfer methods and assays of *CFTR* gene expression. *In vitro* studies examining adenoviral-mediated *CFTR* gene transfer include successful transduction of CF cell lines (Mittereder *et al.*, 1994), polarized epithelial monolayers derived from CF cell lines (Mittereder *et al.*, 1994; Rich *et al.*, 1993; Zabner *et al.*, 1994a) and freshly isolated human nasal and bronchial samples from CF patients (Rosenfeld *et al.*, 1994a). AAV–CFTR vectors have been used to transduce both CF cell lines (Egan *et al.*, 1992) and primary tissue (CF nasal polyps) *in vitro* (Flotte *et al.*, 1993). Transfer of *CFTR* to HeLa and COS-7 cells using the cationic lipids DOTMA and DOTAP, respectively, has been shown to generate cAMP-specific chloride secretion in these cell lines (Hyde *et al.*, 1993; McLachlan *et al.*, 1995), and we have demonstrated that CF nasal cells obtained by brushing can be transfected *in vitro* using the cationic liposome DC-Chol–DOPE (Stern *et al.*, 1995).

11.3.3 In vivo: *CF transgenic animal models*

Transgenic CF mice, which have been generated by a number of laboratories, are now being used in the assessment of gene transfer systems. Initially, 'knockout' models were created using a variety of strategies (Chapter 9; reviewed by Dorin *et al.*, 1994). However, more recently, models mimicking human mutations have been described (Colledge *et al.*, 1995; Delaney *et al.*, 1996; van Doorninck et al., 1995; Zeiher *et al.*, 1995). To assess the effect of gene transfer on the bioelectric properties of these animals, Hyde and co-workers (1993) instilled a cationic liposome (DOTMA–DOPE) complexed with a *CFTR* cDNA into the trachea of the $cftr^{mICam}$ transgenic mice and showed restoration of cAMP-stimulated chloride secretion. We nebulized a liposome (DC-Chol–DOPE)–*CFTR* cDNA complex into $cftr^{mIHGU}$ mice and showed that the CF chloride defect could be corrected by this method in some animals (Alton *et al.*, 1993). However, the relatively large amount of DNA used and the variability of correction suggest that inefficient gene transfer may be a problem with liposome-based systems.

Adenovirus-mediated *CFTR* gene transfer into the nasal cavity of $cftr^{mIUNC}$ transgenic mice has also been studied. Even with very high titres of Ad–CFTR, little change was seen in chloride transport and no correction of the sodium defect was seen (Grubb *et al.*, 1994). Interestingly, Yang and colleagues (1994) have detected CFTR protein expressed from a second-generation, *E2a*-deleted recombinant Ad carrying the *CFTR* cDNA delivered to $cftr^{mIUNC}$ mice. Mice were instilled with 2×10^9 plaque forming units (p.f.u.) of Ad–CFTR; CFTR was detected in 80% of bronchi and for up to 21 days. While a direct comparison of the efficiency of correction of the CF defect using cationic liposomes and adenoviruses has not been undertaken, one group used quantitative RT–PCR to show that the level of exogenous *CFTR* mRNA obtained following nasal instillation of liposome–DNA complexes into BALB/c mice was equivalent to that obtained from 50 multiplicity of infection (MOI) units of adenovirus (Yew *et al.*, 1995). Finally, the lethal intestinal abnormalities associated with the $cftr^{mIUNC}$ mice was rescued, at least in part, by the generation of bi-transgenic animals in which the human *CFTR* gene was expressed specifically in the intestine by placing the human gene under the control of the promoter of the gene for the rat intestinal fatty-acid-binding protein. cAMP-mediated chloride secretion was present in the jejunum and ileum but not the colon, correcting the development of goblet cell hyperplasia observed in the parental null CF transgenics (Zhou *et al.*, 1994). This study suggests gene transfer may be relevant to treatment of some features of the intestinal disease seen in CF patients.

11.3.4 In vivo: *non-transgenic animal models*

Most studies in non-transgenic models have concentrated on assessing the expression of exogenous human *CFTR* by RT–PCR and CFTR protein by immunohistochemistry. A pioneering set of studies by Crystal's group, using both adenovirus-mediated (Rosenfeld *et al.*, 1992) and liposome-mediated gene transfer (Yoshimura *et al.*, 1992) established that the human *CFTR* gene could be expressed in the airways of rodents *in vivo.* In the study by Rosenfeld and co-workers, instillation of an Ad5 vector carrying the *CFTR* cDNA into lungs of cotton rats was followed by appearance of *CFTR* mRNA on day 1, sustained for up to 4 weeks. Human CFTR protein was present in airway epithelial cells 11–14 days post-instillation. Similar results have been obtained in other studies in cotton rats after both single (Yei *et al.*, 1994a) and repetitive administration (Zabner *et al.*, 1994b). Adenoviral-mediated transfer of the *CFTR* gene to primates has shown expression throughout the airways, including the alveoli, but was generally patchy in distribution (Brody *et al.*, 1994; Engelhardt *et al.*, 1993a; Zabner *et al.*, 1994b).

Exogenous *CFTR* RNA and CFTR protein were detectable for up to 6 months in rabbits into which an AAV–CFTR construct had been instilled (Flotte *et al.*, 1993). The same group also reported the presence of AAV vector DNA in airway epithelial cells obtained from Rhesus macaque monkeys whose respiratory tracts had been instilled with an AAV–CFTR vector 3 months previously (Afione *et al.*, 1995), suggesting that long-term gene expression from AAV vectors may be possible *in vivo.*

Delivery of *CFTR* plasmid DNA complexed with the liposome DOTMA–DOPE by intratracheal administration to mice produced human-specific *CFTR* mRNA transcripts in the lung for 4 weeks (Yoshimura *et al.*, 1992). In a similar study using the cationic liposome DMRIE–DOPE, human *CFTR* mRNA was detected in rats instilled 3 days previously (Logan *et al.*, 1995). In most cases, the presence of endogenous CFTR function cannot be distinguished from functional expression of the *CFTR* transgene. However, in the rat, cAMP-mediated chloride conductance is minimal. Thus, as an extension of their experiment showing human *CFTR* mRNA expression, Logan and co-workers instilled into the trachea of normal rats *CFTR* plasmid DNA complexed with DMRIE–DOPE, and measured the ion transport characteristics of the excised rat tracheas in Ussing chambers. Bioelectrical responses consistent with *CFTR* transfer were observed, including cAMP-mediated chloride secretion 3 days after instillation.

11.3.5 Levels of CFTR *gene expression required for therapeutic benefit*

The expression of CFTR is generally low in the adult lung, suggesting that perhaps only low levels of expression will be required for clinical benefit. Recently, data supporting this notion have emerged. Laboratory studies in which different proportions of CF and normal cells were grown in a monolayer showed that no further improvement in the level of corrected chloride transport could be achieved above a mixing of 10% of normal cells with their CF counterparts (Johnson *et al.*, 1992). These data have been confirmed using CF bronchial xenografts grown in *nu/nu* mice. Using an Ad–CFTR vector, an MOI of 100 reconstitutes near-normal levels of cAMP-stimulated chloride transport, even though it was estimated that only 5% of cells in the pseudostratified epithelium were transduced. By contrast, the correction of sodium hyperabsorption was partial and variable (Goldman *et al.*, 1995). In another study, Zabner and co-workers studied Ad2–CFTR transduction of polarized CF epithelial monolayers. At a MOI of 0.1, a partial restoration of chloride secretion was observed and at a MOI of 10, chloride secretion was normal. Parallel studies using a reporter gene showed these MOI values to correspond to 20% and 90% transduction of cells, respectively (Zabner *et al.*, 1994a).

Further observations come from examining different CF genotypes. CF heterozygotes have 50% normal CFTR levels but demonstrate no lung pathology. Furthermore, the chloride conductance of the *CFTR* mutations R347P and R117H has been assessed *in vitro* and has been suggested to be approximately 30% and 15% of normal levels, respectively. Thus, patients who are R347P–ΔF508 and R117H–ΔF508 compound heterozygotes should demonstrate approximately 15% and 7.5% of normal CFTR function, respectively. Interestingly, in the few such patients studied, lung disease is still a prominent feature. Studies in CF mutant mice show that there is an asymptotic relationship between *CFTR* mRNA and chloride transport in both the intestinal and respiratory tracts. Thus, for a relatively small increase in mRNA, there is a disproportionately greater increase in chloride conductance. Interestingly, the relationship with survival (related to intestinal disease in these animals) is even steeper, such that a very small increase in mRNA from the null state can produce almost complete normalization of survival to wild-type levels (Dorin *et al.*, 1996). Overall, these studies are encouraging with respect to gene therapy, suggesting that production of even very low levels of CFTR may be of therapeutic benefit. In these latter studies, CFTR is, of course, produced in every cell, whereas at present the inefficiency of gene transfer is likely to preclude this.

11.3.6 Over-expression of CFTR

Several studies have attempted to address whether there are likely to be any adverse effects due to the over-expression of the CFTR protein as a result of exogenous gene transfer. Transgenic mice expressing the human *CFTR* gene as well as the mouse endogenous gene in the lung were generated using a human *CFTR* gene under control of a lung epithelial cell-specific promoter (from the gene for surfactant protein C). Human CFTR was expressed in distal airway and alveolar cells with no adverse effects in terms of lung weight, morphology or somatic growth (Whitsett *et al.*, 1992). In addition, Rosenfeld and co-workers have shown that increasing expression of the normal human *CFTR* cDNA in a CF pancreatic cell line (CF-PAC-1) results in a progressive increase in the level of CFTR protein but there is a limit to the level of cAMP-stimulated chloride secretion observed (Rosenfeld *et al.*, 1994b). These findings go some way to establishing that over-expression of CFTR is harmless, although one group have suggested this may not be so clear-cut (Schiavi *et al.*, 1993). They reported preliminary findings showing a correlation between high CFTR expression and growth arrest of an epithelial cell line in culture, and that injection of rabbit embryos with *CFTR* cDNA caused the majority of the male offspring to be stillborn.

11.4 Gene transfer studies II: reporter gene expression and safety studies

11.4.1 Reporter gene studies

In addition to studies assessing the effects of *CFTR* transfection, a large number of related reporter gene studies, including the use of the β-galactosidase gene (*β-gal/lacZ*), the luciferase gene (*luc*) or the chloramphenicol acetyltransferase gene (*CAT*), have been performed on airway cells *in vitro* and *in vivo*. Transduction of airway cells *in vitro* and *in vivo* by adenoviral vectors has enabled evaluation of not only the total level of expression but also an estimate of the number of cells transfected, with both being shown to be dose-dependent (Bout *et al.*, 1994; Katkin *et al.*, 1995; Yei *et al.*, 1994a,b). Expression of the *β-gal* reporter gene has been seen in Rhesus monkeys instilled with Ad–β-gal. Large patches of positive staining were observed in the trachea, bronchi and bronchioles 6 days after virus exposure. Basal, mucous, goblet and ciliated cells, and submucosal glands all showed positive staining (Bout *et al.*, 1994). In another study, baboons treated with Ad5–lacZ were assessed at 4 and 21 days for β-galactosidase expression. Transgene expression was predominant in alveolar cells, with only patches in proximal and distal airways (Engelhardt *et al.*, 1993a).

Studies of *in vitro* gene transfer using lipid-mediated gene transfer in different epithelial cells have allowed investigators to estimate the relative efficiency of transfer mediated by different lipids and their relative ability to transfect different respiratory epithelial cell lines (Caplen *et al.*, 1995a; Fasbender *et al.*, 1995; Logan *et al.*, 1995; McLachlan *et al.*, 1995). Tracheal instillation of DOTMA-containing DNA–liposome complexes to mice showed transgene expression in the lung using both the *luc* and *β-gal* reporter genes (Yoshimura *et al.*, 1992). In a similar study by Logan and co-workers using either DOTMA–DOPE or DMRIE–DOPE liposomes, 9 of 11 rats tested demonstrated exogenous *β-gal* expression in approximately 70% of the surface epithelium (bronchi and proximal bronchioles). Estimation of *CAT* expression obtained from whole lung lysates suggests that approximately 1% of all lung cells were transfected (Logan *et al.*, 1995). Interestingly, instillation of plasmid DNA alone into mice trachea has also been shown to produce transgene expression using the *CAT* reporter gene; this level of expression was not enhanced when plasmid DNA was delivered in combination with DOTAP or DOTMA–DOPE liposomes (Meyer *et al.*, 1995). Both the *β-gal* gene and the *CAT* gene have been used to show the efficacy of nebulized DNA–liposome complexes (Alton *et al.*, 1993; McLachlan *et al.*, 1995; Stribling *et al.*, 1992).

Histological analysis of reporter gene expression enables the pattern of gene expression to be assessed, not just with respect to distribution in the lung but also within individual cell types within the respiratory tract. Determination of the exact cell-type transfected within the airway may be critical if it is shown that some cells contribute more significantly to pathology than others. Mastrangeli *et al.* (1993) reported adenoviral transduction of all major cell types of the airway epithelium, including ciliated and basal cells, secretory and undifferentiated, in the cotton rat. The cotton rat was used because its sensitivity to Ad infection is similar to that of humans. However, because the rat has a significantly lower proportion of submucosal glands, these results may not reflect the events that occur in the human airways. One group has suggested that there is a difference in the susceptibility to Ad5 transduction of the airway epithelial cell types *in vivo*: columnar cells show reduced susceptibility as compared with basal cells. In addition, this group showed that mechanical damage which removed columnar cells significantly enhanced transfection (Grubb *et al.*, 1994). This may be related to the differential expression of the principal binding and internalization proteins, the alpha v beta 5 integrins, used by adenovirus. These are expressed at high levels in undifferentiated cells and at very low levels in differentiated columnar cells (Goldman and Wilson, 1995).

Another useful model for examining the cellular distribution of transfer is airway xenografts. Engelhardt and co-workers (1992b) have established a model of the human bronchi by implanting human bronchial xenografts into immune-deficient mice. Initially they showed successful retroviral gene transfer with transduction of between 5% and 10% of the regenerating epithelium. More recently, this group used this model to show successful adenoviral-mediated gene transfer (Engelhardt *et al.*, 1993b); expression was observed in up to 10% of the surface epithelial cells and expression in approximately 40% of submucosal glands, which in this model are retained (Pilewski *et al.*, 1995). Finally, fetal-derived human pulmonary tissue has been used to form xenografts in mice with severe combined immunodeficiency (SCID). Expression of *β-gal* was seen in both epithelial and glandular cells following microinjection of adenovirus (Peault *et al.*, 1994).

Rodent studies using DNA–liposome complexes, whether administered by nasal or tracheal instillation or by nebulization, have shown exogenous expression to be concentrated in the surface epithelium (Alton *et al.*, 1993; Hyde *et al.*, 1993; McLachlan *et al.*, 1995; Yoshimura *et al.*, 1992).

11.4.2 Safety studies: vector–host interactions and gene transfer to the airway

Determination of the safety of any gene delivery system is vital to assessing its viability as a therapeutic agent. With respect to adenoviral vectors, four factors appear to influence the host response to the virus:

  (i)   the continued expression of viral vector genes;
 (ii) *trans*complementation of deleted functions;
(iii) the cellular immune response of the host;
(iv) the humoral immune response of the host.

Reports detecting adenoviral gene expression from vectors are variable but as this may contribute significantly to any host response, the possibility of this occurring has been the subject of much interest. Infrequent expression of *E2a* has been reported from a first-generation Ad vector, although in this study this was not accompanied by expression of the late gene encoding fibre protein (Engelhardt *et al.*, 1993a; Simon *et al.*, 1993). In a different study, *in vitro* assessment of an Ad vector showed that while it did not replicate *in vitro* in HeLa cells, trace vector DNA synthesis was seen and there was expression of Ad late genes for hexon capsid protein in HeLa cells but not in freshly isolated human bronchial epithelial cells (Mittereder *et al.*, 1994). These observations may be relevant in the light of observations suggesting that non-*E1* gene expression from Ad vectors may induce apoptotic cell death and inhibit cell proliferation of human airway epithelial cells even at low titres (Teramoto *et al.*, 1995). Another, as yet theoretical, risk is the possibility of complementation of E1a function by host-derived sequences leading to replication of *E1a*-deleted vectors. Wild-type Ad5 has been shown to mobilize an *E1a*-defective Ad vector carrying *CFTR in vitro* and in a limited manner in cotton rats (Imler *et al.*, 1995). Given that a screen of nasal and bronchial samples from non-CF and CF individuals showed a proportion to be positive for adenoviral *E1a* sequences (21% in non-CF subjects, 13% in CF subjects; Eissa *et al.*, 1994), further mutations may need to be introduced to overcome this potential risk (Imler *et al.*, 1995).

In vivo, one study reported no adenoviral vector replication but the presence of serum neutralizing antibodies in cotton rats was observed after a single application of vector, although in some animals this was transient and levels did not increase after a second administration (Zabner *et al.*, 1994b). In another study assessing Ad–CFTR transfer in cotton rats, gene transfer was accompanied by a dose- and time-dependent inflammation within the airways (Yei *et al.*, 1994a). These findings in cotton rats have been borne out in primates. Wilson *et al.*, (1994) observed a mononuclear cell inflammatory response within the alveolar compartment of primates receiving doses of virus that were required for detectable gene expression. Minimal inflammation was seen at doses of 10^7 and 10^8 but at 10^9 and at 10^{10} p.f.u. per ml, a perivascular infiltrate was also present, with an increase in intensity between 4 and 21 days (Engelhardt *et al.*, 1993a; Simon *et al.*, 1993). A similar dose-dependent increase in inflammatory cells was seen by Brody *et al.* (1994). Microscope analysis showed the infiltrate to consist primarily of lymphocytes, which persisted for at least 2 months in some of the monkeys.

The inflammation seen in these studies may relate to the antigenicity of the adenoviral coat proteins and a cytotoxic T-lymphocyte response. To determine the immunological processes underlying this inflammation, studies have been conducted using mice deficient in different immunological effectors. $CD8^+$ T cells restricted by class I molecules of the major histocompatibility complex (MHC) are activated and lead to destruction of virus-infected cells and thus loss of transgene expression (Yang *et al.*, 1995a). These studies have led to the development of second-generation viruses that inactivate *E2a* and are associated with substantially longer recombinant gene expression and less inflammation (Engelhardt *et al.*, 1994b; Yang *et al.*, 1994). An alternative approach that may overcome this problem is the use of immunosuppressive drugs. Both cyclosporin and dexamethasone, which downregulate the immune response, have been shown to lengthen exogenous expression in the airways of cotton rats (Zsengeller *et al.*, 1995).

However, the most significant limit to the use of adenovirus may be that of the production of neutralizing antibodies, which results in reduced gene transfer following a second administration of virus (Kozarsky *et al.*, 1994; Smith *et al.*, 1993; Yei *et al.*, 1994b). Again using immunologically deficient mice, Yang and co-workers (1995a,b) showed that $CD4^+$ T cells contribute significantly to the formation of neutralizing antibodies in the airway which block subsequent adenovirus-mediated gene transfer. The same group has also shown that interleukin-12 (IL-12) administered at the same time as the first administration of Ad specifically blocked immunoglobulin A (IgA) production and resulted in a 20-fold reduction in neutralizing antibody titre with gene expression seen after re-administration of vector. Similar results were seen following administration of interferon-γ (Yang *et al.*, 1995b).

To date, no large-scale *in vivo* study using DNA–liposome complexes has reported toxicity or evidence of a host response. However, one group has presented preliminary observations describing a minimal to moderate inflammatory response, primarily in the terminal bronchi and alveolar region in BALB/c mice following nasal instillation of DNA–liposome complexes. No specific antibodies nor activation of DNA- or lipid-specific T cells was seen (Marshall *et al.*, 1995). The results of more detailed studies investigating safety aspects of DNA–liposome complexes within the respiratory tract, particularly after repeated administration, are still awaited. Insufflation of the cationic liposome DC-Chol–DOPE into the normal human nose did not alter ion transport measurements, which are themselves a sensitive index of cellular function and integrity (Middleton *et al.*, 1994a).

Finally, there is the question of spread of the administered gene from the respiratory tract. Some spread of adenovirus into the small bowel has been noted in primates, presumably through swallowing at instillation (Zabner *et al.*, 1994b). The possibility of germ-line insertion of the *CFTR* cDNA administered to the airways is likely to be more of a theoretical than a real risk. CFTR is probably expressed in the gonads of individuals without CF, and male CF patients are infertile. In addition, studies so far have shown no germ-line transfer in primates following adenoviral-mediated reporter gene transfer to the lungs (Zabner *et al.*, 1994b), nor in mice nebulized with DNA–liposome complexes (McLachlan *et al.*, 1995).

11.5 Gene transfer studies III: clinical trials

11.5.1 Nasal administration

The nasal cavity represents a useful clinical test area for assessment of the safety and efficacy of *CFTR* gene transfer. Advantages in comparison with the lung include easier access both for gene transfer and measurements of safety and efficacy, as well as the reduced risk in case of the occurrence of side-effects. Furthermore, the nose clearly demonstrates the characteristic ion transport abnormalities of CF. These can be assessed *in vivo* by measurements of potential difference (Knowles *et al.*, 1995a; Middleton *et al.*, 1994b). The baseline potential difference principally relates to sodium transport and this, as well as the response to subsequent perfusion of the sodium-channel blocker amiloride, provides an index of increased sodium transport in CF patients. Both measurements reliably distinguish patients with CF from subjects without disease. To assess chloride transport, the airway epithelium is perfused with a low chloride solution, which provides a driving force to induce chloride secretion. Subsequent perfusion with a β-agonist such as isoprenaline allows for a more selective assessment of cAMP-mediated chloride transport. Again, both these measurements differentiate CF and non-CF subjects.

11.5.2 Nasal studies using adenovirus

In the first published clinical study, Zabner *et al.* (1993) studied adenovirus-mediated *CFTR* cDNA gene transfer to the nose of three volunteers with CF. With respect to safety, a degree of localized inflammation around the site of application was seen, probably related to the method of delivery. *CFTR* mRNA could be demonstrated in two subjects. With respect to correction of the bioelectric abnormalities, baseline potential difference was reduced into the normal range in all three subjects, whereas a β-agonist (terbutaline) produced small changes similar to those seen in subjects without CF following, but not prior to, gene transfer. These changes lasted up to 10 days after the single application, although the study was not designed to assess duration of expression (Zabner *et al.*, 1993). Although these data are encouraging, it is important to note that inflammation can itself reduce baseline potential difference. Furthermore, the crucial measurements of the β-agonist effect were made in the absence of a low chloride concentration, markedly reducing the size of the measured signal. These concerns reduce the degree of confidence that can be attached to these results.

As part of an ongoing clinical protocol conducted by Crystal and co-workers (1994), a sample from one of four CF patients receiving an Ad–CFTR vector to the nasal epithelium displayed *CFTR* mRNA expression and another patient displayed protein expression. Study of the bioelectrical characteristics of nine patients showed an improvement in chloride secretion of approximately 30% over a 2-week period after administration; there was no evidence of vector-induced epithelial damage in any patient (Hay *et al.*, 1995).

In a third trial, which, unlike the previously described ones, was blinded and placebo-controlled, 12 CF patients were recruited to receive one of four different doses of adenovirus or placebo to the nasal epithelium (Boucher *et al.*, 1994). Adenovirus-derived *CFTR* mRNA was demonstrated in five of six patients who received the highest doses. However, no consistent changes in chloride or sodium transport were found. At the lower doses of vector, no toxic effects were seen, but at the highest dose (a titre of 2 x 10^{10} p.f.u., MOI of 1000) there was mucosal inflammation in two of the three patients (Knowles *et al.*, 1995b).

11.5.3 Nasal studies using cationic liposomes

We have recently completed a double-blind, placebo-controlled trial of liposome-mediated *CFTR* cDNA transfer to the nasal epithelium in 15 ΔF508 homozygous subjects with CF (nine CFTR, six placebo) (Caplen *et al.*, 1995b). No safety problems were encountered, either in the routine clinical assessment or by a blinded, semi-quantative analysis of nasal biopsies. Both plasmid DNA and *CFTR* mRNA were detected from the nasal biopsies in five of eight samples available from *CFTR*-treated subjects. Sodium-related measurements (baseline and response to amiloride) were significantly reduced (approximately 20% toward values seen in patients without CF). However, it is important to note that these changes fell within the coefficient of variation of these measurements. More importantly, chloride secretion assessed by perfusion with a solution of low chloride concentration also showed a significant 20% increase toward normal values, a change well outside the variation in these measurements. In two subjects, these chloride responses reached values within the range seen in subjects without CF. These changes in the sodium- and chloride-related measurements paralleled each other, and were no longer present at 7 days.

11.5.4 Lower respiratory tract

One trial has attempted to deliver the *CFTR* cDNA directly to the lung: Crystal *et al.* (1994), as well as applying the adenovirus to the nose, administered an Ad–CFTR vector to the lower respiratory tract by instillation through a bronchoscope. One bronchial sample was found to be positive for normal CFTR protein (Crystal *et al.*, 1994). However, one of the patients (who received the highest dose of 1 x 10^{10} p.f.u.) developed hypotension, fever and respiratory symptoms suggestive of an inflammatory reaction within the lungs. The investigators have suggested that these changes may relate to an increase in IL-6 as a direct result of the adenoviral infection. All clinical symptoms of this event resolved completely over a period of 1 month (McElvaney and Crystal, 1995).

11.5.5 Protocols in progress

At the time of writing, there are several CF gene therapy trials currently in progress or shortly to be initiated. Welsh and co-workers are studying the effects of single and repeated delivery of Ad–CFTR to the nose and single application to the maxillary sinus (Welsh *et al.*, 1995). Several studies are continuing to examine the effect of Ad–CFTR administration to the nose and lung (Crystal *et al.*, 1995a,b; Schatz and Pavirani, 1995; Wilmott *et al.*, 1994; Wilson *et al.*, 1994) and a further study is examining administration to the lung only (D. Meeker, personal communication). Three nasal application studies are also underway, testing the different cationic lipids DMRIE (Sorscher *et al.*, 1994), DC-Chol (D. Gill and S. Hyde, personal communication) and DOTAP (D. Porteous, personal communication). We have recently obtained local and national ethical permission to conduct a trial of liposome-mediated DNA transfer to both the upper and lower respiratory tract of CF individuals. Finally, a trial using AAV to deliver the *CFTR* cDNA to the maxillary sinuses has recently been initiated.

11.5.6 Conclusions from current CF gene therapy trials

With respect to the current state of CF clinical trials, data from both published and unpublished trials can be summarized as follows. In the nasal epithelium, irrespective of the vector system used, approximately 50% of samples show evidence of vector-specific mRNA and there is evidence for functional correction of the chloride defect in about one third of cases. Thus, in the nose it is reasonable to conclude that there is evidence for gene transfer as well as limited evidence of functional correction; neither vector system appears to be superior in this setting. With regard to safety, very few problems have been encountered. In the lung, only adenoviral-mediated data are available. mRNA has been detected in approximately 20% of cases and protein in about 30%. Importantly, functional correction has not been assessed to date. The relationship between dose and toxicity is being defined in the lungs for adenovirus, whereas liposome-mediated transfer has yet to pass this hurdle.

11.6 Problems and future perspectives

11.6.1 Administration

Lung. Nebulization is likely to be the most acceptable delivery system for routine repeated application to the lower airways in humans. Currently available nebulizer technology is able to target the appropriate areas of the lung and has the advantages of widespread deposition. A theoretical disadvantage is the considerable mechanical stress applied to the liquid during the process of nebulization. This might damage the DNA or delivery system, for example by disrupting DNA–liposome complexes. However, nebulization may also have certain advantages, given that one of the major problems with cationic liposome-mediated gene transfer is the aggregation of the liposome–DNA complex into large molecules, which are unlikely to be well suited to efficient gene transfer. The *β-gal* gene has been successfully administered to rodents by aerosolization using both adenovirus and DNA–liposome complexes (Alton *et al.*, 1993; Katkin *et al.*, 1995; McLachlan *et al.*, 1995). In addition, one of the current adenovirus trials uses nebulized delivery (Schatz and Pavirani, 1995); this is also the method of delivery to be used in our forthcoming trial of liposome-mediated *CFTR* transfer.

The presence of mucus and airway surface liquid further complicates predictions of whether delivery to the lower respiratory tract can be achieved. The mucus (gel) layer may impair access to the cell surface, whereas the aqueous (sol) layer may dilute and alter the composition of any topically applied liquid. Furthermore, in the lower airways of patients with CF, the presence of large quantities of infected secretions as well as of endogenous and potentially exogenous nucleases may markedly alter gene transfer efficiency.

Most investigators are concentrating on delivery to the adult respiratory tract through instillation or nebulization. Recently, three studies have investigated the feasibility of transfecting the pulmonary epithelium *in utero*, which may enable tolerance to be induced against the administered vector or transgene. Injection of Ad–β-gal into the amniotic fluid of developing mice (McCray *et al.*, 1995), rats (Sekhon and Larson, 1995), and directly into the trachea of fetal lambs (McCray *et al.*, 1995; Vincent *et al.*, 1995) resulted in X-gal-positive staining predominately in the epithelium in rats and lambs but not in mice, although expression in mice was observed in the fetal epidermis, amniotic membranes and the gastrointestinal tract. Although no adverse reaction was reported in rats, there was more cause for concern in the lambs because histology of lung sections showed evidence of cellular infiltration and lung fluid with high levels of lymphocytes, presumably as a result of an adenovirus-induced immune response. A report of pulmonary epithelial gene transfer following intravenous injection of a liposome–reporter gene complex is also tantalizing and, if repeatable, would suggest the possibility of another administrative route (Zhu *et al.*, 1993). This may be of particular relevance if submucosal glands need to be targeted.

Other targets. While the airways remain the principal target of interest in the development of gene therapy for CF, some studies have started to address the feasibility of using this technology to treat other aspects of the disease. The human lipase cDNA has been transferred *in vivo* to the gallbladder in an attempt to assess the feasibility of *in vivo* gene therapy for exocrine pancreatic insufficiency in CF (Maeda *et al.*, 1994). Two studies have attempted hepatic transfer. One, an *in vitro* study, transduced intrahepatic biliary epithelial cells with an Ad2–CFTR vector. Halide efflux was detected for 31 days, although the number of cells expressing CFTR declined with time (Grubman *et al.*, 1995). Adenovirus transfer into the billary tract by retrograde infusion of vectors carrying either the *lacZ* reporter gene or the *CFTR* gene showed expression in virtually all cells of the intrahepatic bile ducts for up to 21 days (Yang *et al.*, 1993).

11.6.2 Outcome measurements

It is important to point out that, at present, *in vivo* bioelectrical techniques are well established only for measurements of nasal potential difference. To assess functional correction of the lower airway epithelia, new methodologies will have to be developed. Lower airway potential difference is technically more difficult to assess, partly because use of the local anaesthetic lignocaine interferes with the measurements (Alton *et al.*, 1991). To obtain reliable lower airway recordings, it may therefore be necessary to study these patients under some form of general anaesthesia. Although measurements of ion transport provide a useful end-point of CFTR function, at present it is unclear how these relate to pathology. Thus, end-points that are more closely related to susceptibility to infection may also be required. Some (Saiman and Prince, 1993; Schwab *et al.*, 1993), if not all (Plotkowski *et al.*, 1992), studies have indicated that both *Pseudomonas aeruginosa* and *Staphylococcus aureus*, the most common causes of infection in patients with CF, show increased adherence to CF compared with non-CF airway epithelium. Bacterial adherence may therefore provide a further end-point if assessed prior to and following gene transfer within the same patient. Whether a single application to the airways will produce acute changes in either mucus rheology or MCC is debatable, but these parameters are also worth considering. Radiological techniques such as thin-section computed topographic scanning are unlikely to provide a useful measure, at least for short-term assessment of clinical benefit.

11.6.3 New technology

Current gene transfer technologies are inefficient and have known or potential safety issues associated with them. Rapid developments in vector design, particularly of those based on adenovirus, may overcome some of these issues and many new cationic lipids are being developed in an attempt to improve liposome-mediated DNA transfer. Other, more experimental, methods of gene transfer – for example, receptor-mediated DNA delivery – are also likely to come to fruition over the next few years and may allow more precise targeting of particular cell types (Curiel *et al.*, 1992; Ferkol *et al.*, 1993). Similarly, the use of genetic elements which induce more regulated gene expression are more likely to become relevant if cell-type specific expression can be obtained.

11.6.4 Conclusions

The initial excitement in this field has been tempered with the realization that real gene therapy is more difficult to achieve than simple gene transfer. In the pharmaceutical industry, progress from the laboratory to the patient generally takes decades; however, from the cloning of the *CFTR* gene to trials in patients with CF, gene therapy is moving at a much more rapid pace and there are reasons to be optimistic of eventual success in achieving clinical benefit with this strategy.

References

Afione S, Kearns W, Conrad C, Allen S, Reynolds T, Carter B, Guggino W, Cutting G, Flotte T. (1995) Episomal persistence of *rep* deleted AAV-vectors *in vitro* and *in vivo*. *Pediatr. Pulmonol.* **12**: (Suppl.) Abstract 158.

Alton EWFW, Khagani A, Taylor RFH, Logan-Sinclair R, Yacoub M, Geddes DM. (1991) Effect of heart–lung transplantation on airway potential difference in patients with and without cystic fibrosis. *Eur. Respir. J.* **4**: 5–9.

Alton EWFW, Middleton PG, Caplen NJ, Smith SN, Steel DM, Munkonge FM, Jeffery PK, Geddes DM, Hart SL, Williamson R, Fasold KI, Miller AD, Dickinson P, Stevenson BJ, McLachlan G, Dorin JR, Porteous DJ. (1993) Non-invasive liposome-mediated gene delivery can correct the ion transport defect in cystic fibrosis mutant mice. *Nature Genet.* **5**: 135–142.

Bear CE, Li C, Kartner N, Bridges RJ, Jensen TJ, Ramjeesingh M, Riordan JR. (1992) Purification and functional reconstitution of the cystic fibrosis transmembrane regulator (CFTR). *Cell* **68**: 809–818.

Boat TF, Welsh MJ, Beaudet AL. (1989) Cystic fibrosis. In: *The Metabolic Basis of Inherited Diseases*, 6th Edn. (eds CR Scriver, AL Beaudet, WS Sly, D Valle). McGraw-Hill, NY, pp. 2649–2680.

Boucher RC, Stutts MJ, Knowles MR, Cantley L, Gatzy JT. (1986) Na$^+$ transport in cystic fibrosis respiratory epithelia. *J. Clin. Invest.* **78**: 1245–1252.

Boucher RC, Knowles MR, Johnson LG, Olsen JC, Pickles R, Wilson JM, Engelhardt J, Yang Y, Grossman M. (1994) Gene therapy for cystic fibrosis using E1-deleted adenovirus: a Phase I trial in the nasal cavity. *Hum. Gene Ther.* **5**: 615–639.

Bout A, Perricaudet M, Baskin G, Imler J-L, Scholte BJ, Pavirani A, Valerio D. (1994) Lung gene therapy: *in vivo* adenovirus-mediated gene transfer to rhesus monkey airway epithelium. *Hum. Gene Ther.* **5**: 3–10.

Brody SL, Metzger M, Danel C, Rosenfeld MA, Crystal RG. (1994) Acute response of non-human primates to airway delivery of an adenovirus vector containing the human cystic fibrosis transmembrane conductance regulator cDNA. *Hum. Gene Ther.* **5**: 8821–8836.

Caplen NJ, Kinrade E, Sorgi F, Gao X, Gruenert D, Geddes D, Coutelle C, Huang L, Alton EWFW, Williamson R. (1995a) *In vitro* liposome-mediated DNA transfection of epithelial cell lines using the cationic liposome DC-Chol/DOPE. *Gene Ther.* **2**: 603–613.

Caplen NJ, Alton EWFW, Middleton PG, Dorin JR, Stevenson BJ, Gao X, Durham S, Jeffery PK, Hodson ME, Coutelle C, Huang L, Porteous DJ, Williamson R, Geddes DM. (1995b) Liposome-mediated *CFTR* gene transfer to the nasal epithelium of patients with cystic fibrosis. *Nature Med.* **1**: 39–46. Addendum **1**: 272.

Cheng SH, Gregory RJ, Marshall J, Paul S, Souza DW, White GA, O'Riordan CR, Smith AE. (1990) Defective intracellular transport and processing of CFTR is the molecular basis of most cystic fibrosis. *Cell* **63**: 827–834.

Cohn JA. (1994) CFTR localisation: Implications for cell and tissue pathophysiology. In: *Cystic Fibrosis - Current Topics*, Vol. 2 (eds JA Dodge, DJH Brook, JH Widdicombe). John Wiley & Sons, Chichester, pp. 173–191.

Colledge WH, Abella BS, Sothern KW, Ratcliff R, Jiang C, Cheng SH, MacVinish LJ, Anderson JR, Cuthbert AW, Evans MJ. (1995) Generation and characterization of a ΔF508 cystic fibrosis mouse model. *Nature Genet.* **10**: 445–452.

CF Genotype–Phenotype Consortium. (1993) Correlation between genotype and phenotype in patients with cystic fibrosis. *New Engl. J. Med.* **329**: 1308–1313.

Crawford I, Moloney PC, Zeitlin PL, Guggino WB, Hyde SC, Turley H, Gatter KC, Harris A, Higgins C. (1991) Immunocytochemical localization of the cystic fibrosis gene product CFTR. *Proc. Natl Acad. Sci. USA* **88**: 9262–9266.

Crystal RG, McElvaney NG, Rosenfeld MA, Chu C-S, Mastrangeli A, Hay JG, Brody SL, Jaffe HA, Eissa NT, Danel C. (1994) Administration of an adenovirus containing the human *CFTR* cDNA to the respiratory tract of individuals with cystic fibrosis. *Nature Genet.* **8**: 42–51.

Crystal RG, Jaffe A, Brody S, Mastrangeli A, McElvaney NG, Rosenfeld M, Chu C-S, Danel C, Hay J, Eissa T. (1995a) Clinical protocol: A phase I study, in cystic fibrosis patients, of the safety, toxicity, and biological efficacy of a single administration of a replication deficient, recombinant adenovirus carrying the cDNA of the normal cystic fibrosis transmembrane conductance regulator gene in the lung. *Hum. Gene Ther.* **6**: 643–666.

Crystal RG, Mastrangeli A, Sanders A, Cooke J, King T, Gilbert F, Henschke C, Pascal W, Herena J, Harvey B-G, Hirschowitz E, Diaz D, Russi T, Pacheco F, Sikand V, Brion P. (1995b) Clinical Protocol: Evaluation of repeat administration of a replication deficient, recombinant adenovirus containing the normal cystic fibrosis transmembrane conductance regulator cDNA to the airways of individuals with cystic fibrosis. *Hum. Gene Ther.* **6**: 667–703.

Curiel DT, Agarwal S, Rømer MU, Wagner E, Cotten M, Birnstiel ML, Boucher RC. (1992) Gene transfer to respiratory epithelial cells via receptor-mediated endocytosis pathway. *Am. J. Respir. Cell Mol. Biol.* **6**: 247–252.

Delaney SJ, Alton EWFW, Smith SN, Lunn DP, Farley R, Lovelock PK, Thomson SA, Hume DA, Porteous DJ, Dorin JR, Wainwright BJ. (1996) Cystic fibrosis mice carrying the missense mutation G551D replicate human genotype/phenotype correlations. *EMBO J.* **15**: 955–963.

Denning GM, Anderson MP, Amara JF, Marshall J, Smith AE, Welsh MJ. (1992) Processing of mutant cystic fibrosis transmembrane conductance regulator is temperature-sensitive. *Nature* **358**: 761–764.

di Sant' Agensi PA, Darling RC, Perera GA and Shea E. (1953) Abnormal electrolyte composition of sweat in cystic fibrosis of the pancreas. *Pediatrics* **12**: 549–563.

Dorin JR, Dickinson P, Alton EWFW, Porteous DJ. (1994) Animal models of cystic fibrosis. In: *Cystic Fibrosis – Current Topics*, Vol. 2 (eds JA Dodge, DJH Brook, JH Widdicombe). John Wiley & Sons, Chichester, pp. 3–31.

Dorin JR, Farley R, Webb S, Smith SN, Farini E, Delaney SJ, Wainwright B, Alton EWFW, Porteous DJ. (1996) A demonstration using mouse models that successful gene therapy for cystic fibrosis requires only partial gene correction. *Gene Ther.* (in press).

Drumm ML, Pope HA, Cliff WH, Rommens JM, Marvin SA, Tsui L-C, Collins FS, Frizzel RA, Wilson JM. (1990) Correction of the cystic fibrosis defect *in vitro* by retrovirus-mediated gene transfer. *Cell* **62**: 1227–1233.

Egan M, Flotte T, Afione S, Solow R, Zeitlin PL, Carter BJ, Guggino WB. (1992) Defective regulation of outwardly rectifying Cl⁻ channels by protein kinase A corrected by insertion of CFTR. *Nature* **358**: 581–584.

Eissa NT, Chu CS, Crystal RG. (1994) Evaluation of the respiratory epithelium of normals and individuals with cystic fibrosis for the presence of adenovirus E1a sequences relevant to the use of E1a⁻ adenovirus vectors for gene therapy for the respiratory manifestations of cystic fibrosis. *Hum. Gene Ther.* **5:** 1105–1114.

Engelhardt JF, Yankaskas J, Ernst SA, Yang Y, Marino CR, Boucher RC, Cohn JA, Wilson JM. (1992a) Submucosal glands are the predominant site of CFTR expression in the human bronchus. *Nature Genet.* **2:** 240–247.

Engelhardt JF, Yankaskas JR, Wilson JM. (1992b) *In vivo* retroviral gene transfer into human bronchial epithelia of xenografts. *J. Clin. Invest.* **90:** 2598–2607.

Engelhardt JF, Simon RH, Yang Y, Zepeda M, Weber-Pendleton S, Doranz B, Grossman M, Wilson JM. (1993a) Adenovirus-mediated transfer of the *CFTR* gene to lung of non-human primates: biological efficiacy study. *Hum. Gene Ther.* **4:** 759–769.

Engelhardt JF, Yang Y, Stratford-Perricaudet LD, Allen ED, Kozarsky K, Perricaudet M, Yankaskas JR, Wilson JM. (1993b) Direct gene transfer of human CFTR into human bronchial epithelia of xenografts with E1-deleted adenovirus. *Nature Genet.* **4:** 27–34.

Engelhardt JF, Zepeda M, Cohn JA, Yankaskas JR, Wilson JM. (1994a) Expression of the cystic fibrosis gene in adult human lung. *J. Clin. Invest.* **93:** 737–749.

Engelhardt JF, Litzky L, Wilson JM. (1994b) Prolonged transgene expression in cotton rat lung with recombinant adenoviruses defective in E2a. *Hum. Gene Ther.* **5:** 1217–1229.

Fasbender AJ, Zabner J, Welsh MJ. (1995) Optimization of cationic lipid-mediated gene transfer to airway epithelia. *Am. J. Physiol.* **269:** L45–L51.

Ferkol T, Kaetzel CS, Davis PB. (1993) Gene transfer into respiratory epithelial cells by targeting the polymeric immunoglobulin receptor. *J. Clin. Invest.* **92:** 2394–2400.

Flotte TR, Solow R, Owens RA, Afione S, Zeitlin PL, Carter BJ. (1992) Gene expression from adeno-associated virus vectors in airway epithelial cells. *Am. J. Respir. Cell Mol. Biol.* **7:** 349–356.

Flotte TR, Afione SA, Conrad C, McGath SA, Solow R, Oka H, Zeitlin PL, Guggino WB, Carter BJ. (1993) Stable *in vivo* expression of the cystic fibrosis transmembrane conductance regulator with an adeno-associated virus vector. *Proc. Natl Acad. Sci. USA* **90:** 10613–10617.

Gao X, Huang L. (1991) A novel cationic liposome reagent for efficient transfection of mammalian cells. *Biochem. Biophys. Res. Commun.* **179:** 280–285.

Goldman MJ, Wilson JM. (1995) Expression of alpha v beta integrin is neccessary for efficient adenovirus-mediated gene transfer in human airways. *J. Virol.* **69:** 5951–5958.

Goldman MJ, Yang Y, Wilson JM. (1995) Gene therapy in a xenograft model of cystic fibrosis lung corrects chloride transport more effectively than the sodium defect. *Nature Genet.* **9:** 121–131.

Graham A, Hasani A, Alton EWFW, Martin GP, Marriot C, Hodson ME, Clarke SW, Geddes DM. (1993) No added benefit from nebulised amiloride in patients with cystic fibrosis. *Eur. Respir. J.* **6:** 1243–1248.

Gregory RJ, Cheng SH, Rich DP, Marshall J, Paul S, Hehir K, Ostedgaard L, Klinger KW, Welsh MJ, Smith AE. (1990) Expression and characterization of the cystic fibrosis transmembrane conductance regulator. *Nature* **347:** 382–386.

Grubb BR, Pickles RJ, Ye H, Yankaskas JR, Vick RN, Engelhardt JF, Wilson JM, Johnson LG, Boucher RC. (1994) Inefficient gene transfer by adenovirus vector to cystic fibrosis airway epithelia of mice and humans. *Nature* **371:** 802–806.

Grubman SA, Fang SL, Mulberg AE, Perrone RD, Rogers LC, Lee DW, Armentano D, Murray SL, Dorkin HL, Cheng SH. (1995) Correction of the cystic fibrosis defect by gene complementation in human intrahepatic biliary epithelial cell lines. *Gastroenterology* **108:** 584–592.

Hay JG, McElvaney NG, Herena J, Crystal RG. (1995) Modification of nasal epithelial potential differences of individuals with cystic fibrosis consequent to local administration of a normal *CFTR* cDNA adenovirus gene transfer vector. *Hum. Gene Ther.* **6:** 1487–1496.

Hyde S, Gill D, Higgins C, Treizie A, MacVinish LJ, Cuthburt A, Ratcliff R, Evans M, Colledge W. (1993) Correction of the ion transport defect in cystic fibrosis transgenic mice by gene therapy. *Nature* **362:** 250–255.

Imler J-L, Bout A, Dryer D, Dietrlé A, Schultz H, Valerio D, Mehtali M, Pavirani A. (1995) *Trans*-complementation of E1-deleted adenovirus: a new vector to reduce the possibility of codissemination of wild-type and recombinant adenovirus. *Hum. Gene Ther.* **6:** 711–721.

Johnson LG, Olsen JC, Sarkadi B, Moore KL, Swanstrom R, Boucher RC. (1992) Efficiency of gene transfer for restoration of normal airway epithelial function in cystic fibrosis. *Nature Genet.* **2:** 21–25.

Kartner N, Augustinas O, Jensen TJ, Naismith AL, Riordan JR. (1992) Mislocalizaion of ΔF508 CFTR in cystic fibrosis sweat gland. *Nature Genet.* **1**: 321–327.

Katkin JP, Gilbert BE, Langston C, French K, Beaudet AL. (1995) Aerosol delivery of a β-galactosidase adenoviral vector to the lungs of rodents. *Hum. Gene Ther.* **6**: 985–995.

Kerem B-S, Rommens JM, Buchanan JA, Markiewicz D, Cox TK, Chakravarti A, Buchwald M, Tsui L-C. (1989) Identification of the cystic fibrosis gene: Genetic analysis. *Science* **245**: 1073–1080.

Knowles MR, Church NL, Waltner WE, Yankaskas JR, Gilligan P, King M, Edwards LJ, Helms RW, Boucher RC. (1990) A pilot study of aerosolised amiloride for the treatment of cystic fibrosis. *New Engl. J. Med.* **322**: 1189–1194.

Knowles MR, Clarke LL, Boucher RC. (1991) Activation by extracellular nucleotides of chloride secretion in the airway epithelia of patients with cystic fibrosis. *New Engl. J. Med.* **325**: 533–538.

Knowles MR, Paradiso AM, Boucher RC. (1995a) *In vivo* nasal potential difference: techniques and protocols for assessing efficacy of gene transfer in cystic fibrosis. *Hum. Gene Ther.* **6**: 445–455.

Knowles MR, Hohneker KW, Zhou Z, Olsen JC, Noah TL, Hu P-C, Leigh MW, Engelhardt JF, Edwards LJ, Jones KR, Grossman M, Wilson JM, Johnson LG, Boucher RC. (1995b) A controlled study of adenoviral-vector-mediated gene transfer in the nasal epithelium of patients with cystic fibrosis. *New Engl. J. Med.* **333**: 823–831.

Kozarsky KF, McKinley DR, Austin LL, Raper SE, Stratford-Perricaudet LD, Wilson JM. (1994) *In vivo* correction of low denisty lipoprotein receptor deficiency in the Watanabe heritable hyperlipidemic rabbit with recombinant adenoviruses. *J. Biol. Chem.* **269**: 13695–13702.

Kristidis P, Bozon D, Corey M, Markiewicz D, Rommens J, Tsui L-C. (1992) Genetic determination of exocrine pancreatic function in cystic fibrosis. *Am. J. Hum. Genet.* **50**: 1178–1184.

Leigh MW, Kylander JE, Yankaskas JR, Boucher RC. (1995) Cell proliferation in bronchial epithelium and submucosal glands of cystic fibrosis patients. *Am. J. Respir. Cell Mol. Biol.* **12**: 605–612.

Logan JJ, Bebok Z, Walker LC, Peng S, Felgner PL, Siegal GP, Frizzell RA, Dong J, Howard M, Matalon S, Lindsey JR, DuVall M, Sorscher EJ. (1995) Cationic lipids for reporter gene and CFTR transfer to rat pulmonary epithelium. *Gene Ther.* **2**: 38–49.

Maeda H, Danel C, Crystal RG. (1994) Adenovirus-mediated transfer of human lipase complementary DNA to the gallbladder. *Gastroenterology* **106**: 1638–1644.

Marshall J, Lee E, Nietupski J, Siegel C, Nichols M, Rafter P, Cherry M, Kaplan JM, St George JA, Garlick D, Jiang C, Yew NS, Scheule RK, Harris D, Smith AE, Cheng SH. (1995) Development and use of novel synthetic cationic lipids with enhanced gene transfer activity. *Pediatr. Pulmonol.* **12** (Suppl.): Abstract 133.

Mastrangeli A, Danel C, Rosenfeld M, Stratford-Perricaudet L, Perricaudet M, Pavirani A, Lecocq J-P, Crystal RG. (1993) Diversity of airway epithelia cell targets for *in vivo* recombinant adenovirus-mediated gene transfer. *J. Clin. Invest.* **91**: 225–234.

McCray PB, Armstrong K, Zabner J, Miler DW, Koretzky GA, Couture L, Robillard JE, Smith AE, Welsh MJ. (1995) Adenoviral-mediated gene transfer to fetal pulmonary epithelia *in vitro* and *in vivo*. *J. Clin. Invest.* **95**: 2620–2632.

McElvaney NG, Crystal RG. (1995) IL-6 release and air-way administration of human CFTR cDNA adenovirus vector. *Nature Med.* **1**: 182–184.

McLachlan G, Davidson DJ, Stevenson BJ, Dickinson P, Davidson-Smith H, Dorin JR, Porteous DJ. (1995) Evaluation *in vitro* and *in vivo* of cationic liposome-expression construct complexes for cystic fibrosis gene therapy. *Gene Ther.* **2**: 614–622.

Meyer KB, Thompson MM, Levy MY, Barron LG, Szoka FC. (1995) Intratracheal gene delivery to the mouse airway: characterization of plasmid DNA expression and pharmacokinetics. *Gene Ther.* **2**: 450–460.

Middleton PG, Caplen NJ, Gao X, Haung L, Gaya H, Geddes DM, Alton EWFW. (1994a) Nasal application of the cationic liposome DC-Chol:DOPE does not alter ion transport, lung function or bacterial growth. *Eur. Respir. J.* **7**: 442–445.

Middleton PG, Geddes DM, Alton EW. (1994b) Protocols for *in vivo* measurement of ion transport defects in cystic fibrosis nasal epithelium. *Eur. Respir. J.* **7**: 2050–2056.

Mittereder N, Yei S, Bachurski C, Cuppoletti J, Whitsett JA, Tolstoshev P, Trapnell BC. (1994) Evaluation of the efficacy and safety of *in vitro*, adenovirus-mediated transfer of the human cystic fibrosis transmembrane conductance regulator cDNA. *Hum. Gene Ther.* **5**: 719–731.

Olsen JC, Johnson LG, Stutts MJ, Sarkadi B, Yankaskas JR, Swanstrom R, Boucher RC. (1992) Correction of the apical membrane chloride permeability defect in polarized cystic fibrosis airway epithelia following retroviral-mediated gene transfer. *Hum. Gene Ther.* **3**: 253–266.

Peault B, Tirouvanziam R, Sombardier MN, Chen S, Perricaudet M, Gaillard D. (1994) Gene transfer to human fetal pulmonary tissue developed in immunodeficient SCID mice. *Hum. Gene Ther.* **5**: 1131–1137.

Pilewski J, Engelhardt JF, Bavaria JE, Kaiser LR, Wilson JM, Albelda SM. (1995) Adenovirus-mediated gene transfer to human bronchial submucosal glands using xenografts. *Am. J. Physiol.* **268**: L657–L665.

Plotkowski MC, Chevillard M, Pierrot D, Altemayer D, Puchelle E. (1992) Epithelial respiratory cells from cystic fibrosis patients do not possess specific *Pseudomonas aeruginosa*-adhesive properties. *J. Med. Microbiol.* **36**: 104–111.

Ramjeesingh M, Wilschanski MA, Li C, Gymorey K, Kent G, Corey M, Durie P, Tanswell K, Bear CE. (1995) Treatment of the nasal epithelium of CF mice with liposomes containing purified CFTR protein. *Pediatr. Pulmonol.* **13** (Suppl.): Abstract 10.7.

Rich DP, Anderson MP, Gregory RJ, Cheng SH, Paul S, Jefferson DM, McCann JD, Klinger KW, Smith AE, Welsh MJ. (1990) Expression of cystic fibrosis transmembrane conductance regulator corrects defective chloride channel regulation in cystic fibrosis airway epithelial cells. *Nature* **347**: 358–363.

Rich DP, Couture LA, Cardoza LM, Guiggio VM, Armentano D, Espino PC, Hehir K, Welsh MJ, Smith AE, Gregory RG. (1993) Development and analysis of recombinant adenovirus for gene therapy of cystic fibrosis. *Hum. Gene Ther.* **4**: 461–476.

Riordan JR, Rommens JM, Kerem B-S, Alon N, Rozmahel R, Grzelczak Z, Zielenski J, Lok S, Plavsic N, Chou J-L, Drumm ML, Iannuzzi MC, Collins FS, Tsui L-C. (1989) Identification of the cystic fibrosis gene: cloning and characterization of complementary DNA. *Science* **245**: 1066–1073.

Rommens JM, Iannuzzi MC, Kerem B-S, Drumm ML, Melmer G, Dean M, Rozmahel R, Cole JL, Kennedy D, Hidaka N, Zsiga M, Buchwald M, Riordan JR, Tsui L-C, Collins FS. (1989) Identification of the cystic fibrosis gene: chromosome walking and jumping. *Science* **245**: 1059–1065.

Rosenfeld MA, Yoshimura K, Trapnell BC, Yoneyama K, Rosenthal ER, Dalemans W, Fukayama M, Bargon J, Stier LE, Stratford-Perricaudet L, Perricaudet M, Guggino WB, Pavirani A, Lecocq J-P, Crystal RG. (1992) *In vivo* transfer of the human cystic fibrosis transmembrane conductance regulator gene to the airway epithelium. *Cell* **68**: 143–155.

Rosenfeld MA, Chu C-S, Seth P, Danel C, Banks T, Yoneyama K, Yoshimura K, Crystal RG. (1994a) Gene transfer to freshly isolated human respiratory epithelial cells *in vitro* using a replication-deficient adenovirus containing the human cystic fibrosis transmembrane conductance regulator cDNA. *Hum. Gene Ther.* **5**: 331–342.

Rosenfeld MA, Rosenfeld SJ, Danel C, Banks TC, Crystal RG. (1994b) Increasing expression of the normal human *CFTR* cDNA in cystic fibrosis epithelial cells results in a progressive increase in the level of CFTR protein expression, but a limit on the level of cAMP-stimulated chloride secretion. *Hum. Gene Ther.* **5**: 1121–1129.

Rozmahel R, Wischanski M, Matin A, Plyte S, Oliver M, Auerbach W, Moore A, Forstner J, Durie P, Nadeau J, Bear C, Tsui L-C. (1996) Modulation of disease severity in cystic fibrosis transmembrane regulator deficient mice by a secondary genetic factor. *Nature Genet.* **12**: 280–287.

Saiman L, Prince A. (1993) *Pseudomonas aeruginosa* pili bind asialoGM1 which is increased on the surface of cystic fibrosis epithelial cells. *J. Clin. Invest.* **92**: 1875–1890.

Schatz C, Pavirani A. (1995) Gene therapy in mucoviscidosis. *Rev. Pneumol. Clin.* **51**: 201–206.

Scheule RK, Bagley RG, Erickson AL, Wang KX, Fang SL, Vaccaro C, O'Riordan CR, Cheng SH, Smith AE. (1995) Delivery of purified, functional CFTR to epithelial cells *in vitro* using influenza hemagglutinin. *Am. J. Respir. Cell Mol. Biol.* **13**: 330–343.

Schiavi SC, DiTullio P, Abdelkader N, Cunniff M, Reber S, Meade H, Hoppe H, McPherson J, Ebert K, Smith AE, Cheng SH. (1993) Evidence for toxic effects of overexpression of CFTR *in vitro* and *in vivo*. *Pediatr. Pulmonol.* **17**: 69 (Abstract).

Schwab UE, Wold AE, Carson JL, Leigh MW, Cheng PW, Gilligan PH, Boat TF. (1993) Increased adherence of *Staphylococcus aureus* from cystic fibrosis lungs to airway epithelial cells. *Am. Rev. Respir. Dis.* **148**: 365–369.

Sekhon HS, Larson JE. (1995) *In utero* gene transfer into the pulmonary epithelium. *Nature Med.* **1**: 1201–1203.

Simon RH, Engelhardt JF, Yang Y, Zepeda M, Weber-Pendleton S, Grossman M, Wilson JM. (1993) Adenovirus-mediated transfer of the *CFTR* gene to lung of nonhuman primates: toxicity study. *Hum. Gene Ther.* **4:** 771–780.

Smith TAG, Mehaffey MG, Kayda DB, Saunders JM, Yei S, Trapnell BC, McCelland A, Kaleko M. (1993) Adenovirus mediated expression of therapeutic plasma levels of human factor IX in mice. *Nature Genet.* **5:** 397–402.

Sorscher E, Logan JJ, Frizzell RA, Lyrene RK, Bebok Z, Dong JY, Duvall DD, Felgner PL, Matalon S, Walker L, Wiatrak BJ. (1994) Gene therapy for cystic fibrosis using cationic liposome mediated gene transfer: a phase I trial of safety and efficacy in the nasal airway. *Hum. Gene Ther.* **5:** 1259–1277.

Stern M, Munkonge FM, Caplen NJ, Sorgi F, Huang L, Geddes D, Alton EWFW. (1995) Quantitative fluoresecence measurement of chloride secretion in native airway epithelium from CF and non-CF subjects. *Gene Ther.* **2:** 766–774.

Stribling R, Brunette E, Liggitt D, Gaensler K, Debs R. (1992) Aerosol gene delivery *in vivo. Proc. Natl Acad. Sci. USA* **89:** 11277–11281.

Teramoto S, Johnson LG, Huang W, Keigh MW, Boucher RC. (1995) Effect of adenoviral vector infection on cell proliferation in cultured primary human airway epithelial cells. *Hum. Gene Ther.* **6:** 1045–1053.

Trapnell BC, Chu C-S, Paako PK, Banks TC, Yoshimura K, Ferrans VJ, Chernick MS, Crystal R. (1991) Expression of the cystic fibrosis transmembrane regulator gene in the respiratory tract of normal individuals and individuals with cystic fibrosis. *Proc. Natl Acad. Sci. USA* **88:** 6565–6569.

Tsui L-C. (1995) The cystic fibrosis transmembrane conductance regulator gene. *Am. J. Respir. Crit. Care* **151:** S47–S53.

van Doorninck JH, French PJ, Verbeek E, Peters RH, Morreau H, Bijman J, Scholte BJ. (1995) A mouse model for the cystic fibrosis delta F508 mutation. *EMBO J.* **14:** 4403–4411.

Vincent MC, Trapnell BC, Baughman RP, Wert SE, Whitsett JA, Iwamoto HS. (1995) Adenovirus-mediated gene transfer to the respiratory tract of fetal sheep *in utero. Hum. Gene Ther.* **6:** 1019–1028.

Welsh MJ, Smith AE. (1993) Molecular mechanisms of CFTR chloride channel dysfunction in cystic fibrosis. *Cell* **73:** 1251–1254.

Welsh MJ, Anderson MP, Rich DP, Berger HA, Denning GM, Ostegaard LS, Sheppard DN, Cheng SH, Gregory RJ, Smith AE. (1992) Cystic fibrosis transmembrane conductance regulator: a chloride channel with novel regulation. *Neuron* **8:** 821–829.

Welsh MJ, Zabner J, Graham SM, Smith AE, Moscicki R, Wadsworth S. (1995) Adenovirus-mediated gene transfer for cystic fibrosis. Part A: Safety of dose and repeat administration in the nasal epithelium. Part B: Clinical efficacy in the maxillary sinus. *Hum. Gene Ther.* **6:** 205–218.

Whitsett JA, Dey CR, Stripp BR, Wikenheiser KA, Clark JC, Wert SE, Gregory RJ, Smith AE, Cohn JA, Wilson JM, Engelhardt J. (1992) Human cystic fibrosis transmembrane conductance regulator directed to respiratory epithelial cells of transgenic mice. *Nature Genet.* **2:** 13–20.

Wilmott RW, Whitsett JA, Trapnell BC. (1994) Gene therapy of cystic fibrosis utilizing a replication deficient recombinant adenovirus vector to deliver the human cystic fibrosis transmembrane conductance regulator cDNA to the airways. A phase I study. *Hum. Gene Ther.* **5:** 1019–1057.

Wilson JM, Engelhardt JF, Grossman M, Simon RH, Yang Y. (1994) Clinical protocol: Gene therapy of cystic fibrosis lung disease using E1 deleted adenovirus: a phase 1 trial. *Hum. Gene Ther.* **5:** 501–519.

Yang Y, Raper SE, Cohn JA, Engelhardt JF, Wilson JM. (1993) An approach for treating the hepatobiliary disease of cystic fibrosis by somatic gene transfer. *Proc. Natl Acad. Sci. USA* **90:** 4601–4605.

Yang Y, Nunes FA, Berencsi K, Gönczöl E, Engelhardt JF, Wilson JM. (1994) Inactivation of E2a in recombinant adenoviruses improves the prospect for gene therapy in cystic fibrosis. *Nature Genet.* **7:** 362–369.

Yang Y, Li Q, Ertl HC, Wilson JM. (1995a) Cellular and humoral immune responses to viral antigens create barriers to lung-directed gene therapy with recombinant adenoviruses. *J. Virol.* **69:** 2004–2015.

Yang Y, Trinchieri G, Wilson JM. (1995b) Recombinant IL-12 prevents formation of blocking IgA antibodies to recombinant adenovirus and allows repeated gene therapy to mouse lung. *Nature Med.* **1:** 890–893.

Yei S, Mittereder N, Wert S, Whitsett JA, Wilmott RW, Trapnell B. (1994a) *In vivo* evaluation of the safety of adenovirus-mediated transfer of the human cystic fibrosis transmembrane conductance regulator cDNA to the lung. *Hum. Gene Ther.* **5:** 731–744.

Yei S, Mittereder N, Tang K, O'Sullivan C, Trapnell BC. (1994b) Adenovirus-mediated gene transfer for cystic fibrosis: quantitative evaluation of repeated *in vivo* vector administration to the lung. *Gene Ther.* **1:** 192–200.

Yew NS, Wysokenski D, Ziegler R, Cherry M, Rudginsky S, Nichols M, Peterson P, Wan N, Marshall J, Smith AE, Cheng SH. (1995) Plasmid expression vectors for high and sustained expression of CFTR in airway epithelial cells. *Pediatr. Pulmonol.* **12** (Suppl.): Abstract 132

Yoshimura K, Rosenfield MA, Nakamura H, Scherer EM, Pavirani A, Lecocq J-P, Crystal RG. (1992) Expression of the human cystic fibrosis transmembrane conductance regulator gene in the mouse lung after *in vivo* intratracheal plasmid-mediated gene transfer. *Nucleic Acids Res.* **20**: 3233–3240.

Zabner J, Couture LA, Gregory RJ, Graham SM, Smith AE, Welsh MJ. (1993) Adenovirus-mediated gene transfer transiently corrects the chloride transport defect in nasal epithelia of patients with cystic fibrosis. *Cell* **75**: 1–20.

Zabner J, Couture LA, Smith AE, Welsh MJ. (1994a) Correction of cAMP-stimulated fluid secretion in cystic fibrosis airway epithelia: efficiency of adenovirus-mediated gene transfer *in vitro*. *Hum. Gene Ther.* **5**: 585–593.

Zabner J, Petersen DM, Puga AP, Graham SM, Couture LA, Keyes LD, Lukason MJ, George JAS, Gregory RJ, Smith AE, Welsh MJ. (1994b) Safety and efficacy of repetitive adenovirus-mediated transfer of *CFTR* cDNA to airway epithelia of primates and cotton rats. *Nature Genet.* **6**: 75–83.

Zeiher BG, Eichwald E, Zabner J, Smith JJ, Puga AP, McCray PB, Capecchi MR, Welsh MJ, Thomas KR. (1995) A mouse model for the ΔF508 allele of cystic fibrosis. *J. Clin. Invest.* **96**: 2051–2064.

Zhou L, Dey CR, Wert SE, DuVall MD, Frizzell RA, Whitsett JA. (1994) Correction of lethal intestinal defect in a mouse model of cystic fibrosis by human CFTR. *Science* **266**: 1705–1708.

Zhu N, Liggitt D, Liu Y, Debs R. (1993) Systemic gene expression after intravenous DNA delivery into adult mice. Science **261**: 209–211.

Zsengeller ZK, Wert SE, Hull WM, Hu X, Yei S, Trapnell BC, Whitsett JA. (1995) Persistence of replication-deficient adenovirus-mediated gene transfer in lungs of immune-deficent (nu/nu) mice. *Hum. Gene Ther.* **6**: 457–467.

12

Gene therapy for haemophilia B

Kotoku Kurachi and Jian-Min Wang

12.1 Introduction

Both haemophilia A (factor VIII deficiency) and B (factor IX deficiency) have been studied intensively by academic as well as industrial research groups utilizing diverse gene transfer methods (Brownlee, 1995; Kurachi and Yao, 1993; Thompson, 1995). This is due primarily to the fact that these are among the best characterized, recessive genetic diseases (Kurachi *et al.*, 1992, 1993) and can serve as excellent models for other diseases which require an efficient systemic delivery of gene products.

Despite all these efforts, however, none of the gene transfer methods for haemophilias tested to date has yet reached phase I clinical trials in Europe or the USA. An *ex vivo* protocol using skin fibroblasts has been tested with two moderately affected haemophilia B patients in China, and has been reported to have positive effects at least in one of the patients (Lu *et al.*, 1993). The fact that no gene therapy protocol for either of the haemophilias has yet reached the clinical trial phase in Western nations indicates the substantial difficulties involved in developing durable gene therapy for haemophilias. These difficulties are related to the fundamental biology of the blood coagulation system (and were apparently underestimated in the early phase of gene therapy studies) and to the lack of robust gene transfer technologies optimized for haemophilias (Brownlee, 1995; Crystal, 1995; Kurachi and Yao, 1993; Thompson, 1995).

In this chapter, we review the background, gene transfer vectors used and progress made to date, obstacles encountered and future prospects. We focus our discussion mainly on haemophilia B gene therapy, but also comment on factor VIII studies. Previous review articles (Brownlee, 1995; Kurachi and Yao, 1993; Thompson, 1995) also serve as an excellent resource.

12.2 Basic properties of factor IX

The blood coagulation cascade mechanism involves more than two dozen pro-coagulant factors and regulators (Hedner and Davie, 1989; Kurachi *et al.*, 1992, 1993), but critically factor IX and its co-factor, factor VIII, are responsible for activation of factor X. Therefore, deficiency or a substantial reduction in factor IX or factor VIII levels in the circulation, which are due mostly to mutations in their genes, result in typical recessive bleeding disorders (haemophilia B or haemophilia A, respectively) (Kurachi *et al.*, 1992). Since the genes for factor VIII and factor IX are on the X

Gene Therapy, edited by N.R. Lemoine and D.N. Cooper.
© 1996 BIOS Scientific Publishers Ltd, Oxford.

chromosome, clinical presentation of their deficiencies are almost exclusively in males, although they do occur very rarely in females.

Factor IX is a plasma glycoprotein of 415 amino acid residues, post-translationally modified with both *N*- and *O*-linked carbohydrate chains (Kurachi *et al.*, 1993). Factor IX is normally produced by hepatocytes as a single polypeptide chain containing a leader sequence composed of a signal peptide for secretion and a propeptide for γ-carboxylation (an essential post-translational modification) of the amino-terminal region of the mature factor IX protein. Nascent recombinant factor IX polypeptide chains synthesized in the liver or other ectopic tissues or cells must therefore be properly processed by cleavage of the leader sequence, post-translational modification (e.g. γ-carboxylation of the first 12 glutamic acid residues in the amino-terminal region, β-hydroxylation of Asp64, attachment of carbohydrate chains) and other processing required for its full biological activity. The normal plasma concentration of factor IX in humans is 4–5 μg ml^{-1}, and its distribution in the vascular system follows the two-compartment model (Smith and Thompson, 1981; Thompson *et al.*, 1980). Recently, canine factor IX has been isolated and characterized, and its normal plasma concentration was found to be 5.3 mg ml^{-1}, very similar to that of human (Sugahara *et al.*, 1996). This is important because naturally occurring canine mutants with haemophilia B are frequently used as an animal model for testing gene transfer methods and for various haemostatic studies. The reported half-clearance times of human factor IX from the circulation of human, dog and mouse are within the range of 18–23 h (Smith and Thompson, 1981; Thompson *et al.*, 1980; Yao *et al.*, 1994). Significantly shorter half-clearance times, 13.2 and 11.6–17.7 h respectively, have also been reported for recombinant factor IX preparations in rats and dogs (Keith *et al.*, 1995). In the following discussion, we use the well-established half-clearance time of approximately 20 h for human factor IX.

In contrast to factor IX, factor VIII is a large plasma glycoprotein of 2332 amino acid residues, primarily produced by hepatocytes (Kurachi *et al.*, 1992; Zatloukal *et al.*, 1994). The normal plasma concentration of factor VIII is 0.5 mg ml^{-1} in humans, and follows a one-compartment model (with little absorption into the vascular tissue) with a half-clearance time from the circulation of ~10 h (Brownlee, 1995). In the circulation, binding to von Willebrand factor (vWF) is required for its stability. Accordingly, its production by a gene transfer approach must take this into account. The B domain, constituting the central one-third portion of the factor VIII molecule and containing most of the carbohydrate chains, can be deleted with minimal effect on its biological activity. B domain-less factor VIII is often utilized for gene transfer purposes employing various vector systems, including retroviruses and adenoviruses, which have limitations on their insert sizes. The A1 and A2 regions (amino-terminal regions with functional importance) have been reported to possess a substantial silencer activity on expression (Koeberl *et al.*, 1995; Lynch *et al.*, 1993).

12.3 Conditions required for durable haemophilia gene therapy

Basic conditions to be satisfied in establishing a durable gene therapy for the haemophilias include: (i) development of a gene transfer system that can provide highly efficient gene transfer into target cells/tissue; (ii) achievement of stable production of the biologically active transgene products in the circulation at a therapeutic level over a time span of years; (iii) establishment of short- and long-term safety of the method; and (iv) durability and practicality of the method involving only minor surgical procedures, if any.

Gene delivery methods can be categorized into two approaches: *in vivo* direct gene transfer and *ex vivo* gene transfer. Both methods have been tested for factor IX gene transfer into various target tissues, with increasing emphasis on *in vivo* direct gene transfer approaches in recent years. Whatever the method used, transgene products must be secreted efficiently from the cell, and transported into the circulation (*Figure 12.1*). Every step involved in this process, therefore, can substantially affect the systemic production of recombinant factor IX. Blood volume (~8% of body weight for human) alone also presents a formidable problem, demanding a very high systemic production of the transgene. Development of a highly refined and efficient gene delivery system, which can guarantee long-term stable production of factor IX at a therapeutic plasma level, is a challenging task. As a goal, a stable factor IX expression level of 5% of normal or higher (equivalent to ~200 ng ml^{-1} plasma or higher) should be achieved. This level of factor IX would be sufficient to convert severe haemophilia to a mild condition. Ideally, a plasma concentration level of 15% or higher (without exceeding the normal level) is desired. Factor IX protein infused into the vascular system is distributed in two compartments, the circulation and vascular wall tissue retaining about 60 and 40% of the total factor IX, respectively (Smith and Thompson, 1981; Thompson *et al.*, 1980). The kinetics of this process are represented by an equation (Okpako, 1991; Yao and Kurachi, 1992):

$$C_{ss} = K_o T_{1/2} / 0.693 V_d,$$

where C_{ss}, K_o, $T_{1/2}$ and V_d are the steady-state concentration in the circulation, infusion (systemic delivery) rate, half-life and body distribution volume, respectively.

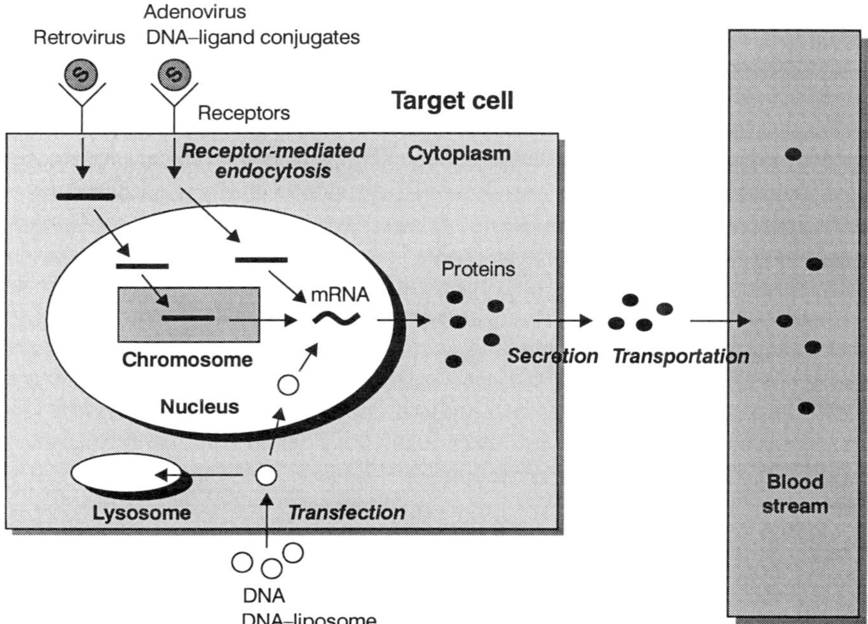

Figure 12.1. Various gene transfer methods and fundamental requirements to be satisfied in haemophilia gene therapy. Factor IX expression cassettes can be transferred into the target cells using viral or non-viral vector systems in either *in vivo* direct or *ex vivo* indirect approaches. Target cells then produce factor IX, which must be secreted efficiently and transported into the circulation.

According to these kinetics, assuming $T_{1/2}$ and V_d are 20 h and 0.25 l kg^{-1}, respectively, a 50 kg man requires a stable systemic production level of factor IX of approximately 1600 µg per 24 h, in order to maintain a plasma factor IX level 5% of the normal level.

Development of a safe and practical gene delivery system that requires no major surgery and which can ensure the stable systemic production of factor IX over years has turned out to be a great challenge.

12.4 Expression vector systems for factor IX gene transfer

Here, we review gene transfer vector systems frequently used to date for factor IX. Since the very early studies of gene transfer, the well-characterized Moloney murine leukaemia virus-derived retroviral vector (Miller, 1992; Mulligan, 1993), has played a major role in factor IX gene transfer (Brownlee, 1995; Kurachi and Yao, 1993; Thompson, 1995). Because its biology potentiates the stable integration of transgenes into the genome of target cells, the retrovirus has been considered to be one of the best choices among various vector systems currently available, with possible long-term stability of the transgene and its expression. Furthermore, with strict quality control of the virus stock used, to exclude potential contamination of helper viruses, the risk of recombinant retroviruses causing random insertional mutations and other detrimental effects in animals as well as humans appears to be minimal (Anderson, 1994; Cornetta *et al.*, 1991). Any possible contamination by helper viruses must be eliminated because lymphoma development in several non-human primates due to helper virus contamination has been reported (Donahue *et al.*, 1992). Some of the limitations of the currently available retroviral vector system include its inability to transfer the transgene into non-proliferating cells and often low titres which can be obtained with highly refined vectors. This limitation causes substantial problems, particularly in using retroviral vectors for direct *in vivo* gene transfer. Secondly, satisfactory *in vivo* expression activity of the transgenes delivered by retrovirus with a titre as low as 10^5 colony forming units (c.f.u.) ml^{-1} has not been achieved. For an *ex vivo* approach, this could be less of a problem, but a high viral titre should make the procedure much more efficient, simpler and perhaps safer. It is, however, important to be aware that an extremely high level of expression of factor IX in a transduced cell may exceed the capacity of individual cells for post-translational modification, particularly γ-carboxylation (a unique and essential modification) (Kurachi, 1991; Kurachi *et al.*, 1993). Our general assumption, which is based on many published reports regarding recombinant factor IX production by cultured cells, is that in order to keep the specific activity of recombinant factor IX sufficiently high ($\geq$90%), the level of factor IX production per 10^6 cells should be kept lower than 10 µg per 24 h. Furthermore, in order to minimize possible insertional mutations due to the recombinant viruses, transgene copy numbers may be better kept as low as 1–2 copies per cells, rather than increasing this to much higher numbers by multiple infections.

The retroviral vector system will continue to maintain its great potential and value for haemophilia gene therapy. Currently, a substantial number of studies are under way to improve the retrovirus system, not only in terms of the expression activity, but also to generate chimeric retrovirus. In the following, some of the important progress which has been made in relation to haemophilia gene therapy is discussed.

Recently, Okuyama *et al.* (1996) reported a remarkable achievement showing that α_1-antitrypsin retrovirus with an internal transcriptional control unit composed of apolipoprotein E enhancer and α_1-antitrypsin promoter can express α_1-antitrypsin at a persistently high level ($\sim$5 μg ml^{-1} plasma) after intraportal infusion into nude mice which have undergone partial hepatectomy. This study indicated that it is possible to have a very strong internal promoter while maintaining a good viral titre ($\geq$10^6 c.f.u. ml^{-1}). This refined retroviral system may be applied to factor IX as well as factor VIII. Because vector sequences [such as long terminal repeats (LTRs)] can apparently have unpredictable interactions with the inserted sequences which substantially affect the overall expression activity, use of any vector system must be tested specifically with the gene sequence of interest (Okuyama *et al.*, 1996). Unfortunately, the approach taken by Okuyama *et al.* (1996) requires a partial hepatectomy to stimulate liver cell proliferation. Such an invasive procedure is not acceptable for the haemophilias, and a much less invasive approach utilizing the vector system must be developed.

We have reported the persistent systemic delivery of factor IX at 10–30 ng ml^{-1} of plasma in mice by using myoblast-mediated gene transfer employing retrovirally transduced primary myoblasts (Yao *et al.*, 1994). This does not require major surgery. Although this factor IX level is not sufficient for correcting haemophilia, we have refined the basic vector structure extensively, and have found that a combination of muscle creatine kinase enhancers, β-actin basal promoter and a factor IX minigene can provide very high and stable factor IX expression in myotubes in culture (Wang *et al.*, 1996). More recently, Naffakh *et al.* (1996) also reported the long-term expression (up to 8 months) of β-glucuronidase by myoblast-mediated gene transfer employing a retrovirus-transduced myoblast, thereby correcting a lysosomal lesion in a mouse disease model. These studies strongly support the rationale of the myoblast-mediated gene transfer approach using a retroviral vector. This approach, however, still requires substantial improvements, particularly with regard to transgene expression levels and retroviral titre, to make it truly durable. It is noteworthy that the MFG-type retroviral vector with B-domain-deleted factor VIII can express and secrete a surprisingly high level of factor VIII (an average of $\sim$80 ng ml^{-1}) from fibroblasts in culture (Dwarki *et al.*, 1995).

In recent years, recombinant adenoviruses have been shown to be highly effective for *in vivo* factor IX gene transfer by intravenous (i.v.) injection, predominantly transferring the transgene into the liver ($\sim$90%). Kay *et al.* (1994) reported a very high, but transient, level of production of recombinant canine factor IX ($\sim$300% of normal) in dogs with haemophilia which lasted for a few days, followed by a rapid decrease to low but still detectable levels over 1–2 months. Adenoviruses were also used to express canine factor IX at a high level for a few weeks in muscles of normal mice (Swiss Webster) as well as in various immune-compromised mice (Dai *et al.*, 1995), again giving a transiently high expression followed by much lower, but prolonged expression in immune-compromised mice for up to a year. The transient expression in normal mice was considered to be due to destruction of the transduced cells by the major histocompatibility complex (MHC) class I/cytotoxic T lymphocyte (CTL) system (Engelhardt *et al.*, 1994; Yang *et al.*, 1994). This immune mechanism was postulated to be due to recognition of viral gene products which are produced by the E1-deleted recombinant adenovirus. As discussed below, however, this mechanism may not be functioning as efficiently as claimed.

Our recent study (Yao *et al.*, 1996) showed that liver cells transduced by factor IX adenovirus (first generation vector with its E1 region replaced with the factor IX expression vector sequence and its E3 region partially deleted) survive well over 6 months (end of the experiment), not only in mice with severe combined immunodeficiency (SCID) and other variously immune-compromised mice, but also in fully immune-competent normal animals (Balb/c). The surviving factor IX transgene is actively expressed in the liver cells, maintaining the plasma factor IX levels at 40–50% of the early maximal level in SCID mice, and in the normal animals as shown by the presence of both human factor IX mRNA and factor IX protein. These observations suggest that the MHC class I/CTL hypothesis proposed to explain the transient expression is not the whole explanation (Dai *et al.*, 1995; Engelhardt *et al.*, 1994; Yang *et al.*, 1994), indicating that the immune mechanism responsible for transient expression is much more complex than previously thought. The successful systemic delivery of human factor VIII (B domain-deleted) at levels well exceeding the normal factor VIII plasma levels for up to 4 weeks (end of the experiment) in C57B/6 mice supports our finding (Connelly *et al.*, 1996). We need, however, to be aware that C57B/6 mice apparently do not develop antibody against the human factor VIII produced. These observations generally support the use of adenovirus as a gene transfer vector even for genetic diseases. However, the factor IX transgene clearance kinetics observed in SCID mice, a system which is considered to represent the maximal possible persistence of the transgene *in vivo*, indicates a substantial decrease of the transgene over time, in accordance with the episomal nature of the adenovirally transferred transgene. The major obstacles to using recombinant adenovirus vectors for haemophilia, therefore, include: (i) how the immune rejection of recombinant adenovirus can be suppressed effectively and safely with repeat viral administration; and (ii) how stable transgene expression can be achieved without gross fluctuations in factor IX expression, and keeping the period of extremely high factor IX level in the circulation as short as possible. The latter may become a major obstacle for expressing pro-coagulant factors like factor IX or factor VIII. Various approaches to the suppression of the immune reaction are in progress. Recently, expression of factor IX at an extremely high level ($\sim$100 μg ml^{-1} plasma) in mice and establishment of immune tolerance against adenovirus in neonate animals was reported (Walter *et al.*, 1996). The absence of immune rejection of adenovirus, however, may invite unwanted infections of wild-type adenovirus causing serious side-effects. Furthermore, this still does not change the transient stability of the transgenes, requiring repeat administration of adenovirus as often as every 3–4 months. Attempts to prolong the period of expression of factor IX at a therapeutic level may result in an unacceptably long period (2–4 weeks or even longer) with a much higher plasma factor IX level than normal. Such a condition, which may tip the balance towards pro-thrombosis, must be avoided in establishing a safe therapy for haemophilia. With substantial modifications and improvements, future generations of adenovirus may become a durable gene delivery vector system for the haemophilias.

The potential of adeno-associated virus (AAV) (Flottee and Carter, 1995), the herpes virus system (Efstathious and Minson, 1995) as well as a recently reported lentivirus vector system (Naldini *et al.*, 1996) is currently not known for haemophilia gene therapy. Because of their ability to transduce non-dividing cells with a wide cell tropism, they may turn out to be useful vector systems in the future.

Direct *in vivo* gene transfer approaches using non-viral vectors, such a vector DNA–ligand complex, have also been tested (Lozier *et al.*, 1994). Major problems invariably encountered in these approaches include very short-lived expression of the transgene (as short as just a few days), low transfection efficiency and/or both. At the current stage of vector development, non-viral vector(s) that can be used for efficient gene transfer and stable *in vivo* expression of the factor IX gene are not viable for clinical application.

Some *ex vivo* approaches combined with non-viral gene expression vectors have made promising progress, as shown by Heartlein *et al.* (1994) for persistent expression of growth hormone over one and a half years in mice. As described below, we also have achieved a long-term stable, safe systemic production of factor IX at a therapeutic level by taking a non-viral *ex vivo* approach.

12.5 Target cells and tissues

The gene for factor IX is expressed in hepatocytes with a stringent tissue specificity. Ideally, therefore, this gene should be targeted to the liver to produce recombinant factor IX at its optimal level. However, because of the significant difficulties involved in this approach, ectopic expression of factor IX by transferring its gene into other tissues which normally do not express the gene has also been tested as an alternative. Provided that factor IX produced by ectopic cells and tissues is similar to that produced by the liver, such approaches should serve as durable methods. Any cells or tissues chosen for ectopic expression of factor IX must be able to carry out all the required co- and post-translational modifications properly without having any significant detrimental effects. Supported by this rationale, various non-hepatic tissues and cells such as skin fibroblasts, vascular tissue and endothelial cells, skeletal muscle, keratinocyte and bone marrow stromal cells have been tested, demonstrating their great potential (Brownlee, 1995; Kurachi and Yao, 1993; Thompson, 1995). Similar ectopic expression systems may also be possible for factor VIII. For factor VIII expression, however, special attention must be paid to its efficient systemic delivery and accessibility to vWF for its stabilization in the circulation.

12.6 Animal models

Naturally occurring canine mutants for both haemophilia A and haemophilia B have served as invaluable resources for haemostasis studies (Dodds, 1988; Giles *et al.*, 1982). Among them, dogs with haemophilia B have been particularly well characterized in terms of the underlying molecular defects in the factor IX gene and factor IX levels in the circulation (Evans *et al.*, 1989; Sugahara *et al.*, 1996). It seems that the mechanisms responsible for the defects are diverse and very similar to those found in the human population. The canine haemophilia B models have been used to test the efficacy and safety of gene transfer approaches (Kay *et al.*, 1994). These canine models can serve as a large animal model which is important to bridge the gap between testing in small animals such as mice and phase I testing in patients. However, difficulties associated with the canine models are due to the intensive maintenance required, including regular transfusions of normal plasma and complicated neonate delivery.

Recently, factor VIII gene-inactivated mice were generated by homologous recombination (Bi *et al.*, 1995). Surprisingly, these mice apparently showed no serious problems, either in daily maintenance (no plasma infusions required) or in husbandry and neonate delivery. Studies to create factor IX-deficient mouse models are also in progress. We can expect that such models will become available for gene transfer studies within a year. Some concerns regarding the use of these mice may include the reliability of the assessment of haemostatic parameters performed with the small animal model, and how well these animals can represent human haemophilia conditions. Another serious concern is how well these animals can tolerate surgery without plasma transfusions. No substantial surgery may be applied to these animals, which greatly restricts their usefulness. Nevertheless, the availability of mouse models is expected to make most of the evaluation process of various gene transfer approaches easier and they promise to be much less costly than canine models.

12.7 Recent developments

Earlier developments with a variety of vector systems and different target cells and tissues tested did not show the superiority of any one method, suggesting a need for the rational design of vector systems (Brownlee, 1995; Thompson, 1995). Various potential gene transfer approaches for haemophilia gene therapy are shown in *Figure 12.2*. Here, we focus on some selected developments over the last few years.

Various gene transfer approaches carried out to date utilizing retroviral vectors have produced high production levels of factor IX *in vitro* (up to ~5 μg 10^6 cells^{-1} 24 h^{-1}) with many different cell types. However, only modest levels (<50 ng ml^{-1} plasma) of persistent expression *in vivo* have been observed to date (Brownlee, 1995; Kurachi and Yao, 1993; Thompson, 1995). The persistent plasma factor IX levels achieved by targeting the canine factor IX gene into the liver of haemophilia B dogs (homogeneous system) by intraportal infusion of virus combined with partial hepatectomy were still less than about 5 ng ml^{-1} plasma (Kay *et al.*, 1993). Other approaches, including keratinocyte

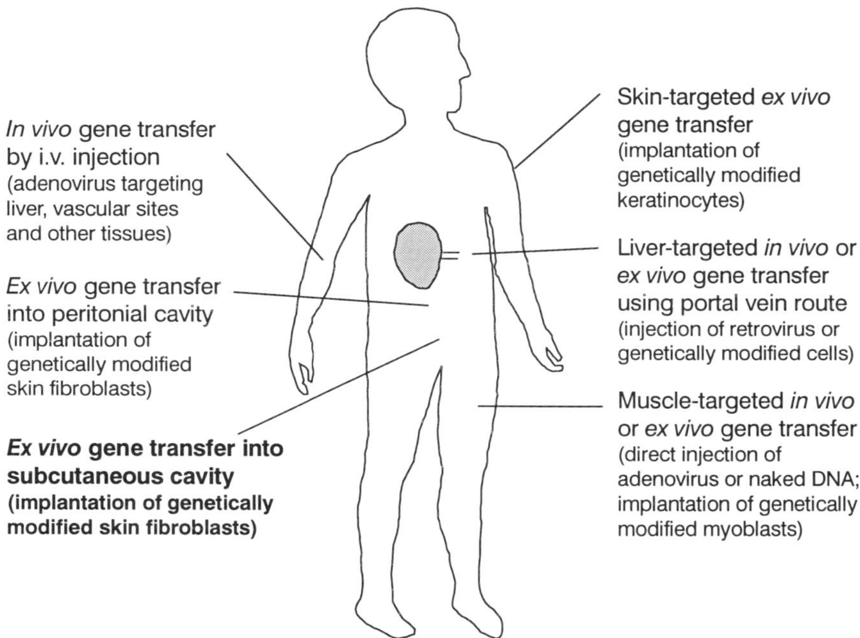

Figure 12.2. Potential gene transfer approaches for haemophilia B gene therapy. Most of the approaches summarized here have been tested with animals including mice, dogs or rabbits (Brownlee, 1995; Kurachi and Yao, 1993; Thompson, 1995). The only method actually tested with haemophilia B patients is that shown in bold letters (*ex vivo* gene transfer into subcutaneous cavity; Lu *et al.*, 1993).

implantations in skin (Fenjves *et al.*, 1996; Gerrard *et al.*, 1993), showed an extremely low level, transient expression. By adopting the mouse primary myoblast-mediated gene transfer approach, we demonstrated the persistent expression of factor IX in the circulation in mice at a level of 10–30 ng ml^{-1} plasma (Yao *et al.*, 1994). As described above, by using the intraportal retrovirus infusion technique, Okuyama *et al.* (1996) reported the successful long-term, stable expression of α_1-antitrypsin (up to 10 months) at a very impressive level of 5 μg ml^{-1} plasma in mice. This approach required a partial hepatectomy, which is not acceptable for haemophilia gene therapy. However, it showed the importance of vector optimization, and the potential application of the α_1-antitrypsin promoter vector system to factor IX as well as factor VIII with an alternative approach requiring no partial hepatectomy.

As mentioned above, all attempts at adenovirus-mediated factor IX gene transfers *in vivo* have resulted in a very high, but only transient, level of expression of factor IX (Dai *et al.*, 1995; Kay *et al.*, 1994; Smith *et al.*, 1993; Yao *et al.*, 1996). Adenovirus-mediated canine factor IX expression in immune-competent haemophilia B dogs showed only a few weeks of high expression followed by a much lower level of expression for up to 1–2 months (Kay *et al.*, 1994). Canine factor IX expression in various immune-compromised mice after intramuscular injection of adenovirus was also shown to be transient, though long-term expression of up to a year at very low levels was also observed (Dai *et al.*, 1995). In our study (Yao *et al.*, 1996), expression of human factor IX in immune-compromised (SCID) mice after a single i.v. administration of recombinant factor IX adenovirus was transient, but showed only a slow decrease in expression over a long period of time, taking many months. In the normal animals (Balb/c), we also found long-term human factor IX production in the liver cells while, as expected, no human factor IX was found in the circulation after about 2 weeks.

Recently, we have demonstrated long-term stable factor IX expression in mice by utilizing the non-viral myoblast-mediated gene transfer approach (Wang *et al.*, 1995). In this approach, SCID mouse primary myoblasts were co-transfected with a refined muscle-specific factor IX expression vector and neomycin-resistant gene vector. After selection with G418, myoblasts (5, 10 and 20 $\times$ 10^6 cells) were injected into hind leg skeletal muscles. Injected cells fused with existing myofibre cells and possibly gave rise to new myofibre cells, and continued expressing recombinant factor IX at levels of approximately 20, 40 and 80 ng ml^{-1} plasma, respectively. With additional cell injections, however, the animals originally injected with 10 $\times$ 10^6 cells elevated their factor IX plasma concentration to 160–200 ng ml^{-1} levels and stayed stable over 8 months (end of the experiment). Other animals which received only the original cell injections maintained their stable expression levels up to 10 months (end of the experiment). Cells transduced with factor IX retrovirus with its LTR as promoter (LIXSN) expressed at about 10 ng ml^{-1}, but the level was stable up to 10 months (end of the experiment). In this study, it was demonstrated that: (i) therapeutic levels of factor IX in the circulation can be stably achieved (no significant fluctuations in expression levels); (ii) repeat therapy is feasible to elevate the factor IX expression level to new higher and stable levels, without having any detrimental effects; and (iii) the promoters used (β-actin promoter linked to muscle-specific enhancer as well as LTR promoter) are not inactivated *in vivo*.

These results indicated that some *ex vivo* approaches are durable for the haemophilias, particularly because these approaches can avoid several serious problems which are associated with the *in vivo* gene transfer approaches utilizing either viral or non-viral vectors. As discussed above, problems including the poor stability of the transgenes and difficulty of repeat therapy can be avoided by adopting some type of *ex vivo* approach, such as myoblast-mediated gene transfer. In addition, any detrimental effects, which may occur after disseminated *in vivo* direct gene transfer utilizing a viral vector may be extremely difficult to abolish, whereas such effects due to an *ex vivo* approach with gene transfer into a restricted area of tissue may be handled much more easily. Furthermore, recently reported *ex vivo* approaches, such as the successful achievement of long-term stable systemic production of growth hormone utilizing transfected fibroblasts (Heartlein *et al.*, 1994), strongly support the rationale of such *ex vivo* approaches. Although *ex vivo* gene transfers require extra steps (isolation of cells in culture, their genetic modification and implantation back into the tissues) and involve the general inconvenient requirement for autologous cells, such an approach can offer very important advantages over the *in vivo* gene transfer approaches. Furthermore, realistically, any major gene therapies for genetic diseases are almost certainly going to be practised at a relatively small number of specialist centres which would be well equipped and well staffed and would provide strict quality control.

12.8 Conclusions

As discussed, the current status of gene therapy studies for the haemophilias indicates no obviously superior method, and there is a need for substantial improvements and optimization of both *in vivo* and *ex vivo* gene transfer methods. Past achievements strongly suggest that it is essential to study the basic biology involved, making certain that every step we take is in balance with the physiological systems involved. We should not forget that we are trying to modify complex natural systems that have evolved through hundreds of millions of years. For example, attempts to establish permanent immune tolerance for adenovirus may result in the unacceptable occurrence of serious side-effects. It is important to evaluate all aspects involved, and not to shortcut some critical issues just to gain one positive aspect. Such shortcuts may generate more serious problems. At the present stage of study, it is wise to pursue our quest by all available avenues including both *in vivo* and *ex vivo* gene transfer approaches by systematic and thorough studies of every step involved, and not to rush into clinical applications with an inadequately developed method. These caveats notwithstanding, we believe that the future prospects for haemophilia gene therapy are great.

References

Anderson WF. (1994) Was it just stupid or are we poor educators? *Hum. Gene Ther.* 5: 791–792.

Bi L, Lawler AM, Antonarakis SE, High KA, Gearhart JD, Kazazian HH, Jr. (1995) Targeted disruption of the mouse factor VIII gene produces a model of hemophilia A. *Nature Genet.* 10: 119–121.

Brownlee GG. (1995) Prospects for gene therapy of hemophilia A and B. *Br. Med. Bull.* 51: 91–105.

Connelly S, Gradner JM, McClelland A, Kaleko M. (1996) High level tissue-specific expressions of functional human factor VIII in mice. *Hum. Gene Ther.* 7: 183–195.

Cornetta K, Morgan RA, Anderson WF. (1991) Safety issues related to retroviral-mediated gene transfer in humans. *Hum. Gene. Ther.* 2: 5–14.

Crystal RG. (1995) Transfer of genes to humans: early lessons and obstacles to success. *Science* 270: 404–410.

Dai Y, Schwarz EM, Gu D, Zhang WW, Sarvetnick N, Verma IM. (1995) Cellular and humoral immune responses to adenoviral vectors containing factor IX gene: tolerization of factor IX and vector antigens allows for long-term expression. *Proc. Natl Acad. Sci. USA* 92: 1401–1405.

Dodds WJ. (1988) Third international registry of animal models of thrombosis and hemorrhagic diseases. *ILAR News* **30**: R1–R32.

Donahue RE, Kessler SW, Bodine D, McDonagh K, Dunbar C, Goodman S, Agricola B, Byrne E, Raffeld M, Moen R, Bacher J, Zsebo KM, Niehuis AW. (1992) Helper virus induced T cell lymphoma in nonhuman primates after retroviral mediated gene transfer. *J. Exp. Med.* **176**: 1125–1135.

Dwarki VJ, Belloni P, Nijjar T, Smith J, Couto L, Rabier M, Clift S, Berns A, Cohen LK. (1995) Gene therapy for hemophilia A: production of therapeutic levels of human factor VIII *in vivo* in mice. *Proc. Natl Acad. Sci. USA* **92**: 1023–1027.

Efstathious S, Minson AC. (1995) Herpes virus-based vectors. *Br. Med. Bull.* **51**: 45–55.

Engelhardt JF, Ye X, Doranz B, Wilson JM. (1994) Ablation of E2A in recombinant adenoviruses improves transgene persistence and decreases inflammatory response in mouse liver. *Proc. Natl Acad. Sci. USA* **91**: 6196–6200.

Evans JP, Brinkhous KM, Brayer GD, Reisner HM, High KA. (1989) Canine hemophilia B resulting from a point mutation with unusual consequences. *Proc. Natl Acad. Sci. USA* **86**: 10095–10099.

Fenjves ES, Yao SN, Kurachi K, Taichman LB. (1996) Loss of expression of a retrovirus-transduced gene in human keratinocytes. *J. Invest. Dermatol.* **106**: 576–578.

Flottee TR, Carter BJ. (1995) Adeno-associated virus vectors for gene therapy. *Gene Ther.* **2**: 357–362.

Gerrard AJ, Hudson DL, Brownlee GG, Watt FM. (1993) Towards gene therapy for haemophilia B using primary human keratinocytes. *Nature Genet.* **3**: 180–183.

Giles AR, Tinlin S, Greenwood R. (1982) A canine model of hemophilic (factor VIII: C deficiency) bleeding. *Blood* **60**: 727–730.

Heartlein MW, Roman VA, Jiang JL, Selers JW, Zulian AM, Treco DA, Selden RF. (1994). Long-term production and delivery of human growth hormone *in vivo*. *Proc. Natl Acad. Sci. USA* **91**: 10967–10971.

Hedner U, Davie EW. (1989) Introduction to hemostasis and the vitamin K-dependent coagulation factors. In: *The Metabolic Basis of Inherited Disease*, Vol. 2, 6th Edn. McGrawHill, New York, pp. 2107–2134.

Kay MA, Rothenberg S, Landen CN, Bellinger DA, Leland F, Toman C, Finegold M, Thompson AR, Read MS, Brinkhous KM, Woo SLC. (1993) *In vivo* gene therapy of hemophilia B: sustained partial correction in factor IX-deficient dogs. *Science* **262**: 117–119.

Kay MA, Landen CN, Rothenberg SR, Taylor LA, Leland F, Wiehle S, Fang B, Bellinger D, Finegold M, Thompson AR, Read M, Brinkhous KM, Woo SLC. (1994) *In vivo* hepatic gene therapy: complete albeit transient correction of factor IX deficiency in hemophilia B dogs. *Proc. Natl Acad. Sci. USA* **91**: 2353–2357.

Keith JC, Jr, Ferranti TJ, Misra B, Frederick T, Rup B, McCarthy K, Faulkner R, Bush L, Schaub RG. (1995) Evaluation of recombinant human factor IX: pharmacokinetic studies in the rat and the dog. *Thromb. Haemost.* **73**: 101–105.

Koeberl DD, Halbert CL, Krumm A, Miller AD. (1995) Sequences within the coding regions of clotting factor VIII and CFTR block transcriptional elongation. *Hum. Gene Ther.* **6**: 469–479.

Kurachi K. (1991) Recombinant antihemophilic factors. In: *Biotechnology of Blood* (ed. J Goldstein). Heinemann, Butterworth, pp. 177–195.

Kurachi K, Yao SN. (1993) Gene therapy of hemophilia B. *Thromb. Haemost.* **70**: 193–197.

Kurachi K, Yao SN, Furukawa M, Kurachi S. (1992) Deficiencies in factors IX and VIII: what is now known. *Hosp. Pract.* **27**: 41–51.

Kurachi K, Kurachi S, Furukawa M, Yao SN. (1993) Biology of factor IX. *Blood Coagul. Fibrinol.* **4**: 953–974.

Lozier JN, Thompson AR, Hu PC, Read M, Brink KM, High KA, Curiel DT. (1994) Efficient transfection of primary cells in a canine hemophilia B model using adenovirus–polylysine DNA complexes. *Hum. Gene Ther.* **5**: 313–322.

Lu DR, Zhou JM, Zheng B, Qiu XF, Xue JL, Wang JM, Meng PL, Han FL, Ming BH, Wang XP, Wang JB, Liang JJ, Jiang ZS. (1993) Stage I clinical trial of gene therapy for hemophilia B. *Sci. China B* **36**: 1342–1351.

Lynch CM, Israel DI, Kaufman RJ, Miller AD. (1993) Sequences in the coding region of clotting factor VIII act as dominant inhibitors of RNA accumulation and protein production. *Hum. Gene Ther.* **4**: 259–272.

Miller AD. (1992) Retroviral vectors. *Curr. Top. Microbiol. Immunol.* **158**: 1–24.

Mulligan RC. (1993) The basic science of gene therapy. *Science* **260**: 926–932.

Naffakh N, Pinset C, Montarras D, Li Z, Paulin D, Danos O, Heard JM. (1996) Long-term secretion of therapeutic proteins from genetically modified skeletal muscles. *Hum. Gene Ther.* **7**: 11–21.

Naldini L, Blömer U, Gallay P, Ory D, Mulligan R, Gage FH, Verma JM, Trono D. (1996) *In vivo* gene delivery and stable transduction of nondividing cells by a lentiviral vector. *Science* 272: 263–267.

Okpako DT. (1991) *Principles of Pharmacology: Topical Approach.* Cambridge University Press, Cambridge, UK, pp. 130–153.

Okuyama T, Huber RM, Bowling W, Pearline R, Kennedy SC, Flye MW, Ponder KP. (1996) Liver directed gene therapy: a retroviral vector with a complete LTR and the ApoE enhancer α_1-antitrypsin promoter dramatically increases expression of human α_1-antitrypsin *in vivo. Hum. Gene Ther.* 7: 637–645.

Smith KJ, Thompson AR. (1981) Labeled factor IX kinetics in patients with hemophilia B. *Blood* 58: 625–629.

Smith TA, Mehaffey MG, Kayda DB, Saunders JM, Yei S, Trapnell BC, McClelland A, Kaleko M. (1993) Adenovirus mediated expression of therapeutic plasma levels of human factor IX in mice. *Nature Genet.* 5: 397–402.

Sugahara Y, Catalfamo J, Brooks M, Hitomi E, Kurachi K. (1996) Isolation and characterization of canine factor IX. *Thromb. Haemost.* 75: 450–455.

Thompson AR. (1995) Progress towards gene therapy for hemophilias. *Thromb. Haemost.* 74: 45–51.

Thompson AR, Forrey AW, Gentry PA, Smith KJ, Harker LA. (1980) Human factor IX in animals: kinetics from isolated, radiolabelled protein and platelet destruction following crude concentrate infusions. *Br. J. Haematol.* 45: 329–342.

Walter J, You O, Hagstrom JN, Sands M, High KA. (1996) Successful expression of human factor IX following repeat administration of an adenoviral vector in mice. *Proc. Natl. Acad. Sci. USA* 93: 3056–3061.

Wang JM, Zheng H, Yao SN, Kurachi K. (1995) High level expression of human factor IX by skeletal myoblast-mediated gene transfer using plasmid vector in mice. *Blood* 86 (Suppl. 1): 240a.

Wang JM, Zheng H, Sugahara Y, Tan J, Yao SN, Olson E, Kurachi K. (1996) Construction of human factor IX expression vectors in retroviral vector frames for optimal production in muscle cells. *Hum. Gene Ther.* 7: 1743–1756.

Yang Y, Nunes FA, Berencsi K, Furth E, Gönczöl E, Wilson JM. (1994) Cellular immunity to viral antigens limits E1-deleted adenoviruses for gene therapy. *Proc. Natl Acad. Sci. USA* 91: 4407–4411.

Yao SN, Kurachi K. (1992) Expression of human factor IX in mice after injection of genetically modified myoblasts. *Proc. Natl Acad. Sci. USA* 89: 3357–3361.

Yao SN, Smith KJ, Kurachi K. (1994) Primary myoblast-mediated gene transfer: persistent expression of human factor IX in mice. *Gene Ther.* 1: 99–107.

Yao SN, Farjo A, Roessler BJ, Davidson BL, Kurachi K. (1996) Adenovirus-mediated transfer of human factor IX gene in immunodeficient and normal mice: evidence for prolonged stability and activity of the transgene in liver. *Viral Immunol.* 9: 141–153.

Zatloukal K, Cotten M, Berger M, Schmidt W, Wagner E, Birnstiel ML. (1994) *In vivo* production of human factor VIII in mice after introsplenic implantation of primary fibroblasts by receptor-mediated, adenovirus-augmented gene delivery. *Proc. Natl Acad. Sci. USA* 91: 5148–5152.

13

Gene therapy for adenosine deaminase deficiency

H.B. Gaspar and C. Kinnon

13.1 Introduction

In 1990, the first clinical trials of somatic gene therapy for inherited disease were started on two girls suffering from adenosine deaminase (ADA) deficiency. For a number of reasons, this rare, autosomal recessive, primary immunodeficiency has acted as a disease model for the development of many of the new technologies associated with gene therapy. To date, the preliminary results of the first four clinical trials have been published.

13.2 Pathophysiology

Deficiency of the enzyme ADA results in the syndrome of severe combined immunodeficiency (SCID). ADA is an enzyme that is expressed in all body tissues and plays an essential role in the purine salvage pathway, catalysing the irreversible deamination of adenosine to inosine and $2'$ deoxyinosine. Its deficiency results in the accumulation of the products of adenosine degradation, characterized by elevated erythrocyte deoxyadenosine triphosphate (dATP), and plasma and urinary deoxyadenosine (dAdo) (Hirschhorn, 1993). Furthermore, *S*-adenosylhomocysteine hydrolase (SAHH) activity is significantly reduced. The mechanism of the immunodeficiency is not clearly understood, though recent evidence shows that raised dATP and decreased SAHH activity block T-lymphocyte differentiation in the thymus (Benveniste *et al.*, 1995). It has also been shown that increased levels of dATP inhibit the enzyme ribonucleotide reductase which is required for DNA synthesis and repair, especially in thymocytes (Cohen *et al.*, 1983). These studies imply that the T-cell deficiency is the primary effect while the B-cell deficiency is secondary.

13.3 Clinical manifestations of ADA-SCID

The condition was first described in 1972 by Giblett and colleagues in Seattle (Giblett *et al.*, 1972). They investigated two unrelated patients with recurrent viral and fungal infections consistent with a T-cell immunodeficiency. On further analysis it was

Gene Therapy, edited by N.R. Lemoine and D.N. Cooper.
© 1996 BIOS Scientific Publishers Ltd, Oxford.

found that both patients were deficient in the ADA enzyme. Thus, ADA-deficient SCID became the first primary immunodeficiency for which the underlying molecular defect was identified. There is considerable heterogeneity in the clinical severity of the disease, ranging from very severely affected infants presenting in the first few months of life through to chronically affected adults presenting later in life (Morgan *et al.*, 1987).

The severely affected phenotype commonly presents with failure to thrive, diarrhoea, recurrent pneumonia and opportunistic infections. Some children present after the first year of life with a less severe degree of immunodeficiency. These have been termed delayed-onset or late-presentation disease. More recently, a small number of adults who had been suffering from an unknown form of cell-mediated immunodeficiency were investigated more fully and found to be ADA deficient (Shovlin *et al.*, 1994). The biochemical and immunological abnormalities generally correlate with the degree of immunodeficiency. The early-onset severe phenotypes have the highest levels of dAdo and dATP, the lowest lymphocyte numbers and the poorest level of lymphocyte response to stimulation. In these patients, there is nearly always an absence of both T and B lymphocytes whereas in the late-onset phenotype there may be some residual immune function (Morgan *et al.*, 1987).

13.4 Therapeutic options

A number of therapeutic options are available for the treatment of ADA deficiency. The treatment of choice is bone marrow transplantation (BMT). Where a related matched donor is available, the success rate for BMT has been reported at > 90% in some studies. However, the mortality and morbidity rates increase significantly in transplantation from a mismatched donor (Fischer *et al.*, 1990). Different treatment strategies have been attempted where a matched related donor is unavailable. Extracellular enzyme replacement therapy was used in the mid-1970s. It was reasoned that since dAdo and dATP are freely diffusible across the cell membrane, the supply of extracellular enzyme would decrease the extracellular level of dAdo and dATP and thus establish a concentration gradient allowing the movement of dAdo and dATP out of cells. Irradiated red blood cell transfusions were initially used as a source of cell-packaged ADA. However, although a few patients showed evidence of immunological improvement and some correction of biochemical parameters, most patients experienced only minimal or no immunological and biochemical benefit (Polmar, 1978). Furthermore, such were the disadvantages of iron overload and infectious complications that this form of therapy was discontinued. Despite this, the concept of enzyme replacement had been successfully demonstrated and this led to the development of an enzyme preparation suitable for parenteral injection, PEG–ADA – a bovine ADA preparation conjugated to polyethylene glycol (Hershfield *et al.*, 1987).

Since the first patient was treated in 1986, over 50 patients have received PEG–ADA. In the majority of patients there has been a marked improvement in biochemical abnormalities followed by an improvement in the immunological parameters. There has been an improvement in both absolute lymphocyte numbers and in their functional responses to mitogenic stimuli (Hershfield *et al.*, 1993). In a number of children there has been such an improvement in humoral immunity that it has been possible to stop intravenous immunoglobulin replacement. Reports suggest that PEG–ADA therapy does not result in completely normal immune parameters longterm. However, it significantly improves the number of peripheral lymphocytes and their function and although mild lymphopaenia may persist, immunity is sufficient to protect against recurrent infection, repeated hospitalization and to allow normal development (Hershfield, 1995). For ethical reasons, all the children treated in the clinical trials of gene therapy undertaken so far, have also received PEG–ADA therapy. This has caused a considerable amount of difficulty in assessing the effect of gene therapy since any clinical improvement may be a result of concomitant PEG–ADA therapy.

13.5 ADA-SCID as a model disease for gene therapy

A number of characteristics have resulted in ADA deficiency becoming an attractive candidate for the initial clinical trials of gene therapy. The *ADA* gene was cloned in 1983 (Valerio *et al.*, 1983) and since this time its function, expression and regulation have been extensively studied. The gene itself is small and the cDNA is approximately 1.1 kb in length. This has been an important consideration since the retroviral constructs used for gene therapy are limited in the amount of DNA that can be accommodated.

The target tissue for gene transfer is well defined in ADA deficiency. Despite being expressed in all body tissues, its deficiency particularly damages the immune system and specifically T lymphocytes. It is also known that ADA-SCID is curable by bone marrow transplantation. In addition, analysis of transplant recipients has revealed that the procedure is curative with full immune reconstitution, even if only donor T lymphocytes engraft (Vossen, 1983). Thus, genetic correction of autologous bone marrow or T lymphocytes might be expected to cure the disease. This target tissue is also easily accessible and can be readily manipulated *ex vivo*, thereby further facilitating gene transfer.

An additional advantage for ADA-SCID as a candidate disorder for gene therapy is that the *ADA* gene is constitutively expressed. The initial trials have mostly used genes driven by constitutively expressing viral promoters, the viral long terminal repeats (LTRs), and not via endogenous promoters, although the clinical trial by Bordignon *et al.* (1995) did use the endogenous *ADA* promoter (see *Figure 13.1*). The use of such relatively unsophisticated first-generation vector systems reflects a general lack of knowledge about the more complicated control systems of tissue-specific, temporally regulated gene products.

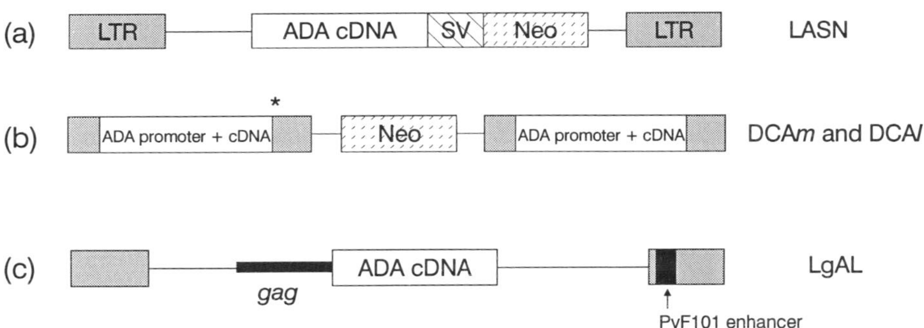

Figure 13.1. Schematic representation of the retroviral vector genomes used in the ADA-SCID clinical trials. LTR, Moloney murine leukaemia virus (MoMLV) long terminal repeat; Neo, neomycin resistance marker; PyF101, polyoma virus enhancer; SV, simian virus 40 early region promoter; *, position of alternative restriction sites for DCA*m* and DCA/retroviruses. Vectors not drawn to scale.

Studies of heterozygote carriers and population screening programmes have shown that individuals with as little as 10% of normal ADA activity have no abnormality of immune function (Daddona *et al.*, 1983). Conversely, individuals have been described with 50 times the normal amount of ADA; they have suffered mild haemolytic anaemia but have not shown any immune system abnormalities (Valentine *et al.*, 1977). Thus, there is a wide range in the level of ADA activity which allows normal immune function but no significant side-effects. This too is an important factor since retroviral-mediated transfer can result in extremely variable levels of gene expression. In the large majority of preclinical studies, only low levels of ADA activity have been achieved using retroviral transfer systems but since as little as 10% of normal activity has been sufficient to allow normal immune function, then retroviral-mediated transfer may suffice as the basis of a curative procedure.

Finally, experience from BMT recipients has shown that there appears to be a growth advantage for ADA-producing (ADA$^+$) lymphoid cells in ADA-deficient (ADA$^-$) patients. In human leukocyte antigen (HLA) genotypically matched transplants to severely affected individuals, ADA$^+$ donor bone marrow cells can be given without the need for prior cytoablation and result in stable engraftment of donor lymphoid cells and restoration of normal immune function. Despite their numerical disadvantage after transplantation into a non-cytoablated environment, ADA$^+$ stem cells are able to proliferate and mature to reconstitute the lymphoid lineage at the expense of ADA$^-$ stem cells (Vossen, 1983). From this, it was assumed that genetic modification of a small percentage of bone marrow stem cells might be sufficient for the progeny of these cells to completely reconstitute the immune system without the need for cytoablative therapy.

13.6 Preclinical studies of gene therapy

The development of gene therapy for ADA deficiency has evolved through a natural succession of *in vitro* and *in vivo* models over a long period of time. Nearly all the studies have used modified retroviral vectors as the vehicle for gene transfer, although a variety of retroviral constructs based on different viruses have been used (see *Figure 13.1*). The target tissues for gene transfer have been divided into studies on transfer into haematopoietic stem cells (HSCs) from mice, primates and humans and those focusing on transfer into T lymphocytes. Since ADA deficiency has been a model disease for gene therapy, a number of studies have been performed by groups not necessarily interested in progressing to clinical trials but in developing novel systems for gene therapy *per se*. In this review, we shall attempt to outline those studies which have been of more direct relevance in developing the clinical protocols used.

13.6.1 Gene transfer into murine HSCs

The target cell initially selected for gene transfer was the HSC. The first studies focused on the transfer of selectable marker genes into murine HSCs using retroviral vectors. Marrow was transduced *ex vivo* under a variety of culture and transduction conditions and transplanted into irradiated recipient mice. The ability of these methods to result in long-term presence and expression of genes was assayed in a number of ways including transplantation of bone marrow cells from primary recipients into secondary recipient mice. Successful transfer and long-term *in vivo* expression of β-globin, hypoxanthine phosphoribosyltransferase (*HPRT*) and neomycin resistance genes after transduction of murine stem cells was initially shown (Dzierzak *et al.*, 1988; Eglitis *et al.*, 1985; Miller *et al.*, 1984).

Following this, a number of groups were able to demonstrate successful retroviral-mediated transfer of the human *ADA* gene into murine HSCs and long-term *in vivo* expression of the transduced gene in the absence of replication-competent retrovirus (Lim *et al.*, 1989; Osborne *et al.*, 1990; van Beusechem *et al.*, 1990; Wilson *et al.*, 1990). Lim *et al.* (1989) used a retroviral vector encoding human ADA driven off an internal phosphoglycerate kinase promoter to transduce prestimulated whole murine marrow prior to transplantation into irradiated recipients. Expression of human ADA was seen in all haemopoietic lineages of primary recipients 4 months after transplantation. Human ADA expression was also shown 6 weeks later in splenic colonies and in the peripheral blood of secondary recipients, thus demonstrating the stability of transfer and expression in early progenitor cells. van Beusechem *et al.* (1990) used an amphotropic retrovirus that transcribed the human *ADA* gene under the control of a hybrid LTR, in which the MoMLV enhancer had been replaced by an F101 polyoma virus enhancer. Expression of human ADA enzyme was detected in lymphoid, myeloid and erythroid lineages 30 days after transplantation. Multi-lineage ADA expression was also seen in secondary and tertiary recipients.

An ADA knockout mouse has recently been generated (Migchielson *et al.*, 1996) in an attempt to create an animal model of ADA deficiency. However, the phenotype of this mouse is very different from human ADA deficiency. ADA-deficient mice died soon after birth with severe hepatocyte degeneration, incomplete lung expansion and small intestinal cell death. Surprisingly, the thymi of neonatal mice appeared histologically normal with no signs of immunodeficiency or infectious disease. In addition, cytometric analysis did not reveal any abnormalities in the major thymocyte subpopulations. Such a clinically different model would pose many difficulties in assessing the benefits of gene transfer for human ADA deficiency.

13.6.2 Gene transfer into large animal HSCs

Having demonstrated the ability of their retrovirus and packaging cell system to transduce early progenitors from murine marrow, van Beusechem *et al.* (1993) used the amphotropic nature of this retrovirus to transduce the bone marrow of rhesus monkeys in an attempt to create a more relevant preclinical model. By co-cultivating the primate marrow with the packaging cell line in the presence of rhesus monkey interleukin-3 (RhIL-3), they were able to show multi-lineage genetic modification over 2 years after gene transfer, again providing evidence that transduction of very early progenitors was possible. Two other groups also showed successful transfer and expression of the *ADA* gene up to 4 months after autologous transplantation of retrovirally transduced rhesus monkey bone marrow (Bodine *et al.*, 1993; Kantoff *et al.*, 1987). However, in all these primate experiments and in those carried out in other large animal models such as dogs and cats (Carter *et al.*, 1992; Lothrop *et al.*, 1991), the level of gene expression was significantly less than that achieved in the murine system. This may be a result of decreased efficiency of transduction of primate marrow by retroviruses, due to less well-defined co-culture conditions, or the lack of experience with large animal bone marrow transplantation and engraftment. It may also be due to a lower density of receptors for amphotropic retroviruses in these species.

13.6.3 Gene transfer into human HSCs

The ultimate target for gene transfer not only for ADA deficiency but for other conditions curable by BMT is the human HSC. Successful transfer of genes into this self-renewing totipotent cell type would allow a once-only procedure for re-population of all lineages with genetically modified cells. Initially, a number of studies demonstrated transfer of genes (including *ADA*) into human haemopoietic cells capable of colony formation up to 2 weeks after gene transfer (Bender *et al.*, 1988; Fink *et al.*, 1990; Hock and Miller, 1986). Although these experiments were able to show the feasibility of retrovirus-mediated transfer into progenitor populations, these early systems were not able to evaluate transfer into more primitive progenitor cells.

The development of long-term bone marrow culture (LTBMC) systems offered the possibility of evaluating gene transfer into more primitive cells. In this system, haemopoiesis can be routinely maintained for 2–3 months and the clonal cell responsible for this, the long-term culture initiating cell (LTCIC) is a more primitive cell than the progenitor cell detected by colony assays (Sutherland *et al.*, 1989). The successful use of cells from LTBMC for autologous BMT (Chang *et al.*, 1989) supports the view that the LTCIC may be functionally related to the primitive stem cells. Using this system, several studies have demonstrated high-frequency transfer of selectable marker genes by retroviruses into LTCICs (Cournoyer *et al.*, 1991; Hughes *et al.*, 1989). These experiments also allowed the evaluation of growth factors leading to more efficient cell transduction. Counoyer *et al.* (1991) used a combination of recombinant (r) IL-6 and rIL-3 to stimulate mononuclear cells during co-cultivation with a high-titre packaging line shedding an ADA-containing retroviral vector. Nine weeks post-infection, 24–44% of clonogenic progenitors from long-term cultures were found to contain the vector sequence. Proviral integration was shown to increase to 65–70% of LTCICs if viral supernatant was added to bone marrow cells maintained on a human bone marrow-derived stromal layer (Moore *et al.*, 1992), suggesting that stromal support may increase stem-cell proliferation and improve retroviral transfer.

More recently, studies have targeted CD34[+] bone marrow cells. The CD34[+] population consists of the earliest haemopoietic progenitors and is thought to contain the HSC. Using conditions similar to those described above, *in vitro* studies have shown retroviral gene transfer into CD34[+] selected populations (Bertolini *et al.*, 1994; Crooks and Kohn, 1993; Xu *et al.*, 1994), although the efficiency of transfer was decreased. All the studies above have used *in vitro* assays to assess transfer into the HSC. Although a system such as LTBMC allows the study of very early progenitor cells, this still remains a surrogate assay.

In an *in vivo* model, van Beusechem *et al.* (1994) transduced CD34[+] CD11b[-] cells from rhesus monkey bone marrow prior to autologous transplantation into irradiated recipients. The presence of transduced cells was observed by PCR in peripheral mononuclear cells and granulocytes cells more than 2 years after transduction, albeit at very low frequency (0.1%) (van Beusechem *et al.*, 1994). More accurate information of effective transduction of truly primitive human HSCs with long-term re-populating ability may come from gene-marking trials (Brenner *et al.*, 1995).

As an alternative source of HSCs, efforts have been directed at transducing progenitor cells derived from human umbilical cord blood. This blood is rich in progenitor cells, as shown by phenotypic and *in vitro* assays but also more convincingly by successful cord blood transplantations for a variety of haemopoietic diseases (Gluckman *et al.*, 1989;

Kernan *et al.*, 1994; Vowels *et al.*, 1993; Wagner *et al.*, 1995). Retroviral gene transfer studies have shown that progenitor and CD34$^+$ cells from human umbilical cord blood can be transduced at frequencies similar to or greater than bone marrow-derived cells (Lu *et al.*, 1993, 1994; Moritz *et al.*, 1993; Williams and Moritz, 1994). These initial studies led to one group optimizing their transfer protocol for a clinical trial of *ADA* gene transfer into cord blood CD34$^+$ cells (Hanley *et al.*, 1994; Kohn *et al.*, 1995).

13.6.4 Gene transfer into T lymphocytes

The difficulty associated with achieving successful retroviral-mediated transfer into human HSCs meant that certain groups targeted alternative cells for the correction of ADA deficiency. The most obvious of these was the T lymphocyte, since this is the cell type primarily affected by the deficiency of ADA. Primary T cells cultured from ADA patients were transduced with the LASN vector (see *Figure 13.1a*), a MoMLV-based vector with the *ADA* gene driven by the 5'LTR and containing a neo^R gene downstream of the *ADA* insert and driven by an internal SV40 promoter (Culver *et al.*, 1991). Transduced cells were shown to grow for a significantly longer period of time in comparison with non-transduced cells and continued to express the *ADA* gene during this time. This suggested that the expression of transduced *ADA* provided a survival advantage to these cells in a culture environment. Further evidence for the benefit of transducing ADA into enzyme-deficient T lymphocytes came from the studies of Ferrari *et al.* (1992). Peripheral blood lymphocytes from *ADA*-deficient patients were transduced with a retroviral vector containing the human *ADA* gene and injected into immunodeficient BNX mice. Only *ADA*-vector-transduced cells were able to show long-term survival in recipient animals. The expression of transduced *ADA* also resulted in human immunoglobulin production and the development of antigen-specific T cells. These experiments suggested that the expression of vector-derived human ADA in the T lymphocytes of ADA$^-$ patients would confer prolonged survival upon these cells, even in a normal ADA environment. Furthermore, the results of the experiment by Ferrari *et al.* (1992) suggested that T-cell modification could reconstitute a certain degree of immune function.

13.7 Clinical trials of gene therapy for ADA deficiency

13.7.1 T-cell gene therapy

In 1990, the first clinical trial of gene therapy for ADA deficiency was started on two girls in the USA (Blaese *et al.*, 1995). Both had presented some years earlier and had commenced PEG–ADA therapy. In each case, there had been an initially good response to PEG–ADA with a subsequent deterioration in total lymphocyte numbers and *in vitro*- and *in vivo*-specific lymphocyte responses. The treatment protocol involved LASN vector supernatant transduction of peripheral T cells apheresed from each patient and stimulated with a combination of rIL-2 and OKT3 (an anti-CD3 complex antibody), both of which serve to increase T-cell proliferation thereby enhancing retroviral infection. The expanded population of T cells were then returned to the patients 9–12 days later by peripheral infusion. It was assumed that despite any prolonged survival that transduction might confer upon the T cells, there would be a progressive decline in the number of transduced cells. Therefore, the procedure was repeated 11–12 times at regular intervals for each patient over a period of 18–24 months.

The efficiency of retroviral gene transfer varied considerably from one round of transduction to another, with a range of 0.1–10% transduced cells. PCR analysis of peripheral cells 2 years after the last infusion showed that in Patient 1, LASN vector sequences were present at 0.3 vector copies per cell whereas in Patient 2, the level of transduced cells was only 0.1–1% of total circulating cells. The level of peripheral blood lymphocyte ADA was in keeping with the PCR findings, with a significant rise in Patient 1 but no change from pre-treatment levels in Patient 2. Immunological reconstitution is difficult to evaluate, partly because of the continued administration of PEG–ADA and also due to the variable results in the two patients. In both patients there was an initial rapid rise in the T-cell count before stabilization in the normal range for Patient 1 and a slight increase from the pre-treatment levels for Patient 2. Cell-mediated immunity was shown to improve in both patients, as assessed by the development of delayed-type hypersensitivity (DTH) skin-test reactivity to a variety of antigens and also by improvement in T-cell immune responses *in vitro*. Humoral immune function appeared to improve with an increased specific antibody response to haemophilus influenza B (Hib) and tetanus immunization and an increase in the level of isohaemagglutinin production, although isohaemagglutinin production decreased considerably with time.

One interesting observation arising from this study is the prolonged survival of the transduced T cells. The percentage of vector-positive cells has remained stable for over 2 years after the last infusion, a far longer survival time than originally predicted. However, the role of exogenous enzyme replacement with PEG–ADA will continue to complicate the outcome of this trial and at present the dose of the drug is being decreased.

A further trial using the same protocol has been commenced on a boy in Japan but as yet very few details of this trial have been disclosed.

13.7.2 Peripheral blood lymphocyte and T cell-depleted bone marrow gene therapy

A slightly different approach to gene therapy was taken by Bordignon and colleagues in 1992 (Bordignon *et al.*, 1995). Having shown previously that gene-modified T cells from ADA-deficient patients could survive for a prolonged period of time and reconstitute immune function in an *in vivo* model, they progressed to a clinical trial of gene therapy. Two structurally identical vectors (see *Figure 13.1b*), expressing the human *ADA* cDNA and distinguishable only by the presence of alternative restriction sites in a non-functional region of the viral LTR, were used to transduce peripheral blood lymphocytes (PBLs) and T cell-depleted bone marrow independently. Thus, it was possible to identify the origin of transduced cells by restriction digest analysis. Prior to co-cultivation with the producer cells, PBLs were expanded with phytohaemagglutinin and rIL-2 while bone marrow was maintained for 3 days in long-term culture with regular addition of vector supernatant. Efficiency of transduction varied from 2% to 40% of PBLs and from 30% to 40% of bone marrow cells.

Two patients were treated by this protocol. Both had followed a similar clinical course, with diagnosis at the age of 2 years and treatment with PEG–ADA. Gene therapy was initiated when administration of PEG–ADA failed to have an effect on immunological parameters. Both patients received infusions of gene-modified PBLs and T cell-depleted bone marrow regularly over a 1–2 year period. Analysis of vector-transduced cells from peripheral blood indicated that initially all vector-positive cells were derived from transduced PBLs. However, 3 years after initiation, and 1 year after discontinuation of gene therapy, peripheral lymphocytes showed the alternative restriction digest pattern, indicating conversion of the circulating transduced lymphocytes from a PBL-derived to a bone marrow-derived population. Further analysis of T-lymphocyte clones, granulocytes and erythrocytes taken at this time also showed a predominantly bone marrow origin. These findings suggested that there had been stable transduction of early haemopoietic progenitors capable of generating multi-lineage progenies. Moreover, the continued presence of transduced bone marrow cells and their differentiation into mature circulating cells appears to confirm the assumption that a small number of ADA$^+$ cells have a survival advantage in an ADA$^-$ environment.

The frequency of transduced cells has varied, with 2–4.7% of T cells and 17–25% of clonable bone marrow progenitors being vector positive. Not surprisingly, given the low frequency of transduced cells in the circulation, the level of total ADA activity in total circulating nucleated cells has remained low (5–18% of normal values). The immune reconstitution in these children appears to be more consistently improved in comparison to the above reported study by Blaese *et al.* (1995). There was an increase in absolute lymphocyte and T-cell numbers in both children into the normal range and there has been an improvement in T-cell proliferative responses to non-specific and specific stimuli. Isohaemagglutinin titre has also improved since the initiation of gene therapy. To study the improvement in T-lymphocyte function further, the T-cell repertoire was analysed by Vβ usage. RT-PCR analysis using Vβ chain-specific primers demonstrated the development of a normal T-cell repertoire.

Although these results appear promising, it must be remembered that both children are still on PEG–ADA therapy, albeit with decreasing doses. The improvements reported may not be sustained with time, as was seen in the initial study by Blaese *et al.* (1995). Furthermore, in both the above trials gene therapy was initiated at a point when PEG–ADA therapy was deemed to have failed. However, this was based on laboratory data and not on clinical well-being since, as both reports point out, all the children were clinically well and free of infection prior to starting therapy.

13.7.3 Bone marrow CD34$^+$ cell gene therapy

The results of two further studies have been reported. Hoogerbrugge *et al.* (1996) performed retroviral-mediated gene transfer into CD34$^+$ bone marrow cells of three children with ADA deficiency in an attempt to effect a cure with a once-only procedure. The retroviral vector (see *Figure 13.1c*) and packaging cell system had been extensively studied in a murine model and also in primates where low-level but prolonged *ADA* gene expression had been reported (van Beusechem *et al.*, 1994). In this study two of the subjects were on enzyme replacement therapy at the start of the trial and in the third child PEG–ADA was only started three months after the gene therapy procedure. The transfer protocol involved a 90 h co-cultivation of the selected cells and packaging cell line with the addition of rIL-3. Gene transfer resulted in a 5–12% transduction frequency in colony-forming cells *in vitro*.

Transduced granulocytes and mononuclear cells persisted in the circulation for 3 months, as shown by nested PCR analysis. In one patient, the transduced gene was detected in the bone marrow 6 months after gene transfer. Subsequently, no evidence of the transduced gene was detectable in either the bone marrow or peripheral blood. In addition, *ADA* gene expression was not detected at any time. A number of reasons might explain the disappointing results. The CD34$^+$ selection meant the number of cells available for transduction was very low, a problem further complicated by the low transduction frequency. The protocol also relied upon a positive proliferative advantage to transfected cells which may not have been borne out by this study. It is also possible that the administration of PEG–ADA may have had a deleterious effect upon the outgrowth of transduced cells by removing their survival advantage. This hypothesis may be supported by the fact that the one child who had not been started on PEG–ADA prior to gene therapy had a positive bone marrow PCR after 6 months.

13.7.4 Umbilical cord blood CD34$^+$ cell gene therapy

The final trial to be published also targeted the CD34$^+$ population but, on this occasion, in cord blood-derived cells (Kohn *et al.*, 1995). Three infants diagnosed prenatally were treated by autologous transplantation of transduced cells. Cord blood was taken from their umbilical cord at birth and CD34$^+$ cells selected prior to transduction with the LASN vector. All three patients were started on PEG–ADA in the first few days of life.

To date, only the results of *ADA* gene presence and expression have been published. No details of immune reconstitution have been formally reported. By progenitor colony assay, it was shown that 12.5%, 19.4% and 21.5% of CD34$^+$ cells had been transduced in the three patients respectively. Analysis of bone marrow was performed 1 year after transplantation. In whole marrow, the LASN vector sequence was present in 1 in 10 000 cells and a myeloid progenitor colony assay showed that 4–6% of CFU-GM colonies contained the vector. The frequency of peripheral lymphocytes containing the LASN vector (as shown by semiquantitative PCR) was 1 in 3000 to 1 in 100 000 cells 18 months after transplantation. At this time, ADA activity in unselected T cells was measured and found to be barely above the levels found in ADA-SCID patients. The results suggest that retroviral-mediated stable transduction of early progenitors in cord blood is possible and that engraftment of these cells without prior cytoablation is possible. However, these patients must be monitored for a much longer period of time before it can be said that permanent engraftment has been achieved. The small numbers of transduced PBLs and the very low level of gene expression in unselected cells suggest that, for the time being at least, immune reconstitution is unlikely and continued PEG–ADA administration will be necessary.

13.7.5 Conclusions

The optimism generated by the first trials of gene therapy has now been tempered with a healthy dose of reality. These initial studies have met with varying degrees of success and no trial can claim to have achieved the objective of clinical cure. The degree of immune reconstitution as a result of gene transfer remains very much in doubt. Very few patients have been treated and the results are inconsistent. It may also be that the improvements so far presented will not be sustained in a similar way to the effects of PEG–ADA on these children. The concurrent administration of PEG–ADA continues to complicate the issue. Although it was felt that PEG–ADA therapy had failed by laboratory-based criteria, these children may still have benefited clinically from its continued administration. It must also be remembered that the peripheral lymphocyte transfer studies initially required PEG–ADA to provide adequate numbers of cells for transduction. Conclusive evidence for the benefit of gene therapy will only be forthcoming on complete PEG–ADA withdrawal. Other concerns of these studies include the fact that all the children treated so far have manifested a delayed-onset and less severe phenotype. It remains to be seen whether gene therapy would be of benefit to the more severely affected population where a greater level of transduced cell engraftment and gene expression may be necessary.

The data from these studies regarding gene expression is poor. When transduced clones or individual vector-positive progenitor colonies have been analysed, the level of ADA activity is normal. However, such is the low frequency of transduced cells in the peripheral blood that the overall level of ADA activity has been barely above pre-treatment levels. In the most encouraging study by Bordignon *et al.* (1995), levels of 5–18% of normal activity in peripheral nucleated cells were seen but these children do not yet have normal immune function. As previously discussed in Section 13.5, it had been assumed that low levels of ADA activity might be sufficient for normal immune function but this may be a simplistic assumption.

Despite these criticisms, a number of significant benefits have been shown. Importantly, none of the children enrolled in these trials have suffered any side-effects as a result of gene transfer therapy. The most significant concern was of replication-competent retrovirus production but in all the studies independent assays for wild-type virus and retroviraemia have been negative. There have also been worries about the random nature of retroviral integration, specifically the risk of insertional mutagenesis, but this has not been borne out. The clinical trials of Bordignon *et al.* (1995) and Kohn *et al.* (1995) demonstrate that it is possible to transduce early progenitor cells capable of generating multi-lineage progeny and that these cells can exist in the marrow and peripheral circulation for a prolonged period of time. The results from Bordignon *et al.* (1995) and Blaese *et al.* (1995) also show that transduced T lymphocytes remain in the circulation for a much longer period of time than originally predicted though whether this is as a result of a positive survival advantage is still unclear.

It must be remembered that these are among the first few clinical trials of a very new technology and it would have been unrealistic to expect complete success immediately. These initial studies have therefore served to underline the feasibility of gene therapy but also to highlight the problems that need to be overcome. Further, they have allowed insights into stem-cell and lymphocyte biology that will need to be more fully understood before gene therapy is ultimately successful.

13.8 Future prospects

The problems facing gene therapy for ADA deficiency mirror the problems facing gene therapy *per se*. The initial results of other clinical trials for inherited disorders and malignancies have also met with limited success. The problems of vector systems and of effective target-cell transduction and gene expression, especially in haemopoietic stem cells, remain the major problems to be overcome. In addition to this, the National Institutes of Health (NIH) in the USA recently voiced concerns that not enough attention is paid to understanding the pathophysiology of targeted diseases prior to gene therapy being attempted. It recommended that further research should be aimed at the mechanisms of disease pathogenesis as well as at better transfer vectors and improved animal models (Touchette, 1996). These recommendations are extremely pertinent to ADA deficiency.

The reasons for the specificity of tissue damage in ADA deficiency still remains unclear. Although there have been recent studies focusing on this problem, there are still a number of questions unanswered. Some children have costochondral abnormalities and there have also been reports of ADA-deficient children suffering neurological problems (Cederbaum *et al.*, 1976; Hirschhorn *et al.*, 1980; Ratech *et al.*, 1985). Thus, the mechanism of disease is complex and requires a greater level of understanding.

The concept of prolonged survival of transduced ADA$^+$ cells in an ADA$^-$ environment remains poorly understood and has not been convincingly confirmed by the gene-transfer studies. Some reports have advocated the use of myeloablation prior to the reinfusion of transduced cells, thus 'making space' for the modified cells to engraft and proliferate (Hoogerbrugge *et al.*, 1996). However, until long-term gene transfer and expression in primitive haemopoietic cells is demonstrated and a convincing animal model has been developed, it is unlikely that such a protocol would become acceptable.

Retroviral vectors as vehicles for gene transfer are limited by their inability to effectively transduce HSCs. Attention has therefore been focused on other delivery systems. Recent studies have described the use of adenoviral vectors (Mitani *et al.*, 1994) and secretable ADA via microcapsules (Hughes *et al.*, 1994) as alternative mechanisms, though both of these systems are in the early stages of development and not without their limitations. Adenovirus-associated viral vectors are less dependent on the cell cycle for transduction and have been shown to transduce quiescent cell populations successfully; however, their ability to transduce HSCs has yet to be established.

The development of these new technologies for gene therapy will go hand in hand with development of gene therapy for ADA deficiency. ADA-SCID remains a paradigm for haemopoietic cell gene therapy and will most likely be one of the first conditions to be successfully treated in this way.

References

Bender MA, Miller AD, Gelinas RE. (1988) Expression of the human beta-globin gene after retroviral transfer into murine erythroleukemia cells and human BFU-E cells. *Mol. Cell Biol.* **8:** 1725–1735.

Benveniste P, Zhu W, Cohen A. (1995) Interference with thymocyte differentiation by an inhibitor of S-adenosylhomocysteine hydrolase. *J. Immunol.* **155:** 536–544.

Bertolini F, de Monte L, Corsini C, Lazzari L, Lauri E, Soligo D, Ward M, Bank A, Malavasi F. (1994) Retrovirus-mediated transfer of the multidrug resistance gene into human haemopoietic progenitor cells. *Br. J. Haematol.* **88:** 318–324.

Blaese RM, Culver KW, Miller AD, Carter C, Fleisher TA, Clerici M, Shearer G, Chang L, Chiang Y, Tolstochev P, Greenblatt JJ, Rosenberg SA, Klein H, Berger M, Mullen CA, Ramsey WJ, Muul L, Morgan RA, Anderson WF. (1995) T lymphocyte-directed gene therapy for ADA-SCID: Initial trial results after 4 years. *Science* **270:** 475–480.

Bodine DM, Moritz T, Donahue RE, Luskey BD, Kessler SW, Martin DI, Orkin SH, Nienhuis AW, Williams DA. (1993) Long-term *in vivo* expression of a murine adenosine deaminase gene in rhesus monkey hematopoietic cells of multiple lineages after retroviral mediated gene transfer into CD34+ bone marrow cells. *Blood* **82:** 1975–1980.

Bordignon C, Notarangelo LD, Nobili N, Ferrari G, Casorati G, Panina P, Mazzolari E, Maggioni D, Rossi C, Servida P, Ugazio AG, Mavilio F. (1995) Gene therapy in peripheral blood lymphocytes and bone marrow for ADA-immunodeficient patients. *Science* **270:** 470–475.

Brenner MK, Cunningham JM, Sorrentino BP, Heslop HE. (1995) Gene transfer into human hemopoietic progenitor cells. *Br. Med. Bull.* **51:** 167–191.

Carter RF, Abrams Ogg AC, Dick JE, Kruth SA, Valli VE, Kamel Reid S, Dube ID. (1992) Autologous transplantation of canine long-term marrow culture cells genetically marked by retroviral vectors. *Blood* **79:** 356–364.

Cederbaum SD, Kaitila I, Rimoin DL, Stiehm ER. (1976) The chondro-osseous dysplasia of adenosine deaminase deficiency with severe combined immunodeficiency. *J. Pediatr.* **89**: 737–742.

Chang J, Morgenstern GR, Coutinho LH, Scarffe JH, Carr T, Deakin DP, Testa NG, Dexter TM. (1989) The use of bone marrow cells grown in long-term culture for autologous bone marrow transplantation in acute myeloid leukaemia: an update. *Bone Marrow Transplant.* **4**: 5–9.

Cohen A, Barankiewicz J, Lederman HM, Gelfand E. (1983) Purine and pyrimidine metabolism in human T lymphocyte regulation of deoxyribonucleotide metabolism. *J. Biol. Chem.* **268**: 12334–12340.

Cournoyer D, Scarpa M, Mitani K, Moore KA, Markowitz D, Bank A, Belmont JW, Caskey CT. (1991) Gene transfer of adenosine deaminase into primitive human hematopoietic progenitor cells. *Hum. Gene Ther.* **2**: 203–213.

Crooks GM, Kohn DB. (1993) Growth factors increase amphotropic retrovirus binding to human CD34+ bone marrow progenitor cells. *Blood* **82**: 3290–3297.

Culver KW, Osborne WR, Miller AD, Fleisher TA, Berger M, Anderson WF, Blaese RM. (1991) Correction of ADA deficiency in human T lymphocytes using retroviral-mediated gene transfer. *Transplant. Proc.* **23**: 170–171.

Daddona PE, Mitchell BS, Meuwissen HJ, Davidson BL, Wilson JM, Koller CA. (1983) Adenosine deaminase deficiency with normal function. *J. Clin. Invest.* **72**: 483–492.

Dzierzak EA, Papayannopoulou T, Mulligan RC. (1988) Lineage-specific expression of a human beta-globin gene in murine bone marrow transplant recipients reconstituted with retrovirus-transduced stem cells. *Nature* **331**: 35–41.

Eglitis MA, Kantoff PW, Gilboa E, Anderson WF. (1985) Gene expression in mice after high efficiency retroviral mediated transfer. *Science* 1395–1398.

Ferrari G, Rossini S, Nobili N, Maggioni D, Garofalo A, Giavazzi R, Mavilio F, Bordignon C. (1992) Transfer of the ADA gene into human ADA-deficient T lymphocytes reconstitutes specific immune functions. *Blood* **80**: 1120–1124.

Fink JK, Correll PH, Perry LK, Brady RO, Karlsson S. (1990) Correction of glucocerebrosidase deficiency after retroviral-mediated gene transfer into hematopoietic progenitor cells from patients with Gaucher disease. *Proc. Natl Acad. Sci. USA* **87**: 2334–2338.

Fischer A, Landais P, Friedrich W, Morgan G, Gerritsen B, Fasth A, Porta F, Griscelli C, Goldman SF, Levinsky RJ, Vossen J. (1990) European experience of bone-marrow transplantation for severe combined immunodeficiency. *Lancet* **336**: 850–854.

Giblett ER, Anderson JE, Cohen F, Pollara B, Meuwissen HJ. (1972) Adenosine-deaminase deficiency in two patients with severely impaired cellular immunity. *Lancet* **2**: 1067–1069.

Gluckman E, Broxmeyer HA, Auerbach AD, Friedman HS, Douglas GW, Devergie A, Esperou H, Thierry D, Socie G, Lehn P, Cooper S, English D, Kurtzberg J, Bard J, Boyse EA. (1989) Hematopoietic reconstitution in a patient with Fanconi's anemia by means of umbilical-cord blood from an HLA-identical sibling. *New Engl. J. Med.* **321**: 1174–1178.

Hanley ME, Nolta JA, Parkman R, Kohn DB. (1994) Umbilical cord blood cell transduction by retroviral vectors: preclinical studies to optimize gene transfer. *Blood Cells* **20**: 539–543.

Hershfield MS. (1995) PEG–ADA: an alternative to haploidentical bone marrow transplantation and an adjunct to gene therapy for adenosine deaminase deficiency. *Hum. Mutat.* **5**: 107–112.

Hershfield MS, Buckley RH, Greenberg ML, Melton AL, Schiff R, Hatem C, Kurtzberg J, Markert ML, Kobayashi RH, Kobayashi AL, Abuchowski A. (1987) Treatment of adenosine deaminase deficiency with polyethylene glycol-modified adenosine deaminase. *New Engl. J. Med.* **316**: 589–596.

Hershfield MS, Chaffee S, Sorensen RU. (1993) Enzyme replacement therapy with polyethylene glycol-adenosine deaminase in adenosine deaminase deficiency: overview and case reports of three patients, including two now receiving gene therapy. *Pediatr. Res.* **33**: S42–47.

Hirschhorn R. (1993) Overview of biochemical abnormalities and molecular genetics of adenosine deaminase deficiency. *Pediatr. Res.* **33**: S35–41.

Hirschhorn R, Paageorgiou PS, Kesarwala HH, Taft LT. (1980) Amelioration of neurologic abnormalities after 'enzyme replacement' in adenosine deaminase deficiency. *New Engl. J. Med.* **303**: 377–380.

Hock RA, Miller AD. (1986) Retrovirus mediated transfer and expression of drug resistance genes in human haematopoietic progenitor cells. *Nature* **320**: 275–277.

Hoogerbrugge PM, van Beusechem VW, Fischer A, Debree M, Le Deist F, Perignon JL, Morgan G, Gaspar HB, Fairbanks LD, Skeoch CH, Moseley A, Harvey M, Levinsky RJ, Valerio D. (1996) Bone marrow gene transfer in three patients with adenosine deaminase deficiency. *Gene Ther.* **3**: 179–183.

Hughes M, Vassilakos A, Andrews DW, Hortelano G, Belmont JW, Chang PL. (1994) Delivery of a secretable adenosine deaminase through microcapsules – a novel approach to somatic gene therapy. *Hum. Gene Ther.* 5: 1445–1455.

Hughes PF, Eaves CJ, Hogge DE, Humphries RK. (1989) High-efficiency gene transfer to human hematopoietic cells maintained in long-term marrow culture [see comments]. *Blood* 74: 1915–1922.

Kantoff PW, Gillio AP, McLachlin JR, Bordignon C, Eglitis MA, Kernan NA, Moen RC, Kohn DB, Yu SF, Karson E, Karlsson S, Zwiebel JA, Gilboa E, Blaese RM, Nienhuis A, O'Reilly RJ, Anderson WF. (1987) Expression of human adenosine deaminase in nonhuman primates after retrovirus-mediated gene transfer. *J. Exp. Med.* 166: 219–234.

Kernan NA, Schroeder ML, Ciavarella D, Preti RA, Rubinstein P, O'Reilly RJ. (1994) Umbilical cord blood infusion in a patient for correction of Wiskott-Aldrich syndrome. *Blood Cells* 20: 245–248.

Kohn DB, Weinberg K, Nolta JA, Heiss LN, Lenarsky C, Crooks GM, Hanley ME, Annett G, Brooks JS, El-Khoureiy A, Lawrence K, Wells S, Moen RC, Bastian J, Williams-Herman DE, Elder M, Wara D, Bowen T, Hershfield M, Mullen CA, Blaese RM, Parkman R. (1995) Engraftment of gene-modified umbilical cord blood cells in neonates with adenosine deaminase deficiency. *Nature Med.* 1: 1017–1023.

Lim B, Apperley JF, Orkin SH, Williams DA. (1989) Long-term expression of human adenosine deaminase in mice transplanted with retrovirus-infected hematopoietic stem cells. *Proc. Natl Acad. Sci. USA* 86: 8892–8896.

Lothrop CD, Jr, al Lebban ZS, Niemeyer GP, Jones JB, Peterson MG, Smith JR, Baker HJ, Morgan RA, Eglitis MA, Anderson WF. (1991) Expression of a foreign gene in cats reconstituted with retroviral vector infected autologous bone marrow. *Blood* 78: 237–245.

Lu L, Xiao M, Clapp DW, Li ZH, Broxmeyer HE. (1993) High efficiency retroviral mediated gene transduction into single isolated immature and replateable CD34(3+) hematopoietic stem/progenitor cells from human umbilical cord blood. *J. Exp. Med.* 178: 2089–2096.

Lu L, Xiao M, Clapp DW, Li ZH, Broxmeyer HE. (1994) Stable integration of retrovirally transduced genes into human umbilical cord blood high-proliferative potential colony-forming cells (HPP-CFC) as assessed after multiple HPP-CFC colony replatings *in vitro*. *Blood Cells* 20: 525–530.

Migchielson AAJ, Breuer ML, van Roon MA, te Riele H, Zurcher C, Ossendorp F, Toutain S, Hershfield MS, Berns A, Valerio D. (1996) Adenosine-deaminase-deficient mice die perinatally and exhibit liver-cell degeneration, atelectasis and small intestinal cell death. *Nature Genet.* 10: 279–287.

Miller AD, Eckner RJ, Jolly DJ, Friedmann T, Verma IM. (1984) Expression of a retrovirus encoding human HPRT in mice. *Science* 225: 630–632.

Mitani K, Graham FL, Caskey CT. (1994) Transduction of human bone marrow by adenoviral vector. *Hum. Gene Ther.* 5: 941–948.

Moore KA, Deisseroth AB, Reading CL, Williams DE, Belmont JW. (1992) Stromal support enhances cell-free retroviral vector transduction of human bone marrow long-term culture-initiating cells. *Blood* 79: 1393–1399.

Morgan G, Levinsky RJ, Hugh Jones K, Fairbanks LD, Morris GS, Simmonds HA. (1987) Heterogeneity of biochemical, clinical and immunological parameters in severe combined immunodeficiency due to adenosine deaminase deficiency. *Clin. Exp. Immunol.* 70: 491–499.

Moritz T, Keller DC, Williams DA. (1993) Human cord blood cells as targets for gene transfer: potential use in genetic therapies of severe combined immunodeficiency disease. *J. Exp. Med.* 178: 529–536.

Osborne WR, Hock RA, Kaleko M, Miller AD. (1990) Long-term expression of human adenosine deaminase in mice after transplantation of bone marrow infected with amphotropic retroviral vectors. *Hum. Gene Ther.* 1: 31–41.

Polmar SH. (1978) Enzyme replacement and other biochemical approaches to the therapy of adenosine deaminase deficiency. *Ciba Found. Symp.* 68: 213–230.

Ratech H, Greco MA, Gallo G, Rimoin DL, Kamino H, Hirschhorn R. (1985) Pathologic findings in adenosine deaminase-deficient severe combined immunodeficiency. I. Kidney, adrenal, and chondro-osseous tissue alterations. *Am. J. Pathol.* 120: 157–169.

Shovlin CL, Simmonds HA, Fairbanks LD, Deacock SJ, Hughes JM, Lechler RI, Webster AD, Sun XM, Webb JC, Soutar AK. (1994) Adult onset immunodeficiency caused by inherited adenosine deaminase deficiency. *J. Immunol.* 153: 2331–2339.

Sutherland HJ, Eaves CJ, Eaves AC, Dragowska W, Lansdorp PM. (1989) Characterization and partial purification of human marrow cells capable of initiating long-term hematopoiesis *in vitro*. *Blood* 74: 1563–1570.

Touchette N. (1996) Gene therapy: Not ready for prime time. *Nature Med.* 2: 7–8.

Valentine WN, Paglia DE, Tartaglia AP, Gilsanz F. (1977) Hereditary haemolytic anaemia with increased red cell adenosine deaminase (45- to 70-fold) and decreased adenosine triphosphates. *Science* **195**: 783–785.

Valerio D, Duyvesteyn MG, Meera Khan P, Geurts van Kessel A, de Waard A, van der Eb AJ. (1983) Isolation of cDNA clones for human adenosine deaminase. *Gene* **25**: 231–240.

van Beusechem VW, Kukler A, Einerhand MP, Bakx TA, van der Eb AJ, van Bekkum DW, Valerio D. (1990) Expression of human adenosine deaminase in mice transplanted with hemopoietic stem cells infected with amphotropic retroviruses. *J. Exp. Med.* **172**: 729–736.

van Beusechem VW, Bakx TA, Kaptein LC, Bart Baumeister JA, Kukler A, Braakman E, Valerio D. (1993) Retrovirus-mediated gene transfer into rhesus monkey hematopoietic stem cells: the effect of viral titers on transduction efficiency. *Hum. Gene Ther.* **4**: 239–247.

van Beusechem VW, Bart Baumeister JA, Bakx TA, Kaptein LC, Levinsky RJ, Valerio D. (1994) Gene transfer into nonhuman primate CD34+CD11b– bone marrow progenitor cells capable of repopulating lymphoid and myeloid lineages. *Hum. Gene Ther.* **5**: 295–305.

Vossen JM. (1983) Bone marrow transplantation in the treatment of primary immunodeficiencies. *Ann. Clin. Res.* **19**: 285–292.

Vowels MR, Tang RL, Berdoukas V, Ford D, Thierry D, Purtilo D, Gluckman E. (1993) Brief report: correction of X-linked lymphoproliferative disease by transplantation of cord-blood stem cells [see comments]. *New Engl. J. Med.* **329**: 1623–1625.

Wagner JE, Kernan NA, Steinbuch M, Broxmeyer HE, Gluckman E. (1995) Allogeneic sibling umbilical-cord-blood transplantation in children with malignant and non-malignant disease. *Lancet* **346**: 214–219.

Williams DA, Moritz T. (1994) Umbilical cord blood stem cells as targets for genetic modification: new therapeutic approaches to somatic gene therapy. *Blood Cells* **20**: 504–515.

Wilson JM, Danos O, Grossman M, Raulet DH, Mulligan RC. (1990) Expression of human adenosine deaminase in mice reconstituted with retrovirus-transduced hematopoietic stem cells. *Proc. Natl Acad. Sci. USA* **87**: 439–443.

Xu LC, Young HA, Blanco M, Kessler S, Roberts AB, Karlsson S. (1994) Poor transduction efficiency of human hematopoietic progenitor cells by a high-titer amphotropic retrovirus producer cell clone. *J. Virol.* **68**: 7634–7636.

Cardiovascular disease

B.A. French

14.1 Introduction

A great deal of basic and applied research is focused upon the cardiovascular system and it is therefore not surprising that many of the recent advances in gene therapy have been made in this same field. Early work using *ex vivo* gene transfer, with the implantation of transduced cells to peripheral arteries (Nabel *et al.*, 1989; Wilson *et al.*, 1989), quickly gave way to methods of direct *in vivo* gene transfer (Nabel *et al.*, 1990). This field has advanced to the point where approval has now been given for a clinical protocol for gene therapy in the treatment of peripheral vascular disease (Isner *et al.*, 1995). This chapter provides an overview of research to date, an assessment of current limitations, and forecasts potential applications of gene therapy in the cardiovascular system (*Table 14.1*).

Table 14.1. Gene therapy for cardiovascular disease

Target tissue	Current limitations	Gene therapy for:
Peripheral vessels	Safe and effective vectors	Atherosclerosis Thrombosis Metabolism
Coronary arteries	Safe and effective vectors Local delivery systems	Atherosclerosis Thrombosis Restenosis
Myocardium	Safe and effective vectors Local delivery systems	Ischaemia Transplant rejection
Liver	Safe and effective vectors	Atherosclerosis Metabolism

Nearly every conceivable method of direct gene transfer has been evaluated in cardiovascular tissues. Early studies employed retroviral vectors or complexes of cationic liposomes with plasmid DNA to accomplish gene transfer to the vasculature, but the efficiency of recombinant adenovirus has made this the method of choice for contemporary research. However, it should be noted that several studies have demonstrated that the first generation of recombinant adenovirus provokes a vigorous inflammatory response in host tissues following direct *in vivo* gene transfer. Recent results from human trials of gene therapy for cystic fibrosis (CF) mediated by adenovirus type 5 (Ad5) suggest that it may not always be possible to apply an effective dose of recom-

Gene Therapy, edited by N.R. Lemoine and D.N. Cooper.
© 1996 BIOS Scientific Publishers Ltd, Oxford.

binant adenovirus without provoking an inflammatory reaction (Knowles *et al.*, 1995). Second-generation Ad5 vectors have been developed to address the problem of inflammation, but it will be prudent to evaluate carefully the pro-inflammatory potential of these vectors before additional clinical trials are undertaken.

14.2 Gene therapy for the peripheral vasculature

14.2.1 Research to date

Direct gene transfer to the vasculature is conceptually appealing because peripheral vessels are easily accessed and the endothelium is in direct contact with the bloodstream. Since many of the pioneering studies of gene therapy were first performed in the peripheral vasculature, this system serves as a useful paradigm to illustrate some of the essential points regarding *in vivo* gene transfer. The earliest studies to demonstrate gene transfer to the vasculature made use of retroviral vectors, since these were the first gene-transfer vectors with the efficiency necessary to make such experiments possible. However, retroviruses are rapidly inactivated by complement in human serum (Rother *et al.*, 1995), so that in order to obtain gene transfer it was necessary to remove the vascular cells, culture them *in vitro*, infect them with recombinant retrovirus, and then seed the cells back into the intact vessel. Implantation into the vascular wall was accomplished using either direct instillation of cells (Nabel *et al.*, 1989) or by means of a vascular graft (Wilson *et al.*, 1989). Although these studies were successful in demonstrating gene transfer, the clinical application of this approach was limited due to the inefficiency and technical complexity of this type of *ex vivo* gene transfer.

A far less cumbersome method of accomplishing gene transfer is to introduce genetic material directly into the tissue of interest *in vivo*. This concept of direct *in vivo* gene transfer was pursued by Nabel *et al.* (1990) in a pioneering study which employed both retroviral vectors and liposome–DNA complexes. Peripheral vessels of experimental swine were surgically isolated, and the target segment was cleared of blood using a specialized double-balloon catheter. Solutions carrying the retroviral vector (or liposome–DNA complexes) were then introduced through the same double-balloon catheter. Direct transfer of the *lac*Z gene from *Escherichia coli* was demonstrated 3 days later using a histochemical stain for β-galactosidase activity. The efficiency of the gene-transfer process was low, but subsequent studies showed that even low-level expression of cDNAs encoding potent growth factors could alter the morphology of peripheral vessels (Nabel *et al.*, 1993a,b). These experiments illustrate the point that the expression of potent, diffusible growth factors from a limited number of vascular cells can have a measurable impact upon vessel morphology. On the other hand, it would be unreasonable to expect that low-level expression of a structural protein might have a similar impact.

The use of the strategy employing diffusible growth factors has recently been demonstrated in a study which used plasmids carrying the cDNA for vascular endothelial growth factor (VEGF) (Takeshita *et al.*, 1993). Although the efficiency of plasmid-mediated gene transfer to vascular tissue is extremely low, this method has a safety advantage since plasmid DNA itself is fairly innocuous and can readily be purified to pharmaceutical standards. Extremely sensitive methods must be employed to detect gene expression resulting from the direct introduction of plasmid DNA. Nevertheless, VEGF is such a potent molecule that even a single intra-arterial bolus of the purified growth factor will stimulate collateral formation in ischaemic rabbit limbs (Takeshita *et al.*, 1994). These investigations served as a foundation for the first clinical study of vascular gene therapy to win the approval of the Recombinant DNA Advisory Committee (RAC; Isner *et al.*, 1995). This protocol illustrates well the manner in which the strengths and weaknesses of a particular disease, gene therapy, and gene-transfer vector can be combined in a complementary fashion to achieve the desired clinical outcome. The fact that peripheral vascular disease can be treated with trace amounts of VEGF has been exploited by using a gene-therapy vector (plasmid DNA) which has a favourable safety profile. The low transfection efficiency of plasmid-mediated gene transfer is not a drawback in this particular application, because even low levels of gene expression are adequate to obtain the desired end-point.

Although the gene therapy protocol for peripheral vascular disease represents an important milestone along the road to gene therapy for cardiovascular disease, it should also be recognized as an exception to the general rule that gene transfer must be efficient in order to be effective. Clearly, the level of gene expression necessary to achieve a desired clinical outcome will vary widely depending upon the particular disease and the nature of the gene employed for its treatment. It follows that an efficient method of gene transfer will be more effective against a broader spectrum of disease than an inefficient method of gene transfer. To be precise, the efficiency of recombinant gene expression in a given tissue is dependent not only upon the ratio of transfected to untransfected cells, but also upon the level of gene expression in each transfected cell. The proportion of transfected cells is a function of gene transfer, whereas the level of gene expression in each cell is largely determined by gene copy number and promoter strength. The proportion of transfected cells achieved using plasmid DNA or retroviral vectors may be adequate to combat a few diseases, but far more effective methods of gene transfer will be required to address the broader field of cardiovascular disease. Therefore, the biggest hurdle facing gene therapy for the vasculature remains the same – the problem of getting enough genes into enough cells to achieve clinically significant physiological change in the tissue of interest.

The overall efficiency of a particular gene-transfer procedure is actually determined by taking the product of the 'efficiency' of delivering the vector to the tissue of interest and the 'efficiency' with which the therapeutic DNA is introduced into individual cells comprising the target tissue. Effective delivery techniques were developed for the peripheral vasculature in the earliest studies of direct gene transfer (Lim *et al.*, 1991; Nabel *et al.*, 1990). The technique of arterial ligation followed by vector instillation might be considered the 'gold standard' for work in the peripheral vasculature, but the double-balloon catheter employed by Nabel and colleagues is a close approximation, particularly when used in isolated arteries under direct visualization (Nabel *et al.*, 1990, 1993a,b). Of course, almost any catheter-based, local delivery device (Riessen and Isner, 1994) can be used in peripheral vessels, and the delivery characteristics of many such devices have been examined (Rome *et al.*, 1994, Willard *et al.*, 1994). In short, each catheter-based device has its own delivery characteristics, but the most effective and reproducible methods of gene transfer in peripheral vessels are based upon isolating a segment of artery so that high concentrations of vector can be infused at a constant pressure over extended periods of time (up to 30 min). Given this reasonably effective method for achieving local delivery, the overall efficiency of gene transfer in such vessels is then primarily determined by the gene-transfer efficiency of the vector itself.

This in turn explains much of the current interest in adenoviral vectors, since recombinant adenovirus is one of the most efficient recombinant gene transfer vectors known. Under ideal conditions in cell culture, there is a nearly a one-to-one correspondence between the number of infectious viral particles added to the cell culture dish and the number of cells that ultimately express the recombinant gene of interest. Recombinant adenovirus has proved to be the vector of choice in animal studies demonstrating that neointimal formation after arterial injury can be prevented by the expression of conditional toxins (Ohno *et al.*, 1994), or cytostatic molecules (Chang *et al.*, 1995). However, there are a number of technical difficulties which prevent these promising animal experiments from maturing into clinically relevant treatment modalities.

14.2.2 Current limitations

As mentioned in Section 14.1, the inflammatory response of the host to the first generation of recombinant Ad5 vectors has limited their effectiveness in preliminary human trials (Knowles *et al.*, 1995). The fact that inflammation was not noted in many of the early animal studies might be attributed to variability in the host response between species, and even variability between strains of a single species (Barr *et al.*, 1995). That humans respond to recombinant adenovirus in the same way that they respond to wild-type adenovirus is not surprising when it is remembered that a recombinant Ad5 is still 90–95% wild-type in its genetic makeup. However, there are several approaches to this problem which hold out promise (Engelhardt *et al.*, 1994, Wang *et al.*, 1995) and it is reasonable to assume that the problem of inflammation will be overcome within a matter of years.

Animal studies have revealed a second potential limitation to Ad5 which may limit its use in humans. Exposing an animal to Ad5 (recombinant or wild-type) will prompt the immune system to raise circulating antibodies, which can in turn effectively neutralize Ad5 during subsequent administration (Mittal *et al.*, 1993; Yang *et al.*, 1995). An analogous situation exists in the human population, where most children contract Ad5 during the first few years of life. As a result, the vast majority of potential gene therapy patients carry measurable titres of neutralizing antibody against Ad5 and this natural immunity of the human population may limit the efficiency of Ad5-mediated gene transfer in clinical trials of gene therapy.

It will therefore be important to determine the titre of neutralizing antibody present in patients before administering virus-mediated gene therapy. The titre of antibody against Ad5 may be negligibly low in some patients, whereas in others a moderate level of neutralizing antibody might be overcome by increasing the dose of Ad5. Patients with a high titre of antibody against Ad5 may not be candidates for this gene therapy vector, and it may be necessary to use other recombinant viruses carrying different epitopes. Other strains of adenovirus may prove useful in this regard, or it may be possible to use recombinant DNA technology to re-define the epitopes displayed by Ad5 to the immune system. Ultimately, sera from potential gene therapy patients will most likely be screened against a panel of the available gene therapy vectors in order to select the most appropriate vector system. This same approach would apply to patients seeking a second treatment of gene therapy, since they may well have developed antibody against the vector used during the first course of treatment.

The ultimate goal of gene therapy is to develop recombinant vectors that are both safe and effective. Although vectors derived from Ad5 are currently recognized as the most efficient, many questions remain regarding their safety. Conversely, naked plasmid DNA is relatively innocuous but the efficiency of plasmid-mediated gene transfer is so low as to be negligible in most tissues. Ultimately, it may be possible to create 'synthetic viruses' which combine the safety of plasmid DNA with the efficiency of a virus. Important progress has been made towards this goal (Cristiano *et al.*, 1993; Wagner *et al.*, 1992). However, many years of research may be required to develop a 'synthetic virus' which matches the efficiency of Ad5. In the meantime, the number of diseases which can be treated with a particular vector system will grow as a function of the efficiency of that system.

14.2.3 Potential applications

Given the assumption that a reasonably safe, effective and durable method of gene transfer will be developed in the near future, it is interesting to speculate as to how such a vector might be employed in the peripheral vasculature. The vasculature is fairly accessible via a catheter, the endothelium can be targeted at high frequency by Ad5 (Lemarchand *et al.*, 1993; Willard *et al.*, 1994), and protein products secreted by the endothelium will be rapidly disseminated by the bloodstream. Gene therapy could therefore be focused upon specific vascular segments, not only for the treatment of local disease but also for distribution to the rest of the body via the circulatory system.

There are numerous examples in which gene therapy might be effective in treating local vascular disease. Recurrent thrombosis in injured arteries might be addressed by activating the fibrinolytic system with tissue-type plasminogen activator (tPA) or the urokinase-type plasminogen activator (uPA). Similarly, thrombin might be inhibited with recombinant hirudin or antithrombin III. Alternatively, the generation of thrombin might be inhibited with specific coagulation factor antagonists such as tick anticoagulant peptide (which inhibits factor Xa). Moreover, vascular tone might be improved, and the adherence of platelets to injured vessel surfaces might be reduced by increasing nitric oxide production via the expression of nitric oxide synthase. Note, however, that gene therapy strategies involving such recombinant molecules would be totally useless in acute settings, because it takes 24 h for gene expression to reach half-maximal levels after direct gene transfer to the vasculature (French *et al.*, 1994a,b). The 12–24 h delay between the administration of gene therapy and the actual expression of recombinant genes is easily understood in terms of the time it takes for DNA to penetrate the plasma membrane, traverse the cytoplasm and enter the nucleus. Once in the nucleus, a recombinant gene must be transcribed to mRNA, the mRNA must exit the nucleus to be translated by ribosomes, and the resulting protein will usually need to be processed in the Golgi apparatus before secretion from the cell. For this reason, gene therapy involving fibrinolytic and thrombolytic strategies will only be useful in the chronic setting of recurrent fibin deposition or thrombosis. The use of such an approach would be enhanced by catheter-based techniques to localize gene therapy to the injured vessel, thus avoiding the potential problem of adverse systemic events.

The problem of atherosclerosis in peripheral vessels might also be addressed by gene therapy. For example, disrupting the interaction between vascular cell adhesion molecule 1 (VCAM-1) on the surface of endothelial cells and very late antigen 4 (VLA4) on the surface of circulating monocytes might prevent the latter from traversing the endothelial barrier and maturing into the macrophages which contribute to atherogenesis. Indeed, Chen *et al.* (1994) have reported the construction of a recombinant Ad5 vector which directs the expression and secretion of a soluble form of VCAM-1. The secreted molecule still binds VLA4, and thus has the potential to act as a competitive inhibitor of this mechanism of the cell adhesion system. The local expression of therapeutic protein within atherosclerotic vessels has potential, but the problem of atherosclerosis is probably best addressed by altering lipid profiles.

Interestingly, the peripheral vasculature might also prove useful in treating disease not directly related to the cardiovascular system. For example, many disorders in human metabolism are secondary to mutations in genes that code for hepatic enzymes. Although it would at first seem logical to direct gene therapy for such diseases to the liver, it may actually be advantageous in certain applications to target gene therapy to specific arterial segments instead. Gene therapy to peripheral arteries has a number of advantages, including ease of access, and the possibility of later removal should it become necessary to terminate gene therapy. Consider, for example, the case in which it might become necessary to reduce (or terminate) a patient's gene therapy at some future date. This could entail major surgery in the case of liver gene therapy, but the endothelial cells lining a specific arterial segment could easily be removed by simple abrasion of the artery with an angioplasty balloon. A molecular alternative for such a contingency plan is to include a conditionally lethal gene (such as that for herpes simplex virus thymidine kinase) in the gene transfer vector, so that transfected cells could be destroyed as required (with ganciclovir treatment).

14.3 Gene therapy for the coronary vasculature

14.3.1 Research to date

Direct gene transfer to the coronary vasculature was first demonstrated by Lim *et al.* (1991) in a study where a coronary bypass was used so that liposome–DNA complexes could be infused into a ligated segment of a canine coronary artery. The feasibility of using perforated balloon catheters to deliver liposome–DNA complexes to coronary arteries of dogs was demonstrated by Chapman *et al.* (1992). Similar techniques were later used to deliver recombinant adenovirus to pig coronary arteries (French *et al.*, 1994a). Recombinant Ad5 has also been introduced via catheter into the coronary arterial cirulation of rabbits, resulting in gene transfer to both the vasculature and the myocardium (Barr *et al.*, 1994). Although many of the principles governing direct gene transfer to peripheral arteries also apply to coronary arteries, direct comparison between the two systems is often complicated by the technical challenges posed by local delivery to coronary arteries.

14.3.2 Current limitations

The clinical implementation of coronary gene therapy is contingent not only upon highly efficient gene transfer vectors, but also upon advanced catheter-based delivery devices capable of safe but effective delivery. Given that suitable gene-transfer vectors will likely be developed within a period of years, the most daunting challenge facing coronary gene therapy is one of local delivery. Coronary arteries present a more challenging target for gene therapy than do peripheral vessels, owing to the increased velocity of blood flow and the critical importance of maintaining distal perfusion. Double-balloon catheters similar to the ones employed in peripheral arteries by Nabel *et al.* (1994a,b) have been modified to allow distal perfusion. However, the general use of such devices is limited by the highly branched nature of human coronary arteries. In practice, these branches make it difficult to confine Ad5 (or any other gene transfer vector) within a defined target segment of coronary artery.

14.3.3 Potential applications

Gene therapy for the prevention of coronary restenosis is a conceptually appealing application of gene therapy. The pathophysiology of clinical restenosis is actually the product of a number of complex molecular, biological and physiological processes. Nevertheless, each of the complex mechanisms that might contribute to restenosis can conceivably be modulated by gene transfer, whether they involve thrombosis, vascular smooth muscle cell activation, or even vascular remodelling. Each of these mechanisms is known to involve the modulation of gene expression, so it follows that direct *in vivo* gene transfer might be a logical approach to the problem. Separate gene therapy strategies might be developed to control thrombosis, to modulate cytokine production or signalling, to inhibit vascular smooth muscle cell proliferation or migration, to neutralize the local production of oxygen-derived free radicals, to promote re-endothelialization, to inhibit the elaboration of extracellular matrix or to influence vascular remodelling. Moreover, combination therapies targeting each of these mechanisms could be developed and employed as necessary to achieve lesion-specific gene therapy.

The widespread application of recombinant tPA for thrombolysis stands as testimony to the potency of recombinant protein, and gene therapy might be considered as a method by which affected tissues can be programmed to produce their own specific recombinant pharmaceutical product. The localized expression of recombinant genes in a specific target tissue is also appealing since this approach would serve to minimize adverse systemic effects. Although a number of innovative, catheter-based, local delivery devices have been developed and approved for clinical use, their use is limited by the fact that the beneficial effects of conventional drugs are extremely transient whereas clinical restenosis develops over a period of months. When considered in this light, direct gene transfer is one of the few methods by which a single local delivery procedure might give rise to a long-term therapeutic effect.

Access to the coronary arteries for the purpose of gene transfer might readily be obtained with only minor modifications to standard catheterization and angioplasty techniques. It is, however, important to consider that the only reasonable opportunity to perform gene therapy is immediately after balloon angioplasty, when the site of balloon injury has been established and the target is still readily accessible via a catheter. The pathology of atherosclerotic coronary lesions after balloon injury is extremely complex. Not only is the fractured atherosclerotic plaque still present but the ballooned artery has been denuded of endothelium with dissection planes extending deep into the media and serving as substrates for thrombus formation. Given the fact that balloon angioplasty removes endothelium and the observation that balloon injury enhances gene transfer to smooth vascular muscle cells (Lee *et al.*, 1993), the dissection planes created by balloon angioplasty should serve as channels to facilitate gene transfer to the media (Landau *et al.*, 1995). This might actually prove advantageous in channelling gene therapy to the site where it is most needed, since neointimal formation later on is most prominent in these same dissection planes.

Given the many studies in which Ad5-mediated gene therapy has inhibited neointimal formation in peripheral arteries (Chang *et al.*, 1995; Guzman *et al.*, 1994; Ohno *et al.*, 1994), there is little doubt that gene therapy for coronary restenosis would work in the clinical setting, if only it were possible to deliver the appropriate concentration and volume of the appropriate Ad5 vector to the appropriate place at the appropriate pressure. The problem is that each of these parameters is critical and unforgiving. Too low a concentration of Ad5 might not give adequate levels of gene transfer, whereas too high a concentration might lead to cytopathogenicity from viral capsid proteins (Schulick *et al.*, 1995). Too little volume might not be adequate to fill dissection planes while too much volume might propagate dissections. Too little pressure might be inadequate to force gene transfer reagent down dissection planes into the media, whereas too much pressure would easily propagate these dissections. The problem is further complicated by the fact that each of these variables can vary widely from patient to patient and from lesion to lesion. Current clinical methods of arterial visualization and local delivery are for the most part inadequate to support the lesion-specific approach necessary to customize gene therapy. In the final analysis, the economic and engineering challenges facing gene therapy for coronary restenosis may be more significant than the genetic or immunological challenges.

Although a reliable and cost-effective gene therapy against coronary restenosis may take some time to develop, there are a number of other interesting potential applications of coronary gene therapy. Intracoronary stents are gaining favour among interventionalists since the 'early gain' they provide in luminal diameter serves to offset the 'late loss' of restenosis, as demonstrated recently by the BENESTENT (BElgium NEtherlands STENT) and STRESS (STent REStenosis Study) trials. At present, there do not appear to be any significant long-term side-effects associated with implanting metallic coils within human coronary arteries, and current research in the field has therefore focused upon inhibiting the neointimal formation which inevitably results from stent deployment. This problem is particularly acute in smaller coronary arteries, which by their very nature are less able to accommodate 'late-loss'. Some of the innovative approaches under development include stents which have been made radioactive, stents which have been covered with therapeutic coatings, and stents composed of biocompatible or biodegradable polymers. It may be possible to incorporate gene-transfer vectors into coated or biocompatible stents to reduce their thrombogenicity or to inhibit neointimal formation. However, these approaches must each be optimized individually before they might be evaluated combinatorially, and the practical use of stent-based gene therapy will depend greatly upon whether adequate results can be obtained using more conventional technology.

Another potential application of coronary gene therapy is in the setting of coronary artery bypass grafts. The same antiproliferative gene therapy strategies developed for restenosis might prove beneficial in internal mammary artery bypass grafts. Other genes might prove useful in preventing vasculitis and/or prolonging graft survival. Provocative studies undertaken in rat carotid arteries indicate that genetic modification of interposition vein grafts with antisense oligonucleotides can promote medial thickening and even inhibit the accelerated atherosclerosis responsible for graft failure (Mann *et al.*, 1995). It is doubtful that genetically modified vein grafts will ever be preferred to internal mammary grafts but such a modification would integrate well into standard surgical practice, provided that it was cost-effective and free of complications.

Perhaps the most exciting, useful and practical gene therapy for coronary arteries will not involve coronary gene transfer at all but rather gene transfer to the liver. Animal studies have already demonstrated that genetic deficiencies in the low-density lipoprotein (LDL) receptor can be corrected by gene therapy to the liver (Ishibashi *et al.*, 1993; Wilson *et al.*, 1992). It may be possible to use similar methods to manipulate cholesterol levels in humans. For example, high-density lipoprotein (HDL) cholesterol levels might be raised with apolipoprotein AI, or cholesterol clearance enhanced by over-expression of the LDL receptor. There are a number of reasons why this type of approach might be one of the first cardiovascular gene therapies to find widespread application. Among these is the magnitude of the clinical problem and the pivotal role of cholesterol metabolism in cardiovascular disease, coupled with increased awareness of the value of preventive medicine. It is not difficult to appreciate that an effective gene therapy for manipulating lipid levels would significantly reduce the rate of atherosclerosis along with all of its clinical *sequelae* (angina, stroke, myocardial infarction and ultimately even restenosis). Recent studies with the statin family of 3-hydroxy-3-methylglutaryl coenzyme A (HMG-CoA) reductase inhibitors have demonstrated that favourable changes in lipid profiles can even lead to the regression of atherosclerosis (Gotto, 1995).

The relative ease with which genes can be targeted to the liver is another reason why this organ is likely to become an important early candidate for cardiovascular gene therapy. The structure and function of the liver make it the primary target organ for gene transfer following the intravenous administration of Ad5, or even of synthetic complexes formed of liposomes and plasmid DNA.

14.4 Gene therapy for the myocardium

14.4.1 Research to date

Early work demonstrating gene transfer to cardiomyocytes in living animals was carried out using the direct injection of plasmid DNA (Acsadi *et al.*, 1991; Lin *et al.*, 1990). This technique has proved valuable for studying the regulation of myocardial gene expression *in vivo*. However, the potential of plasmid DNA as a vector for myocardial gene therapy is limited by the fact that gene uptake and expression is limited to the few cardiomyocytes that line the needle track of the injection. It would therefore be necessary to perforate the myocardium at extremely high density in order to transfect even a small percentage of the total cardiomyocytes in a given region of the heart. From a practical viewpoint, there are relatively few clinical procedures that provide open-chest access to the heart and fewer still that are performed on an elective basis in such a manner that gene therapy by direct injection might safely be employed. This does not exclude the possibility that isolated applications might develop. For example, the direct infusion of a plasmid encoding the gene for a potent secreted mitogen such as VEGF might well stimulate collateral formation in ischaemic regions of the left ventricle, provided that the myocardium was still viable and that a suitable opportunity for direct gene transfer presented itself. Although it may be reasonable to deliver this type of gene therapy on an adjunctive basis, it is doubtful that many interventional or surgical procedures would be undertaken primarily for the purpose of accomplishing gene transfer.

14.4.2 Current limitations

The challenge facing gene therapy to the myocardium is ultimately the same as that facing any other form of gene therapy: the familiar problem of getting enough genes into enough cells in the tissue of interest. The demonstration of widespread, long-term expression in skeletal muscle and heart following intravenous injection in neonatal mice was therefore a turning point in gene transfer to the myocardium (Stratford-Perricaudet *et al.*, 1992). Although the animal model system was inadequate to expose all of the technical shortcomings subsequently associated with first-generation Ad5 vectors, this study demonstrated for the first time a reasonably efficient method of direct *in vivo* gene transfer to the heart. The use of catheter-based techniques for Ad5-mediated gene transfer to the heart was demonstrated by Barr *et al.* (1994). Subsequently, a direct comparison of direct injection of plasmid DNA versus Ad5 into the left ventricle of pigs demonstrated that Ad5 was some 140 000 times more efficient than naked plasmid DNA on a molar basis (French *et al.*, 1994b). However, this study also revealed that the increased efficiency of Ad5 came at the price of a vigorous inflammatory response directed against the infected (recombinant) cardiomyocytes. Recent clinical studies of Ad5-mediated gene therapy for CF indicate that the host inflammatory response to the first generation of Ad5 vectors may preclude their effective use in humans (Knowles *et al.*, 1995).

14.4.3 Potential applications

It is anticipated that within a matter of years improvements in recombinant adenoviral vectors will overcome the technical shortcomings which presently limit their clinical application, and no doubt the efficiency of alternative gene transfer vectors will be improved substantially. Using these improved vectors in conjunction with catheter-based techniques, it may soon be possible to reprogramme gene expression within ischaemic beds of myocardium in conjunction with percutaneous transluminal coronary angioplasty (PTCA). This opens up the possibility of using Ad5-mediated gene therapy to improve myocardial recovery and performance. For example, the revival of myocardium might be hastened by gene transfer of transcriptional activators of myocardial differentiation or collateral formation might be stimulated by genes encoding fibroblast growth factor or VEGF. As the mechanisms underlying the phenomenon of ischaemic preconditioning are elucidated, it may become possible to enhance myocardial survival during future rounds of ischaemic insult. There are many potential applications which might be expected to improve recovery following myocardial infarction, but such adjunctive therapies will be extremely difficult to evaluate in randomized clinical trials. It may take years to implement adjunctive gene therapy because it will first be necessary to demonstrate that the benefit clearly offsets the substantial safety concerns.

The safety of gene therapy will always be an issue of paramount concern, particularly when viral vectors are involved. It is no coincidence that most of the clinical gene therapy protocols approved by the Recombinant DNA Advisory Committee involve cancer because the prognosis in such cases is often grim and the small risk inherent in the gene therapy protocols is easily offset by the potential gain. As gene therapy vectors improve and clinical experience accumulates, the inherent risk will decrease but the potential for therapeutic success will continue to be a deciding factor in equations governing the implementation of gene therapy. Therefore, the earliest applications of gene therapy in the cardiovascular system will be those which can clearly demonstrate significant therapeutic success in the face of reasonable risk. In addition, it is clear that the markets available to gene therapy will be limited to areas where available pharmaceutical therapies are clearly inadequate. Furthermore, the earliest applications of gene therapy will be those in which it can be administered in concert with conventional medical practice, since physicians will be understandably reluctant to modify established clinical protocols.

Given the current limitations for the local delivery of gene therapy vectors and the realities of modern medical practice summarized above, it is apparent that cardiac transplantation should be one of the most fertile fields for the early application of gene therapy in the cardiovascular system. The prognosis of a patient on the cardiac transplant list is by definition very poor, and the long-term survival of the patients completing transplant clearly needs to be improved. Conventional pharmaceutical products used to control transplant rejection are far from optimal and the end-points commonly used to assess the success of a cardiac transplant are hard end-points indeed. Most compelling, however, is the fact that gene therapy will fit seamlessly into the transplant arena. Following explant from the donor, these hearts are commonly maintained in preservation solution for up to 6 h. During this period of time, the organ would be an ideal substrate for gene therapy. Not only could defined doses of gene therapy be applied under carefully controlled conditions of temperature, pressure and concentration but this approach would also prevent the systemic dissemination of gene therapy which might otherwise be prohibitive.

There are a number of genes which might be expected to improve the long-term survival of cardiac transplants. Local production of interleukin-10 might control the activation of leukocytes. Similarly, the expression of cell adhesion molecules or major histocompatibility complex antigens might be altered to control rejection. Multiple proteins could be encoded by a single gene therapy vector so that several molecular strategies could be pursued in parallel to achieve optimal results. As molecular strategies for the control of transplant rejection are perfected, it may even become feasible to apply similar techniques in pig xenografts. Of course, a very high transfection rate (>90%) would be required to make such strategies viable but it should prove possible to achieve such high rates under ideal *ex vivo* conditions.

14.5 Summary

The technical barriers impeding practical gene therapy for cardiovascular disease are likely to fall within a matter of years. When they do, the fields of atherosclerosis and cardiac transplant may be two of the areas in which gene therapy will have its earliest widespread impact. It is important to note that the human genome project will be maturing during the same period of time, and that gene therapy will be the most direct route by which the new genetic discoveries will find immediate clinical application. The anticipated synergy between gene therapy technology and the human genome project should provide important impetus in the evolution of modern medicine. It is important to realize, however, that the current academic trend towards generalized training (vs. specialization) and economic trends toward managed care (vs. best care at any cost) run counter to the clinical implementation of gene therapy. It may therefore take decades before gene therapy can expand beyond a limited set of early applications to become a preferred treatment modality for the majority of cardiovascular disease states.

References

Acsadi G, Jiao S, Jani A, Duke D, Williams P, Chong W, Wolff JA. (1991) Direct gene transfer and expression into rat heart *in vivo*. *New Biol.* **3**: 71–81.

Barr D, Tubb J, Ferguson D, Scaria A, Lieber A, Wilson C, Perkins J, Kay MA. (1995) Strain related variations in adenovirally mediated transgene expression from mouse hepatocytes *in vivo*: comparisons between immunocompetent and immunodeficient inbred strains. *Gene Ther.* **2**: 151–155.

Barr E, Carroll J, Kalynych AM, Tripathy SK, Kozarsky K, Wilson JM, Leiden JM. (1994) Efficient catheter-mediated gene transfer into the heart using replication-defective adenovirus. *Gene Ther.* **1:** 51–58.

Chang MW, Barr E, Seltzer J, Jiang YQ, Nabel GJ, Nabel EG, Parmacek MS, Leiden JM. (1995) Cytostatic gene therapy for vascular proliferative disorders with a constitutively active form of the retinoblastoma gene product. *Science* **267:** 518–522.

Chapman GD, Lim CS, Gammon RS, Culp SC, Desper JS, Bauman RP, Swain JL, Stack RS. (1992) Gene transfer into coronary arteries of intact animals with a percutaneous balloon catheter. *Circ. Res.* **71:** 27–33.

Chen SJ, Wilson JM, Muller DW. (1994) Adenovirus-mediated gene transfer of soluble vascular cell adhesion molecule to porcine interposition vein grafts. *Circulation* **89:** 1922–1928.

Cristiano RJ, Smith LC, Woo SL. (1993) Hepatic gene therapy: adenovirus enhancement of receptor-mediated gene delivery and expression in primary hepatocytes. *Proc. Natl Acad. Sci. USA* **90:** 2122–2126.

Engelhardt JF, Ye X, Doranz B, Wilson JM. (1994) Ablation of E2A in recombinant adenoviruses improves transgene persistence and decreases inflammatory response in mouse liver. *Proc. Natl Acad. Sci. USA* **91:** 6196–6200.

French BA, Mazur W, Ali NM, Geske RS, Finnigan JP, Rodgers GP, Roberts R, Raizner AE. (1994a) Percutaneous transluminal *in vivo* gene transfer by recombinant adenovirus in normal porcine coronary arteries, atherosclerotic arteries, and two models of coronary restenosis. *Circulation* **90:** 2402–2413.

French BA, Mazur W, Geske RS, Bolli R. (1994b) Direct *in vivo* gene transfer into porcine myocardium using replication-deficient adenoviral vectors. *Circulation* **90:** 2414–2424.

Gotto AM, Jr. (1995) Lipid lowering, regression, and coronary events. A review of the Interdisciplinary Council on Lipids and Cardiovascular Risk Intervention, Seventh Council meeting. *Circulation* **92:** 646–656.

Guzman RJ, Hirschowitz EA, Brody SL, Crystal RG, Epstein SE, Finkel T. (1994) *In vivo* suppression of injury-induced vascular smooth muscle cell accumulation using adenovirus-mediated transfer of the herpes simplex virus thymidine kinase gene. *Proc. Natl Acad. Sci. USA* **91:** 10732–10736.

Ishibashi S, Brown MS, Goldstein JL, Gerard RD, Hammer RE, Herz J. (1993) Hypercholesterolemia in low density lipoprotein receptor knockout mice and its reversal by adenovirus-mediated gene delivery. *J. Clin. Invest.* **92:** 883–893.

Isner JM, Walsh K, Symes J, Pieczek A, Takeshita S, Lowry J, Rossow S, Rosenfield K, Weir L, Brogi E. (1995) Arterial gene therapy for therapeutic angiogenesis in patients with peripheral artery disease. *Circulation* **91:** 2687–2692.

Knowles MR, Hohneker KW, Zou Z, Olsen JC, Noah TL, Hu P-C, Leigh MW, Engelhardt JF, Edwards LJ, Jones KR, Grossman M, Wilson JM, Johnson LG, Boucher RC. (1995) A controlled study of adenoviral-vector-mediated gene transfer in the nasal epithelium of patients with cystic fibrosis. *New Engl. J. Med.* **333:** 823–831.

Landau C, Pirwitz MJ, Willard MA, Gerard RD, Meidell RS, Willard SE. (1995) Adenoviral mediated gene transfer to atherosclerotic arteries after balloon angioplasty. *Am. Heart J.* **129:** 1051–1057.

Lee SW, Frapnell BC, Rade JJ, Virmani R, Dichek DA. (1993) *In vivo* adenoviral vector-mediated gene transfer into balloon-injured rat carotid arteries. *Circ. Res.* **73:** 797–807.

Lemarchand P, Jones M, Yamada I, Crystal RG. (1993) *In vivo* gene transfer and expression in normal uninjured blood vessels using replication-deficient recombinant adenovirus vectors. *Circ. Res.* **72:** 1132–1138.

Lim CS, Chapman GD, Gammon RS, Muhlestein JB, Bauman RP, Stack RS, Swain JL. (1991) Direct *in vivo* gene transfer into the coronary and peripheral vasculatures of the intact dog. *Circulation* **83:** 2007–2011.

Lin H, Parmacek MS, Morle G, Bolling S, Leiden JM. (1990) Expression of recombinant genes in myocardium *in vivo* after direct injection of DNA. *Circulation* **82:** 2217–2221.

Mann MJ, Gibbons GH, Kernoff RS, Diet FP, Tsao PS, Cooke JP, Kaneda Y, Dzau VJ. (1995) Genetic engineering of vein grafts resistant to atherosclerosis. *Proc. Natl Acad. Sci. USA* **92:** 4502–4506.

Mittal SK, McDermott MR, Johnson DC, Prevec L, Graham FL. (1993) Monitoring foreign gene expression by a human adenovirus-based vector using the firefly luciferase gene as a reporter. *Virus Res.* **28:** 67–90.

Nabel EG, Plautz G, Boyce FM, Stanley JC, Nabel GJ. (1989) Recombinant gene expression within endothelial cells of the arterial wall. *Science* **244:** 1342–1344.

Nabel EG, Plautz G, Nabel GJ. (1990) Site-specific gene expression *in vivo* by direct gene transfer into the arterial wall. *Science* **249:** 1285–1288.

Nabel EG, Shum L, Pompili V, Yang ZY, San H, Shu HB, Liptay S, Gold L, Gordon D, Derynck R. (1993a) Direct transfer of transforming growth factor beta 1 gene into arteries stimulates fibrocellular hyperplasia. *Proc. Natl Acad. Sci. USA* **90**: 10759–10763.

Nabel EG, Yang ZY, Plautz G, Forough R, Zhan X, Haudenschild CC, Maciag T, Nabel GJ. (1993b) Recombinant fibroblast growth factor-1 promotes intimal hyperplasia and angiogenesis in arteries *in vivo*. *Nature* **362**: 844–846.

Ohno T, Gordon D, San H, Pompili VJ, Imperiale MJ, Nabel GJ, Nabel EG. (1994) Gene therapy for vascular smooth muscle cell proliferation after arterial injury. *Science* **265**: 781–784.

Riessen R, Isner JM. (1994) Prospects for site-specific delivery of pharmacologic and molecular therapies. *J. Am. Coll. Cardiol.* **23**: 1234–1244.

Rome JJ, Shayani V, Flugelman MY, Newman KD, Farb A, Virmani R, Dichek DA. (1994) Anatomic barriers influence the distribution of *in vivo* gene transfer into the arterial wall. Modeling with microscopic tracer particles and verification with a recombinant adenoviral vector. *Arterioscler. Thromb.* **14**: 148–161.

Rother RP, Squinto SP, Mason JM, Rollins SA. (1995) Protection of retroviral vector particles in human blood through complement inhibition. *Hum. Gene Ther.* **6**: 429–435.

Schulick AH, Newman KD, Virmani R, Dichek DA. (1995) *In vivo* gene transfer into injured carotid arteries. Optimization and evaluation of acute toxicity. *Circulation* **91**: 2407–2414.

Stratford-Perricaudet LD, Makeh I, Perricaudet M, Briand P. (1992) Widespread long-term gene transfer to mouse skeletal muscles and heart. *J. Clin. Invest.* **90**: 626–630.

Takeshita S, Zheng LP, Asahare T, Riessen R, Brogi E, Ferrara N, Symes JF, Isner JM. (1993) *In vivo* evidence of enhanced angiogenesis following direct arterial gene transfer of the plasmid encoding vascular endothelial growth factor. *Circulation* **88**: I-476 (Abstract).

Takeshita S, Zheng LP, Brogi E, Kearney M, Pu LQ, Bunting S, Ferrara N, Symes JF, Isner JM. (1994) Therapeutic angiogenesis. A single intraarterial bolus of vascular endothelial growth factor augments revascularization in a rabbit ischemic hind limb model. *J. Clin. Invest.* **93**: 662–670.

Wagner E, Plank C, Zatloukal K, Cotten M, Birnstiel ML. (1992) Influenza virus haemagglutinin HA-2 N-terminal fusogenic peptides augment gene transfer by transferrin–polysine–DNA complexes: towards a synthetic virus-like gene-transfer vehicle. *Proc. Natl Acad. Sci. USA* **89**: 7934–7938.

Wang, Q, Jia X-C, Finer MH. (1995) A packaging cell line for propagation of recombinant adenovirus vectors containing two lethal gene-region deletions. *Gene Ther.* **2**: 775–783.

Willard JE, Landau C, Glamann DB, Burns D, Jessen ME, Pirwitz MJ, Gerard RD, Meidell RS. (1994) Genetic modification of the vessel wall. Comparison of surgical and catheter-based techniques for delivery of recombinant adenovirus. *Circulation* **89**: 2190–2197.

Wilson JM, Birinyi LK, Salomon RN, Libby P, Callow AD, Mulligan RC. (1989) Implantation of vascular grafts lined with genetically modified endothelial cells. *Science* **244**: 1344–1346.

Wilson JM, Grossman M, Cabrera JA, Wu CH, Chowdhury NR, Wu GY, Chowdhury JR. (1992) Hepatocyte-directed gene transfer *in vivo* leads to transient improvement of hypercholesterolemia in low density lipoprotein-deficient rabbits. *J. Biol. Chem.* **267**: 963–967.

Yang Y, Li Q, Ertl HC, Wilson JM. (1995) Cellular and humoral immune responses to viral antigens create barriers to lung-directed gene therapy with recombinant adenoviruses. *J. Virol.* **69**: 2004–2015.

15

Cancer gene therapy I: genetic intervention strategies

Lesley-Ann Martin and Nicholas R. Lemoine

15.1 Introduction

The aim of current cancer treatments is to destroy as much of the malignancy as possible without damaging normal tissue. However, despite continuing advances in surgical techniques and improvements in radiotherapeutic and chemotherapeutic regimes, this often proves difficult when tumours are inaccessible or widely disseminated throughout the body (Sikora, 1993). The continuing characterization of genes contributing to the development of cancer presents an opportunity to utilize these genetic elements and their products as targets for the treatment and (in the long term) prevention of the disease. In this chapter, we review some of the gene therapy strategies that are being developed for the treatment of cancer together with a consideration of how genetic intervention could be used in the context of the hereditary predisposition syndromes.

15.2 Somatic and germ-line therapy

Gene therapy can be defined as the introduction and expression of exogenous genetic material into human cells to provide therapeutic benefit (Lemoine, 1994). This can be achieved by correcting an existing abnormality or by providing the cells with a new function. At present, the permanent correction of an inherited genetic condition by gene transfer to germ cells cannot be contemplated (Gutierrez *et al.*, 1992; Lemoine, 1994; Wivel and Walters, 1993).

15.3 Prevention of hereditary predisposition syndromes

Prevention of cancer currently relies on epidemiological studies to identify factors involved in the aetiology of the disease and the subsequent modification of behavioural patterns of the general public. However, these studies require extensive research and the implementation of new health education measures, both of which can take decades to come to fruition (Gullick and Handyside, 1994).

In recent years, it has become evident that some individuals are predisposed to neoplasia at predictable sites due to inherited genetic mutation. Unfortunately, until

Gene Therapy, edited by N.R. Lemoine and D.N. Cooper.
© 1996 BIOS Scientific Publishers Ltd, Oxford.

preventative measures or effective cures are developed, such individuals will live with the prospect of developing cancer in mid-life or even childhood. There are currently few methods for the prevention of cancer, and those that do exist are often unacceptable; for instance, carriers of the *BRCA1* gene are offered prophylactic mastectomy and oophorectomy even though it is still difficult to quantify the actual risk of disease (Bilimoria and Morrow, 1995). With the progress reported for the prediction of inherited genetic diseases such as cystic fibrosis (CF) and muscular dystrophy, it has been suggested that such prenatal diagnostic techniques could be offered to couples with genetic traits predisposing to cancer, thereby preventing transmission of the gene to any offspring (Gullick and Handyside, 1994). However, termination of a pregnancy after chorionic villus sampling (CVS) may prove unethical to many couples when the baby would be otherwise normal. A more acceptable alternative may be *in vitro* fertilization (IVF) coupled with pre-implantation diagnosis. This has been used successfully to prevent inherited X-linked recessive diseases (Handyside *et al.*, 1990) as well as CF (Handyside *et al.*, 1992), Lesch–Nyhan syndrome, Tay–Sachs disease, haemophilia A and Duchenne muscular dystrophy (DMD) (Delhanty *et al.*, 1994). However, a prerequisite for the success of such an approach is the identification and characterization of the genes. There are essentially three categories of genetic cancer predisposition:

(i)   those where mutation of an identified gene locus affecting the families is known, for example *APC* in familial adenomatous polyposis (*Table 15.1*);
(ii) those where a defined chromosomal linkage is known but the gene has yet to be identified, for example Beckwith–Wiedemann syndrome (*Table 15.1*);
(iii) those where there is a familial site-specific cancer phenotype without definite linkage to a chromosomal locus, for example prostate cancer.

Important factors in implementing interventional strategies are the sheer scale of the problem of screening large genes with diverse mutation sites, and the realization that any one gene may be just one of many that can contribute to susceptibility to a given tumour type (Friend, 1996).

The diagnosis of genetic mutations after IVF involves the removal of one or two cells from the embryo and the subsequent analysis of their DNA by polymerase chain reaction (PCR) amplification techniques. Such methods could be applied to mutations resulting in cancer predisposition where the mutation is well characterized, and efforts are now directed toward development of multiplex PCR systems to allow analysis of several exons or several genes simultaneously. However, where linkage data form the sole evidence for a mutation, such techniques would be impractical. Fluorescence *in situ* hybridization (FISH) with labelled Y probes has been used to sex embryos, allowing the selection of female embryos where X-linked recessive diseases have been diagnosed (Harper *et al.*, 1994). This technology could be offered to those women carrying the *BRCA1* mutation so that only male embryos are selected, although there is some evidence that males carrying the gene have a higher risk of prostate and other cancers (Gullick and Handyside, 1994).

Although it is technically possible that the regimes described could be applied to cancer screening, the question as to whether such schemes would ever be feasible depends on their cost effectiveness. If an individual inherits a genetic mutation, this does not automatically result in the development of a life-threatening cancer, because of interactions with modifier genes and environmental factors which influence the

Table 15.1. Familial inherited genes predisposing to cancer

	Gene	Location
Mapped locations		
Renal cell carcinoma	*RCC*	3pl4
Beckwith–Wiedemann syndrome	*BWS*	11p15
Neuroblastoma	*NB*	1p36
Identified genes		
Familial breast cancer	*BRCA2*	13q12
Familial breast/ovarian cancer	*BRCA1*	17q21
Male breast cancer	*AR*	X
Multiple endocrine neoplasia type 1	*MEN1*	11q13
Familial melanoma	*CDKN2/P16*	9q21
Tuberous sclerosis 1	*TSC1*	9q34
Tuberous sclerosis 2	*TSC2*	16p13
Retinoblastoma	*RB1*	13q14
Li–Fraumeni syndrome	*TP53*	17q13
Neurofibromatosis type 1	*NF1*	17q11
Neurofibromatosis type 2	*NF2*	22q12
Familial adenomatous polyposis	*APC*	5q21
Lynch family syndrome 2	*MSH2*	2p16
Lynch family syndrome 2	*MLH1*	3p21
Von Hippel–Lindau syndrome	*VHL*	3p25
Multiple endocrine neoplasia type 2A	*RET*	10q11
Multiple endocrine neoplasia type 2B	*RET*	10q11
Basal cell nevus syndrome	*BCNS*	9q31

phenotype. Current statistical methods that are used to determine the probability of cancer development do not, in general, take into account such modifier genes which may influence the course of action (Friend, 1996). However, many scientists feel that individuals have the right to know if they are carrying a genetic mutation associated with cancer, and only a dialogue between the biomedical and general communities will decide how this can be best directed (Goodman, 1996). Already a test for the *BRCA1* 185delAG mutation is available commercially to the public, and it is likely that there will be expansion to other candidate genes. However, although we have the technology to screen for genetic mutations, we have to ask whether we have enough knowledge of gene–gene and gene–environment interactions to give good advice about risk to individuals concerned.

15.4 Strategies for somatic genetic intervention

Somatic gene therapy involves the insertion of genes into the diploid cells of an individual where the genetic material is not passed onto the subject's progeny. The transfer is achieved by physical means or by using virus-mediated delivery, and is conventionally envisaged for application in patients with established cancer (Lemoine, 1994).

In essence, there are five major approaches to genetic intervention for cancer therapy (replacement or augmentation of gene expression, blockade of gene expression by antisense technology, genetic prodrug activation therapy, genetic immodulation and polynucleotide vaccination), each with distinct advantages and disadvantages for clinical application.

15.5 Replacement of tumour suppressor gene function

The majority of cancer gene therapy protocols have concentrated on the tumour suppressor gene, *TP53*. The protein product of this gene, p53, plays a pivotal role in arresting cell growth in response to DNA damage so that repair or apoptosis can occur. Abnormalities in the *TP53* gene have been reported in over 50% of human tumours (Levine *et al.*, 1994). Studies have demonstrated that *ex vivo* introduction of a wild-type *TP53* gene into tumour cells expressing mutant p53 can suppress malignant growth both *in vitro* and *in vivo* (Cai *et al.*, 1993; Chen *et al.*, 1990; Takahashi *et al.*, 1992). A study using replication-incompetent adenovirus to deliver a *TP53* minigene into p53 mutant tumours established in nude mice has been described by Wills *et al.* (1994). Tumour growth was suppressed in the p53-treated animals as compared with the controls, although the p53-treated mice still exhibited some tumour burden which could represent shut-off of the cytomegalovirus (CMV) promoter preventing expression of p53 or failure to transduce all tumour cells due to low transduction efficiencies. Interestingly, the mice injected with the parental adenovirus showed more tumour suppression than the buffer-treated controls, which may be the result of an immune response involving tumour necrosis factor and interleukins. A p53 construct under control of the CMV promoter (AdCMV.p53) was used to treat microscopic residual head and neck squamous carcinoma in nude mice; autopsy showed that the construct had inhibited development of tumours as compared with control animals (Clayman *et al.*, 1995). Yang *et al.* (1995) used a similar construct for *in vivo* studies with nude mice inoculated subcutaneously with prostate tumour cells and demonstrated that, after 3 weeks, tumour suppression was evident in the treated mice. A clinical trial to test the efficacy of p53 augmentation has recently been approved; patients with lung cancer positive for p53 mutations are to be treated with a replication-defective retrovirus expressing wild-type p53 delivered via intratracheal installation (Zhang and Roth, 1994).

A related tumour suppressor gene of interest is the p53-inducible *Waf-1* which encodes p21^{Waf}. This binds to and inhibits cyclin-dependent kinases (cdks) by forming a quaternary structure with cdks, cyclins and proliferating cellular nuclear antigen (PCNA) (Harper *et al.*, 1993). In the vast majority of tumours, the activity of cdks is uncontrolled due to loss of p21^{Waf} expression. El-Deiry *et al.* (1994) demonstrated that over-expression of p21^{Waf} in p53-deficient cells suppressed tumour growth. Interestingly, use of truncated versions of p21^{Waf} has demonstrated that only the N-terminal domain of this protein is required to act as a suppressor. Two truncated subunits of p21^{Waf} (from amino acids 1–80 or 1–89) showed an increased activity against tumour cell development compared with the wild-type p21^{Waf}. These studies look promising for the development of high-efficiency synthetic suppressor genes for replacement therapies.

MTSI/p16^{ink4} is another candidate suppressor gene that could be exploited in augmentation therapies. This gene encodes p16 which inhibits cdk4 in complex with cyclin D1. Deletions in exons 1 and 2 of *MTSI*/p16^{ink4} have been found in a high proportion of pancreatic cancers and oesophagus squamous cancer cell lines (Caldas *et al.*, 1994; Liu *et al.*, 1995). Studies by Serrano *et al.* (1995) have shown that ectopic expression of p16 blocks entry into S phase of the cell cycle (induced by *ras* or c-*myc* oncogenes) if the cells expressed functional retinoblastoma protein. Since this condition is satisfied in the majority of pancreatic cancers, a p16^{ink4} replacement strategy might be feasible (Lemoine, 1994).

The retinoblastoma (*RB1*) gene encodes a phosphoprotein that is constitutively expressed in normal cells. Mutations in this gene have been implicated in several cancers including retinoblastoma, prostate carcinoma, osteosarcoma, soft-tissue sarcoma, breast carcinoma and small lung carcinoma (Bookstein *et al.*, 1990; Shew *et al.*, 1989; T'Ang *et al.*, 1989). Transduction of cell lines derived from *RB1*-negative tumours with a retroviral vector expressing a functional retinoblastoma (Rb) protein suppresses tumorigenicity, suggesting that this could be an exploitable target gene in this context (Bookstein *et al.*, 1990; Huang *et al.*, 1988; Tanaka *et al.*, 1991).

Two further suppressor genes of interest are *APC* which is defective in the germ line of patients with predispositions for adenomatous polyposis as well as a large proportion of sporadic colorectal tumours (Smith *et al.*, 1993), and the *DPC4* suppressor gene (which is inactivated in pancreatic carcinoma, colorectal carcinoma and a small proportion of other tumours; Hahn *et al.*, 1996). There have been reports of a liposomal delivery system used to deliver the human *APC* gene under control of a constitutive promoter to rodent colonic epithelium cells by rectal infusion (Westbrook *et al.*, 1994). The gene was transiently expressed for up to 3 weeks at one-tenth the level of the endogenous *APC* gene with no side-effects. However, expression of the gene was only transient, and repeated doses would be required to suppress the tumour burden.

Although current preclinical models suggest that expression of a single tumour suppressor gene can produce dramatic anti-tumour effects, there are a number of problems associated with gene augmentation/replacement as a clinical strategy. By the time a tumour has been diagnosed, the cells have accumulated a number of genetic mutations contributing to complete transformation, and there may be separate subclones which have evolved by different pathways with the resulting problem of heterogeneity of molecular profile (Goyette *et al.*, 1992; Vogelstein *et al.*, 1988, 1989). It is also possible that individual tumour suppressor genes may be important only at particular points in tumour evolution, so that there may be only a limited window of opportunity for therapeutic intervention.

15.6 Blockade of dominant gene expression by antisense technology

Naturally occurring antisense interactions are known to modulate gene expression in prokaryotes, and similar mechanisms are thought to occur also in eukaryotes (Murray and Crockett, 1992; Nellen and Lichtenstein, 1993; Thomas, 1992). As a consequence, there has been an explosion in the development of antisense technology to target specifically cellular transcription and translation. Initial attempts to downregulate genes *in vitro* using antisense oligonucleotides appeared successful, and there has been increasing interest in their potential therapeutic application *in vivo* (Barton and Lemoine, 1995; Carter and Lemoine, 1993).

There are four strategies for the use of antisense agents which are all based on complementary base pairing. The most widely documented involves the exogenous introduction of short complementary single-stranded oligonucleotides (normally DNA ~15–20 nucleotides in length) which are thought to act by binding specifically to their corresponding cellular mRNA partner blocking translation (Carter and Lemoine, 1993; Murray and Crockett, 1992; Nellen and Lichtenstein, 1993; Prins *et al.*, 1993; Toulmé, 1992). These antisense oligonucleotides may target several different stages of the translation pathway. Cleavage by RNase H specifically degrades the RNA subunit in the DNA–RNA hybrid. Antisense oligonucleotides which target the 5′-untranslated region may hinder sterically the binding of the 40S ribosomal subunit. Similarly, those binding to or close to the initiation codon (AUG) could prevent assembly of the translation initiation complex. In some instances, it has been suggested that antisense oligonucleotides which bind to the nuclear localized precursor mRNA (hnRNA) may provide a better target, as the size of the hrRNA pool is smaller and so less antisense agent would be required to downregulate gene expression. It has also been suggested that targeting hrRNA intron–exon junctions may afford a higher degree of specificity (Carter and Lemoine, 1993).

The second approach involves specific binding of oligonucleotides to gene targets prior to transcription by forming a triple helix. This can be manipulated to produce irreversible binding at a genetic locus or even selective cleavage, resulting in loss of gene expression and cell death. The third approach targets the transcription machinery. Certain proteins including transcription factors, DNA polymerases and RNA polymerases have the ability to recognize nucleic acid motifs. These factors, which are essential for transcriptional initiation, can be sequestered away from the genetic site by providing the cell with double-stranded oligodeoxynucleotides which act as 'traps' (Carter and Lemoine, 1993). Harel-Bellan *et al.* (1989) have reported the use of this approach to modulate the expression of the human T-cell heat shock protein by introducing a 14-bp synthetic deoxyoligonucleotide (mapped in the enhancer region of the gene) into cells. The fourth approach uses ribozymes, small oligoribonucleotides with a specific base sequence resulting in a self-splicing activity (*Figure 15.1*). Although their mode of action is significantly different from the other three antisense technologies described above, their targets and the outcome of their activity is similar. Ribozyme activity can be directed against various RNAs by the incorporation of an antisense region into the ribozyme, although at present this requires that the target has the consensus sequence GUC, excluding some genes as targets. Cai *et al.* (1995) reported the development of a ribozyme strategy for the suppression of lung cancer cell growth. A retroviral construct was used to deliver an anti-p53 ribozyme able to cleave p53 pre-mRNA at codon 187 close to the intron 5–exon 6 boundary. When

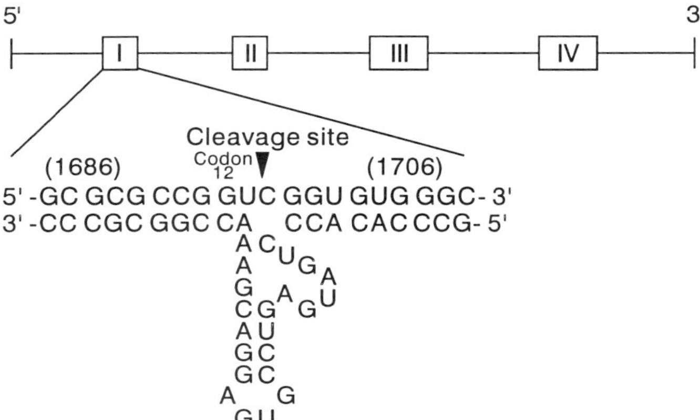

Figure 15.1. The structure of a ribozyme targeting H-*ras* mutated at codon 12. The complementary H-*ras* RNA (1688–1714) is shown with the GUC cleavage site found in a mutated but not normal H-*ras* allele. Reproduced from Carter and Lemoine (1993) Antisense technology for cancer therapy: does it make sense? *Br. J. Cancer* **67**: 869–876, with permission from Macmillan Press Ltd.

transduced into human lung carcinoma cells, the level of mutant p53 was markedly reduced and a control ribozyme with a mutation preventing cleavage of p53 pre-mRNA confirmed the ribozyme's specific activity *in vitro*.

Szczylik *et al.* (1991) reported one of the first clinical applications for antisense oligonucleotides in the treatment of chronic myeloid leukaemia with the Philadelphia chromosome (Ph) translocation t(9;22) giving rise to a fusion gene *BCR–ABL*, the target for the antisense oligonucleotide in this trial. When leukaemic blast cells derived from patients with Ph-positive leukaemia were exposed to either of two complementary deoxyoligonucleotides (18-mers) targeted to the *BCR–ABL* junction, colony formation was suppressed whilst control granulocyte–macrophages were unaffected. When the two populations of cells were mixed and treated with the antisense oligonucleotides, the majority of surviving cells were of granulocyte–macrophage phenotype. *In vivo* studies by Skorski *et al.* (1994) demonstrated that mice with severe combined immunodeficiency (SCID) injected with the chronic myeloid leukaemia-blast crisis cell line BV173 developed a disease similar to leukaemia patients. When these mice were treated with a 26-mer BCR–ABL antisense oligonucleotide (1 mg day^{-1} over 9 days), the clonogenic leukaemic cells disappeared, there was a marked decrease in the *BCR–ABL* cellular mRNA and the treated animals survived over 10 weeks longer than controls. Similar studies using antisense oligonucleotides targeted to different points of the *BCR–ABL* junction have also reported successful inhibition of clonogenic leukaemic cells *in vitro* (Mahon *et al.*, 1995a,b).

A further study has used the *ERBB2* antisense oligonucleotides (varying in length between 15 and 20 nucleotides) to inhibit cell proliferation in breast cancer (Colomer *et al.*, 1994). An *ERBB2* antisense oligonucleotide (used at 20 mM) inhibited the proliferation of breast cancer cell lines (*ERBB2*-positive) by 60% relative to controls, and ERBB2 protein expression decreased in a dose-dependent manner. When the antisense *ERBB2* oligonucleotide was screened in breast cancer cells with an *ERBB2*-negative expression status, it was shown to have no effect on cell proliferation. Although the study was successful in its selective targeting potential, parallel studies highlighted some of the problems associated with stability of antisense deoxyoligonucleotides that might be encountered *in vivo*.

Akhatar *et al.* (1991) studied the stability of structurally unmodified oligodeoxynucleotides (D-oligo) compared with analogues containing phosphorothioate (S-oligo), methylphosphonate (MP-oligo) and alternating methylphosphonate and phosphodiester internucleoside linkages (Alt-MP-oligo) in HeLa cell extracts and sera (*Figure 15.2*). They concluded that in any medium screened, D-oligos were the least stable as compared with the others and that the MP- and Alt-MP-oligos were more resistant to phosphatase activity as compared with S- and D-oligos. The D-oligos are readily taken up by cells, are non-toxic and form excellent hybrids that induce RNAse H activity, but they have very low stability; the MP-oligos have good stability, low toxicity and are readily taken up by cells but exhibit poor hybridization characteristics. The S-oligos appear to have a combination of the best characteristics of both D- and MP-oligos (Carter and Lemoine, 1993). However, they have been shown to exhibit some non-specific effects that will be discussed later.

Some investigators have tried to overcome the instability problems using recombinant plasmids. For instance, there have been reports of successful inhibition of *KRAS* expression in lung cancer cells using a recombinant plasmid containing a 2 kb genomic fragment of this oncogene in antisense orientation. The construct was trans-

(a)   (b)   (c)

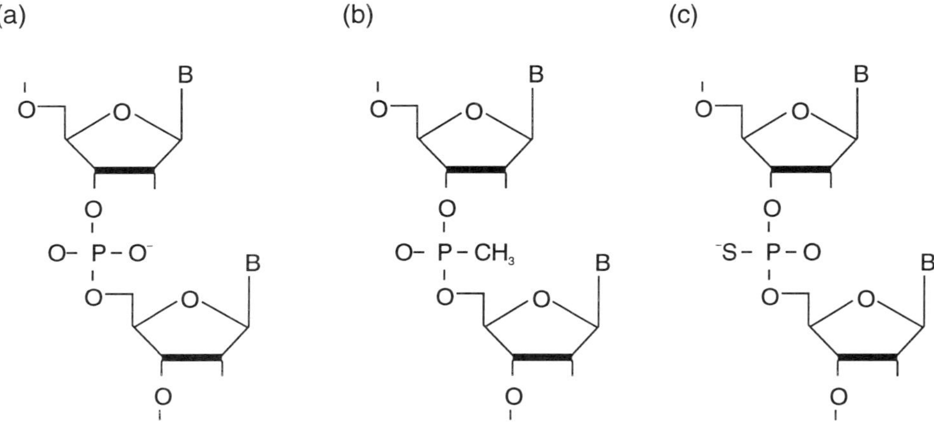

Figure 15.2. Oligonucleotide structure showing common analogues achieved by substitution at the internucleotide phosphate. B = purine or pyrimidine base. (a) 'Natural' phosphodiester linkage; (b) methylphosphonate linkage; (c) phosphorothioate linkage. Reproduced from Carter and Lemoine (1993) Antisense technology for cancer: does it make sense? *Br. J. Cancer* **67**: 869–876, with permission from Macmillan Press Ltd.

fected into a number of cell lines and shown to inhibit translation of KRAS but not HRAS nor NRAS. When cells expressing the mutant gene were treated with the antisense construct there was a threefold downregulation of the gene and a partial reduction in cell viability. *In vivo* studies of xenografts in nude mice treated with the construct demonstrated substantial reductions in both *KRAS* expression and tumour growth (Mukhopadhyay *et al.*, 1991). A subsequent study using the same construct to test *in vivo* treatment of an orthotopic human lung cancer model in nude mice gave similar results. In this series of experiments, the construct was delivered to cells via a replication-deficient retrovirus by intratracheal instillation. At postmortem, only 13% of the animals treated with the antisense construct had residual tumours as compared with 90% of the control animals. Further studies showed that the tumour suppression was dose-dependent (Georges *et al.*, 1993).

Antisense reduction of *BCL2* gene expression has been reported to reverse chemoresistance in B-cell lymphoma lines. *BCL2* is transcriptionally deregulated in non-Hodgkin's lymphoma due to a translocation which juxtaposes it to an immunoglobin heavy chain locus forming a fusion gene, and its overexpression results in failure of apoptosis. It has been demonstrated that human lymphoma cell lines treated with either an antisense *BCL2* oligonucleotide or an inducible expression plasmid containing the sequence in an antisense orientation had reduced *BCL2* expression levels and were sensitized to cytosine arabinoside (araC) and methotrexate (MTX). This approach may provide a novel mechanism for improving chemotherapeutic treatment of cancer (Kitada *et al.*, 1994).

Antisense technology in its present state has many deficiencies which preclude its general clinical application. One example reported by Barton and Lemoine (1995) demonstrated that antisense oligonucleotides directed against p53 (including some in clinical trial; Bishop *et al.*, 1996) did show an anti-proliferative effect, but this was unrelated to effects on p53. Phosphorothioate oligonucleotides have been used extensively in antisense technologies but major problems have been reported with their use. For instance, it has been shown that although these oligonucleotides activate RNAse H activity when bound to mRNA, at high concentration they will bind to and inhibit the catalytic action of the enzyme (Gao *et al.*, 1992). Some of the inhibitory effects noted may be the result of binding to molecules other than RNA. Phosphorothioate oligonucleotides are polyanions and can interact with other polyanions such as heparin. The non-sequence-specific binding with other proteins makes it difficult to determine exactly what is causing an inhibitory effect. For instance, phosphorothioate oligonucleotides have been shown to compete with human immunodeficiency virus (HIV) for a soluble CD4 factor, not in a sequence-specific manner but instead in a length-related manner (Yakubov *et al.*, 1993). Another example of non-sequence-dependent inhibition has been noted with basic fibroblast growth factor (bFGF) in a manner resembling suramin and pentosan polysulphate, two polyanions currently in cancer trials (Guvakova *et al.*, 1995). Sequence-specific antisense blockade of c-*myb* in a rat carotid model of restenosis has been reported by Simons *et al.* (1992), but subsequent studies by Burgess *et al.* (1995) has suggested that the anti-proliferative effect of antisense oligonucleotides to c-*myb* and c-*myc* in smooth muscle cells is actually a non-antisense mechanism, and Stein (1995) has suggested that it may have been the result of a non-specific interaction of the phosphothioate oligonucleotide with bFGF as described by Guvakova *et al.* (1995). The motif CpG in a nucleic acid sequence is thought to be a potential immunomodulator, as demonstrated by Krieg *et al.* (1995). Mice injected intraperitoneally with phosphorothioates show a dramatic increase in immunoglobulin secretion accompanied by increased expression of major histocompatibility complex (MHC) class II markers, and oligonucleotides containing the CpG motif can induce interferons and augment natural killer cells (Tamamoto *et al.*, 1994; Yamamoto *et al.*, 1992). Such responses bring into question data that have been reported from murine tumour models. It would seem that greater care in the design and screening of antisense oligonucleotides is required before conclusions can be drawn about their sequence-specific blockade. Stein and Krieg (1994) recently reviewed some of the anomalies associated with the use of antisense technology, the controls that should be incorporated and suggestions on how to interpret data. It is hoped that this will aid in the standardization of antisense technology. Even though such problems exist, several phase I clinical trials have been approved for the use of phosphorothioate antisense oligonucleotides (Bayever *et al.*, 1993; Spinolo *et al.*, 1992). Ironically, although many of the antisense systems described may be inhibitory as a result of non-sequence-specific interactions, they may in the future prove to be as beneficial in the treatment of cancer as those demonstrated to be sequence-specific.

15.7 Genetic prodrug activation therapy

Genetic prodrug activation therapy (GPAT) utilizes the transcriptional differences between normal and neoplastic cells to drive the selective expression of a metabolic suicide gene to confer sensitivity to a prodrug (*Figure 15.3*).

There are a number of genes encoding enzymes for specific prodrug activation (*Table 15.2*), the major target being cellular nucleic acid pathways. Three enzymes have become the prototypes for the GPAT system. Herpes simplex virus thymidine kinase (HSV-TK) converts ganciclovir (guanosine analogue) to its monophosphate form which is then further metabolized by cellular kinases to di- and triphosphates to inhibit DNA polymerase α (Reid *et al.*, 1988) and cause chain termination (Mar *et al.*,

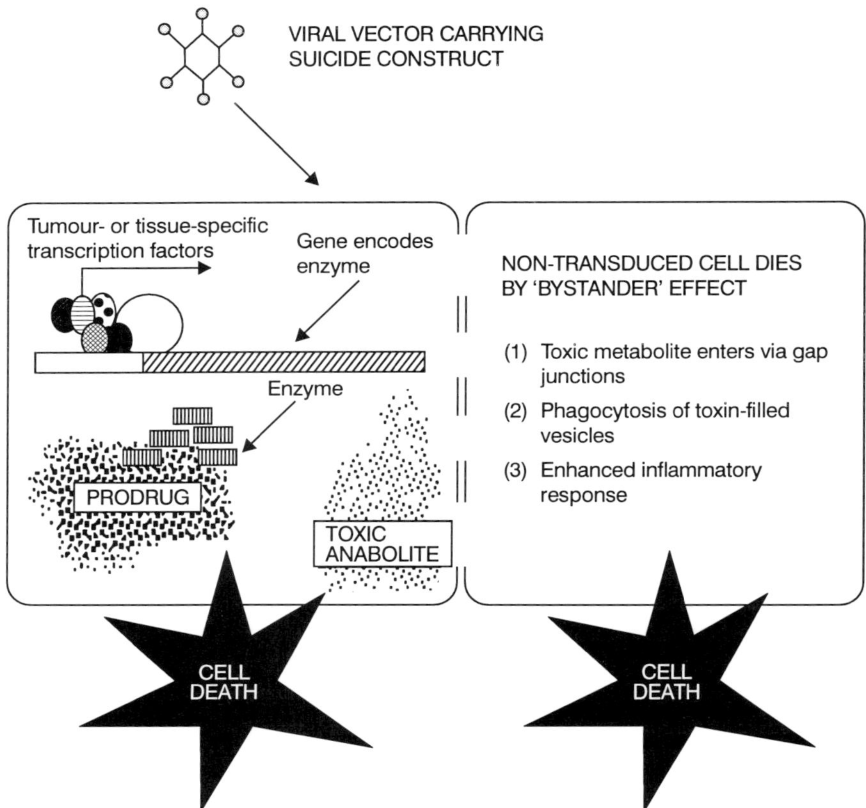

Figure 15.3. Schematic diagram showing genetic prodrug activation therapy and accompanying bystander effect in neoplastic cells.

1985). Varicella zoster virus thymidine kinase (VZV-TK) converts 9-(β-D-arabino-furanosyl)-6-methoxy-9H-purine (araM) to its monophosphorylated form which is further converted by cellular enzymes to adenine arabinonucleoside triphosphate (araATP) (Huber *et al.*, 1991). Finally, cytosine deaminase (Danielsen *et al.*, 1992) deaminates 5-fluorocytosine (5-FC) to 5-fluorouracil (5-FU) which is then converted to 5-fluorouridine 5′-triphosphate and 5-fluoro-2′deoxyuridine 5′monophosphate, obstructing DNA and RNA synthesis.

A major requirement for any suicide gene strategy is to target specifically the toxin gene expression to malignant tissue. At present, there are two approaches to this problem. Transduction targeting, which relies on the preferential delivery of genes to actively dividing cells usually by means of a viral vector, is suited to types of cancer arising in quiescent tissues like the brain and nervous system rather than those in organs like the gastrointestinal tract where there is a continual renewal of cells. Transcriptional targeting relies on unique tissue-specific or tumour-specific transcriptional elements to drive the expression of the toxic protein only in those cells that contain transcription factors capable of activating the promoter elements. As a consequence, for such a system to be effective, the regulatory elements of the promoter/enhancer need to be fully characterized (Garver *et al.*, 1994). These two methods can be combined to provided an

Table 15.2. Some of the enzyme–prodrug systems studied

Enzyme	Prodrug	Drug
Viral thymidine kinase	Ganciclovir 6-Methoxypurine arabino- nucleoside (araM)	Ganciclovir triphosphate Adenine arabinonucleoside triphosphate (araATP)
Cytosine deaminase	5-Fluorocytosine	5-Fluorouracil
Cytochrome P-450	Cyclophosphamide	Phosphoamide mustard
Carboxypeptidase A	Methotrexate-alanine	Methotrexate
Carboxypeptidase G2	Benzoic acid mustard glucuronide	Benzoic acid mustard
Nitroreductase	CB 1954 (5-aziridine 2,4 dinitrobenzamidine)	5-Aziridine 2,4-hydroxyamino 2- nitrobenzamidine-4-hydroxyl
β-Glucuronidase	Adriamycin glucuronide Daunomycin glucuronide Epirubicin glucuronide	Adriamycin Daunomycin Eprirubicin
Alkaline phosphatase	Doxorubicin phosphate Mitomycin phosphate Etoposide phosphate	Doxorubicin Mitomycin Etopside
Linamarase	Amygdalin	Cyanide
β-Glucosidase	Amygdalin	Cyanide
Xanthine oxidase	Xanthine	Oxygen radicals
β-Lactamase	Cephalosporin mustard carbamate	Nitrogen mustard

improved targeting system, but each system has its own pitfalls which are described in more detail in the following sections.

15.7.1 Transduction targeting

The first report of a GPAT system was published by Moolten (1986) who described the transduction of a mouse cell line with an *HSV-tk* construct which became sensitive to ganciclovir *in vitro* and *in vivo*. *In vivo* studies on tumours in syngeneic animals resulted in total eradication of neoplastic cells. However, further development of the technique was hampered by poor delivery systems, and later studies have concentrated on their improvement.

A specialized delivery system has been described by Culver *et al.* (1992) for the treatment of intracerebal gliomas. Retrovirus producer cell lines were used to deliver replication-defective retroviruses carrying the *HSV-tk* gene (under the control of a constitutive promoter) into rat brains containing small intracerebal gliomas. Animals were then treated 7 days later with ganciclovir over a 5-day period. Postmortem examination showed that 75% of the tumours had regressed in the treated animals as compared with none in untreated controls. The study also indicated the presence of an intriguing 'bystander effect'. Analysis of the tumour cells 4 weeks post-injection showed that even with levels of transduction as low as 10%, significant tumour regression occurred. The bystander effect is of fundamental importance for the success of GPAT systems and will be discussed in more detail in Section 15.7.3.

One of the advantages of this protocol is the use of vector-secreting cell lines which increase the chance of cell transduction by maintaining a relatively high titre of virus around the tumour prior to treatment with the prodrug. At present, several clinical trials have been approved to evaluate this system (Klatzmann *et al.*, 1996a,b; Oldfield *et al.*, 1993). Oldfield *et al.* (1993) have an ongoing study which uses stereotactic apparatus fixed to a patient's skull to allow several deposits of vector-producing cells (secreting the replication-defective retrovirus carrying the *HSV-tk* gene) to be made into tumour sites. Patients are then treated a week later with a twice daily dose (5 mg kg^{-1}) of ganciclovir over a 2-week period. Current reports suggest that while tumour regression is initially evident in several patients, recurrences are usual (Mullen, 1994). The less impressive results in the clinical trial as compared with the preclinical model may be because human tumours generally have a lower growth index compared with animal models. As a consequence, infectivity with the retroviral vector may not be as efficient. Also, the ratio of vector-producing cells to tumour cells was between 10 and 100:1 in the initial animal study, and this could not be achieved in the human investigation.

A similar trial is under way for the treatment of metastatic malignant melanoma (Klatzmann *et al.*, 1996a). In this system, patients have been divided into groups receiving either 1×10^8, 2×10^8 or 3×10^8 cells secreting a replication-defective retrovirus construct expressing the *HSV-tk* gene. The cells are injected either intracutaneously, subcutaneously or intralymphatically. Patients are treated after 7 days with ganciclovir at 10 mg kg^{-1} each day for 14 days. There are potential problems with this system, similar to those outlined above. There may be side-effects to the injection of producer cells, resulting in the undesirable expression of transgenes in normal tissues, even though there will be a dilution effect for any 'escaped' cells.

Several studies have investigated the use of adenovirus vectors for delivery of metabolic suicide genes. A model for the treatment of human malignant mesothelioma and lung cancer has been reported (Hwang *et al.*, 1995; Smythe *et al.*, 1994). In this system, the *HSV-tk* gene was placed under control of the Rous sarcoma virus (RSV) long terminal repeat (LTR) and delivered by a replication-defective adenovirus (Ad.RSV*tk*). *In vitro* studies showed that REN mesothelioma cells were effectively transduced by the adenovirus vector and that infected cells were more sensitive to ganciclovir (100- to 1000-fold) than uninfected parent cells (Smythe *et al.*, 1994). Similar results were shown in a SCID mouse model of mesothelioma. Of the control animals, 19 out of 20 had large amounts of tumour in the peritoneal cavity compared with only one of the 10 animals treated with the Ad.RSV*tk* construct and ganciclovir. A very strong bystander effect was also noted, such that only 10% of the cells needed to be transduced *in vivo*. The protocol was subsequently shown to be effective *in vitro* and *in vivo* for the treatment of a lung cell line (Hwang *et al.*, 1995).

Clinical trials using a replication-defective adenovirus system to deliver the *HSV-tk* gene under the control of the RSV LTR have been approved for the treatment of malignant glioma (Albelda, 1995). The virus is to be introduced using stereotactic apparatus followed by treatment with ganciclovir. Samples of tissue are to be removed for analysis of *HSV-tk* expression and signs of inflammation (Eck and Alavi, 1995). The same construct has also been approved for treatment of patients with mesothelioma. Patients are to receive varying doses of the adenovirus (between 10^9 and 10^{12} viral particles) followed by ganciclovir treatment.

A further system has been reported for the treatment of colonic carcinomas (Hirschowitz *et al.*, 1995). The suicide gene, in this instance the gene for cytosine deaminase, was placed under the control of the CMV constitutive promoter and delivered to HT29 colon carcinoma cells by a replication-defective adenovirus. Transfer of the construct was measured *in vitro* by the ability of cell lysates to convert [³H]5-FC to 5-FU. The growth of the HT29 cells was suppressed in the presence of 5-FC in a dose-dependent manner. *In vivo* studies in nude mice demonstrated that injection of the adenovirus construct into subcutaneous HT29 tumours rendered then sensitive to 5-FC and that there was a fourfold reduction in tumour size as compared with the untreated control animals after 15 days.

Although transduction targeting is proving surprisingly successful, there are still deficiencies in the technology. For example, each of the systems described so far relies on constitutive promoters for expression of the suicide gene. This is not so much of a problem for the treatment of brain tumours (for instance where the tumour cells are rapidly dividing in a relatively static background), but the treatment of gastrointestinal and lung carcinomas could result in normal stem cells also being transduced. Even direct injection of viruses to the tumour site does not eliminate this problem. Efforts are being made to improve specificity by means of transcriptional targeting (discussed in Section 15.7.2) and by modifying the natural tropism of retroviruses and adenoviruses (see Chapters 3 and 4).

15.7.2 Transcriptional targeting

Transcriptional targeting exploits tumour-specific promoter elements to drive the expression of a toxic protein only in those cells that contain transcription factors able to activate the promoter (Miller and Vile, 1995). Hence, for such a system to work, the regulatory transcriptional elements need to be fully characterized.

The first reported system harnessed one of two promoters, either the hepatoma-associated α-fetoprotein (AFP) or normal liver-associated albumin to drive a VZV-TK suicide gene. *In vitro* it was possible to engineer expression restricted to the malignant hepatoma cells using the AFP promoter and expression restricted to normal liver cells using the albumin promoter (Huber *et al.*, 1991). Subsequent studies to test the concept in transgenic mice were disappointing as recombinant retrovirus constructs failed to express the suicide gene; this was attributed to silencing of the retroviral sequences (Richards and Huber, 1993).

Melanomas have provided the target for another tissue-specific transcriptional suicide gene strategy (Vile and Hart, 1993a,b). In this system, the *HSV-tk* gene was placed under control of the tyrosinase promoter. Melanin (which is produced exclusively by melanocytes and their derivatives) is synthesized by the hydroxylation of tyrosine by tyrosinase which is upregulated in melanomas (Kluppel *et al.*, 1991). Transfected murine melanoma cells expressing *HSV-tk* under the control of the tyrosinase promoter were injected into immunocompetent syngeneic mice, and tumour regression was noted in 29 of the 30 animals subsequently treated with ganciclovir. Although the system looks promising, there may be problems with expression of tyrosinase in other cell types, for example spinal ganglia, astrocytes and Schwann cells, which may result in normal cells being sensitized to ganciclovir. Once again this emphasizes the need to be able to target delivery as well as expression.

Harris *et al.* (1994) have exploited the promoter elements of the proto-oncogene *ERBB2*. This has been found to be over-expressed in a number of cancers (Gullick, 1991) including pancreatic and breast carcinomas. The transcriptional subunits implicated in its upregulation in breast cancer are well documented (Hollywood and

Hurst, 1993). One site in particular, situated in the proximal promoter, has been shown to bind an AP2-like moiety designated OB2-1 which is directly linked to *ERBB2* overexpression in breast cancer (Bosher *et al.*, 1995). This promoter segment was used in a retroviral construct to drive the gene for cytosine deaminase. A panel of cells with varying *ERBB2* status were transduced and treated with 5-FC. The proportion of cell death was directly related to the *ERBB2* expression level of the cells. Because of the success of this system, a clinical trial to test its efficacy has been approved in which women with cutaneous intralesional breast tumours that have an *ERBB2*-positive status are being injected with up to 400 μg of the suicide gene construct followed by a 48-h infusion of 5-FC.

In recent years, there have been several reports of the development of chimeric promoters in an attempt to improve transcriptional targeting. One study has exploited the use of the *myc–max* oncogene which has been found to be upregulated in a number of cancers. The protein heterodimer is able to bind the motif CACGTG which switches on transcriptional activity. A construct was prepared which contained four repeats of this motif upstream of a minimal promoter driving a luciferase reporter gene (Fujita *et al.*, 1994). myc-positive cells transfected with the construct showed a four- to sixfold increase in expression of the luciferase gene compared with transfected myc-negative cells. Established myc-positive tumours in nude mice were injected with a suicide construct, encoding the *Pseudomonas* enterotoxin. There was a 50% decrease in tumour burden in the toxin-treated animals as compared with the controls (Fujita *et al.*, 1994). Subsequent studies replacing the gene for the toxin with the *HSV-tk* gene have proved similarly successful (Kumagai *et al.*, 1996).

One of the first examples describing the rearrangement of transcriptional elements to produce a synthetic tissue-specific promoter was described by Huber and co-workers utilizing elements of the gene for carcinoembryonic antigen (CEA) (Austin and Huber, 1993; Richards *et al.*, 1993, 1995). CEA is a tumour-associated cell surface antigen that is expressed in normal tissue but upregulated in tumour tissue such as lung and colorectal cancer (Schrewe *et al.*, 1990). Artificial mini-genes composed of various fragments from the 5′-flanking region of the *CEA* gene driving expression of a luciferase reporter gene were screened in cell lines to determine which sequences were essential for specific expression of CEA. The first generation constructs (containing 10 800 bp to 299 bp of the 5′ sequence) showed only a 5- to 10-fold increase in expression in CEA-positive cells compared with CEA-negative cells, but further analysis of multimers of positively acting elements produced a 10- to 30-fold increase in expression. By using these multimers, the expression level of the reporter construct was shown to be tissue-specific and two to four times stronger than the constitutive simian virus 40 promoter. Su *et al.* (1996) have reported the development of a chimeric promoter composed of the albumin promoter element fused to the AFP enhancer to drive the *HSV-tk* gene. Cell lines positive or negative for albumin and/or AFP were transduced with an adeno-associated virus carrying the chimeric promoter. The albumin- and AFP-positive cell lines were shown to be sensitive to ganciclovir, whereas the negative cell lines were resistant.

It would appear that development of such chimeric promoters is a promising approach for transcriptional targeting. However, there have been reports indicating that delivery of these minigenes in viral delivery systems and subsequent expression are not always efficient. This may be attributed variously to partial loss of tissue specificity from the internal promoter, promoter interference and reduced virus titres due to the large inserts, often arranged in the opposite orientation to the viral LTR (Richards *et al.*, 1993; Vile *et al.*, 1994a; Vile *et al.*, 1995). To overcome these problems, Vile *et al.* (1995) have found that by replacing the Moloney murine leukaemia virus enhancer (in the 3' LTR) with the murine tyrosinase promoter/enhancer (directing expression of interleukin-2) tissue specificity can be maintained. With further developments in the design of 'tailor-made' transcriptional elements and modifications of viral vector systems, GPAT may provide a feasible treatment for cancer.

15.7.3 'Bystander effects'

One of the phenomena that has been noted by several groups studying the GPAT system is that treatment of mixed populations of transduced and non-transduced cells with prodrug can result in unexpectedly high levels of cell death. This has been termed 'bystander' killing, a feature which is extremely advantageous for GPAT systems where the transduction efficiency is poor. Studies by Culver *et al.* (1992) demonstrated that only 10–70% of glioma cells were transduced with the construct, and yet almost complete tumour destruction was observed. Similar studies showed that in xenografts composed of a mixed population where only 10% of cells carried the suicide gene, over half the tumours were eradicated after prodrug treatment (Moolten *et al.*, 1992). Bi *et al.* (1993) have investigated the mechanism of bystander killing. They co-cultured two populations of cells at different densities. One cell line expressed the gene for β-galactosidase and the other an *HSV-tk* suicide gene. When cells at low densities were treated with ganciclovir only the *HSV-tk*-transduced cells died, but when confluent, 100% cell death of both populations was produced. It was postulated that the bystander effect was due to metabolic co-operation between cells, the result of trafficking of small molecules (mol. wt <1000 Da) between cells via gap junctions. Other investigators have suggested that it could also result from apoptotic vesicles containing the *HSV-tk* gene derived from dying cells being phagocytosed (Freeman *et al.*, 1993), and more recent work has suggested that a T-cell-mediated response could contribute. Vile *et al.* (1994b) demonstrated that lung tumours transduced with an *HSV-tk* gene in immunocompetent mice which were then treated with ganciclovir showed a 90% reduction in tumour size as compared with the saline-treated controls. However, the same study repeated in nude mice (T-cell-deficient) resulted in no change in tumour burden between the ganciclovir-treated and the saline-treated control animals. Findings by Mullen *et al.* (1994) are in accord with these results. In their *in vivo* studies, tumours derived from different cell lines transduced with a gene for cytosine deaminase and treated with 5-FC showed growth suppression as compared with control animals. When the tumours were excised and the animals challenged with wild-type tumour, they exhibited a type-specific resistance which the investigators linked to a strong T-cell response to dying cells or cytosine deaminase acting as a super-antigen. Freeman *et al.* (1995) have sought to exploit the bystander effect in a clinical trial to treat ovarian cancer. Women with stage III ovarian cancer are to have irradiated ovarian cancer cells transduced with an *HSV-tk* suicide construct delivered intraperitoneally followed by treatment with ganciclovir.

Although the initial results for the GPAT system look promising, there are still some technical difficulties associated with its use. Effective delivery systems are still required as, even under optimized conditions *in vitro*, only 10–70% of cells are transduced and *in vivo* the level of transduction can be much lower. Also, even efficient transduction does not guarantee good gene expression, because of problems such as gene silencing (Richards and Huber, 1993). There is also a need to develop stronger promoters to drive tissue- or tumour-specific expression. Attempts to improve this situation are already under way in the design of chimeric promoters. Improvements in targeted retrovirus and adenovirus systems are being made by modifying their tropism, but these crippled hybrids often have low infectivity which, in turn, results in low transduction efficiencies.

From the data presented so far, it appears that the GPAT system is the most effective of those reviewed for the treatment of cancer. Antisense technologies have many problems not only in the interpretation of data but also in experimental design, and until such anomalies are fully rectified, it is hard to see their rational use for cancer gene therapy. Similarly, at present, the multiple genetic disorders that occur during transformation of a cell make it difficult to treat by augmentation or knockout of single genes, and such treatment may need to be applied during the pre-neoplastic phase.

15.8 The future for cancer gene therapy

A plethora of information is currently being gathered which is enhancing our understanding of the human genome and its transcriptional control. The further recognition of positive acting factors leading to cell growth and negative factors leading to suppression and apoptosis will lead to more rational development of therapeutic agents for cancer treatment. This, together with improvements in viral and synthetic delivery systems, development of chimeric promoters for targeted transcription and the prospect of producing designer minigenes, makes the future for cancer gene therapy an exciting one.

References

Akhatar S, Kole R, Juliano RL. (1991) Stability of antisense DNA oligonucleotide analogs in cellular extracts and sera. *Life Sci.* **49**: 1793–1801.

Albelda SM. (1995) Treatment of advanced mesothelioma with recombinant adenovirus H5.020RSVTK: a Phase I trial. *Hum. Gene Ther.* **6**: 509–514.

Austin EA, Huber BH. (1993) A first step in the development of gene therapy for colorectal carcinoma: cloning, sequencing, and expression of *Escherichia coli* cytosine deaminase. *Mol. Pharmacol.* **43**: 380–387.

Barton CM, Lemoine NR. (1995) Antisense oligonucleotides directed against p53 have antiproliferative effects unrelated to effects on p53 expression. *Br. J. Cancer* **71**: 429–437.

Bayever E, Iversen PL, Bishop MR, Sharp JG, Tewary HK, Arneson MA, Pirruccello SJ, Ruddon RW, Smith LJ, Zon G, Kessinger A. (1993) Systemic administration of a phosphorothioate oligonucleotide complementary to p53 in acute myelogenous leukaemia and myelodysplastic syndrome: phase I clinical trial. *Cancer Gene Ther.* **1** (Suppl.): 11A.

Bi WL, Parysek LM, Warnick R, Stambrook PJ. (1993) *In vitro* evidence that metabolic cooperation is responsible for the bystander effect observed with HSVtk retroviral gene therapy. *Hum. Gene Ther.* **4**: 725–732.

Bilimoria MM, Morrow M. (1995) The woman at increased risk for breast cancer: evaluation and management strategies. *Cancer J. Clin.* **45**: 263–278.

Bishop MR, Iversen PL, Bayever E, Sharp JG, Greiner TC, Copple BL, Ruddon R, Zon G, Spinolo J, Arneson M, Armitage JO, Kessinger A. (1996) Phase I trial of an antisense oligonucleotide OL(1)p53 in hematologic malignancies. *J. Clin. Oncol.* **14**: 1320–1326.

Bookstein R, Shew JY, Chen PL, Scully P, Lee WH. (1990) Suppression of tumorigenicity of human prostate carcinoma cells by replacing a mutated RB gene. *Science* **247**: 712–715.

Bosher JM, Williams T, Hurst HC. (1995) The developmentally regulated transcription factor AP-2 is involved in c-erbB-2 overexpression in human mammary carcinoma. *Proc. Natl Acad. Sci. USA* **92**: 744–747.

Burgess T, Fisher EF, Ross SL, Bready JV, Qian YX, Bayewitch LA, Cohen AM, Herrera CJ, Hu SSF, Kramer TB, Lott FD, Martin FH, Pierce GF, Simonet L, Farrell CL. (1995) The antiproliferative activity of c-myb and c-myc antisense oligonucleotides in smooth muscle cells caused by a nonantisense mechanism. *Proc. Natl Acad. Sci. USA* **92**: 4051–4055.

Cai DW, Mukhopadhyay T, Lui T, Fujiwara T, Roth JA. (1993) Stable expression of the wild-type p53 gene in human lung cancer cells after retrovirus-mediated gene transfer. *Hum. Gene Ther.* **4**: 617–624.

Cai DW, Mukhopadhyay T, Roth JA. (1995) Suppression of lung cancer cell growth by ribozyme-mediated modification of p53 pre-mRNA. *Cancer Gene Ther.* **2**: 199–205.

Caldas C, Hahn SA, da Costa LT, Redston MS, Schutte M, Seymour AB, Weinstein CL, Hruban RH, Yeo CJ, Kern SE. (1994) Frequent somatic mutations and homozygous deletions of the p16 (MTSI) gene in pancreatic adenocarcinoma. *Nature Genet.* **8**: 27–32

Carter G, Lemoine NR. (1993) Antisense technology for cancer therapy: does it make sense? *Br. J. Cancer* **67**: 869–876.

Chen PL, Chen Y, Bookstein R, Lee WH. (1990) Genetic mechanisms of tumor suppression by the human p53 gene. *Science* **250**: 1576–1580.

Clayman GL, El-Nagger AK, Roth JA, Zhang WW, Goepfert H, Taylor DL, Liu TJ. (1995) *In vivo* molecular therapy with p53 adenovirus for microscopic residual head and neck squamous carcinoma. *Cancer Res.* **55**: 1–6.

Colomer R, Lupu R, Bacus SS, Gelmann EP. (1994) *ERBB-2* antisense oligonucleotides inhibit the proliferation of breast carcinoma cells with *ERBB-2* oncogene amplification. *Br. J. Cancer* **70**: 819–825.

Culver KW, Ram Z, Wallbridge S, Ishii H, Oldfield EH, Blaese RM. (1992) *In vivo* gene transfer with retroviral vector-producer cells for treatment of experimental brain tumors. *Science* **256**: 1550–1552.

Danielsen S, Kilstrup M, Barilla K, Jochimsen B, Neuhard J. (1992) Characterization of the *Escherichia coli codBA* operon encoding cytosine permease and cytosine deaminase. *Mol. Microbiol.* **6**: 1335–1344.

Delhanty JD, Handyside AH, Winston RM. (1994) Preimplantation diagnosis. *Lancet* **343**: 1569–1570

Eck SL, Alavi JB. (1995) Treatment of advanced CNS malignancy with the recombinant Adenovirus H5.020RSVTK: a Phase I trial. *Hum. Gene Ther.* **6**: 509–512.

El-Deiry WS, Harper JW, O'Connor PM, Velculescu VE, Canman CE, Jackman J, Pietenpol JA, Burrell M, Hill DE, Wang Y, Wiman KG, Mercer WE, Kastan MB. (1994) *WAF1/CIP1* is induced in p53-mediated G1 arrest and apoptosis. *Cancer Res.* **54**: 1169–1174.

Freeman S, Abboud CN, Whartenby KA, Packman CH, Koeplin DS, Moolten FL, Abraham GN. (1993) The 'bystander effect': tumor regression when a fraction of the tumor mass is genetically modified. *Cancer Res.* **53**: 5274–5283.

Freeman SM, McCune C, Robinson W, Abbound CN, Abraham GN, Angel C, Marrogi A. (1995) The treatment of ovarian cancer with a gene modified cancer vaccine: a phase I study. *Hum. Gene Ther.* **6**: 927–939.

Friend SH. (1996) Breast cancer susceptibility testing: realities in the post-genomic era. *Nature Genet.* **13**: 16–17.

Fujita K, Sugaya S, Ishii S, Tanaka K. (1994) Specific expression of *Pseudomonas* exotoxin in tumour cells promoted by myc binding sequence. *Cancer Gene Ther.* **1** (Suppl. 1): 3.

Gao W, Han F, Storm C, Egan W, Cheng YC. (1992) Phosphorothioate oligo-nucleotides are inhibitors for human DNA polymerases and RNAse H: implications for antisense technology. *Mol. Pharmacol.* **41**: 223–229.

Garver RI, Goldsmith KT, Rodu B, Hu P-C, Sorscher EJ, Curiel DT. (1994) Strategy for achieving selective killing of carcinomas. *Gene Ther.* **1**: 46–50.

Georges RN, Mukhopadhyay T, Zhang Y, Yen N, Roth JA. (1993) Prevention of orthotopic human lung cancer growth by intratracheal instillation of a retroviral antisense K-*ras* construct. *Cancer Res.* **53**: 1743–1746.

Goodman L. (1996) Box: breast cancer mutation screening. *Nature Genet.* **13**: 17.

Goyette MC, Cho K, Fasching CL, Levy DB, Kinzler KW, Paraskeva C, Vogelstein B, Stanbridge EJ. (1992) Progression of colorectal cancer is associated with multiple tumor suppressor gene defects but inhibition of tumorigenicity is accomplished by correction of any single defect via chromosome transfer. *Mol. Cell. Biol.* **12**: 1387–1395.

Gullick WJ. (1991) Prevalence of abberrant expression of the epidermal growth factor receptor in human cancers. *Br. Med. Bull.* **47**: 87–98.

Gullick WJ, Handyside A. (1994) Pre-implantation diagnosis of inherited predisposition to cancer. *Eur. J. Cancer* **30A**: 2030–2032.

Gutierrez AA, Lemoine NR, Sikora K. (1992) Gene therapy for cancer. *Lancet* **339**: 715–721.

Guvakova MA, Yakubov LA, Volodavsky I, Tonkinson JL, Stein CA. (1995) Phosphorothioate oligonucleotides bind to basic fibroblast growth factor, inhibit its binding to cell surface receptors, and remove it from low affinity binding sites on extracellular matrix. *J. Biol. Chem.* **270**: 2620–2627.

Hahn SA, Schutte M, Hoque ATMS, Moskaluk CA, da Costa LT, Rozenblum E, Weinstein CL, Fischer A, Yeo J, Hruban RH, Kern SE. (1996) *DPC4*, a candidate tumor suppressor gene at human chromosome 18q21.1. *Science* **271**: 350–353.

Handyside AH, Kontogaianni EH, Hardy K, Winston RM. (1990) Pregnancies from biopsied human preimplantation embryos sexed by Y-specific DNA amplification. *Nature* **344**: 768–770.

Handyside AH, Lesko JG, Tarinn JJ, Winston RM, Hughes MR. (1992) Birth of a normal girl after *in vitro* fertilization and preimplantation diagnostic testing for cystic fibrosis. *New Engl. J. Med.* **327**: 905–909.

Harel-Bellan A, Durum S, Muegge K, Abbas A, Farrar WL. (1988) Specific inhibition of lymphokine biosynthesis and autocrine growth using antisense oligonucleotides in Th1 and Th2 helper T cell clones. *J. Exp. Med.* **168**: 2309–2318.

Harper JC, Coonen E, Ramaekers FCS, Delhanty JD, Handyside AH, Winston RM, Hopman AH. (1994) Identification of the sex of human preimplantation embryos in two hours using an improved spreading method and fluorescent *in-situ* hybridisation (FISH) using directly labelled probes. *Hum. Reprod.* **9**: 721–724.

Harper JW, Adami GR, Wei N, Keyomarsi K, Elledge SJ. (1993) The p21 Cdk-interacting protein Cip1 is a potent inhibitor of G1 cyclin-dependent kinases. *Cell* **75**: 805–816.

Harris JD, Gutierrez AA, Hurst HC, Sikora K, Lemoine NR. (1994) Gene therapy for cancer using tumour-specific prodrug activation. *Gene Ther.* **1**: 170–175.

Hirschowitz EA, Ohwada A, Pascal WR, Russi TJ, Crystal RG. (1995) *In vivo* adenovirus-mediated gene transfer of the *Escherichia coli* cytosine deaminase gene to human colon carcinoma-derived tumors induces chemosensitivity to 5-flurocytosine. *Hum. Gene Ther.* **6**: 1055–1063.

Hollywood DP, Hurst H. (1993) A novel transcription factor, OB21, is required for overexpression of the proto-oncogene *c-erbB-2* in mammary tumour lines. *EMBO J.* **12**: 2369–2375.

Huang HJ, Yee JK, Shrew JY, Chen PL, Bookstein R, Friedmann T, Lee EY, Lee WH. (1988) Suppression of the neoplastic phenotype by replacement of the *RB* gene in human cancer cells. *Science* **242**: 1563–1566.

Huber BE, Richards CA, Krenitsky TA. (1991) Retroviral-mediated gene therapy for the treatment of hepatocellular carcinoma: an innovative approach for cancer therapy. *Proc. Natl Acad. Sci. USA* **88**: 8039–8043.

Hwang CH, Smythe WR, Elshami AS, Kucharczuk JC, Amin KM, Williams JP, Litzky LA, Kaiser LR, Albelda SM. (1995) Gene therapy using adenovirus carrying the herpes simplex-thymidine kinase gene to treat *in vivo* models of human malignant mesothelioma and lung cancer. *Am. J. Respir. Cell Mol. Biol.* **13**: 7–16.

Kitada S, Takayama S, de Reil K, Tanaka S, Reed JC. (1994) Reversal of chemoresistance of lymphoma cells by antisense-mediated reduction of *bcl-2* gene expression. *Antisense Res. Devel.* **4**: 71–79.

Klatzmann D, Herson S, Cherin P, Chosidow O, Baillet F, Bensimon G, Boyer O, Salzmann J-L. (1996a) Gene therapy for metastatic malignant melanoma: evaluation of tolerance to intratumoral injection of cells producing recombinant retroviruses carrying the herpes simplex virus type 1 thymidine kinase gene, to be followed by ganciclovir administration. *Hum. Gene Ther.* **7**: 255–267.

Klatzmann D, Philippon J, Valery CA, Bensimon G, Salzmann J-L. (1996b) Clinical protocol. Gene therapy for glioblastoma in adult patients: safety and efficacy evaluation of an *in situ* injection of recombinant retroviruses producing cells carrying the thymidine kinase gene of the herpes simplex type 1 virus, to be followed with the administration of ganciclovir. *Hum. Gene. Ther.* **7**: 109–126.

Kluppel M, Beermann F, Puppert S, Schmid E, Hummler E, Schutz G. (1991) The mouse tyrosinase promoter is sufficient for expression in melanocytes and in the pigmented epithelium of the retina. *Proc. Natl Acad. Sci. USA* **88**: 3777–3781.

Krieg AM, Yi A-K, Matson S, Waldschmidt TJ, Bishop GA, Teasdale R, Koretzky GA, Kilman DM. (1995) CpG motifs in bacterial DNA trigger direct B-cell activation. *Nature* **374**: 546–549.

Kumagai T, Tanio Y, Osaki T, Hosoe S, Tachibana I, Ueno K, Kijima T, Horai T, Kishimoto, T. (1996) Eradication of *myc*-overexpressing small cell lung cancer cells transfected with herpes simplex virus thymidine kinase gene containing *myc–max* response elements. *Cancer Res.* **56**: 354–358.

Lemoine NR. (1994) Genetic intervention for therapy and prevention of pancreatic cancer. *Dig. Surg.* **11:** 170–177.

Levine AJ, Perry ME, Chang A, Silver A, Dittmer D, Wu M, Welsh D. (1994). The role of the p53 tumour-suppressor gene in tumorigenesis. *Br. J. Cancer* **69:** 409–416.

Liu Q, Yan YX, McClure M, Natagawa H, Fujimura F, Rustgi AK. (1995) *MTS1* (*CDKN2*) tumor suppressor gene deletions are a frequent event in esophagus squamous cancer and pancreatic adenocarcinoma cell lines. *Oncogene* **10:** 619–622.

Mahon FX, Belloc F, Vianes I, Bardot C, Boiron JM, Cowen D, Lacombe F, Brizard A, Bilhou-Nabera C, Bernard P, Broustet A, Reiffers J. (1995a) Specific oligomer anti BCR–ABL junctions in chronic myeloid leukaemia: a cell cycle analysis and CFU-GM study. *Leukaemia Lymphoma* **19:** 423–429.

Mahon FX, Ripoche J, Pigeonnier V, Jazwiec B, Pigneux A, Moreau JF, Reiffers J. (1995b) Inhibition of chronic myelogenous leukaemia cells harbouring a *BCR–ABL* B3A2 junction by antisense oligonucleotides targeted at the B2A2 junction. *Exp. Hematol.* **23:** 1606–1611.

Mar EC, Chiou JF, Cheng YC, Huang ES. (1985) Inhibition of cellular DNA polymerase alpha and human cytomegalovirus-induced DNA polymerase by triphosphates of 9-(2-hydroxyethoxymethyl) guanine and 9-(1,3-dihydroxy-2-propoxymethyl) guanine. *J. Virol.* **53:** 776–780.

Miller N, Vile R. (1995) Targeted vectors for gene therapy. *FASEB J.* **9:** 190–199.

Moolten FL. (1986) Tumour chemosensitivity conferred by inserted herpes thymidine kinase genes: paradigm for a prospective cancer control strategy. *Cancer Res.* **46:** 5276–5281.

Moolten FL, Wells JM, Mroz PJ. (1992) Multiple transduction as a means of preserving ganciclovir chemosensitivity in sarcoma cells carrying retrovirally transduced herpes thymidine kinase genes. *Cancer Lett.* **64:** 257–263.

Mukhopadhyay T, Tainsky M, Cavender AC, Roth JA. (1991) Specific inhibition of K-*ras* expression and tumorigenicity of lung cancer cells by antisense RNA. *Cancer Res.* **51:** 1744–1748.

Mullen CA. (1994) Metabolic suicide genes in gene therapy. *Pharmacol. Ther.* **63:** 199–207.

Mullen CA, Coale MM, Lowe R, Blaese RM. (1994) Tumors expressing the cytosine deaminase suicide gene can be eliminated *in vivo* with 5-fluorocytosine and induce protective immunity to wild type tumor. *Cancer Res.* **54:** 1503–1506.

Murray J, Crockett N. (1992) Antisense techniques: an overview. In: *Antisense RNA and DNA* (ed. J Murray). Wiley-Liss, New York, pp. 1–49.

Nellen W, Lichtenstein C. (1993) What makes an mRNA anti-sense-itive? *Trends Biochem. Sci.* **18:** 419–423.

Oldfield EH, Ram Z, Culver KW, Blaese RM, DeVroom HL, Anderson WF. (1993) Gene therapy for the treatment of brain tumors using intra-tumoral transduction with the thymidine kinase gene and intravenous ganciclovir. *Hum. Gene Ther.* **4:** 39–69.

Prins J, de Vries E, Mulder N. (1993) Antisense of oligonucleotides and the inhibition of oncogene expression. *Clin. Oncol.* **5:** 245–252.

Reid R, Mar EC, Huang ES, Topal MD. (1988) Insertion and extension of acyclic, dideoxy and ara nucleotides by herpesviridae, human alpha and human beta polymerases. A unique inhibition mechanism for 9-(1,3-dihydroxy-2-propoxymethyl)guanine triphosphate. *J. Biol. Chem.* **263:** 3898–3904.

Richards CA, Huber BE. (1993) Generation of a transgenic model for retrovirus-mediated gene therapy for hepatocellular carcinoma is thwarted by lack of transgene expression. *Hum. Gene Ther.* **4:** 143–150.

Richards CA, Wolberg AS, Huber BE. (1993) The transcriptional control region of the human carcinoembryonic antigen gene: DNA sequence and homology studies. *DNA Sequence* **4:** 185–196.

Richards CA, Austin EA, Huber BE. (1995) Transcriptional regulatory sequences of carcinoembryonic antigen: identification and use with cytosine deaminase for tumor-specific gene therapy. *Hum. Gene Ther.* **6:** 881–893.

Schrewe H, Thompson J, Bona M, Hefta LJF, Maruya A, Hassauer M, Shively JE, Von Kleist S, Zimmerman W. (1990) Cloning of the complete gene for carcinoembryonic antigen: analysis of its promoter indicates a region conveying cell-type specific expression. *Mol. Cell. Biol.* **10:** 2738–2748.

Serrano M, Gomez-Lahoz E, De Pinho RA, Beach D, Bar-Sagi D. (1995) Inhibition of *Ras*-induced proliferation and cellular transformation by p16[INK4]. *Science* **267:** 249–252.

Shew JY, Ling N, Yang XM, Fodstad O, Lee WH. (1989) Antibodies detecting abnormalities of the restinoblastoma susceptibility gene product (pp110RB) in osteosarcomas and synovial sarcomas. *Oncogene Res.* **4:** 205–214.

Sikora K. (1993) Gene therapy for cancer. *Trends Biotechnol.* **11:** 197–201.

Simons M, Edelman E, DeKeyser J, Langer R, Rosenberg R. (1992) Antisense c-*myb* oligonucleotides inhibit intimal arterial smooth muscle cell accumulation *in vivo*. *Nature* 359: 67–70.

Skorski T, Nieborowska-Skorska M, Nicolaides NC, Szczylik C, Iversen P, Iozzo RV, Zon G, Calabretta B. (1994) Suppression of Philadelphia leukaemia cell growth in mice by BCR–ABL antisense oligonucleotide. *Proc. Natl Acad. Sci. USA* 91: 4504–4508.

Smith K, Johnson K, Bryan T, Hill D, Markowitz S, Willson J, Paraskeva C, Petersen G, Hamilton S, Vogelstein B, Kinzler K. (1993) The *APC* gene product in normal and tumor cells. *Proc. Natl Acad. Sci. USA* 90: 2846–2850.

Smythe WR, Hwang HC, Amin KM, Eck SJ, Wilson JM, Kaiser LR, Albelda SM. (1994) Recombinant adenovirus transfer of the HSV-thymidine kinase gene to thoracic neoplasms: an effective *in vitro* drug sensitization system. *Cancer Res.* 54: 2055–2059.

Spinolo J, Iversen P, Smith L, Pirucello SJ, Haines KM, Norton SE, Kay HD, Arneson MS, Cook P, Zon G, Bayever E. (1992) Antisense p53 oligodeoxynucleotide for systemic antileukaemic therapy. *Hum. Gene Ther.* 3: 24A.

Stein CA. (1995) Does antisense exist? It may, but only under very special circumstances. *Nature Med.* 1: 1119–1121.

Stein CA, Krieg A. (1994) Problems in interpretation of data derived from *in vitro* and *in vivo* use of antisense oligonucleotides. *Antisense Res. Devel.* 4: 67–69.

Su H, Chang JC, Xu SM, Kan YW. (1996) Selective killing of AFP-positive hepatocellular carcinoma cells by adeno-associated virus transfer of the herpes simplex virus thymidine kinase gene. *Hum. Gene Ther.* 7: 463–470.

Szczylik C, Skorski T, Nicolaides NC, Manzella L, Malaguarnera L, Venturelli D, Gewirtz AM, Calabretta B. (1991) Selective inhibition of leukaemia cell proliferation by *BCR–ABL* antisense oligonucleotides. *Science* 253: 262–265.

T'Ang A, Wu KJ, Hashimoto T, Liu WY, Takahashi R, Shi XH, Mihara K, Zhang FH, Chen YY, Du C. (1989) Genomic organization of the human retinoblatoma gene. *Oncogene* 4: 401–407.

Takahashi T, Carbone D, Nau MM, Hida T, Linnoila I, Ueda R, Minna JD. (1992) Wild-type but not mutant p53 suppresses the growth of human lung cancer cells bearing multiple genetic lesions. *Cancer Res.* 52: 2340–2343.

Tamamoto T, Yamamoto S, Kataoka T, Tokunaga T. (1994) Ability of oligonucleotides with certain palindromes to induce interferon production and augment natural killer cell activity is associated with their base length. *Antisense Res. Devel.* 4: 119–122.

Tanaka K, Oshimura M, Kikuchi R, Seki M, Hayashi T, Miyaki M. (1991) Suppression of tumorigenicity in human colon carcinoma cells by introduction of normal chromosome 5 or 18. *Nature* 349: 340–342.

Thomas J-J. (1992) Regulation of gene expression and function by antisense RNA in bacteria. In: *Antisense RNA and DNA* (ed. J Murray). Wiley-Liss, New York, pp. 51–77.

Toulmé JJ. (1992) Artificial regulation of gene expression by complementary oligonucleotides – an overview. In: *Antisense RNA and DNA* (ed. J Murray). Wiley-Liss, New York, pp. 175–194.

Vile RG, Hart IR. (1993a) *In vitro* and *in vivo* targeting of gene expression to melanoma cells. *Cancer Res.* 53: 962–967.

Vile RG, Hart IR. (1993b) Use of tissue-specific expression of the herpes simplex virus thymidine kinase gene to inhibit growth of established murine melanomas following direct injection of DNA. *Cancer Res.* 53: 3860–3864.

Vile R, Miller N, Chernajovsky Y, Hart I. (1994a) A comparison of the properties of different retroviral vectors containing the murine tyrosinase promoter to achieve transcriptionally targeted expression of the HSVtk or IL-2 genes. *Gene Ther.* 1: 307–316.

Vile RG, Nelson JA, Castleden S, Chong H, Hart IR. (1994b) Systemic gene therapy of murine melanoma using tissue specific expression of the HSVtk gene involves an immune component. *Cancer Res.* 54: 6223–6234.

Vile RG, Diaz RM, Miller N, Mitchell S, Russell SJ. (1995) Tissue-specific gene expression from Mo-MLV retroviral vectors with hybrid LTRs containing the murine tyrosinase enhancer/promoter. *Virology* 214: 307–313.

Vogelstein B, Fearon ER, Hamilton SR, Kern SE, Preisinger AC, Leppert M, Nakamura Y, White R, Smits AM, Bos JL. (1988) Genetic alterations during colorectal-tumor development. *N. Engl. J. Med.* 319: 525–532.

Vogelstein B, Fearon ER, Kern SE, Hamilton SR, Preisinger AC, Nakamura Y, White R. (1989) Allelotype of colorectal carcinomas. *Science* 244: 207–211.

Westbrook CA, Chumura SJ, Arenas RB, Kim SY, Otto G. (1994) Human *APC* gene expression in rodent colonic epithelium *in vivo* using liposomal gene delivery. *Hum. Mol. Genet.* **3**: 2005–2010.

Wills KN, Maneval DC, Menzel P, Harris MP, Sutjipto S, Vaillancourt M-T, Huang W-M, Johnson DE, Anderson SC, Wen SF, Bookstein R, Shepard HM, Gregory RJ. (1994) Development and characterization of recombinant adenoviruses encoding human p53 for gene therapy of cancer. *Hum. Genet. Ther.* **5**: 1079–1088.

Wivel NA, Walters L. (1993) Germline gene modification and disease prevention: some medical and ethical perspectives. *Science* **262**: 533–538.

Yakubov L, Khaled Z, Zhang LM, Truneh A, Vlassov V, Stein CA. (1993) Oligonucleotides interact with recombinant CD4 at multiple sites. *J. Biol. Chem.* **268**: 18818–18823.

Yamamoto S, Yamamoto T, Shimada S, Kuramoto E, Yano O, Kataoka T, Tokunaga T. (1992) DNA from bacteria, but not vertebrates, induces interferons, activates natural killer cells and inhibits tumor growth. *Microbiol. Immunol.* **36**: 983–987.

Yang C, Cirielli C, Capogrossi MC, Passaniti A. (1995) Adenovirus-mediated wild-type expression induces apoptosis and suppresses tumorigenesis of prostatic tumor cells. *Cancer Res.* **55**: 4210–4213.

Zhang WW, Roth JA. (1994) Anti-oncogene and tumor suppressor gene therapy – examples from a lung cancer animal model. *In Vivo* **8**: 755–769.

16

Cancer gene therapy II: immunomodulation strategies

Joanna Galea-Lauri and Joop Gäken

16.1 Introduction

Immune-gene therapy has become a new approach in the treatment of cancer. This involves active immunization with gene-modified tumour cells to stimulate specific anti-tumour immunity leading to tumour rejection. The effects exerted on tumour cells by these genetic manipulations are usually pleiomorphic, either direct (e.g. cytolysis or upregulation of immune modulatory molecules) or indirect (e.g. recruitment of the effector immune cells to the tumour site), or both. Several tumour antigens have already been defined but it appears that these antigens alone are unable to stimulate an effective immune response (Kuiper *et al.*, 1995; van den Eynde and Brichard, 1995). This means that most tumours are not immunogenic; that is, they lack the molecules necessary for presentation of their tumour antigens and are thus able to avoid attack by the immune system. To overcome this immune failure, investigators in the field have therefore attempted to engineer tumour cells either to secrete cytokines or to express co-stimulatory immune molecules on their surfaces leading to the recruitment of immune effector cells at the tumour site. Both approaches can incite potent anti-tumour immunity and, in some experimental models, have resulted in rejection of tumour cells already established in the host. This chapter will review progress made using this immunotherapy strategy and discuss future therapeutic implications.

16.2 Activation of immune cells

Understanding the principles underlying the induction of an immune response is pivotal to the design of new strategies for the eradication of tumours by enhancing our immune defence mechanisms. An effective immune response is only possible if appropriate activation of several types of immune cell, including T cells, B cells, antigen-presenting cells (APCs) and natural killer (NK) cells occurs. T cells play a crucial role in several events of the immune system (Julius *et al.*, 1993). Not only are they involved in the recognition of foreign antigens and the destruction of virally infected cells but they also provide help to B cells for the production of antibodies involved in

Gene Therapy, edited by N.R. Lemoine and D.N. Cooper.
© 1996 BIOS Scientific Publishers Ltd, Oxford.

the neutralization and destruction of foreign antigens. Two subsets of T cells have been defined: CD4[+] T cells are generally considered to be helper cells (Th), providing help in the form of cytokines for other cells to perform their function; whereas CD8[+] T cells are cytotoxic and are involved in the killing of target cells such as virally infected and tumour cells (Mosmann and Coffman, 1989). In mice, CD4[+] T cells have been classified into Th1 and Th2 subsets according to the pattern of cytokines they produce (Street *et al.*, 1990). Typically, Th1 clones secrete interleukin-2 (IL-2) and interferon-γ (IFN-γ) and augment cell-mediated immunity, whereas Th2 clones secrete IL-4, IL-5, IL-6 and IL-10 and play an auxiliary role in antibody production. Human T-cell clones display a Th1-like or Th2-like cytokine profile that is distinct from that described in mice (Seder and Paul, 1994).

16.2.1 Recognition of antigen by T cells

It is now well established that in order to respond to their target antigen, T cells must receive signals from the cell presenting the antigen to them. These cells are usually APCs such as dendritic cells, macrophages, Langerhans cells and B cells. APC cells use immune molecules known as major histocompatibility complex (MHC) molecules in mouse, or human leukocyte antigens (HLA) in humans, to form surface complexes with the antigen in a form that is recognized by the T-cell receptor (TcR) (also known as CD3 complex) on the lymphocyte (Germain, 1994). There are two classes of MHC molecules, both of which are highly polymorphic. MHC class I is mainly involved in the presentation of endogenous protein fragments derived either from self antigens or from viral gene products (Howard, 1995). MHC class II molecules present fragments of proteins that have entered the cell by phagocytosis or endocytosis, mainly derived from infectious organisms such as bacteria and parasites (Harding, 1995). In order to recognize the MHC–antigen complex on the APC, T lymphocytes use their unique TcR which, together with their CD8 or CD4 co-receptors, bind to MHC class I or II, respectively (Davis and Bjorkman, 1988).

TcR occupancy alone is not sufficient to activate T lymphocytes to respond; in addition, these cells need to receive antigen-independent signals obtained by engagement of some of their other cell surface molecules with complementary ligands expressed on the surface of the APC (van Seventer *et al.*, 1991b). Depending on their nature, some of these molecules stabilize the interaction of the T lymphocyte with the APCs and others transduce co-stimulatory signals leading to cytokine secretion, proliferation and induction of their effector function. Although several cell surface molecules have been implicated as potential co-stimulatory molecules, a large number of recent studies show that an important co-stimulatory activity required for IL-2-driven proliferation of T lymphocytes is provided by members of the CD28 family on the T-lymphocyte surface (Linsley and Ledbetter, 1993; van Seventer *et al.*, 1991b). The ligands for these molecules are members of the B7 family, namely B7-1, B7-2 and B7-3 (Boussiotis *et al.*, 1993) and are present mainly on the surface of the APC. Details of these molecules are provided below. In the absence of co-stimulatory signals, T cells enter a state of unresponsiveness called anergy, and despite occupancy of their TcR, these cells cannot become activated (Gimmi *et al.*, 1993; Harding *et al.*, 1992; Lenschow and Bluestone, 1993; Linsley and Ledbetter, 1993; Schwartz, 1993). Details of the signal-transduction pathways leading to the activation of T cells have recently been reviewed (Galea-Lauri *et al.*, 1996).

Once the T lymphocyte is activated to respond to a particular antigen, it is then capable of recognition and/or destruction of other cells that display the same antigen. Whilst the activation of T lymphocytes requires the co-stimulatory function, the effector phase (i.e. destruction of the target cell) does not usually require interaction with co-stimulatory molecules on the target cells. However, in special circumstances the presence of these co-stimulatory molecules may alter the nature and/or potency of T lymphocyte-mediated lysis of target cells.

16.2.2 NK cells

NK cells constitute a population of lymphocytes distinguishable from other lymphocytes by the absence of receptors found on B and T cells such as surface immunoglobulin molecules and the TcR, respectively (Yokoyama, 1995). Unlike cytotoxic T cells, NK cells can spontaneously and non-specifically kill certain tumours and virus-infected cells without prior sensitization, in a process called natural killing. In addition, NK cells do not need target cell expression of MHC class I molecules for activation. It is now postulated that NK cells survey tissues for normal expression of MHC class I molecules and lyse targets when MHC class I expression is aberrant or absent, as is the case in some tumours. Thus, cytotoxic T lymphocytes (CTLs) and NK cells have opposing requirements for target-cell MHC class I expression thus providing a fail-safe system for target cells that have escaped destruction by CTLs due to, for example, downregulation of MHC.

Unlike T-cell activation, models of NK-cell activation are less well understood. Recognition of susceptible targets and activation of NK cells appear to be mediated by a NK-cell receptor NKR-P1 that binds carbohydrate determinants on target cells and initiates target cell lysis (Bezouska et al., 1994a,b). NK receptors, such as Ly-49, that bind to MHC class I have also been identified but, unlike the NKR-P1 receptor, these deliver a negative signal to the NK cell and prevent their activation (Ljunggren and Karre, 1990). However, discrepancies exist as to the requirement for NK activation. For example, monoclonal antibodies against human NKR-P1 do not stimulate NK cells (Lanier et al., 1994) and do not block natural killing (Chambers et al., 1989; Karlhofer and Yokoyama, 1991; Lanier et al., 1994).

16.2.3 Mechanisms for lymphocyte-mediated cytotoxicity

Two models of cell death have been molecularly defined to describe the mechanisms used by lymphocytes to mediate killing of the target cell (Podack, 1995). The first model involves granzymes (perforin and associated granule proteases) which are responsible for the enforced death of non-co-operative targets critical for the elimination of tumour and parasitized cells (Podack, 1985; Tschopp and Nabholz, 1990). The second model involves the cell surface receptors Fas and Fas ligand which are responsible for induced suicide in compliant cells and are essential for the control of cell populations as part of the developmental and homeostatic mechanisms (Rouvier et al., 1993). NK cells and CD8+ MHC class I restricted CTLs have been shown to lyse their targets by the same lytic machinery involving perforin (Henkart, 1994; Kagi et al., 1994). Details of the killing process have been reviewed by Podack and Kupfer (1991) and will not be discussed here.

16.3 Co-stimulatory molecules

16.3.1 CD28 and CTLA-4 receptor family

The CD28 receptor, originally called Tp44, is a homodimeric glycoprotein composed of two disulphide-linked 44 kDa subunits; it belongs to the immunoglobulin supergene family (Aruffo and Seed, 1987). CD28 is expressed on the surface of about 95% of CD4$^+$ and 50% of CD8$^+$ human peripheral blood T lymphocytes and on all mouse T lymphocytes (Aruffo and Seed, 1987; Gross *et al.*, 1992). Furthermore, its expression can be increased 4–8-fold by treatment with phorbol esters or with anti-CD3 antibodies (Linsley and Ledbetter, 1993). CD28 has also been shown to be expressed on murine NK cells (Nandi *et al.*, 1994), human NK cell lines (Azuma *et al.*, 1992) but not adult human peripheral blood NK cells (Nagler *et al.*, 1989). CD28 bears structural homology to CTLA-4, an activation molecule that is expressed on the surface of both CD4$^+$ and CD8$^+$ peripheral T lymphocytes but only after activation and then at very low levels (Freeman *et al.*, 1992). The overall level of sequence similarity between CTLA-4 and CD28 is approximately 20% (Dariavach *et al.*, 1988). The region of greatest homology is near the carboxyl-terminal end and contains the peptide MYPPPY, which is completely conserved in CD28 and CTLA-4 molecules from all species studied. The upregulation of CTLA-4 on activated T lymphocytes correlates with the relatively slow induction of B7-1 expression on B lymphocytes, with peak induction achieved 48–72 h after their stimulation (Linsley *et al.*, 1991). The role of CTLA-4 has until recently remained obscure, primarily due to lack of antibodies. In CD28-deficient mice, CTLA-4 can still be induced but there appears to be no B7-dependent T-cell proliferation, suggesting that CTLA-4, at least by itself, cannot substitute for CD28 in providing co-stimulatory signals (Green *et al.*, 1994). A recent report by Krummel and Allison (1995) demonstrates that cross-linking of CTLA-4 together with the TcR and CD28 results in profound inhibition of IL-2 production and proliferation. This suggests that CTLA-4 and CD28 have opposing effects, with CTLA-4 having an inhibitory effect on T-cell activation (Walunas *et al.*, 1994). This might be a reflection of a differential affinity for B7 molecules because CTLA-4 binds B7 ligands with a much higher avidity than CD28 (Mondino and Jenkins, 1994). The role of CTLA-4 might be to dampen and limit T-cell responses by terminating IL-2 production, thereby promoting the generation of memory T cells.

16.3.2 B7 receptor family

As described in Section 16.1.1, CD28 and CTLA-4 bind to three structurally related proteins, B7-1, B7-2 and B7-3, defined by different monoclonal antibodies and the fusion protein CTLA-4Ig [consisting of the extracellular domain of CTLA-4 fused to the constant domain of the human immunoglobulin (Ig) G1] which recognizes all three (Boussiotis *et al.*, 1993). B7-1 (Freeman *et al.*, 1989, 1991) and B7-2 (Freeman *et al.*, 1993a,b) but not B7-3 have been cloned and their nucleotide sequences determined. Both are members of the immunoglobulin supergene family, with extracellular regions containing two Ig-like domains, a transmembrane domain and a short cytoplasmic tail. B7-1 and B7-2 are about 25% identical at the amino acid level and both are heavily glycosylated with molecular masses of 55–60 kDa and 70 kDa, respectively. B7-1 is expressed on dendritic cells and on other APCs but only after activation. For example, it is not expressed on resting B cells but its expression on these cells can be induced by a variety of stimuli such as lipopolysaccharide (LPS), cAMP and cross-linking of surface molecules including surface immunoglobulin, CD40 and MHC class II (Azuma *et al.*, 1993; Yokochi *et al.*, 1982). The induction of B7-1 on B cells is relatively slow with peak induction being achieved 48–72 h after stimulation. Resting monocytes express little or no B7-1, but it is induced by IFN-γ treatment (Freedman *et al.*, 1991).

In contrast to B7-1, B7-2 is rapidly induced on resting B cells with peak expression achieved by 24 h. B7-2 is constitutively expressed on monocytes and its expression is upregulated by addition of IFN-γ. B7-2 is the most prevalent B7 form expressed on dendritic cells. B7-1 and B7-2 are also induced on human T cells stimulated for long periods with anti-CD3 plus IL-2, although the level of expression is very low (Azuma *et al.*, 1993). T cells must receive co-stimulatory signals from APCs within the first 12 h of TcR stimulation for maximal IL-2 production to occur (Mondino and Jenkins, 1994). The rapid appearance of B7-2 on B cells after activation suggests that it is the critical co-stimulatory molecule used by B cells to initiate IL-2 production. Subsequently, B7-1 may provide signals important for continued lymphokine production. B7-2 appears to have a similar binding affinity for CD28 and CTLA-4, as does B7-1, but with different kinetics of binding from B7-1 (Linsley *et al.*, 1994). Emerging evidence suggests that B7-1 and B7-2 might not be equivalent in their effects on T-cell co-stimulation. Data from two different groups suggest that these molecules might direct the differentiation of T cells into Th cells, with B7-1 inducing Th1 and B7-2 inducing Th2 subsets (Freeman *et al.*, 1995; Kuchroo *et al.*, 1995). More studies are needed to confirm these findings, especially since two other groups have reported no differences in the outcome of co-stimulation by B7-1 and B7-2 (Lanier *et al.*, 1995; Levine *et al.*, 1995).

16.3.3 Other co-stimulatory molecules involved in T-lymphocyte activation

In addition to CD28, various other molecules also appear to transduce co-stimulatory signals important for the activation of T lymphocytes (*Figure 16.1*). For example, CD28-deficient mice mount normal delayed-type hypersensitivity and cytotoxic T-cell responses following viral infection (Shahinian *et al.*, 1993). In addition, Johnson and Jenkins (1994) have shown that the monocytic cell line U937, as well as CD14$^+$ monocytes, possess a novel membrane-bound co-stimulatory activity which is independent of CD28–B7 interaction. Two other members of the immunoglobulin supergene family, intercellular cell adhesion molecule 1 (ICAM-1) and leukocyte function-associated antigen 3 (LFA-3), are expressed on the surface of APCs and have been shown to co-stimulate T lymphocytes after interaction with LFA-1 and CD2 (another member of the immunoglobulin supergene family), respectively (van Seventer *et al.*, 1991a). Certain combinations of anti-CD2 antibodies can also enhance proliferation of T lymphocytes stimulated with co-immobilized anti-CD3 antibodies but without detectable production of IL-2 (Harding *et al.*, 1992). CD2 signalling involves association with the TcR and induces the same signal-transduction pathways activated by the TcR engagement (Kanner *et al.*, 1992; Moingeon *et al.*, 1992). LFA-1 (CD11a/CD18) is a member of the β$_2$ integrin family and is expressed by resting T lymphocytes in a conformational state that binds with low avidity to its ligands ICAM-1, ICAM-2 and ICAM-3 (Dustin and Springer, 1989). After TcR signalling, the affinity of LFA-1 for ICAM-1 increases rapidly and transiently, allowing the T lymphocyte to adhere more tightly to the APC (Hibbs *et al.*, 1991). Immobilized, purified ICAM-1 molecules enhance T-lymphocyte

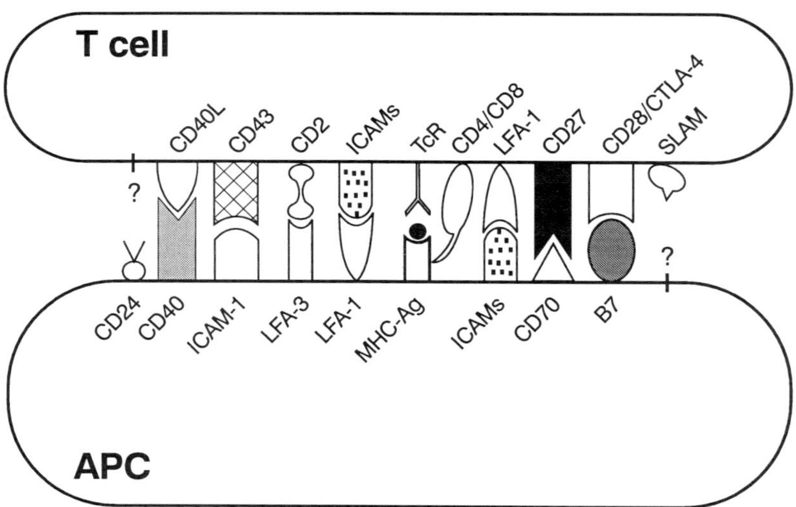

Figure 16.1. Cell surface molecules involved in the adhesion and activation of T cells by APCs. TcR together with the co-receptor (CD4 or CD8) bind to the MHC-antigen (Ag) complex and deliver an antigen-dependent signal to the T cell. In addition, the T cell receives antigen-independent signals from engagement of its co-stimulatory ligands CD28, CTLA-4, CD27 and CD43 with complementary ligands on the APC. The ligand/s for SLAM and for CD24 have not yet been identified but both molecules have been shown to provide additional co-stimulation to the T cell. Molecules such as LFA-1, CD-2 and ICAM-1 are necessary to stabilize the T cell–APC interaction. Activation of the T cell by engagement of its surface molecules results in proliferation and lymphokine release necessary for their function.

proliferation in response to anti-CD3 antibodies but do not induce IL-2 production (van Seventer *et al.*, 1992). Furthermore, transfection of ICAM-1 cDNA into fibroblasts that express low levels of MHC enables them to activate specific T lymphocytes (Altmann *et al.*, 1989). Similar to the TcR and CD2, LFA-1 molecules transduce signals by activating the inositol phospholipid pathway (van Seventer *et al.*, 1992). CD2 and LFA-1 may therefore enhance T-cell proliferation by contributing to the total pool of TcR-derived second messengers, thereby mimicking a higher level of TcR occupancy. Recent work by Dubey *et al.* (1995) has provided new evidence that ICAM-1 and B7-1 are both required for optimum co-stimulation of naive T lymphocytes. Supporting this data, Cavallo *et al.* (1995) have shown that co-expression of B7-1 and ICAM-1 on tumours is required for both tumour rejection and the establishment of a memory response. This emphasizes the involvement of several different molecules in the activation of T cells.

Another pathway that seems to be involved in the co-stimulation of naive T cells involves interaction between CD27 and its ligand CD70 (Hintzen *et al.*, 1994a). CD27, a member of the tumour necrosis factor (TNF)/nerve growth factor (NGF) receptor family, is a 120 kDa transmembrane homodimeric molecule expressed on most T lymphocytes, B lymphocytes and NK cells. CD27-mediated, T-lymphocyte proliferation does not seem to be dependent on the IL-2/IL-2R system. However, it does involve the activation of protein tyrosine kinase and protein kinase C (Kobata *et al.*, 1994). CD70 is a type II transmembrane molecule with significant structural homology to TNF-α, TNF-β, lymphotoxin β and CD40L. CD70 expression is restricted to B cells in some germinal centres, stromal cells in the thymic medulla, and scattered T cells in tonsils, skin and gut (Bowman *et al.*, 1994; Hintzen *et al.*, 1994b). Using transfected mouse fibroblasts expressing human CD70, it has been shown that interaction of CD27 with its ligand provides a potent second signal for cytokine production, induction of activation antigens and proliferation of unprimed T cells (marked by CD45RA expression), and to a lesser extent, of primed peripheral blood T cells (marked by CD45RO expression) (Kobata *et al.*, 1994). In contrast to co-stimulatory signals delivered via B7-1, CD70 was found to induce relatively low levels of IL-2, IL-4 and IL-10 but comparable levels of TNF-α secretion (Hintzen *et al.*, 1995). Furthermore, CD27-CD70 interactions have also been shown to enhance alloantigen-induced proliferation and cytolytic activity in CD8$^+$ T lymphocytes (Brown *et al.*, 1995). The finding that CD70 and B7-1 co-operate in the induction of T-cell proliferation again indicates that several different molecules may be essential for optimal T-cell stimulation. The interaction between CD27 and its ligand CD70 might be of particular importance for the recruitment of T cells from the unprimed T-cell pool. Moreover, since CD70 expression is confined to activated B and T lymphocytes, only a limited set of APCs are able to generate this specific second signal for T-cell expansion.

Liu *et al.* (1992b) have shown that an antibody to the heat-stable antigen (CD24 in humans) expressed on B lymphocytes, dendritic cells and immature thymocytes blocks the co-stimulatory function of dendritic cells and T lymphocyte-depleted splenocytes. Furthermore, transfection of murine CD24 cDNA confers co-stimulatory function upon CHO cells that lacked this previously (Liu *et al.*, 1992a). Relevant to T-cell co-stimulation and tumour immune responses, Wang *et al.* (1995) showed that expression of this heat-stable antigen on murine melanoma K1735ML cells provides co-stimulation for tumour-specific T-cell proliferation and cytotoxicity in mice.

Other molecules recently reported as providing co-stimulatory function include CD43 (leukosialin) and SLAM. The co-stimulatory function of CD43 was described using T lymphocytes from CD28-deficient mice where monoclonal antibody R2/60 to CD43 synergizes with TcR engagement in inducing proliferation (Sperling *et al.*, 1995). CD43 is a large mucin-like protein containing one *N*-linked and 70–90 *O*-linked oligosaccharides (Mentzer *et al.*, 1987). The nature of the mechanism by which CD43 increases T-lymphocyte proliferation is unknown. If CD43 functions by initiating a signal transduction cascade, it will be important to identify its ligand and the cells on which it is expressed. A previous report shows that ICAM-1 binds to CD43 and can therefore be a potential ligand (Rosenstein *et al.*, 1991). SLAM belongs to the immunoglobulin supergene family and is constitutively expressed on memory T lymphocytes, T-lymphocyte clones, a proportion of B lymphocytes and is rapidly induced on naive T lymphocytes after activation. Antibody-induced cross-linking of SLAM enhances antigen-specific proliferation of CD4$^+$ T lymphocytes and cytokine production, particularly of IFN-γ but of not IL-4 or IL-5. Furthermore, in the absence of any other stimuli and without CD28 involvement, engagement of SLAM induces the proliferation of T-lymphocyte clones (Cocks *et al.*, 1995).

In addition to the above co-stimulatory molecules, Zuckerman *et al.* (1995) have described a unique co-stimulatory molecule called VAM-1 on a UV-induced mouse class II$^+$ fibrosarcoma (6130-VAR1) which is capable of stimulating both cytokine production and proliferation in Th1 clones. These fibrosarcoma cells lack any of the known co-stimulatory molecules, including members of the B7 family, and do not bind either CD28 or CTLA-4.

16.3.4 Relevance of co-stimulation in the induction of tumour immunity

The critical importance of both CD28–B7 and CTLA-4–B7 interactions in T-cell activation has been confirmed *in vivo*. This has been achieved by the introduction of B7 into tumour cells and by use of agents that block CD28–B7 interactions, respectively (reviewed by Guinan *et al.*, 1994). The inhibitory role of CTLA-4 on T-cell activation *in vitro* has been recently confirmed *in vivo* by the demonstration that blockade of CTLA-4 by antibodies leads to the enhancement of tumour immunity (Leach *et al.*, 1996). Blocking CTLA-4 might affect T-cell responses in two non-exclusive ways. First, removal of inhibitory signals may lower the overall threshold of T-cell activation and allow normally unreactive T cells to become activated. Alternatively, CTLA-4 blockade might sustain proliferation of activated T cells by removing inhibitory signals that would normally terminate the response, thus allowing for greater expansion of tumour-specific T cells. The importance of such co-stimulatory factors was illustrated previously by the fact that, in their absence, TcR recognition of antigen appears to induce tolerance or anergy in that clone of T lymphocytes (Harding *et al.*, 1992; Schwartz, 1993). This involves a block both in IL-2 production and responsiveness to IL-4. The induction of T-cell anergy may be one of the means by which neoplasms evade recognition by the immune system. The anergic state can be prevented by co-stimulation with B7 (Harding *et al.*, 1992) and, at least in model systems, reversed by administration of high doses of IL-2, resulting in the gradual recovery of the activated state (Essery *et al.*, 1988; Schwartz, 1990). The kinetics of recovery indicate that the reversal of the anergic state is not due to the emergence of a subpopulation of anergy-resistant lymphocytes growing out, but rather is the result of reversal of the anergic state and the subsequent activation of most anergized T lymphocytes (Schwartz, 1990).

16.4 Gene therapy of cancer

Understanding how the immune system works has been critical in the design and engineering of tumour cells to stimulate an anti-tumour response. To this end, investigators have been introducing genes into cancer cells that make them secrete cytokines which then help recruit and activate immune cells at the tumour site. In addition, tumour cells are also being genetically engineered to act as APCs, thereby ensuring presentation of their tumour antigens to the immune system and activation of the cells that lead to their destruction.

16.4.1 Cytokine gene-modified tumour cells

Cytokines represent a heterogeneous group of somatic cell polypeptides involved in immunity and inflammation. They control the extent, amplitude and kinetics of the immune response. Their potency in the induction and maintenance of the immune response has led earlier investigators to demonstrate the effectiveness of high doses of systemic IL-2 on the tumour rejection of both murine and human tumours (Rosenberg *et al.*, 1985, 1987). However, because of dose-limiting toxicities associated with systemic cytokine administration, tumour cells have been genetically engineered to secrete these molecules locally and thus stimulate the immune system to reject the tumour without systemic toxicity. Recent investigations have demonstrated that the growth of murine tumours can be inhibited by the insertion of cDNAs encoding several distinct cytokines (summarized in *Table 16.1*).

Table 16.1. Cytokine gene transfer for anti-cancer therapy

Cytokine	Anti-tumour effect	References
IL-2	Expression of high levels of IL-2 in transduced tumour cells results in tumour regression; a systemic immunity is induced in some tumour models. These anti-tumour responses are mainly due to CD8[+] T cells	Connor *et al.* (1993); Fearon *et al.* (1990); Gansbacher *et al.* (1990); Salvadori *et al.* (1994)
IL-4	Expression of high levels of IL-4 in transduced tumour cells results in tumour regression; a systemic immunity is induced in some tumour models. Both CD8[+] and CD4[+] T cells are involved in these anti-tumour effects	Li *et al.* (1990); Ohira *et al.* (1994); Tepper *et al.* (1989)
IL-6	IL-6-transfected carcinoma cells show reduced tumorigenicity and suppression of metastatic competence in syngeneic mice	Porgador *et al.* (1992, 1993)
IL-7	IL-7-transduced tumour cells are less tumorigenic and systemic anti-tumour responses have been reported. IL-7 transduction results in a marked inhibition of the immune-suppressor TGF-β	Dubinett *et al.* (1995); McBride *et al.* (1992); Miller *et al.* (1993)
IL-10	CHO cells transfected with IL-10 lose tumorigenicity and suppress the growth of co-injected untransduced CHO cells. Tumour infiltration of macrophages is decreased	Richter *et al.* (1993)
IL-12	IL-12-transduced tumour cells are less tumorigenic and systemic immunity can be induced. NK cells and the induction of IFN-γ by IL-12 play important roles in the early phase of the anti-tumour response but both CD4[+] and CD8[+] T cells play a major role in tumour rejection	Martinotti *et al.* (1995); Tahara *et al.* (1995)
IL-13	IL-13-transduced P815 tumour cells are rejected and induce long-lasting immunity, probably due to pleiotropic effects including improved stimulation of immune-specific, anti-tumour effector cells	Richter *et al.* (1993)
IL-15	Activity is comparable to that of IL-2 but is more potent. It induces lymphokine-activated killer (LAK) cells and tumour-infiltrating lymphocytes (TIL) with increased levels of IFN-γ and GM–CSF secretion	Gamero *et al.* (1995); Lewko *et al.* (1995); Munger *et al.* (1995)
IFN-α, IFN-γ	Transduction of tumour cells with either IFN-α or IFN-γ results in tumour regression and induction of systemic immunity in certain animal models. Effects of interferon are thought to be mediated by induction of MHC molecules used to present tumour antigens by the tumour cells	Ferrantini *et al.* (1993); Kaido *et al.* (1995); Restifo *et al.* (1992); Rosenthal *et al.* (1994); Watanabe *et al.* (1989)
TNF-α	TNF-α-transduced tumour cells are sometimes rejected in syngeneic animals but do not lead to enhanced systemic immunity. It is unclear how TNF-α exerts its *in vivo* effects but, *in vitro*, TNF-α inhibits cell proliferation	Asher *et al.* (1991); Blankenstein *et al.* (1991); Qin *et al.* (1993)

16.4.2 Significance of B7 in T cell-mediated killing of tumour cells

Exciting studies have shown that expression of the gene for B7-1 in the weakly immunogenic K1735-M2 MHC class I[+] and II[+] mouse melanoma cell line induces an immune response by CD8[+] T lymphocytes. Chen, L. *et al.* (1992) showed that K1735-M2 tumour cells transfected with the gene for B7-1 and the *E7* gene from human papilloma virus (HPV) 16 were immunogenic, while cells transfected with *E7* only grew progressively in immune-competent mice. Furthermore, inoculation of B7-1 and E7[+] cells could prevent the growth of simultaneously inoculated B7-1[-] but E7[+] cells. In addition, pre-established tumours were rejected in 40% of mice provided that vaccinations with B7[+]–E7[+] tumour cells were started within 4 days of inoculation of the normal tumour cells. CTLs could be isolated from mice treated with B7-1[+] and E7[+] cells and were shown to lyse B7-1[-] K1735 melanoma cells *in vitro* in an MHC class I-restricted manner. Depletion of CD8[+] T lymphocytes with specific antibodies completely abrogated tumour rejection. In contrast, depletion of CD4[+] lymphocytes had no effect, indicating that tumour rejection was mainly due to CD8[+] CTLs (Chen, L. *et al.*, 1992). Townsend and Allison (1993) reported similar findings with the same K1735 mouse melanoma except that they could induce rejection and protect against re-challenge with unmodified tumour cells without the need for co-expression of transfected exogenous E7 antigen. These differences may be due to slight variations in the cell lines used and the way in which the experiments were performed using cultured cells (Chen, L. *et al.*, 1992) or fragments of tumours propagated in nude mice (Townsend and Allison, 1993). Nevertheless, both groups agree that an antigen (tumour or viral) presented in the MHC class I of the melanoma cells is recognized by CD8[+] CTLs. B7-1 expressed by these melanoma cells then provides the co-stimulatory signal to its counter-receptor CD28, inducing IL-2 expression, T-lymphocyte proliferation and clonal expansion, resulting in CD8[+]-mediated killing of the tumour cells.

In contrast, Baskar *et al.* (1993) showed exclusive involvement of CD4[+] T lymphocytes in the rejection of the MHC class I[+], class II[-], weakly immunogenic, SaI mouse sarcoma cell line. SaI cells transfected with a full-length syngeneic class II molecule are immunologically rejected by the autologous host and the mice become protected against a challenge with unmodified cells. In contrast, SaI cells transfected with a class II molecule with a truncated cytoplasmic domain are as tumorigenic as wild-type SaI cells, suggesting that the cytoplasmic region is required to induce immunity. However, transfection of a cytoplasmic domain-truncated class II MHC molecule together with B7-1 prevented tumour formation and protected against subsequent challenge with unmodified tumour cells. CD4[+] and CD8[+] T-lymphocyte depletion studies indicated that CD4[+] lymphocytes are critical for tumour rejection. However, the possible involvement of CD8[+] T lymphocytes could not be ruled out in this study (Baskar *et al.*, 1993). In a follow-up study, this group showed that SaI cells transfected with *B7-1* and autologous A[k] MHC class II genes were rejected in syngeneic mice and conferred protective immunity against a subsequent challenge with wild-type SaI cells or SaI tumour fragments. Furthermore, large established SaI sarcoma tumours could be successfully treated by immunization with SaI cells transfected with MHC class II and *B7-1* genes. CD4 and CD8 depletion studies showed that CD4[+] T cells are important for tumour rejection in naive animals while the rejection of established tumours seemed to depend mainly on CD8[+] T cells (Baskar *et al.*, 1995). These results support the hypothesis that MHC class I[+], II[+], B7-1[+] tumour cells can activate both CD4[+] and CD8[+] tumour-specific T cells. Similar results were obtained in MHC class II[+] K1735 cells transfected with B7-1 and the human-melanoma-associated antigen p97. Mice with established wild-type tumours were cured when treated with gene-modified cells but only if treatment was started before the tumours reached a critical size. Both CD4[+] and CD8[+] T cells were involved in the regression of these established tumours (Li *et al.*, 1994).

Ramarathinam *et al.* (1994) reported that expression of B7-1 by the MHC class I[+], class II$^-$ mouse plasmacytoma cell line J558 could induce rejection of tumour formation mediated by CD8$^+$ T lymphocytes but failed to induce cross-protection against challenge with parental J558 cells. This observation is similar to our own findings, which show that expression of B7-1 and IL-2 on a poorly immunogenic class I$^+$, class II$^-$ adenocarcinoma (NC) reduces their tumorigenicity. Furthermore, treatment with irradiated B7-1–IL-2 NC cells induces the rejection of previously established NC tumours but fails to induce cross-protection against a subsequent challenge (Gäken *et al.*, 1995; Hollingsworth *et al.*, 1995). Chen *et al.* (1994) have investigated the effect of B7-1 expression on the tumorigenicity of a number of tumour cell lines. Their findings indicate that B7-1 prevents tumour formation and protects against challenge with parental cells in the immunogenic tumour cell lines RMA, E6B2, P815 and EL4, whilst the non-immunogenic cell lines B16, Ag104, MCA101 and MCA102 remained tumorigenic after transduction with the gene for B7-1 (Chen *et al.*, 1994). These results show that the effective rejection of tumours as a result of B7 expression is dependent on the inherent immunogenicity of the cell lines used. However, a recent reassessment of the role of B7-1 expression in tumour rejection of B16 murine melanoma cells showed that the immunogenicity of these cells depends on the level of B7-1 expression (Wu *et al.*, 1995). B16 cells with high levels of B7-1 did not grow in syngeneic mice, whereas wild-type cells and cells with low B7-1 expression grew progressively. Furthermore, antibody depletion *in vivo* showed that NK1.1 and CD8$^+$ T cells, but not CD4$^+$ T cells, were responsible for the tumour rejection *in vivo*. This direct killing of the tumour cells did not result in the induction of systemic immunity (Wu *et al.*, 1995).

Recent findings have established that B7-1 co-stimulation by tumour cells is not only effective against solid tumours but could also be a useful means of treating leukaemia. The murine non-leukaemic myeloid cell line 32D was transformed by introduction of the human *BCR/abl* gene, generating a subline that was leukaemic and lethal in syngeneic animals (Matulonis *et al.*, 1995). Transduction of this cell line with the gene for B7-1 and inoculation of syngeneic animals with these cells resulted in only a transient leukaemia whilst these animals were also protected against re-challenge with unmodified leukaemic cells. Furthermore, treatment with B7-1$^+$ leukaemic cells prolonged the survival of animals previously injected with a lethal dose of B7-1$^-$ leukaemic cells. Surprisingly, this treatment only worked when the injections were given intravenously as opposed to subcutaneously (Matulonis *et al.*, 1995). This difference could be due to the preferential migration of the injected cells to lymphoid organs such as the spleen where they might interact directly with reactive T cells, as opposed to inoculation under the skin to where naive T cells do not migrate (Springer, 1994).

B7-1$^+$ K1735 murine melanoma cells have been shown not only to induce protective immune responses against re-challenge with B7-1$^-$ K1735 cells but also against an unrelated syngeneic melanoma cell line CM519 and a UV-induced fibrosarcoma cell line UV5498–4 (Townsend *et al.*, 1994). Similar results were obtained in the EL-4 murine thymoma model: immunization with EL-4 resulted in cross-protection against a subsequent challenge with a syngeneic thymoma C6VLB (Townsend *et al.*, 1994). These results suggest the presence of shared tumour antigens recognized by tumour-specific T cells, a finding which might lead to the possible development of broad-spectrum, anti-tumour vaccines.

Other exciting studies have shown that the anti-tumour effects induced by B7-1$^+$

tumour cells can be amplified by the co-expression of cytokine genes. Salvadori *et al.* (1995) have shown that expression of both B7-1 and IL-2 in the murine fibrosarcoma CMS5 induces the rejection of these cells in syngeneic animals whilst expression of either of these genes alone results in the rapid formation of tumours. Similarly, we find that whereas the expression of either B7-1 or IL-2 alone has little effect on the tumorigenicity of the poorly immunogenic NC adenocarcinoma cell line, the combined expression of B7-1 and IL-2 induces potent anti-tumour responses including the rejection of established tumours (Gäken *et al.*, 1995; Hollingsworth *et al.*, 1995). A possible explanation for the enhancement of anti-tumour responses observed with the combined expression of B7-1 and IL-2 might be the reversal of anergy. The absence of co-stimulation in tumour-bearing animals has been shown to induce anergy in the potentially reactive T cells (Bonneville *et al.*, 1990; Rammensee *et al.*, 1989; Ramsdell *et al.*, 1989). However, in a number of examples, high-level expression of IL-2 has been shown to prevent the onset of anergy and to reverse established anergy (Essery *et al.*, 1988; Schwartz, 1990). Furthermore, IL-2 secretion by tumour cells overcomes the abnormal signal transduction (Salvadori *et al.*, 1994) often seen in T lymphocytes from tumour-bearing mice (Mizoguchi *et al.*, 1992). These findings indicate the importance of studies into the therapeutic potential of the combined expression of different co-stimulators together with immunomodulatory cytokines (*Figure 16.2*).

Despite the biochemical differences between B7-1 and B7-2, early reports seemed to indicate that the anti-tumour effect of B7-2 co-stimulation was indistinguishable from B7-1. Transfection of the immunogenic mouse mastocytoma cell line P815 with B7-2 resulted in regression of these tumours in syngeneic animals and immune protection against a subsequent challenge with wild-type cells (Yang *et al.*, 1995). Depletion studies showed a requirement for CD8$^+$ but not CD4$^+$ T cells. In contrast, transfection of B7-2 alone or in combination with B7-1 had no effect on the tumorigenicity of the non-immunogenic fibrosarcoma MCA102 (Yang *et al.*, 1995). In a different study, MC38 murine colonic adenocarcinoma cells infected with vaccinia virus vectors containing the gene for either B7-1 or B7-2 were rejected to an equal extent in syngeneic mice (Hodge *et al.*, 1994). A recent report by Matulonis *et al.* (1996) showed that B7-1 is superior to B7-2 co-stimulation in the induction and maintenance of T-cell-mediated anti-leukaemia immunity. Further studies using different tumour models are needed to establish the potency of B7-1 and B7-2 in anti-tumour immune responses, and to elucidate the molecular mechanisms of rejection.

16.4.3 Importance of B7-1 in NK-mediated killing of tumour cells

Insights into the biology of NK cells indicate that these cells are also important mediators of the immune surveillance of cancer (Whiteside and Herberman, 1995). These are cells with a large granular morphology endowed with the selective ability to kill a variety of cell types, including a wide range of tumour cells and virus-infected cells (Gidlund *et al.*, 1981). NK cells kill these targets spontaneously, without the need for prior sensitization and without class I MHC restriction (Seaman *et al.*, 1987; Welsh *et al.*, 1991). However, NK cells appear to require the expression of CD28 for their cytotoxic activity. For example, the YT2C2 human NK cell line expresses CD28 and can spontaneously kill both murine and human cell lines expressing B7 proteins (Azuma *et al.*, 1992). The participation of CD28 and B7 in the MHC-unrestricted killing of these targets was demonstrated by correlation of target sensitivity with levels of B7

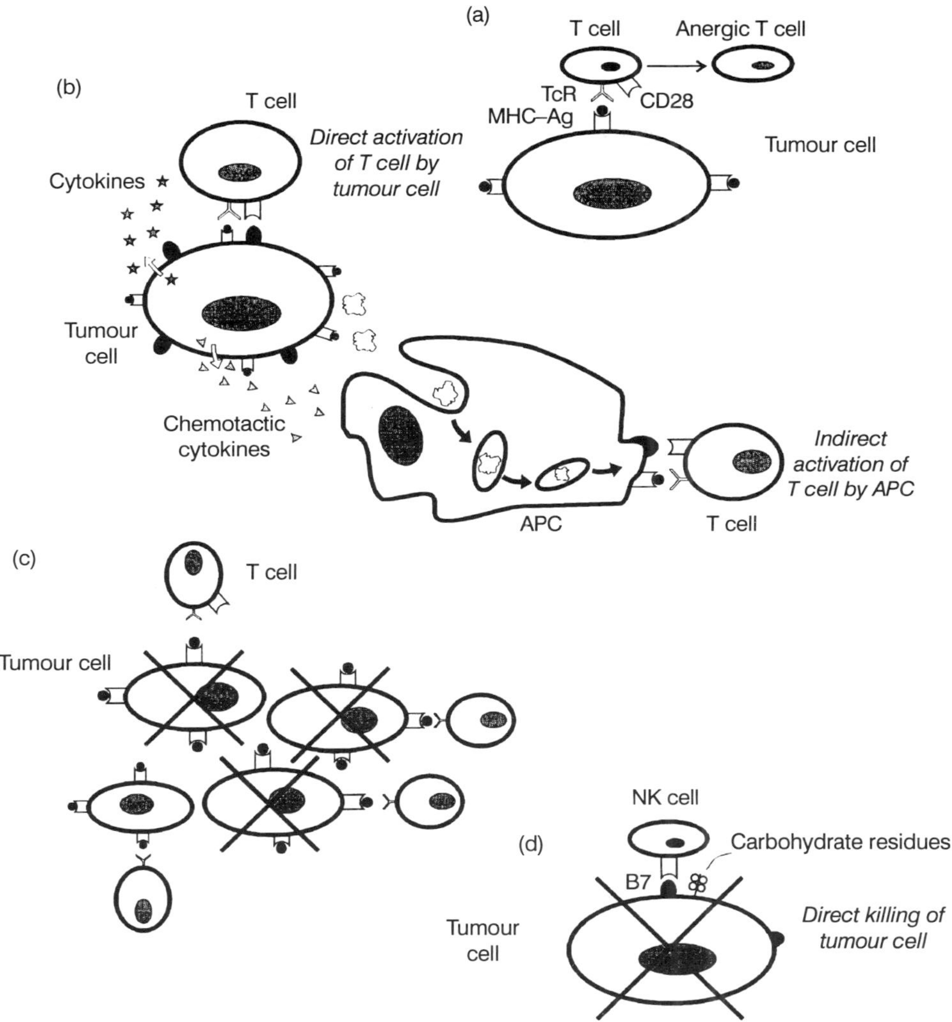

Figure 16.2. Modulation of tumour cells to evoke an anti-tumour immune response. (a) Most tumour cells lack co-stimulatory molecules but can express low levels of MHC on their surface. TcR engagement of the MHC–Ag complex in the absence of co-stimulation leads to the induction of anergy in those T cells. (b) Tumour cells can be modulated by gene transfer to express surface co-stimulators such as B7, which are necessary for T-cell activation, or to secrete cytokines such as IL-2, which help to activate the T cells. Alternatively, tumour cells can be modified to secrete cytokines which attract APCs to the tumour site. The APCs take up the exogenous tumour fragments, process them and present their tumour antigens with MHC class II molecules, thus leading to indirect activation of T cells. (c) Once activated, both CD4 and CD8 T cells can recognize other tumour cells (of the same origin) and destroy them without the need for the expression of the co-stimulatory molecules on target cells. (d) Furthermore, B7 expression can render tumour cells targets for NK cells, thus leading to their direct killing. Other molecules involved in NK recognition of tumour cells include carbohydrate structures, but very little is known about this.

expression, inhibition of killing by anti-CD28 or B7 antibodies, and by making both murine and human cell lines susceptible to NK-mediated lysis by transfection of B7 (Azuma *et al.*, 1992). However, CD28–B7 interactions on their own are not sufficient for NK-mediated lysis of target cells; other adhesion molecules, in particular LFA-1 and ICAM-1, play a significant role. In addition, van den Driessche *et al.* (1994) have shown that the experimental metastatic potential of BW5147-derived T-lymphoma cells is determined by the level of expression of the MHC class I antigen H-2D^k, resulting in resistance of cells expressing high levels of H-2D^k to NK-mediated lysis. In view of the possible role of B7-1 in NK-cell activation, this group transfected BW5147-derived tumour cells with B7-1 and showed that this modulates their sensitivity toward NK effector cells (Geldhof *et al.*, 1995). It is therefore possible that transduction of tumour cells with co-stimulatory molecules allows the participation of NK cells in tumour rejection.

16.4.4 Alternative immunomodulation strategies

DNA vaccination against tumour antigens. Recent studies have shown that direct injection of plasmids expressing a variety of genes leads to T-cell and antibody response *in vivo* against these proteins, thus providing a novel means of specific immunization (reviewed by Spooner *et al.*, 1995). Expression of DNA encoding tumour-specific proteins or fragments of these proteins has been shown to be an effective means of breaking immunological tolerance and generating tumour-specific immune responses. For example, the idiotypic Ig of the malignant B cell in B-cell lymphomas provides a suitable target but this requires a vaccine for each patient. The strategy used for making such vaccines is to clone the variable genes of the idiotypic Ig by polymerase chain reaction (PCR) and to inject the patient with an expression vector containing this *V* gene. This approach has been successfully adopted in a murine model where it has been shown that intramuscular idiotypic DNA vaccination induces low levels of anti-idiotypic antibody in serum (Stevenson *et al.*, 1995). Levels can be increased dramatically by co-injection of DNA plasmids encoding either IL-2 or granulocyte–macrophage colony-stimulating factor (GM–CSF), and specific proliferative anti-idiotypic T cells are induced. A preliminary small trial of DNA vaccines for chemotherapy-resistant patients with B-cell lymphomas has now been started by the same investigators.

Direct injection of plasmid DNA encoding the full-length cDNA for human carcinoembryonic antigen (CEA) induced humoral and cellular immune responses specific for human CEA in immunized mice, and fully protected the animals against tumour challenge with syngeneic, CEA-expressing, colon carcinoma cells (Conry *et al.*, 1994, 1995).

Similarly, mutant oncogenes and fusion proteins, the products of chromosomal translocations such as p210$^{bcr/abl}$ and PML/RAR, could provide novel tumour antigens. For example, synthetic peptides corresponding to the joining region of p210$^{bcr/abl}$ protein elicit murine CD4$^+$ class II restricted T cells that can proliferate in response to BCR/ABL protein (Chen, W. *et al.*, 1992). Furthermore, peptides based on the amino acid sequence of p21^{ras} mutated at codons 12 and 13 are presented by class II molecules and induce CD4$^+$ T-cell responses (Gedde-Dahl *et al.*, 1992). These findings suggest that vaccines based on immunization with plasmid DNAs that express epitopes of appropriate tumour-specific antigens may be a good way forward for the future, not only for virally induced human cancers but also for cancers with unique tumour-specific antigens.

Newcastle disease virus tumour therapy. Several viruses have been evaluated as potential agents for cancer treatment, using either their oncolytic properties to lyse cancer cells or their potential augmenting effects on the immune responses to tumours. One of the viruses studied is the Newcastle disease virus (NDV), an avian paramyxovirus which is a potent inducer of interferon in a number of cell lines (Slattery *et al.*, 1980). Heicappell *et al.* (1986) showed that infection of ESb tumour cells with NDV led to expression of viral antigens and to an increase in immuno-genicity of this cell line. Furthermore, an effective anti-metastatic therapy was achieved in a murine tumour model that combined surgery with post-operative immunotherapy using NDV-modified autologous tumour cells. Experiments with xenografts of human neuroblastoma and sarcomas in athymic mice demonstrated that intra-tumour injection of NDV resulted in complete regression of the tumour, which did not recur for the duration of the study (Lorence *et al.*, 1994).

Infection of tumour cells with low amounts of NDV leads to the activation of tumour-specific CD4$^+$ and CD8$^+$ T cells after sensitization *in vivo* and re-stimulation *in vitro* (Schild *et al.*, 1989). Further studies showed that transfection of murine tumour cells with the viral gene for haemagglutinin-neuraminidase augmented the generation of specific CTLs as well as infection with NDV (Ertel *et al.*, 1993). These findings have led to clinical trials to increase the immunogenicity of cancer vaccines for the post-operative immunotherapy of metastasis (Schlag *et al.*, 1992).

16.5 Conclusions

Tumour immunity has been largely ignored until recently because there was little evidence for the presence of an immune reaction against the majority of spontaneously arising animal neoplasms. The conventional assays involved the immunization of syngeneic animals with irradiated tumour cells and subsequent challenge with viable tumour cells. If tumour rejection antigens were present, it was argued, this approach would prevent tumour formation when challenged with viable tumour cells. The question as to what extent human tumours are immunogenic is still controversial but results obtained in recent years indicate the presence of specific antigens on a number of human tumours and the ability of the immune system to recognize and react against these antigens. Boon and co-workers were the first to develop a method for the isolation of tumour antigens and to demonstrate their recognition by CD8$^+$ T lymphocytes in human malignant melanoma (De Plaen *et al.*, 1988; van den Eynde *et al.*, 1991). This approach has been used in several laboratories and a number of human melanoma-associated tumour antigens have already been isolated (Bakker *et al.*, 1994; Boël *et al.*, 1995; Brichard *et al.*, 1993; Gaugler *et al.*, 1994; Kawakami *et al.*, 1994; van den Eynde *et al.*, 1995; van der Bruggen *et al.*, 1991). New biochemically based procedures have also been developed by Mandelboim *et al.* (1994) and Cox *et al.* (1994) for the identification of novel tumour antigens. The presence of tumour-specific antigens is well established, increasing the feasibility of an immunological approach to cancer therapy.

Although it is clear that several molecules play a vital role in the induction of T-lymphocyte activation and co-stimulation, it is still unknown which combination of these molecules would be appropriate to achieve this most effectively. Characterization of these molecules opens up the possibility of introducing them into tumour cells in order to enhance their immunogenicity. By engineering tumour cells to express such co-stimulatory molecules, they would become capable not only of presenting their tumour antigens but also of delivering co-stimulatory signals, thus mimicking APCs with the potential of stimulating the appropriate T lymphocytes to recognize and eradicate the unmodified tumour cells. This approach offers the advantage that not all of the tumour cells will need to be transduced for recognition and destruction by the immune system. However, it should be pointed out that the immunogenicity of transfected tumours might still be impaired by insufficient expression of other molecules required for efficient T-cell activation, such as MHC and adhesion molecules. One way to overcome these problems would be to fuse tumour cells with 'professional' APCs and use the resulting hybrid cell as a tumour vaccine. Using this approach, fusing rat BERH-2 hepatocarcinoma cells with syngeneic activated B cells resulted in hybrid cells that expressed B7, LFA-1 and elevated levels of MHC class I and II. Inoculation of syngeneic animals with these hybrid cells led to their rejection and a T cell-dependent immunity to the parental tumour cells (Guo *et al.*, 1994). A similar rejection of established tumours was observed after injection of hybrid cells obtained from the fusion of mouse EL4 thymoma cells with A20 B lymphoma cells (Stuhler and Walden, 1994). This strategy may be useful in cases where gene modification of tumour cells is not possible by more conventional methods. Although there are obstacles to overcome, advances in our understanding of the role of co-stimulatory signals in the regulation of immune responses offer encouragement that, in the near future, immunotherapy may become an effective clinical tool for the treatment of cancer.

Acknowledgements

We would like to thank Dr F. Farzaneh for critical reading of the manuscript and for helpful discussions, and I. Giddings, M. Kuiper and S. Vallian for assistance with the illustrations. The Immune Gene Therapy Group is supported by the Lewis Family Charitable Trust, the King's Medical Research Trust grants 11 and 171, the Joint Research Council, the Leukemia Research Fund and the Cancer Research Campaign.

References

Altmann DM, Hogg N, Trowsdale J, Wilkinson D. (1989) Cotransfection of ICAM-1 and HLA-DR reconstitutes human antigen-presenting cell function in mouse L cells. *Nature* 338: 512–514.

Aruffo A, Seed B. (1987) Molecular cloning of a CD28 cDNA by a high-efficiency COS cell expression system. *Proc. Natl Acad. Sci. USA* 84: 8573–8577.

Asher AL, Mule JJ, Kasid A, Restifo NP, Salo JC, Reichert CM, Jaffe G, Fendly B, Kriegler M, Rosenberg SA. (1991) Murine tumor cells transduced with the gene for tumor necrosis factor α. Evidence for paracrine immune effects of tumor necrosis factor against tumors. *J. Immunol.* 146: 3227-3234.

Azuma M, Cayabyab M, Buck D, Phillips JH, Lanier LL. (1992) Involvement of CD28 in MHC-unrestricted cytotoxicity mediated by a human natural killer leukemia cell line. *J. Immunol.* 149: 1115–1123.

Azuma M, Ito D, Yagita H, Okumura K, Phillips JH, Lanier LL, Somoza C. (1993) B70 antigen is a second ligand for CTLA-4 and CD28. *Nature* 366: 76–79.

Bakker AB, Schreurs MW, de Boer AJ, Kawakami Y, Rosenberg SA, Adema GJ, Figdor CG. (1994) Melanocyte lineage-specific antigen gp100 is recognized by melanoma-derived tumor-infiltrating lymphocytes. *J. Exp. Med.* 179: 1005–1009.

Baskar S, Ostrand-Rosenberg S, Nabavi N, Nadler LM, Freeman GJ, Glimcher LH. (1993) Constitutive expression of B7 restores immunogenicity of tumor cells expressing truncated major histocompatibility complex class II molecules. *Proc. Natl Acad. Sci. USA* 90: 5687-5690.

Baskar S, Glimcher L, Nabavi N, Jones RT, Ostrand-Rosenberg S. (1995) Major histocompatibility complex class II⁺ B7-1⁺ tumor cells are potent vaccines for stimulating tumor rejection in tumor-bearing mice. *J. Exp. Med.* **181**: 619–629.

Bezouska K, Vlahas G, Horvath O, Jinochova G, Fiserova A, Giorda R, Chambers WH, Feizi T, Pospisil M. (1994a) Rat natural killer cell antigen, NKR-P1, related to C-type animal lectins is a carbohydrate-binding protein. *J. Biol. Chem.* **269**: 16945–16952.

Bezouska K, Yuen CT, O'Brien J, Childs RA, Chai W, Lawson AM, Drbal K, Fiserova A, Pospisil M, Feizi T. (1994b) Oligosaccharide ligands for NKR-P1 protein activate NK cells and cytotoxicity. *Nature* **372**: 150–157.

Blankenstein T, Qin ZH, Uberla K, Muller W, Rosen H, Volk HD, Diamantstein T. (1991) Tumor suppression after tumor cell-targeted tumor necrosis factor α gene transfer. *J. Exp. Med.* **173**: 1047–1052.

Boël P, Wildmann C, Sensi ML, Brasseur R, Renauld JC, Coulie P, Boon T, van der Bruggen P. (1995) BAGE: a new gene encoding an antigen recognized on human melanomas by cytolytic T lymphocytes. *Immunity* **2**: 167–175.

Bonneville M, Ishida I, Itohara S, Verbeek S, Berns A, Kanagawa O, Haas W, Tonegawa S. (1990) Self-tolerance to transgenic γδ T cells by intrathymic inactivation. *Nature* **344**: 163–165.

Boussiotis VA, Freeman GJ, Gribben JG, Daley J, Gray G, Nadler LM. (1993) Activated human B lymphocytes express three CTLA-4 counterreceptors that costimulate T-cell activation. *Proc. Natl Acad. Sci. USA* **90**: 11059–11063.

Bowman MR, Crimmins MA, Yetz Aldape J, Kriz R, Kelleher K, Herrmann S. (1994) The cloning of CD70 and its identification as the ligand for CD27. *J. Immunol.* **152**: 1756–1761.

Brichard V, Van Pel A, Wolfel T, Wolfel C, De Plaen E, Lethe B, Coulie P, Boon T. (1993) The tyrosinase gene codes for an antigen recognized by autologous cytolytic T lymphocytes on HLA-A2 melanomas. *J. Exp. Med.* **178**: 489–495.

Brown GR, Meek K, Nishioka Y, Thiele DL. (1995) CD27-CD27 ligand/CD70 interactions enhance alloantigen-induced proliferation and cytolytic activity in CD8⁺ T lymphocytes. *J. Immunol.* **154**: 3686–3695.

Cavallo F, Martin-Fontecha A, Bellone M, Heltai S, Gatti E, Tornaghi P, Freschi M, Forni G, Dellabona P, Casorati G. (1995) Co-expression of B7-1 and ICAM-1 on tumors is required for rejection and the establishment of a memory response. *Eur. J. Immunol.* **25**: 1154–1162.

Chambers WH, Vujanovic NL, Deleo AB, Olszowy MW, Herberman RB, Hiserodt JC. (1989) Monoclonal-antibody to a triggering structure expressed on rat natural-killer cells and adherent lymphokine-activated killer cells. *J. Exp. Med.* **169**: 1373–1389.

Chen L, Ashe S, Brady WA, Hellström I, Hellström KE, Ledbetter JA, McGowan P, Linsley PS. (1992) Costimulation of antitumor immunity by the B7 counterreceptor for the T lymphocyte molecules CD28 and CTLA-4. *Cell* **71**: 1093–1102.

Chen L, McGowan P, Ashe S, Johnston J, Li Y, Hellström I, Hellström KE. (1994) Tumor immunogenicity determines the effect of B7 costimulation on T cell-mediated tumor immunity. *J. Exp. Med.* **179**: 523–532.

Chen W, Peace DJ, Rovira DK, You SG, Cheever MA. (1992) T-cell immunity to the joining region of p210^bcr-abl protein. *Proc. Natl Acad. Sci. USA* **89**: 1468–1472.

Cocks BG, Chang CC, Carballido JM, Yssel H, de Vries JE, Aversa G. (1995) A novel receptor involved in T-cell activation. *Nature* **376**: 260–263.

Connor J, Bannerji R, Saito S, Heston W, Fair W, Gilboa E. (1993) Regression of bladder tumors in mice treated with interleukin 2 gene-modified tumor cells. *J. Exp. Med.* **177**: 1127–1134.

Conry RM, LoBuglio AF, Kantor J, Schlom J, Loechel F, Moore SE, Sumerel LA, Barlow DL, Abrams S, Curiel DT. (1994) Immune response to a carcinoembryonic antigen polynucleotide vaccine. *Cancer Res.* **54**: 1164–1168.

Conry RM, LoBuglio AF, Loechel F, Moore SE, Sumerel LA, Barlow DL, Curiel DT. (1995) A carcinoembryonic antigen polynucleotide vaccine has *in vivo* antitumour activity. *Gene Ther.* **2**: 59–65.

Cox AL, Skipper J, Chen Y, Henderson RA, Darrow TL, Shabanowitz J, Engelhard VH, Hunt DF, Slingluff CLJ. (1994) Identification of a peptide recognized by five melanoma-specific human cytotoxic T cell lines. *Science* **264**: 716–719.

Dariavach P, Mattei MG, Golstein P, Lefranc MP. (1988) Human Ig superfamily CTLA-4 gene: chromosomal localization and identity of protein sequence between murine and human CTLA-4 cytoplasmic domains. *Eur. J. Immunol.* **18**: 1901–1905.

Davis MM, Bjorkman PJ. (1988) T-cell antigen receptor genes and T-cell recognition. *Nature* **334**: 395–402.

De Plaen E, Lurquin C, Van Pel A, Mariame B, Szikora JP, Wolfel T, Sibille C, Chomez P, Boon T. (1988) Immunogenic (TUM−) variants of mouse tumor P815: cloning of the gene of TUM− antigen P91A and identification of the tum− mutation. *Proc. Natl Acad. Sci. USA* **85:** 2274–2278.

Dubey C, Croft M, Swain SL. (1995) Costimulatory requirements of naive CD4$^+$ T cells. ICAM-1 or B7-1 can costimulate naive CD4 T cell activation but both are required for optimum response. *J. Immunol.* **155:** 45–57.

Dubinett SM, Huang M, Dhanani S, Economou JS, Wang J, Lee P, Sharma S, Dougherty GJ, McBride WH. (1995) Down-regulation of murine fibrosarcoma transforming growth factor-β1 expression by interleukin 7. *J. Natl Cancer Inst.* **87:** 593–597.

Dustin ML, Springer TA. (1989) T-cell receptor cross-linking transiently stimulates adhesiveness through LFA-1. *Nature* **341:** 619–624.

Ertel C, Millar NS, Emmerson PT, Schirrmacher V, von Hoegen P. (1993) Viral hemagglutinin augments peptide-specific cytotoxic T cell responses. *Eur. J. Immunol.* **23:** 2592–2596.

Essery G, Feldmann M, Lamb JR. (1988) Interleukin-2 can prevent and reverse antigen-induced unresponsiveness in cloned human T lymphocytes. *Immunology* **64:** 413–417.

Fearon ER, Pardoll DM, Itaya T, Golumbek P, Levitsky HI, Simons JW, Karasuyama H, Vogelstein B, Frost P. (1990) Interleukin-2 production by tumor cells bypasses T helper function in the generation of an antitumor response. *Cell* **60:** 397-403.

Ferrantini M, Proietti E, Santodonato L, Gabriele L, Peretti M, Plavec I, Meyer F, Kaido T, Gresser I, Belardelli F. (1993) α1-interferon gene transfer into metastatic Friend leukemia cells abrogated tumorigenicity in immunocompetent mice: antitumor therapy by means of interferon-producing cells. *Cancer Res.* **53:** 1107-1112.

Freedman AS, Freeman GJ, Rhynhart K, Nadler LM. (1991) Selective induction of B7/BB-1 on interferon γ stimulated monocytes: a potential mechanism for amplification of T cell activation through the CD28 pathway. *Cell Immunol.* **137:** 429–437.

Freeman GJ, Freedman AS, Segil JM, Lee G, Whitman JF, Nadler LM. (1989) B7, a new member of the Ig superfamily with unique expression on activated and neoplastic B cells. *J. Immunol.* **143:** 2714–2722.

Freeman GJ, Gray GS, Gimmi CD, Lombard DB, Zhou LJ, White M, Fingeroth JD, Gribben JG, Nadler LM. (1991) Structure, expression, and T cell costimulatory activity of the murine homologue of the human B lymphocyte activation antigen B7. *J. Exp. Med.* **174:** 625–631.

Freeman GJ, Lombard DB, Gimmi CD, Brod SA, Lee K, Laning JC, Hafler DA, Dorf ME, Gray GS, Reiser H, June CH, Thompson CB, Nadler LM. (1992) CTLA-4 and CD28 mRNA are coexpressed in most T cells after activation. Expression of CTLA-4 and CD28 mRNA does not correlate with the pattern of lymphokine production. *J. Immunol.* **149:** 3795–3801.

Freeman GJ, Borriello F, Hodes RJ, Reiser H, Gribben JG, Ng JW, Kim J, Goldberg JM, Hathcock K, Laszlo G. (1993a) Murine B7-2, an alternative CTLA4 counter-receptor that costimulates T cell proliferation and interleukin 2 production. *J. Exp. Med.* **178:** 2185–2192.

Freeman GJ, Gribben JG, Boussiotis VA, Ng JW, Restivo VA, Jr, Lombard LA, Gray GS, Nadler LM. (1993b) Cloning of B7-2: a CTLA-4 counter-receptor that costimulates human T cell proliferation. *Science* **262:** 909–911.

Freeman GJ, Boussiotis VA, Anumanthan A, Bernstein GM, Ke XY, Rennert PD, Gray GS, Gribben JG, Nadler LM. (1995) B7-1 and B7-2 do not deliver identical costimulatory signals, since B7-2 but not B7-1 preferentially costimulates the initial production of IL-4. *Immunity* **2:** 523–532.

Gäken J, Darling D, Hollingsworth S, Hirst W, Kuiper M, Towner P, Mufti G, Farzaneh F. (1995) Combined expression of B7-1 and IL-2: synergy in tumor rejection. *J. Cell. Biochem.* **S21A:** 421.

Galea-Lauri J, Farzaneh F, Gäken J. (1996) Novel costimulators in the immune gene therapy of cancer. *Cancer Gene Ther.* **3:** 202–214.

Gamero AM, Ussery D, Reintgen DS, Puleo CA, Djeu JY. (1995) Interleukin 15 induction of lymphokine-activated killer cell function against autologous tumor cells in melanoma patient lymphocytes by a CD18-dependent, perforin-related mechanism. *Cancer Res.* **55:** 4988–4994.

Gansbacher B, Zier K, Daniels B, Cronin K, Bannerji R, Gilboa E. (1990) Interleukin 2 gene transfer into tumor cells abrogates tumorigenicity and induces protective immunity. *J. Exp. Med.* **172:** 1217-1224.

Gaugler B, Van den Eynde B, van der Bruggen P, Romero P, Gaforio JJ, De Plaen E, Lethe B, Brasseur F, Boon T. (1994) Human gene MAGE-3 codes for an antigen recognized on a melanoma by autologous cytolytic T lymphocytes. *J. Exp. Med.* **179:** 921–930.

Gedde-Dahl T, III, Eriksen JA, Thorsby E, Gaudernack G. (1992) T-cell responses against products of oncogenes: generation and characterisation of human T-cell clones specific for p21 ras-derived synthetic peptides. *Hum. Immunol.* 33: 266–274.

Geldhof AB, Raes G, Bakkus M, Devos S, Thielemans K, De Baetselier P. (1995) Expression of B7-1 by highly metastatic mouse T lymphocytes induces optimal natural killer cell-mediated cytotoxicity. *Cancer Res.* 55: 2730–2733.

Germain RN. (1994) MHC-dependent antigen processing and peptide presentation: providing ligands for T lymphocyte activation. *Cell* 76: 287-299.

Gidlund M, Örn A, Pattengale PK, Jansson M, Wigzell H, Nilsson K. (1981) Natural killer cells kill tumour cells at a given stage of differentiation. *Nature* 292: 848–850.

Gimmi CD, Freeman GJ, Gribben JG, Gray G, Nadler LM. (1993) Human T-cell clonal anergy is induced by antigen presentation in the absence of B7 costimulation. *Proc. Natl Acad. Sci. USA* 90: 6586–6590.

Green JM, Noel PJ, Sperling AI, Walunas TL, Gray GS, Bluestone JA, Thompson CB. (1994) Absence of B7-dependent responses in CD28-deficient mice. *Immunity* 1: 501–508.

Gross JA, Callas E, Allison JP. (1992) Identification and distribution of the costimulatory receptor CD28 in the mouse. *J. Immunol.* 149: 380–388.

Guinan EC, Gribben JG, Boussiotis VA, Freeman GJ, Nadler LM. (1994) Pivotal role of the B7:CD28 pathway in transplantation tolerance and tumor immunity. *Blood* 84: 3261–3282.

Guo Y, Wu M, Chen H, Wang X, Liu G, Li G, Ma J, Sy MS. (1994) Effective tumor vaccine generated by fusion of hepatoma cells with activated B cells. *Science* 263: 518–520.

Harding CV. (1995) Phagocytic processing of antigens for presentation by MHC molecules. *Trends Cell Biol.* 5: 105–109.

Harding FA, McArthur JG, Gross JA, Raulet DH, Allison JP. (1992) CD28-mediated signalling co-stimulates murine T cells and prevents induction of anergy in T-cell clones. *Nature* 356: 607-609.

Heicappell R, Schirrmacher V, von Hoegen P, Ahlert T, Appelhans B. (1986) Prevention of metastatic spread by postoperative immunotherapy with virally modified autologous tumor cells. Parameters for optimal therapeutic effects. *Int. J. Cancer* 37: 569–577.

Henkart PA. (1994) Lymphocyte-mediated cytotoxicity: two pathways and multiple effector molecules. *Immunity* 1: 343–346.

Hibbs ML, Xu H, Stacker SA, Springer TA. (1991) Regulation of adhesion of ICAM-1 by the cytoplasmic domain of LFA-1 integrinβ subunit. *Science* 251: 1611–1613.

Hintzen RQ, de Jong R, Lens SM, van Lier RA. (1994a) CD27: marker and mediator of T-cell activation? *Immunol. Today* 15: 307-311.

Hintzen RQ, Lens SM, Koopman G, Pals ST, Spits H, van Lier RA. (1994b) CD70 represents the human ligand for CD27. *Int. Immunol.* 6: 477-480.

Hintzen RQ, Lens SM, Lammers K, Kuiper H, Beckmann MP, van Lier RA. (1995) Engagement of CD27 with its ligand CD70 provides a second signal for T cell activation. *J. Immunol.* 154: 2612–2623.

Hodge JW, Abrams S, Schlom J, Kantor JA. (1994) Induction of antitumor immunity by recombinant vaccinia viruses expressing B7-1 or B7-2 costimulatory molecules. *Cancer Res.* 54: 5552–5555.

Hollingsworth S, Gäken J, Darling D. (1995) Induction of tumour rejection by combination B7.1/IL-2 expressing tumour cells. *Cancer Gene Ther.* 2: 240.

Howard JC. (1995) Supply and transport of peptides presented by class I MHC molecules. *Curr. Opin. Immunol.* 7: 69–76.

Johnson JG, Jenkins MK. (1994) Monocytes provide a novel costimulatory signal to T cells that is not mediated by the CD28/B7 interaction. *J. Immunol.* 152: 429–437.

Julius M, Maroun CR, Haughn L. (1993) Distinct roles for CD4 and CD8 as co-receptors in antigen receptor signalling. *Immunol. Today* 14: 177-183.

Kagi D, Ledermann B, Burki K, Seiler P, Odermatt B, Olsen KJ, Podack ER, Zinkernagel RM, Hengartner H. (1994) Cytotoxicity mediated by T cells and natural killer cells is greatly impaired in perforin-deficient mice. *Nature* 369: 31–37.

Kaido T, Bandu MT, Maury C, Ferrantini M, Belardelli F, Gresser I. (1995) IFN-α1 gene transfection completely abolishes the tumorigenicity of murine B16 melanoma cells in allogeneic DBA/2 mice and decreases their tumorigenicity in syngeneic C57BL/6 mice. *Int. J. Cancer* 60: 221–229.

Kanner SB, Damle NK, Blake J, Aruffo A, Ledbetter JA. (1992) CD2/LFA-3 ligation induces phospholipase-C γ1 tyrosine phosphorylation and regulates CD3 signaling. *J. Immunol.* 148: 2023–2029.

Karlhofer FM, Yokoyama WM. (1991) Stimulation of murine natural killer (NK) cells by a monoclonal antibody specific for the NK1.1 antigen. IL-2-activated NK cells possess additional specific stimulation pathways. *J. Immunol.* **146**: 3662–3673.

Kawakami Y, Eliyahu S, Delgado CH, Robbins PF, Rivoltini L, Topalian SL, Miki T, Rosenberg SA. (1994) Cloning of the gene coding for a shared human melanoma antigen recognized by autologous T cells infiltrating into tumor. *Proc. Natl Acad. Sci. USA* **91**: 3515–3519.

Kobata T, Agematsu K, Kameoka J, Schlossman SF, Morimoto C. (1994) CD27 is a signal-transducing molecule involved in CD45RA⁺ naive T cell costimulation. *J. Immunol.* **153**: 5422–5432.

Krummel MF, Allison JP. (1995) CD28 and CTLA-4 have opposing effects on the response of T cells to stimulation. *J. Exp. Med.* **182**: 459–465.

Kuchroo VK, Das MP, Brown JA, Ranger AM, Zamvil SS, Sobel RA, Weiner HL, Nabavi N, Glimcher LH. (1995) B7-1 and B7-2 costimulatory molecules activate differentially the Th1/Th2 developmental pathways: application to autoimmune disease therapy. *Cell* **80**: 707-718.

Kuiper M, Peakman M, Farzaneh F. (1995) Ovarian tumour antigens as potential targets for immune gene therapy. *Gene Ther.* **2**: 7-15.

Lanier LL, Chang C, Phillips JH. (1994) Human NKR-P1A. A disulfide-linked homodimer of the C-type lectin superfamily expressed by a subset of NK and T lymphocytes. *J. Immunol.* **153**: 2417-2428.

Lanier LL, O'Fallon S, Somoza C, Phillips JH, Linsley PS, Okumura K, Ito D, Azuma M. (1995) CD80 (B7) and CD86 (B70) provide similar costimulatory signals for T cell proliferation, cytokine production, and generation of CTL. *J. Immunol.* **154**: 97-105.

Leach DR, Krummel MF, Allison JP. (1996) Enhancement of antitumor immunity by CTLA-4 blockade. *Science* **271**: 1734–1736.

Lenschow DJ, Bluestone JA. (1993) T cell co-stimulation and *in vivo* tolerance. *Curr. Opin. Immunol.* **5**: 747-752.

Levine BL, Ueda Y, Craighead N, Huang ML, June CH. (1995) CD28 ligands CD80 (B7-1) and CD86 (B7-2) induce long-term autocrine growth of CD4⁽⁺⁾ T cells and induce similar patterns of cytokine secretion *in vitro*. *Int. Immunol.* **7**: 891–904.

Lewko WM, Smith TL, Bowman DJ, Good RW, Oldham RK. (1995) Interleukin-15 and the growth of tumor derived activated T-cells. *Cancer Biother.* **10**: 13–20.

Li WQ, Diamantstein T, Blankenstein T. (1990) Lack of tumorigenicity of interleukin 4 autocrine growing cells seems related to the anti-tumor function of interleukin 4. *Mol. Immunol.* **27**: 1331–1337.

Li Y, McGowan P, Hellström I, Hellström KE, Chen L. (1994) Costimulation of tumor-reactive CD4⁺ and CD8⁺ T lymphocytes by B7, a natural ligand for CD28, can be used to treat established mouse melanoma. *J. Immunol.* **153**: 421–428.

Linsley PS, Ledbetter JA. (1993) The role of the CD28 receptor during T cell responses to antigen. *Annu. Rev. Immunol.* **11**: 191–212.

Linsley PS, Brady W, Urnes M, Grosmaire LS, Damle NK, Ledbetter JA. (1991) CTLA-4 is a second receptor for the B cell activation antigen B7. *J. Exp. Med.* **174**: 561–569.

Linsley PS, Greene JL, Brady W, Bajorath J, Ledbetter JA, Peach R. (1994) Human B7-1 (CD80) and B7-2 (CD86) bind with similar avidities but distinct kinetics to CD28 and CTLA-4 receptors. *Immunity* **1**: 793–801.

Liu Y, Jones B, Aruffo A, Sullivan KM, Linsley PS, Janeway CAJ. (1992a) Heat-stable antigen is a costimulatory molecule for CD4 T cell growth. *J. Exp. Med.* **175**: 437-445.

Liu Y, Jones B, Brady W, Janeway CA, Jr, Linsley PS. (1992b) Co-stimulation of murine CD4 T cell growth: cooperation between B7 and heat-stable antigen. *Eur. J. Immunol.* **22**: 2855–2859.

Ljunggren HG, Karre K. (1990) In search of the 'missing self': MHC molecules and NK cell recognition. *Immunol. Today* **11**: 237-244.

Lorence RM, Katubig BB, Reichard KW, Reyes HM, Phuangsab A, Sassetti MD, Walter RJ, Peeples ME. (1994) Complete regression of human fibrosarcoma xenografts after local Newcastle disease virus therapy. *Cancer Res.* **54**: 6017-6021.

Mandelboim O, Berke G, Fridkin M, Feldman M, Eisenstein M, Eisenbach L. (1994) CTL induction by a tumour-associated antigen octapeptide derived from a murine lung carcinoma. *Nature* **369**: 67-71.

Martinotti A, Stoppacciaro A, Vagliani M, Melani C, Spreafico F, Wysocka M, Parmiani G, Trinchieri G, Colombo MP. (1995) CD4 T cells inhibit *in vivo* the CD8-mediated immune response against murine colon carcinoma cells transduced with interleukin-12 genes. *Eur. J. Immunol.* **25**: 137-146.

Matulonis UA, Dosiou C, Lamont C, Freeman GJ, Mauch P, Nadler LM, Griffin JD. (1995) Role of B7-1 in mediating an immune response to myeloid leukemia cells. *Blood* **85**: 2507-2515.

Matulonis UA, Dosiou C, Freeman GJ, Lamont C, Mauch P, Nadler LM, Griffin JD. (1996) B7-1 is superior to B7-2 costimulation in the induction and maintenance of T cell-mediated antileukemia immunity. Further evidence that B7-1 and B7-2 are functionally distinct. *J. Immunol.* **156:** 1126–1131.

McBride WH, Thacker JD, Comora S, Economou JS, Kelley D, Hogge D, Dubinett SM, Dougherty GJ. (1992) Genetic modification of a murine fibrosarcoma to produce interleukin 7 stimulates host cell infiltration and tumor immunity. *Cancer Res.* **52:** 3931–3937.

Mentzer SJ, Remold OD, Crimmins MA, Bierer BE, Rosen FS, Burakoff SJ. (1987) Sialophorin, a surface sialoglycoprotein defective in the Wiskott-Aldrich syndrome, is involved in human T lymphocyte proliferation. *J. Exp. Med.* **165:** 1383–1392.

Miller AR, McBride WH, Dubinett SM, Dougherty GJ, Thacker JD, Shau H, Kohn DB, Moen RC, Walker MJ, Chiu R. (1993) Transduction of human melanoma cell lines with the human interleukin-7 gene using retroviral-mediated gene transfer: comparison of immunologic properties with interleukin-2. *Blood* **82:** 3686–3694.

Mizoguchi H, O'Shea JJ, Longo DL, Loeffler CM, McVicar DW, Ochoa AC. (1992) Alterations in signal transduction molecules in T lymphocytes from tumor-bearing mice. *Science* **258:** 1795–1798.

Moingeon P, Lucich JL, McConkey DJ, Letourneur F, Malissen B, Kochan J, Chang HC, Rodewald HR, Reinherz EL. (1992) CD3 ζ dependence of the CD2 pathway of activation in T lymphocytes and natural killer cells. *Proc. Natl Acad. Sci. USA* **89:** 1492–1496.

Mondino A, Jenkins MK. (1994) Surface proteins involved in T cell costimulation. *J. Leuk. Biol.* **55:** 805–815.

Mosmann TR, Coffman RL. (1989) Th1 and Th2 cells: different patterns of lymphokine secretion lead to different functional properties. *Annu. Rev. Immunol.* **7:** 145–173.

Munger W, DeJoy SQ, Jeyaseelan RS, Torley LW, Grabstein KH, Eisenmann J, Paxton R, Cox T, Wick MM, Kerwar SS. (1995) Studies evaluating the antitumor activity and toxicity of interleukin-15, a new T cell growth factor: comparison with interleukin-2. *Cell Immunol.* **165:** 289–293.

Nagler A, Lanier LL, Cwirla S, Phillips JH. (1989) Comparative studies of human FcRIII-positive and negative natural killer cells. *J. Immunol.* **143:** 3183–3191.

Nandi D, Gross JA, Allison JP. (1994) CD28-mediated costimulation is necessary for optimal proliferation of murine NK cells. *J. Immunol.* **152:** 3361–3369.

Ohira T, Ohe Y, Heike Y, Podack ER, Olsen KJ, Nishio K, Nishio M, Miyahara Y, Funayama Y, Ogasawara H, Arioka H, Kunikane H, Fukuda M, Kato H, Saijo N. (1994) *In vitro* and *in vivo* growth of B16F10 melanoma cells transfected with interleukin-4 cDNA and gene therapy with the transfectant. *J. Cancer Res. Clin. Oncol.* **120:** 631–635.

Podack ER. (1985) The molecular mechanism of lymphocyte-mediated tumor-cell lysis. *Immunol. Today* **6:** 21–27.

Podack ER. (1995) Functional significance of two cytolytic pathways of cytotoxic T lymphocytes. *J. Leuk. Biol.* **57:** 548–552.

Podack ER, Kupfer A. (1991) T-cell effector functions: mechanisms for delivery of cytotoxicity and help. *Annu. Rev. Cell Biol.* **7:** 479–504.

Porgador A, Tzehoval E, Katz A, Vadai E, Revel M, Feldman M, Eisenbach L. (1992) Interleukin 6 gene transfection into Lewis lung carcinoma tumor cells suppresses the malignant phenotype and confers immunotherapeutic competence against parental metastatic cells. *Cancer Res.* **52:** 3679–3686.

Porgador A, Tzehoval E, Vadai E, Feldman M, Eisenbach L. (1993) Immunotherapy via gene therapy: comparison of the effects of tumor cells transduced with the interleukin-2, interleukin-6, or interferon-γ genes. *J. Immunother.* **14:** 191–201.

Qin Z, Kruger Krasagakes S, Kunzendorf U, Hock H, Diamantstein T, Blankenstein T. (1993) Expression of tumor necrosis factor by different tumor cell lines results either in tumor suppression or augmented metastasis. *J. Exp. Med.* **178:** 355–360.

Ramarathinam L, Castle M, Wu Y, Liu Y. (1994) T cell costimulation by B7/BB1 induces CD8 T cell-dependent tumor rejection: an important role of B7/BB1 in the induction, recruitment, and effector function of antitumor T cells. *J. Exp. Med.* **179:** 1205–1214.

Rammensee HG, Kroschewski R, Frangoulis B. (1989) Clonal anergy induced in mature Vβ6$^+$ T lymphocytes on immunizing Mls-1^b mice with Mls-1^a expressing cells. *Nature* **339:** 541–544.

Ramsdell F, Lantz T, Fowlkes BJ. (1989) A nondeletional mechanism of thymic self tolerance. *Science* **246:** 1038–1041.

Restifo NP, Spiess PJ, Karp SE, Mule JJ, Rosenberg SA. (1992) A nonimmunogenic sarcoma transduced with the cDNA for interferon γ elicits CD8$^+$ T cells against the wild-type tumor: correlation with antigen presentation capability. *J. Exp. Med.* **175:** 1423–1431.

Richter G, Kruger Krasagakes S, Hein G, Huls C, Schmitt E, Diamantstein T, Blankenstein T. (1993) Interleukin 10 transfected into Chinese hamster ovary cells prevents tumor growth and macrophage infiltration. *Cancer Res.* **53:** 4134–4137.

Rosenberg SA, Mulé JJ, Spiess PJ, Reichert CM, Schwarz S. (1985) Regression of established pulmonary metastasis and subcutaneous tumor mediated by the systemic administration of high-dose recombinant interleukin 2. *J. Exp. Med.* **161:** 1169–1188.

Rosenberg SA, Lotze MT, Muul LM, Chang AE, Avis FP, Leitman S, Linehan WM, Robertson CN, Lee RE, Rubin JT. (1987) A progress report on the treatment of 157 patients with advanced cancer using lymphokine activated killer cells and interleukin-2 or high dose interleukin-2 alone. *New Engl. J. Med.* **316:** 889–897.

Rosenstein Y, Park JK, Hahn WC, Rosen FS, Bierer BE, Burakoff SJ. (1991) CD43, a molecule defective in Wiskott-Aldrich syndrome, binds ICAM-1. *Nature* **354:** 233–235.

Rosenthal FM, Cronin K, Bannerji R, Golde DW, Gansbacher B. (1994) Augmentation of antitumor immunity by tumor cells transduced with a retroviral vector carrying the interleukin-2 and interferon γ cDNAs. *Blood* **83:** 1289–1298.

Rouvier E, Luciani MF, Golstein P. (1993) Fas involvement in Ca$^{(2+)}$-independent T cell-mediated cytotoxicity. *J. Exp. Med.* **177:** 195–200.

Salvadori S, Gansbacher B, Pizzimenti AM, Zier KS. (1994) Abnormal signal transduction by T cells of mice with parental tumors is not seen in mice bearing IL-2-secreting tumors. *J. Immunol.* **153:** 5176–5182.

Salvadori S, Gansbacher B, Wernick I, Tirelli S, Zier K. (1995) B7-1 amplifies the response to interleukin-2-secreting tumor vaccines *in vivo*, but fails to induce a response by naive cells *in vitro*. *Hum. Gene Ther.* **6:** 1299–1306.

Schild H, von Hoegen P, Schirrmacher V. (1989) Modification of tumor cells by a low dose of Newcastle disease virus. Augmented tumor-specific T cell responses as a result of CD4$^+$ and CD8$^+$ immune T cell cooperation. *Cancer Immunol. Immunother.* **28:** 22–28.

Schlag P, Manasterski M, Gerneth T, Hohenberger P, Dueck M, Herfarth C, Liebrich W, Schirrmacher V. (1992) Active specific immunotherapy with Newcastle-disease-virus-modified autologous tumor cells following resection of liver metastases in colorectal cancer. *Cancer Immunol. Immunother.* **35:** 325–330.

Schwartz RH. (1990) A cell culture model for T lymphocyte clonal anergy. *Science* **248:** 1349–1356.

Schwartz RH. (1993) T cell anergy. *Sci. Am.* **269:** 48–54.

Seaman WE, Sleisenger M, Eriksson E, Koo GC. (1987) Depletion of natural killer cells in mice by monoclonal antibody to NK-1.1. Reduction in host defense against malignancy without loss of cellular or humoral immunity. *J. Immunol.* **138:** 4539–4544.

Seder AR, Paul WE. (1994) Acquisition of lymphokine-producing phenotype by CD4$^+$ T cells. *Annu. Rev. Immunol.* **12:** 635–673.

Shahinian A, Pfeffer K, Lee KP, Kundig TM, Kishihara K, Wakeham A, Kawai K, Ohashi PS, Thompson CB, Mak TW. (1993) Differential T cell costimulatory requirements in CD28-deficient mice. *Science* **261:** 609–612.

Slattery E, Taira H, Broeze R, Lengyel P. (1980) Mouse interferons: production by Ehrlich ascites tumour cells infected with Newcastle disease virus and its enhancement by theophylline. *J. Gen. Virol.* **49:** 91–96.

Sperling AI, Green JM, Mosley RL, Smith PL, DiPaolo RJ, Klein JR, Bluestone JA, Thompson CB. (1995) CD43 is a murine T cell costimulatory receptor that functions independently of CD28. *J. Exp. Med.* **182:** 139–146.

Spooner RA, Deonarain MP, Epenetos AA. (1995) DNA vaccination for cancer treatment. *Gene Ther.* **2:** 173–180.

Springer TA. (1994) Traffic signals for lymphocyte recirculation and leukocyte emigration: the multistep paradigm. *Cell* **76:** 301–314.

Stevenson FK, Zhu D, King CA, Ashworth LJ, Kumar S, Thompsett A, Hawkins RE. (1995) A genetic approach to idiotypic vaccination for B cell lymphoma. *Ann. N.Y. Acad. Sci.* **772:** 212–226.

Street NE, Schumacher JH, Fong TA, Bass H, Fiorentino DF, Leverah JA, Mosmann TR. (1990) Heterogeneity of mouse helper T cells. Evidence from bulk cultures and limiting dilution cloning for precursors of Th1 and Th2 cells. *J. Immunol.* **144:** 1629–1639.

Stuhler G, Walden P. (1994) Recruitment of helper T cells for induction of tumour rejection by cytolytic T lymphocytes. *Cancer Immunol. Immunother.* **39:** 342–345.

Tahara H, Zitvogel L, Storkus WJ, Zeh HJ, McKinney TG, Schreiber RD, Gubler U, Robbins PD, Lotze MT. (1995) Effective eradication of established murine tumors with IL-12 gene therapy using a polycistronic retroviral vector. *J. Immunol.* **154:** 6466–6474.

Tepper RI, Pattengale PK, Leder P. (1989) Murine interleukin-4 displays potent anti-tumor activity *in vivo. Cell* **57:** 503–512.

Townsend SE, Allison JP. (1993) Tumor rejection after direct costimulation of CD8[+] T cells by B7-transfected melanoma cells. *Science* **259:** 368–370.

Townsend SE, Su FW, Atherton JM, Allison JP. (1994) Specificity and longevity of antitumor immune responses induced by B7-transfected tumors. *Cancer Res.* **54:** 6477-6483.

Tschopp J, Nabholz M. (1990) Perforin-mediated target cell lysis by cytolytic T lymphocytes. *Annu. Rev. Immunol.* **8:** 279–302.

van den Driessche T, Geldhof A, Bakkus M, Toussaint Demylle D, Brijs L, Thielemans K, Verschueren H, De Baetselier P. (1994) Metastasis of mouse T lymphoma cells is controlled by the level of major histocompatibility complex class I H-2D^k antigens. *Int. J. Cancer* **58:** 217-225.

van den Eynde B, Brichard VG. (1995) New tumour antigens recognised by T cells. *Curr. Opin. Immunol.* **7:** 674–681.

van den Eynde B, Lethe B, Van Pel A, De Plaen E, Boon T. (1991) The gene coding for a major tumor rejection antigen of tumor P815 is identical to the normal gene of syngeneic DBA/2 mice. *J. Exp. Med.* **173:** 1373–1384.

van den Eynde B, Peeters O, De Backer O, Gaugler B, Lucas S, Boon T. (1995) A new family of genes coding for an antigen recognized by autologous cytolytic T-lymphocytes on a human-melanoma. *J. Exp. Med.* **182:** 689–698.

van der Bruggen P, Traversari C, Chomez P, Lurquin C, De Plaen E, Van den Eynde B, Knuth A, Boon T. (1991) A gene encoding an antigen recognized by cytolytic T lymphocytes on a human melanoma. *Science* **254:** 1643–1647.

van Seventer GA, Shimizu Y, Horgan KJ, Luce GE, Webb D, Shaw S. (1991a) Remote T cell co-stimulation via LFA-1/ICAM-1 and CD2/LFA-3: demonstration with immobilized ligand/mAb and implication in monocyte-mediated co-stimulation. *Eur. J. Immunol.* **21:** 1711–1718.

van Seventer GA, Shimizu Y, Shaw S. (1991b) Roles of multiple accessory molecules in T-cell activation: bilateral interplay of adhesion and costimulation. *Curr. Opin. Immunol.* **3:** 294–303.

van Seventer GA, Bonvini E, Yamada H, Conti A, Stringfellow S, June CH, Shaw S. (1992) Costimulation of T cell receptor/CD3-mediated activation of resting human CD4[+] T cells by leukocyte function-associated antigen-1 ligand intercellular cell adhesion molecule-1 involves prolonged inositol phospholipid hydrolysis and sustained increase of intracellular Ca^{2+} levels. *J. Immunol.* **149:** 3872–3880.

Walunas TL, Lenschow DJ, Bakker CY, Linsley PS, Freeman GJ, Green JM, Thompson CB, Bluestone JA. (1994) CTLA-4 can function as a negative regulator of T cell activation. *Immunity* **1:** 405–413.

Wang YC, Zhu L, McHugh R, Sell KW, Selvaraj P. (1995) Expression of heat-stable antigen on tumor cells provides co-stimulation for tumor-specific T cell proliferation and cytotoxicity in mice. *Eur. J. Immunol.* **25:** 1163–1167.

Watanabe Y, Kuribayashi K, Miyatake S, Nishihara K, Nakayama E, Taniyama T, Sakata T. (1989) Exogenous expression of mouse interferon γ cDNA in mouse neuroblastoma C1300 cells results in reduced tumorigenicity by augmented anti-tumor immunity. *Proc. Natl Acad. Sci. USA* **86:** 9456–9460.

Welsh RM, Brubaker JO, Vargas Cortes M, O'Donnell CL. (1991) Natural killer (NK) cell response to virus infections in mice with severe combined immunodeficiency. The stimulation of NK cells and the NK cell-dependent control of virus infections occur independently of T and B cell function. *J. Exp. Med.* **173:** 1053–1063.

Whiteside TL, Herberman RB. (1995) The role of natural killer cells in immune surveillance of cancer. *Curr. Opin. Immunol.* **7:** 704–710.

Wu T, Huang AY, Jaffee EM, Levitsky HI, Pardoll DM. (1995) A reassessment of the role of B7-1 expression in tumor rejection. *J. Exp. Med.* **182:** 1415–1421.

Yang G, Hellström KE, Hellström I, Chen L. (1995) Antitumor immunity elicited by tumor cells transfected with B7-2, a second ligand for CD28/CTLA-4 costimulatory molecules. *J. Immunol.* **154:** 2794–2800.

Yokochi T, Holly RD, Clark EA. (1982) B lymphoblast antigen (BB-1) expressed on Epstein-Barr virus-activated B cell blasts, B lymphoblastoid cell lines, and Burkitt's lymphomas. *J. Immunol.* **128:** 823–827.

Yokoyama WM. (1995) Natural killer cell receptors. *Curr. Opin. Immunol.* **7:** 110–120.

Zuckerman LA, Sant AJ, Miller J. (1995) Identification of a unique costimulatory activity for murine T helper 1 T cell clones. *J. Immunol.* **154:** 4503–4512.

The muscular dystrophies

Caroline A. Sewry and Terence A. Partridge

17.1 Introduction

The muscular dystrophies are genetically determined disorders characterized by progressive weakness and degeneration of muscle. Traditionally they have been subdivided according to their severity, clinical presentation and mode of inheritance, but it is now apparent that similar clinical phenotypes can result from different genetic defects. As more genes and gene products are identified, the classification of the muscular dystrophies is changing to take account of this (see *Journal of Neuromuscular Disorders* for a regular update).

Many of the proteins that are defective in the muscular dystrophies are not only expressed in skeletal muscle but also in cardiac and smooth muscle, the central and peripheral nervous systems and visceral tissues. A clinical effect may not be apparent in all tissues, but therapeutic strategies and targeting methods may need to be carefully devised to accommodate the diverse expression of some proteins.

Research into gene therapy in the muscular dystrophies is presently concentrating on the recessive disorders, with the idea that these might be treated by complementation of the defective genes with an extra copy. Although a few dominant genes have been mapped (e.g. facioscapulohumeral dystrophy, Bethlem myopathy) and in some cases cloned (myotonic dystrophy), the precise mechanisms governing the relationship between the genetic defect, protein expression and the clinical phenotype has yet to be defined and, in consequence, no general strategy has been proposed for dominant disorders. This may relate, in part, to the lack of suitable animal models for the dominant muscular dystrophies.

The various muscular dystrophies provide perhaps the widest range of known gene defects within a single tissue as targets for gene therapy. Of those identified so far, the majority are components of the sarcolemmal dystrophin-associated complex (*Figure 17.1*), probably reflecting a bias towards genes associated with the first discovered, dystrophin, as likely candidates for autosomal dystrophies.

17.2 Duchenne and Becker muscular dystrophy

The gene responsible for X-linked Duchenne muscular dystrophy (DMD) was the first defective gene to be isolated in the muscular dystrophies [see Emery, 1993; Partridge, 1993, for reviews on the molecular aspects of DMD and Becker muscular dystrophy (BMD) and dystrophin]. Since then, considerable research efforts have been made to understand the role and interactions of the gene product, dystrophin,

Gene Therapy, edited by N.R. Lemoine and D.N. Cooper.
© 1996 BIOS Scientific Publishers Ltd, Oxford.

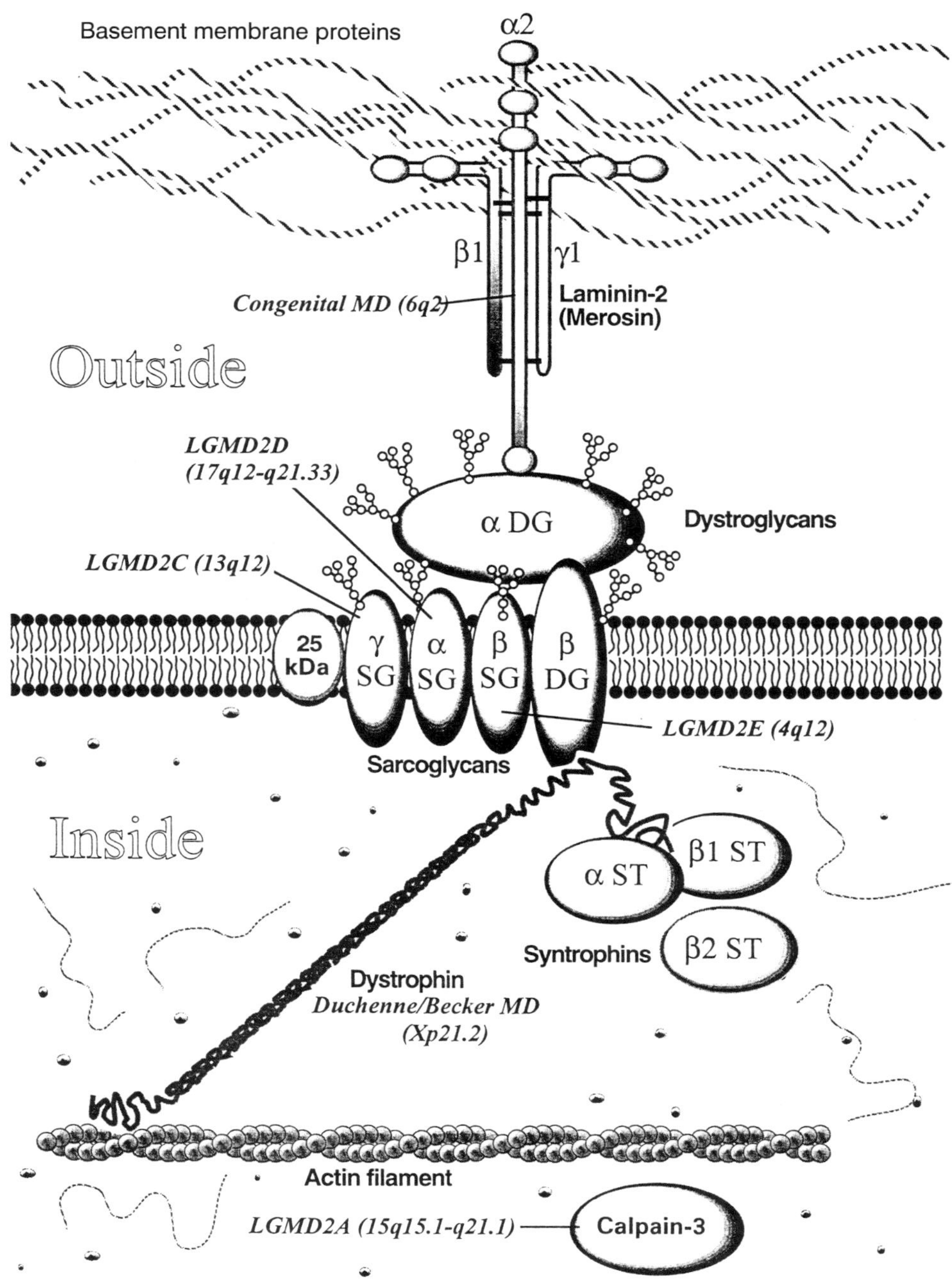

Figure 17.1. Diagram of the hypothetical relationships between the various components of the dystrophin–transmembrane complex. Inherited muscular dystrophies known to be associated with particular components and their chromosomal locations are indicated in italic text. Also indicated is the calpain-3-associated disease. Emerin, the protein which is defective in Emery–Dreifuss muscular dystrophy (MD), is not shown on this diagram. DG, dystroglycan; SG, sarcoglycan; ST, syntrophin.

and to develop therapeutic strategies. These largely involve animal models with defects in the dystrophin gene, in particular the *mdx* mouse (Bulfield *et al.*, 1984; Sicinski *et al.*, 1989) and transgenic mice (Phelps *et al.*, 1995; Wells *et al.*, 1995). Clinical trials in DMD patients using myoblast transfer (see below) have so far proved unsuccessful (Gussoni *et al.*, 1992; Huard *et al.*, 1992; Karpati *et al.*, 1993; Mendell *et al.*, 1995; Morandi *et al.*, 1994).

The defective gene (*DMD*) responsible for Duchenne dystrophy, and its milder allelic form, BMD, is extremely large and complex, being composed of 2.5 Mb of genomic sequence arranged in 79 exons. There are at least seven promoters that give rise to variably sized transcripts and which manifest tissue-specific expression (Ahn and Kunkel, 1993; Lidov *et al.*, 1995). In skeletal muscle, the most abundant and largest isoform, dystrophin, has a molecular mass of approximately 427 kDa. The amino acid sequence of dystrophin predicts four distinct structural domains with an N-terminal actin-binding region, a rod domain with homology to β-spectrin, a cysteine rich domain and a unique C-terminal region that shows considerable splice variation. This structure and interaction with actin has led to the belief that dystrophin is a cytoskeletal protein that may help to stabilize the membrane during contraction. Recent evidence also suggests that components of a transmembrane complex associated with dystrophin (*Figure 17.1*) may have a role in signal transduction and cytoskeletal organization (Campanelli *et al.*, 1994; Gee *et al.*, 1994; Yang *et al.*, 1995).

In normal muscle, dystrophin is evenly distributed on the cytoplasmic face of the sarcolemma. Studies of thin sections, however, show that it has a costameric distribution, similar to other cytoskeletal proteins (Porter *et al.*, 1992; Straub *et al.*, 1992). In DMD, and the *mdx* mouse and canine models, dystrophin is absent from the sarcolemma of most fibres because of mutations that disrupt the reading frame (Monaco *et al.*, 1988; Sharp *et al.*, 1992; Sicinski *et al.*, 1989). Approximately 65–70% of patients have a detectable deletion, 5–10% have a duplication and the remainder have point mutations, many of which have not been identified. Although mutations can arise in various parts of the gene, most occur in two 'hot-spots', one N-terminal involving exon 7, and the other in the rod domain, involving exon 44. A variety of alternative splicing events occur in the gene, some of which restore the reading frame and give rise to the expression of minor transcripts. Small amounts of protein may, therefore, be identified by immunoblotting or immunolabelling (Nicholson *et al.*, 1990). Such splicing events and restoration of the reading frame are also believed to give rise to 'revertant fibres' in muscle from DMD patients (Klein *et al.*, 1992; Sherratt *et al.*, 1993). These show almost normal immunolabelling with antibodies to dystrophin and occur as isolated fibres or small clusters. Assessment of gene therapy studies has to take account of these inherent dystrophin-positive fibres.

Maintenance of the reading frame also occurs in the majority of cases of BMD, and the dystrophin expressed is of smaller molecular mass and usually of lower abundance. The presence of the cysteine-rich region is thought to be essential for the correct localization of dystrophin (Arahata *et al.*, 1991). No correlation has been found between the size of deletions and severity in either DMD or BMD. One of the largest deletions detected was in a BMD patient who lacks almost 50% of the rod domain (England *et al.*, 1990). The mild phenotype in this patient is being exploited in gene therapy studies, by the transfer of this 'mini-gene' into model systems (Kumar-Singh and Chamberlain, 1996; Phelps *et al.*, 1995; Wells *et al.*, 1995).

A variable degree of non-progressive intellectual impairment is a feature of many DMD patients (Dubowitz, 1995). The brain isoform of dystrophin, driven by a 5′ promoter, has a similar molecular mass to that in skeletal muscle and is absent in the brain of *mdx* mice (see Ahn and Kunkel, 1993). No correlation between the size or position of deletions and intellectual impairment has been found. There has been a report, however, of a severe clinical phenotype and mental retardation in patients with mutations affecting Dp116, the C-terminal transcript with an initiation site upstream of exon 56 (Comi *et al.*, 1995). This isoform is expressed in Schwann cells, and demyelination has been found in patients with an absence of this isoform, in addition to the absence of full-length transcripts. Motor nerve conduction velocity in these patients, however, is not affected. It has also been suggested that the absence of Dp140, the isoform initiated in exon 44, may correlate with reduced intellectual impairment. Gene therapy directed towards the nervous system in muscular dystrophies has not been addressed, and most efforts are attempting to improve the quality of life by improving motor function.

Another small transcript of the dystrophin gene, Dp71 or apo-dystrophin-1, is the major transcript in several non-muscle tissues, including brain and liver. The promoter for this isoform lies between exons 62 and 63, and alternatively spliced isoforms have been cloned (Austin *et al.*, 1995). This transcript is barely detectable in normal skeletal or cardiac muscle, but expression occurs in the heart in some X-linked cardiomyopathic patients (Muntoni *et al.*, 1995). These patients have minimal skeletal muscle weakness because of a splicing event in the 5′ promoter region that results in near normal expression of dystrophin in skeletal muscle. In the heart, this event does not take place and the absence of dystrophin leads to a severe cardiomyopathy. Gene therapy in these patients will, therefore, only need to target cardiac muscle.

Interestingly, an autosomal homologue of dystrophin, called utrophin (Love *et al.*, 1989), has very strong homology at the protein level and is found associated with similar transmembrane glycoproteins in many tissues. Further interest in this protein has been generated by the finding that it is widely expressed in developing muscle, becoming restricted to the folds of the neuromuscular junction as the fibres mature, and that its expression is elevated in dystrophic muscle fibres. Together, these observations have led to the proposal that utrophin is a developmental isoform of dystrophin and that, if its expression in dystrophic muscle could be increased artificially, it might serve to partially replace the function of dystrophin (Tinsley and Davies, 1993).

17.3 Recessive limb-girdle muscular dystrophies

Dystrophin is associated with a large oligomeric complex of sarcolemmal proteins and glycoproteins (*Figure 17.1*; see Campbell, 1995; Matsumura and Campbell 1994; Ozawa *et al.*, 1995). Five components are transmembrane (α-, β-, γ-sarcoglycan, β-dystroglycan and a 25 kDa glycoprotein), one is extracellular (α-dystroglycan) and the syntrophin triplet is cytoplasmic (α-, β₁-, β₂-syntrophin). α- and β-dystroglycan are post-translational products from a single mRNA from a gene on chromosome 3p21. α-Dystroglycan binds to the G domain of the α2 chain of laminin-2 (merosin) and to the transmembrane component, β-dystroglycan. It is also a ligand for agrin. β-Dystroglycan has been shown to bind to the cysteine-rich region of dystrophin, which in turn binds to F-actin filaments. The syntrophin complex is associated with the C-terminal domain of dystrophin, but the precise association with the sarcoglycan complex has not yet been identified. Thus, the dystrophin-associated glycoprotein complex links the extracellular matrix with the cytoskeleton of the muscle fibre. It is believed to have a stabilizing effect on the sarcolemma and may provide protection against contraction-induced stress. Any disruption of this link may render the muscle fibre susceptible to damage and lead to necrosis.

In DMD, the absence of dystrophin is accompanied by reduced expression of all the dystrophin-associated proteins, although the mRNAs are apparently normal. Similarly in BMD, there is reduced expression of dystrophin and of the proteins of the complex. In carriers of DMD, fibres with an absence of dystrophin also show reduced expression of the dystrophin-associated proteins (Sewry *et al.*, 1994). Restoration of dystrophin expression in transgenic *mdx* mice results in normal sarcolemmal expression of the dystrophin-associated proteins (Wells *et al.*, 1995). Thus, gene therapy of one component of the sarcolemmal complex should restore the complete complex.

No disease-related defects have yet been identified in the dystroglycan or syntrophin complexes. Mutations in each component of the sarcoglycan complex, however, result in variants of autosomal recessive limb-girdle muscular dystrophy (*Figure 17.1*) (Bonneman *et al.*, 1995; Lim *et al.*, 1995; Noguchi *et al.*, 1995; Roberds *et al.*, 1994). Recent studies suggest that the sarcoglycan complex acts as a unit and that a primary deficiency in one component is associated with a secondary deficiency of the other two components. It is assumed that gene therapy for one defect will restore expression of the other two components, in a situation analogous to the restoration of the glycoproteins in transgenic *mdx* mice.

The first glycoprotein to be associated with a muscular dystrophy was the 50 kDa component, α-sarcoglycan (previously known as adhalin). A deficiency of this protein was found in a group of North African patients with a severe, childhood, autosomal recessive muscular dystrophy (SCARMD) that has some similarity to DMD (see Campbell, 1995; Ozawa *et al.*, 1995). The defect was later mapped to chromosome 13q but the gene encoding α-sarcoglycan was subsequently located on chromosome 17q. Thus, the reduced expression of α-sarcoglycan in the North African patients is a secondary effect. The gene (*LGMD2C*) on chromosome 13q is now known to code for the 35 kDa glycoprotein, γ-sarcoglycan. A variety of mutations have been identified in the genes for α- and γ-sarcoglycan and both are known to have a wide geographical distribution (Noguchi *et al.*, 1995; Roberds *et al.*, 1994). The gene (*LGMD2B*) for the 43 kDa glycoprotein, β-sarcoglycan, has also been mapped recently, to chromosome 4q, and mutations have been observed in an inbred American Amish population and in one isolated case (Bonneman *et al.*, 1995; Lim *et al.*, 1995).

17.4 Congenital muscular dystrophies

The 156 kDa glycoprotein, α-dystroglycan, has been shown to be the ligand for the α2 chain of laminin-2 (merosin) (Ervasti and Campbell, 1993). This laminin subunit is deficient in about 50% of patients with congenital muscular dystrophy (CMD), and linkage and mutational analysis indicate that this is caused by mutation in the gene for α2 laminin (*LAMA2*) on chromosome 6q22 (Helbling-Leclerc *et al.*, 1995; Sewry *et al.*, 1995; Tomé *et al.*, 1994). Thus, another defect in the link between the extracellular matrix and the cytoskeleton is associated with a muscular dystrophy. Patients with a defect in α2 laminin (often referred to as merosin-deficient CMD) have a severe phenotype that involves skeletal and cardiac muscle and the peripheral and central nervous systems, all tissues in which the protein is expressed (Mercuri *et al.*, 1995; Philpot *et al.*, 1995; Shorer *et al.*, 1995). It is not yet known if restoration of protein expression in all these tissues is required to improve the phenotype, but studies of two allelic models, the *dy/dy* and *dy^{2J}/dy^{2J}* mouse, both with a primary deficiency of the α2 chain of laminin, may prove informative (Sunada *et al.*, 1994; Xu *et al.*, 1994).

Patients with a pronounced reduction of α2 laminin have muscle weakness and hypotonia from birth, or within the first few months of life. They rarely achieve independent ambulation; they have reduced motor nerve conduction velocities, abnormal evoked visual responses, some have a cardiomyopathy and most have white matter abnormalities on magnetic resonance imaging (MRI) of the brain. However, they have a normal intellect, in contrast to patients with the Fukuyama form of CMD, prevalent in Japan. These patients show a secondary reduction of α2 laminin as the disorder is linked to chromosome 9q31–33 (Toda *et al.*, 1993). Evidence from a few cases suggests that the muscle–eye–brain form of CMD may also map to the same region on 9q, but a Finnish group of patients with muscle–eye–brain disease are not allelic with Fukuyama CMD and do not map to 9q31–33.

Several forms of laminin have been identified, all composed of one heavy chain (α) and two light chains (β and γ) arranged in a cross-like arrangement. Homologues of each chain have been sequenced, and various combinations of these give rise to different heterotrimeric forms of laminins. Expression of these different laminins is tissue-specific and developmentally regulated. The predominant form of laminin on muscle fibres, laminin-2, is composed of α2–β1–γ1 chains. In addition to the deficiency of α2 laminin, some CMD patients also show a secondary reduction in the β1 chain (Sewry *et al.*, 1995). A reduction of β1 laminin has also been reported in some patients with a deficiency of α-sarcoglycan (Yamada *et al.*, 1995). Recent evidence also suggests that there may be a secondary reduction both of the β2 laminin chain and of α-sarcoglycan on muscle fibres in patients with the Walker–Warburg form of CMD (Wewer *et al.*, 1995). The data therefore suggest an association between laminin and α-sarcoglycan.

In some disorders, the deficiency of protein is accompanied by over-expression of a homologous protein. For example, CMD patients with a deficiency of the α2 chain of laminin show an over-expression of the homologous α1 chain on muscle fibres (Sewry *et al.*, 1995; Tomé *et al.*, 1994), whilst in limb-girdle patients with a reduction of α-sarcoglycan and β1 laminin, there appears to be over-expression of β2 laminin (Yamada *et al.*, 1995). This may be a compensatory effect, but it is not known if it will interfere with gene therapy attempts that target existing fibres.

All patients with a deficiency of α2 laminin studied to date have been shown to link to the region of the *LAMA2* gene on chromosome 6. There is, therefore, no current evidence to suggest genetic heterogeneity. It is apparent, however, that the amount of protein detected may vary from total absence, to slight traces, or only a moderate reduction. This suggests that some mutations may be out-of-frame, whereas others are in-frame, giving rise to expression of a truncated protein. This may be analogous to the *dy^{2J}* mutant mouse in which a splicing event in the α2 laminin gene results in reduced abundance of a truncated portion of α2 laminin (Sunada *et al.*, 1995; Xu *et al.*, 1994).

No gene therapy studies with the α2 laminin gene in humans have been attempted, probably because the structure of the gene is not fully known. However, cell transplantation therapy, which is not dependent on such detailed knowledge, has been attempted in the *dy* mouse model (Vilquin *et al.*, 1996). Like dystrophin, the $\alpha 2$ laminin gene is large, with a transcript of at least 10 kb, making cDNA viral vectors difficult to construct. Also, the protein has several important sites for binding to other proteins such as integrins and other extracellular matrix proteins, and the essential parts of the gene required to restore the complete laminin heterotrimer are not fully understood.

17.5 Other muscular dystrophies

Recent molecular analysis has identified defective proteins that are not structural, and not directly linked with the dystrophin-associated glycoprotein complex. A different strategy for gene therapy may therefore have to be designed. Emery–Dreifuss muscular dystrophy is characterized by skeletal muscle weakness, joint contractures and a cardiomyopathy with arrhythmia. One form of the disease is X-linked and mutations in the gene (*EMD*) have been identified (Bione *et al.*, 1994, 1995). The gene codes for a small protein, emerin, of 254 amino acids with a hydrophobic C-terminal region, similar to that found in membrane proteins of the secretory pathway involved in vesicular transport. It has been shown recently to localize to the nuclear membrane and is expressed ubiquitously (Manilal *et al.*, 1996).

The recessive limb-girdle muscular dystrophy, LGMD2A, is caused by mutations in a gene (*CAPN3*) on chromosome 15q15 that codes for the muscle-specific calcium-activated neutral protease, calpain-3 (Richard *et al.*, 1995). The localization of this enzyme in muscle has not been reported, possibly because of rapid autolysis, but transfection in COS cells and myoblasts have localized it to the nucleus. It has been suggested that calpain-3 may be involved in cell signalling and it is another example of a mutation that does not influence the sarcolemmal cytoskeleton directly. Difficulties in gene therapy in LGMD2A may also be compounded by a possible digenic inheritance in which expression of the mutations in calpain-3 may be dependent on genetic background.

Linkage of other forms of muscular dystrophies is known, such as the recessive limb-girdle dystrophy on chromosome 2p, a dominant limb-girdle dystrophy on 5q and dominant facioscapulohumeral dystrophy, but the genes involved have not been isolated, nor the relevant protein localized.

17.6 General principles of therapy

For recessive dystrophies, there is a rationale for genetic therapy based on complementation of the defective gene with an extra copy. This would entail the targeting of an expression construct encoding the competent gene to the great majority of muscle fibres; a feat which we cannot achieve presently with any useful level of efficiency. An exception to this may be dystrophies where the defective protein is located outside the muscle fibre, in the extracellular matrix, as in merosin-deficient CMD. Since the basement membrane is in part dependent on fibroblastic cells (Swasdison and Mayne, 1992), it may be more feasible to target these cells, rather than the muscle cells. In fact, a recent report of the partial restoration of the $\alpha 2$-laminin chain by cell transplantation in the merosin-deficient *dy/dy* mouse suggests that not all myogenic cells nor all fibroblastic cells are able to synthesize this protein in muscle, with the implication that some unidentified subset of one of these categories of cell is able to do so (Vilquin *et al.*, 1996).

A more elegant prospect than genetic complementation would be a discrete correction of the defective gene, but this is too complex to be worth consideration for the moment, since, in addition to the presently unattainable objective of introducing the construct into a high proportion of muscle cells in the first place, it entails accurate and highly efficient targeting of the reparative construct to the genetic defect in the genome of muscle nuclei.

Genetic therapies for dominantly inherited dystrophies such as dystrophia myotonica and facioscapulohumoral dystrophy, where the relationship between the mutant gene and the physiological dysfunction is ill understood, are yet more difficult to envisage. Certainly, it seems unlikely that genetic complementation could be of use in such diseases, and gene repair, were it achievable, looks to be the only rational option.

17.7 Gene targeting

A fundamental problem for genetic therapies in general is the efficient and preferably selective delivery of the genetic construct to the cells to be modified. For some tissues, a degree of selectivity may be provided to an otherwise non-selective process of gene delivery by virtue of the relative ease of access of the target tissue, for instance via the vascular system (e.g. liver, bone marrow), the airways (e.g. lung lining cells) or other compartmental body spaces (e.g. the retina). However, skeletal muscle does not enjoy any of these advantages and, because it is the most abundant single tissue in the body and is widely dispersed, the possibility of vascular dissemination of genetic material is regarded as the most feasible route of delivery of modifying genes into the majority of muscle fibres.

In the absence of a mechanism for targeting the therapeutic construct to muscle, it is possible nonetheless to obtain tissue-specific gene expression by using a muscle-specific promoter to restrict expression to muscle cells (Dickson and Dunckley, 1993).

17.8 Myoblast transplantation

Up to now, cell transplantation has been the only successful form of genetic therapy for any inherited disease, in the form of transplantation of normal bone marrow for genetic diseases of the blood. Long-term efficacy of this approach relies upon the presence of multipotential haematopoietic stem cells in the transplant which are able to regenerate the full complement of genetically sound mature blood cells within the recipient. By analogy, it had been argued that genetically normal myogenic precursor cells might be able to repair and reconstitute the damaged muscle fibres of muscular dystrophy patients. We were able to validate the principle by implantation of myogenic cells derived from normal mouse muscle into muscles of the *mdx* dystrophic mouse; the most successful examples producing up to 30% of dystrophin-positive fibres from an injection of half a million cells into the tibialis anterior muscle (Partridge *et al.*, 1989). A similar experiment in which normal human myogenic cells were xenografted into *mdx* mice resulted in a rather lower yield of dystrophin-positive fibres (Karpati *et al.*, 1989).

Subsequent attempts to transfer this therapy to humans have, in those studies which have been well monitored, proven totally unsuccessful (Karpati *et al.*, 1993; Morandi *et al.*, 1994) or have produced no significant biochemical or functional improvement (Gussoni *et al.*, 1992; Huard *et al.*, 1992a,b). Only the most recent report (Mendell *et al.*, 1995) has demonstrated unequivocally that the transplanted myoblasts have produced dystrophin protein within fibres of the host muscle, and this in no more than 10% of the biopsy sample. One of the likely causes of the general failure is immune rejection. This was presaged as a potential problem in earlier mouse studies (Grounds *et al.*, 1980; Morgan *et al.*, 1987) and has been confirmed by those human myoblast transplantation trials which were designed to examine this phenomenon (Huard *et al.*, 1992b). Loss of myogenicity during extensive tissue culture of human myogenic cells derived from biopsies of adult muscle may also have played some part in the failure of these trials (Karpati *et al.*, 1993) and recent experiments suggest that exposure of myoblasts to high levels of basic fibroblast growth factor (bFGF) prior to grafting may partially counteract this (Kinoshita *et al.*, 1995).

17.9 Genetic therapies

A number of more direct strategies for genetic therapy of muscular dystrophies are currently being explored. All of the following have shown some promise in experimental models but none has yet been tested in human trials.

17.9.1 Plasmid insertion

Direct injection. It came as a surprise that simple injection of mammalian expression plasmids, including those encoding the dystrophin gene, into skeletal muscle resulted in uptake and long-term expression of the construct within skeletal muscle fibres in the region of the injection (Acsadi *et al.*, 1991; Wolff *et al.*, 1990). However, only a small number of fibres could be transfected in this way, even by injection of very large amounts of DNA and, although provocation of regeneration in the muscle was found to improve efficiency (Davis *et al.*, 1993), the final level of genetic conversion was still small and this approach now seems to have been abandoned as a likely route to therapy for primary genetic disease of muscle.

Targeted plasmid insertion. The goal of genetic delivery systems is to direct the gene in question accurately to the tissue of choice. This might be achieved by linking the plasmid to a ligand or antibody for a receptor which is, ideally exclusively but at least preferentially, expressed on the surface of the target cell. Many receptors become internalized when they bind to their ligand and, where the adenoviral coat is used for this purpose, this seems to augment the efficiency with which the plasmid gains access to the cytoplasmic and eventually the nuclear compartments of the cell (Curiel *et al.*, 1992).

17.9.2 Recombinant viral vectors

One of the more heavily researched areas of attempted gene therapy in skeletal muscle as for other tissues is the use of the innate cell entry mechanisms of genetically modified viruses to introduce expression constructs into the cell (Dickson and Dunckley, 1993; Karpati and Acsadi, 1993; reviewed in Turner *et al.*, 1996). None of the vectors used to date show any tropism for skeletal muscle, so specificity of expression must be conferred by use of tissue-specific promoters if expression of the gene in non-muscle tissues is to be avoided.

Retroviral vectors. Recombinant retroviral vectors were the first to be used to introduce expression markers into muscle cells. Although they do not have the capacity to carry the full-length dystrophin gene, a 'mini-dystrophin' gene of approximately half the normal size was inserted into a packageable retroviral vector and was shown to be able to transduce dystrophic mouse myoblasts in culture (Dunckley *et al.*, 1992) and, when injected into dystrophic mouse muscles *in vivo* (Dunckley *et al.*, 1993), it gave rise to muscle fibres expressing the mini-dystrophin product. This minigene was derived from a mild Becker dystrophy patient (England *et al.*, 1990) and would be expected to confer a great amelioration of the disease when expressed in dystrophic muscle fibres. This has proved to be the case, for animals made transgenic for this and other similar mini-dystrophin genes were as effectively protected against disease as those made transgenic for the full-length dystrophin gene. Such constructs are thus an important feature of the retroviral strategy for DMD (Kumar-Singh and Chamberlain, 1996; Phelps *et al.*, 1995; Wells *et al.*, 1995).

The advantages of the retrovirus as a vector are that it becomes permanently integrated into the genome of the cells it infects and so offers the prospect of lifelong effectiveness. It also has the advantage over other vectors that the viral genes are not expressed within the transduced cells, thus minimizing the likelihood that these will become the target of an immune response. The main problem for this vector is that it requires division of the target cell in order to become integrated, thus limiting its range *in vivo* to proliferating myogenic cells in growing or regenerating muscle. Since myogenic cells are dividing asynchronously in both situations and sporadically in the case of regenerating muscle, genetic conversion of significant amounts of skeletal muscle would entail the chronic presence of high titres of the recombinant retrovirus within the intercellular spaces of the muscles; attempts to achieve this by implanting mitomycin-treated retroviral producer cells have indeed raised the levels of transduction in the muscles of the recipient but this requires the abrogation of the host immune system for the duration of the infection period (G. Dickson and D.J. Wells, personal communication).

Adenoviral vectors. A non-integrating, recombinant viral vector is not an immediately obvious choice for introducing genetic material into skeletal muscle. However, adenoviruses do exhibit a number of advantages for this purpose. They infect a wide range of cells via a common integrin receptor, they are relatively easy to concentrate to high titre and potentially they could be engineered to package up to about 30 kb of expression plasmid; this is more than sufficient to accommodate a full-length dystrophin cDNA (13 kb) plus a muscle-specific promoter element such as the muscle creatine kinase (MCK) or skeletal muscle α actin promoter. Such use of tissue-specific promoters to confer specificity of expression to the plasmid carried by the adenovirus may be required to compensate the broad spectrum of infectivity of this class of vector.

In experiments on *mdx* mice, recombinant adenoviral vectors have had the most striking effects to date of any system for introducing the mini-dystrophin gene into *mdx* skeletal muscle. Direct injection of the virus into muscle resulted in the conversion of a large proportion of the fibres to dystrophin positivity (Ragot *et al.*, 1993) and significant numbers can be converted by the intravascular route or even by intragastric introduction of the virus (Huard *et al.*, 1995b). However, the effectiveness of infection drops rapidly with the age of the animal, being most effective in newborn animals and barely measurable with mice of a few weeks old (Acsadi *et al.*, 1994). This seems to reflect, in part, an intrinsic loss of availability of viral receptors on muscle as it develops, for regenerating muscle regains the ability to be infected. However, there also seems to be a strong immune component to this phenomenon for, in older animals, muscle fibres of immunodeficient mice remain more readily infectable than those of immunocompetent mice. Indeed, it is now apparent that muscle fibres transduced by adenoviral vectors beyond the first few postnatal days, when tolerance can still be induced in the recipient mice, are subject to a cell-mediated immune rejection both of the adenoviral proteins and the protein encoded by the construct being carried by the vector (Turner *et al.*, 1996).

Herpes vectors. Development of recombinant herpes viruses as vectors has been directed largely at transduction of cells of neural origin but these viruses have been show to infect myoblasts and muscle fibres efficiently enough to be of interest (Huard *et al.*, 1995a).

As with adenoviral vectors, these seem to infect only myoblasts and immature muscle fibres with any useful level of efficiency and they carry a large and complex viral genome which is a potential target for immune attack. The great advantage of this vector is that it has the potential to carry large plasmid inserts.

17.10 Secondary targets of therapy

Genetic therapies for inherited muscle diseases have concentrated attention on the primary genetic defect. In many cases, this may not be the only appropriate target for therapy, for the myodegenerative component of the disease which is the immediate consequence of this primary genetic defect appears not to be the direct cause of the major loss of function of the muscle. Rather, this loss of function seems to be associated with the progressive fibrotic and fatty degenerative changes which accumulate as a consequence of the chronic degenerative and inflammatory processes. These changes, which are probably themselves pathogenic and not spontaneously reversible, together with the mechanisms which underlie them, are as important a group of targets for therapy in the muscular dystrophies as is the primary defect. It may be that genetic methods for modulating cytokine expression could be used to suppress or reverse these secondary and tertiary pathological changes, but pharmacological approaches seem to be at least as appropriate. Indeed, given the known and foreseeable problems of genetic therapies, it is unwarranted at present to neglect the pharmacological routes to therapy for all stages in the pathogenic cascade between primary genetic defect and end-stage degenerative change.

References

Acsadi G, Dickson G, Love DR, Jani A, Walsh FS, Gurusinghe A, Wolff JA, Davies KE. (1991) Human dystrophin expression in mdx mice after intramuscular injection of DNA constructs. *Nature* **352**: 815–818.

Acsadi G, Jani A, Massie B, Simoneau M, Holland P, Blaschuk K, Karpati G. (1994) A differential efficiency of adenovirus-mediated *in vivo* gene transfer into skeletal muscle cells of different maturity. *Hum. Mol. Genet.* **3**: 579–584.

Ahn AH, Kunkel LM. (1993) The structural and functional diversity of dystrophin. *Nature Genet.* **3**: 238–291.

Arahata K, Beggs A, Honda H, Ito S, Ishiura S, Tsukahara T, Ishigura T, Eguchi C, Orimo S, Arikawa E, Kaido M, Nonaka I, Sugita H, Kunkel LM. (1991) Preservation of the C-terminus of dystrophin molecule in the skeletal muscle from Becker muscular dystrophy. *J. Neurol. Sci.* **101**: 148–156.

Austin RC, Howard PL, D'Souza VND, Klamut HJ, Ray PN. (1995) Cloning and characterization of alternatively spliced isoforms of Dp71. *Hum. Mol. Genet.* **9**: 1475–1483.

Bione S, Maestrini E, Rivella S, Mancini M, Regus S, Romeo G, Toniolo, D. (1994) Identification of a novel X-linked gene responsible for Emery–Dreifuss muscular dystrophy. *Nature Genet.* **8**: 323–327.

Bione S, Small K, Aksmanovic VMA, D'Urso M, Ciccodicula A, Merlini L, Morandi L, Kress W, Yates JRW, Warren ST, Toniolo D. (1995) Identification of new mutations in Emery–Dreifuss muscular dystrophy and evidence for genetic heterogeneity of the disease. *Hum. Mol. Genet.* **4:** 1859–1863.

Bonneman CG, Modi R, Noguchi S, Mizuno Y, Yoshida M, Gussoni E, McNally EM, Duggan DJ, Angelini C, Hoffman EP, Ozawa E, Kunkel LM. (1995) β-Sarcoglycan (A3b) mutations cause autosomal recessive muscular dystrophy with the loss of the sarcoglycan complex. *Nature Genet.* **11:** 266–272.

Bulfield G, Siller WG, Wight PAL, Moore KJ. (1984) X chromosome-linked muscular dystrophy (*mdx*) in the mouse. *Proc. Natl Acad Sci. USA* **81:** 1189–1192.

Campanelli JT, Roberds SL, Campbell KP, Scheller RH. (1994) A role for dystrophin-associated glycoproteins and utrophin in agrin-induced AChR clustering. *Cell* **77:** 663–674.

Campbell KP. (1995) Three muscular dystrophies: loss of cytoskeletal–extracellular matrix linkage. *Cell* **80:** 675–679.

Comi GP, Ciafaloni E, de Silva HAR, Prelle A, Bardoni A, Rigoletto C, Robotti M, Bresolin N, Moggio M, Fortunato F, Ciscato P, Turconi A, Rose AD, Scarlato G. (1995) A $G^{+1} \to$ A transversion at the 5' splice site of intron 69 of the dystrophin gene causing the absence of peripheral nerve Dp116 and severe clinical involvement in a DMD patient. *Hum. Mol. Genet.* **4:** 2171–2174.

Curiel DT, Agarwal S, Romer MU, Wagner E, Cotten M, Birnstiel ML, Boucher RC. (1992) Gene transfer to respiratory epithelial cells via the receptor-mediated endocytosis pathway. *Am. J. Respir. Cell Mol. Biol.* **6:** 247–252.

Davis HL, Whalen RG, Demeneix BA. (1993) Direct gene transfer into skeletal muscle *in vivo*: factors affecting efficiency of transfer and stability of expression. *Hum. Gene Ther.* **4:** 151–159.

Dickson G, Dunckley M. (1993) Human dystrophin gene transfer: genetic correction of dystrophin deficiency. In: *Molecular and Cell Biology of Muscular Dystrophy* (ed. TA Partridge). Chapman & Hall, London, pp. 283–302.

Dubowitz V. (1995) *Muscle Disorders in Childhood*, 2nd Edn. WB Saunders, London.

Dunckley MG, Love DR, Davies KE, Walsh FS, Morris GE, Dickson G. (1992) Retroviral-mediated transfer of a dystrophin minigene into *mdx* mouse myoblasts *in vitro*. *FEBS Lett.* **296:** 128–134.

Dunckley MG, Wells DJ, Walsh FS, Dickson G. (1993) Direct retroviral-mediated transfer of a dystrophin minigene into *mdx* mouse muscle *in vivo*. *Hum. Mol. Genet.* **2:** 717–723.

Emery AEH. (1993) *Duchenne Muscular Dystrophy*, 2nd Edn. Oxford Monographs in Medical Genetics, No. 15, Oxford Medical, Oxford.

England SB, Nicholson LVB, Johnson MA, Forest SM, Love DR, Zubrzycka-Gaarn EE, Bulman DE, Harris JB, Davies KE. (1990) Very mild muscular dystrophy associated with the deletion of 46% of dystrophin. *Nature* **343:** 180–182.

Ervasti JM, Campbell KP. (1993) A role for dystrophin–glycoprotein complex as a transmembrane linker between laminin and actin. *J. Cell Biol.* **122:** 809–823.

Gee SH, Montanaro F, Lindenbaum MH, Carbonetto S. (1994) Dystroglycan-alpha, a dystrophin-associated glycoprotein, is a functional agrin receptor. *Cell* **77:** 675–686.

Grounds MD, Partridge TA, Sloper JC. (1980) The contribution of exogenous cells to regenerating skeletal muscle: an isoenzyme study of muscle allografts in mice. *J. Pathol.* **132:** 325–341.

Gussoni E, Pavlath GK, Lanctot AM, Sharma KR, Miller RG, Steinman L, Blau HM. (1992) Normal dystrophin transcripts detected in Duchenne muscular dystrophy patients after myoblast transplantation. *Nature* **356:** 435–438.

Helbling-Leclerc A, Xu Zhang, Topaloglu H, Cruaud C, Tesson F, Weissenbach J, Tomé FMS, Schwartz K, Fardeau M, Tryggvason K, Guicheney P. (1995) Mutations in the laminin α2-chain gene (*LAMA2*) cause merosin deficient congenital muscular dystrophy. *Nature Genet.* **11:** 216–218.

Huard J, Bouchard JP, Roy R, Malouin F, Dansereau G, Labrecque C, Albert N, Richards CL, Lemieux B, Tremblay JP. (1992a) Human myoblast transplantation: preliminary results of 4 cases. *Muscle Nerve* **15:** 550–560.

Huard J, Roy R, Bouchard JP, Malouin F, Richards CL, Tremblay JP. (1992b) Human myoblast transplantation between immunohistocompatible donors and recipients produces immune reactions. *Transplant. Proc.* **24:** 3049–3051.

Huard J, Goins WF, Glorioso JC. (1995a) Herpes simplex virus type I vector-mediated gene transfer to muscle. *Gene Ther.* **2:** 385–392.

Huard J, Lochmuller H, Acsadi G, Jani A, Massie B, Karpati G. (1995b) The route of administration is a major determinant of the transduction efficiency of rat tissues by adenoviral recombinants. *Gene Ther.* **2:** 107–115.

Karpati G, Acsadi G. (1993) The potential for gene therapy in Duchenne muscular dystrophy and other genetic muscle diseases. *Muscle Nerve* **16:** 1141–1153.

Karpati G, Pouliot Y, Zubrzycka GE, Carpenter S, Ray PN, Worton RG, Holland P. (1989) Dystrophin is expressed in *mdx* skeletal muscle fibers after normal myoblast implantation. *Am. J. Pathol.* **135:** 27–32.

Karpati G, Ajdukovic D, Arnold D, Gledhill RB, Guttmann R, Holland P, Koch PA, Shoubridge E, Spence D, Vanasse M, Watters GV, Abrahamowicz M, Duff C, Worton RG. (1993) Myoblast transfer in Duchenne muscular dystrophy. *Ann. Neurol.* **34:** 8–17.

Kinoshita I, Vilquin JT, Tremblay JP. (1995) Pretreatment of myoblast cultures with basic fibroblast growth factor increases the efficacy of their transplantation in *mdx* mice. *Muscle Nerve* **18:** 834–841.

Klein CJ, Coovert DD, Bulman DE, Ray PN, Mendell JR, Burghes AHM. (1992) Somatic reversion/suppression in Duchenne muscular dystrophy (DMD): evidence supporting a frame-restoring mechanism in rare dystrophin-positive fibers. *Am. J. Hum. Genet* **50:** 950–959.

Kumar-Singh R, Chamberlain JS. (1996) Encapsidated adenovirus minichromosomes allow delivery and expression of a 14 kb dystrophin cDNA to muscle cells. *Hum. Mol. Genet.* **5:** 913–921.

Lidov HG, Selig S, Kunkel LM. (1995) Dp140: a novel 140 kDa CNS transcript from the dystrophin locus. *Hum. Mol. Genet.* **4:** 329–325.

Lim LE, Duclos F, Broux O, Bourg N, Sunada Y, Allamand V, Meyer J, Richard I, Moomaw C, Slaughter C, Tomé FMS, Fardeau M, Jackson CE, Beckmann JS, Campbell KP. (1995) β-Sarcoglycan: characteristics and role in limb-girdle muscular dystrophy linked to 4q12. *Nature Genet.* **11:** 257–265.

Love DR, Hill DF, Dickson G, Spurr NK, Byth BC, Marsden RF, Walsh FS, Edwards YH, Davies KE. (1989) An autosomal transcript in skeletal muscle with homology to dystrophin. *Nature* **339:** 55–58.

Manilal S, Nguyen thi M, Sewry CA, Morris G. (1996) The Emery–Dreifuss muscular dystrophy protein, emerin, is a nuclear membrane protein. *Hum. Mol. Genet.* **5:** 801–808.

Matsumura K, Campbell KP. (1994) Dystrophin–glycoprotein complex: its role in the molecular pathogenesis of muscular dystrophy. *Muscle Nerve* **17:** 2–15.

Mendell JR, Kissel JT, Amato AA, King W, Signore L, Prior TW, Sahenk Z, Benson S, McAndrew PE, Rice R, Nagaraja H, Stephens R, Lantry L, Morris G, Burghes AHM. (1995) Myoblast transfer in the treatment of Duchenne's muscular dystrophy. *New Engl. J. Med.* **333:** 832–838.

Mercuri E, Muntoni F, Berardinelli A, Pennock J, Sewry C, Philpot J, Dubowitz V. (1995) Somatosensory and visual evoked potentials in congenital muscular dystrophy: correlation with MRI changes and muscle merosin status. *Neuropediatrics* **26:** 3–7.

Monaco AP, Bertelson CJ, Liechti GS, Moser H, Kunkel LM. (1988) An explanation for the phenotypic differences between patients bearing partial deletions of the DMD locus. *Genomics* **2:** 90–95.

Morandi L, Bernasconi P, Gebbia M, Mora M, Crosti F, Mantegazza R, Cornelio F. (1994) Lack of mRNA and dystrophin expression in DMD patients three months after myoblast transfer. *Neuromusc. Disord.* **5:** 291–295.

Morgan JE, Coulton GR, Partridge TA. (1987) Muscle precursor cells invade and repopulate freeze-killed skeletal muscles. *J. Muscle Res. Cell Motil.* **8:** 386–396.

Muntoni F, Wilson L, Marrosu G, Marrosu MG, Cianchetti C, Mestroni L, Ganau A, Dubowitz V, Sewry C. (1995) A mutation in the dystrophin gene selectively affecting dystrophin expression in the heart. *J. Clin. Invest.* **96:** 693–699.

Nicholson LVB, Johnson MA, Gardner-Medwin D, Bhattacharya S, Harris JB. (1990) Heterogeneity of dystrophin expression in patients with Duchenne and Becker muscular dystrophy. *Acta Neuropathol.* **80:** 239–250.

Noguchi S, McNally EM, Othmane KB, Hagiwara Y, Mizuno Y, Yoshida M, Yamamoto H, Bonnemann CG, Gussoni E, Denton PH, Kyriakides T, Middleton L, Hentati F, Hamida MB, Nonaka I, Vance JM, Kunkel LM, Ozawa I. (1995) Mutations in the dystrophin-associated protein γ-sarcoglycan in chromosome 13 muscular dystrophy. *Science* **270:** 819–822.

Ozawa E, Yoshida M, Suzuki A, Mizuno Y, Hagiwara Y, Noguchi S. (1995) Dystrophin-associated proteins in muscular dystrophy. *Hum. Mol. Genet.* **4:** 1711–1716.

Partridge T. (1993) *Molecular and Cell Biology of Muscular Dystrophy.* Chapman & Hall, London.

Partridge TA, Morgan JE, Coulton GR, Hoffman EP, Kunkel LM. (1989) Conversion of *mdx* myofibres from dystrophin-negative to -positive by injection of normal myoblasts. *Nature* **337:** 176–179.

Phelps SF, Hauser MA, Cole NM, Rafael JA, Hinkle RT, Faulkner JA, Chamberlain JS. (1995) Expression of full-length and truncated dystrophin mini-genes in transgenic *mdx* mice. *Hum. Mol. Genet.* **4:** 1251–1258.

Philpot J, Sewry C, Pennock J, Dubowitz, V. (1995) Clinical phenotype in congenital muscular dystrophy: correlation with expression of merosin in skeletal muscle. *Neuromusc. Disord.* **5:** 301–305.

Porter GA, Dmytrenko GM, Winkelmann JC, Bloch RJ. (1992) Dystrophin co-localizes with β-spectrin in distinct subsarcolemma domains in mammalian skeletal muscle. *J. Cell Biol.* **117:** 997–1005.

Ragot T, Vincent N, Chafey P, Vigne E, Gilgenkrantz H, Couton D, Cartaud J, Briand P, Kaplan JC, Perricaudet M, Kahn A. (1993) Efficient adenovirus-mediated transfer of a human minidystrophin gene to skeletal muscle of *mdx* mice. *Nature* **361:** 647–650.

Richard I, Broux O, Allamand V, Fougerousse F, Chiannilkulchaii N, Bourg N, Brenguier Devaud C, Pasturaud P, Roudaut C, Hillaire D, Passos-Bueno Zatz M, Tischfield Fardeau M, Jackson CE, Cohen D, Beckmann JS. (1995) Mutations in the proteolytic enzyme calpain-3 cause limb-girdle muscular dystrophy type 2A. *Cell* **81:** 1–20.

Roberds SL, Leturcq F, Allamand V, Piccolo F, Jeanpierre M, Anderson RD, Lim LE, Lee JC, Tomé FMS, Romero NB, Fardeau M, Beckmann JS, Kaplan JC, Campbell KP. (1994) Missense mutations in the adhalin gene linked to autosomal recessive muscular dystrophy. *Cell* **78:** 625–633.

Sewry CA, Matsumura K, Campbell KP, Dubowitz, V. (1994) Expression of dystrophin-associated glycoproteins in carriers of Duchenne muscular dystrophy. *Neuromusc. Disord.* **4:** 406–409.

Sewry CA, Philpot J, Mahoney D, Wilson LA, Muntoni F, Dubowitz V. (1995) Expression of laminin subunits in congenital muscular dystrophy. *Neuromusc. Disord.* **5:** 307–316.

Sharp NJ, Kornegay JN, Van Camp CS, Herbstreith MH, Secore SL, Kettle S, Hung WY, Constantinou CD, Dykstra MJ, Roses AD, Bartlett RJ. (1992) An error in dystrophin mRNA processing in golden retriever muscular dystrophy, an animal homologue of Duchenne muscular dystrophy. *Genomics* **13:** 115–121.

Sherratt TG, Vulliamy T, Dubowitz V, Sewry CA, Strong PN. (1993) Exon skipping and translation in patients with frameshift deletions in the dystrophin gene. *Am. J. Hum. Genet.* **53:** 1007–1015.

Shorer Z, Philpot J, Muntoni F, Sewry C, Dubowitz V. (1995) Peripheral nerve involvement in congenital muscular dystrophy. *J. Child Neurol.* **10:** 472–475.

Sicinski P, Geng Y, Ryder-Cook AS, Barnard EA, Darlison MG, Barnard PJ. (1989) The molecular basis of muscular dystrophy in the *mdx* mouse: a point mutation. *Science* **244:** 1578–1580.

Straub V, Bittner RE, Leger JJ, Voit T. (1992) Direct visualization of the dystrophin network on skeletal muscle fiber membrane. *J. Cell Biol.* **119:** 1183–1191.

Sunada Y, Bernier SM, Kozak CA, Yamada Y, Campbell KP. (1994) Deficiency of merosin in dystrophic *dy* mice and genetic linkage of laminin M chain to the *dy* locus. *J. Biol. Chem.* **269:** 13729–13732.

Sunada Y, Bernier SM, Utani A, Yamada Y, Campbell KP. (1995) Identification of a novel mutant transcript of laminin alpha 2 chain gene responsible for muscular dystrophy and dysmyelination in *dy²ᴶ* mice. *Hum. Mol. Genet.* **4:** 1055–1061.

Swasdison S, Mayne R. (1992) Formation of highly organized skeletal muscle fibers *in vitro*: comparison with muscle development *in vivo*. *J. Cell Sci.* **102:** 643–652.

Tinsley JM, Davies KE. (1993) Utrophin: a potential replacement for dystrophin? *Neuromusc. Disord.* **3:** 537–539.

Toda T, Segawa M, Nomura Y, Nonaka I, Masuda K, Ishihara T, Suzuki M, Tomita I, Origuchi Y, Ohno K, Misugi N, Sasaki Y, Takeda K, Kawai M, Otani K, Murakami T, Saito K, Fukuyama Y, Shimizu T, Kanazawa I, Nakamura Y. (1993) Localization of a gene for Fukuyama-type congenital muscular dystrophy to chromosome 9q31–33. *Nature Genet.* **5:** 283–286.

Tomé, FMS, Evangelista T, Leclerc A, Sunada Y, Manole E, Estournet Barois A, Campbell KP, Fardeau M. (1994) Congenital muscular dystrophy with merosin deficiency. *C.R. Acad. Sci. Paris* **317:** 351–357.

Turner G, Dunkley MG, Dickson G. (1996) Gene therapy of Duchenne muscular dystrophy. In: *Dystrophin: Gene, Protein and Cell* (eds S Brown, J Lucy). Cambridge University Press, Cambridge (in press).

Vilquin JT, Kinishita I, Roy B, Goulet M, Engvall E, Tomé F, Fardeau M, Tremblay J. (1996) Partial laminin α2 chain restoration in the α2 chain-deficient *dy/dy* mouse by primary muscle cell culture transplantation. *J. Cell. Biol.* **133:** 185–197.

Wells DJ, Wells KE, Asante EA, Turner G, Sunada Y, Campbell KP, Walsh FS, Dickson G. (1995) Expression of human full-length and mini-dystrophin in transgenic *mdx* mice: implications for gene therapy of Duchenne muscular dystrophy. *Hum. Mol. Genet.* **4:** 1245–1250.

Wewer UM, Durkin ME, Zhang X, Laursen H, Nielsin NH, Towfighi J, Engvall E, Albrechtsen R. (1995) Laminin beta 2 chain and adhalin deficiency in the skeletal muscle of Walker–Warburg syndrome (cerebro-ocular dysplasia-muscular dystrophy). *Neurology.* 45: 2099–2101.

Wolff JA, Malone RW, Williams P, Chong W, Acsadi G, Jani A, Felgner PL. (1990) Direct gene transfer into mouse skeletal muscle *in vivo. Science* 247: 1465–1468.

Xu H, Wu XR, Wewer U, Engvall E. (1994) Murine muscular dystrophy caused by a mutation in the laminin α2 (LAMA-2) gene. *Nature Genet.* 8: 297–301.

Yamada H, Tomé FMS, Higuchi I, Kawai H, Azibi K, Chaouch M, Roberds SL, Tanaka T, Fujita S, Mitsui T, Fukunaga H, Miyoshi K, Osama M, Fardeau M, Kaplan JC, Shimuzu T, Campbell KP, Matsumura K. (1995) Laminin abnormality in severe childhood autosomal recessive muscular dystrophy. *Lab. Invest.* 72: 715–721.

Yang B, Jung D, Motto D, Meyer J, Koretzky G, Campbell KP. (1995) SH3 domain-mediated interaction of dystroglycan and Grb2. *J. Biol. Chem.* 270: 11711–11714.

18

Recent developments in gene therapy for neurological disorders

P.R. Lowenstein

> Scientific discoveries are not the proper subject for newspaper scoops and all media of mass communication should have equal opportunity for simultaneous access to the information. In the future, we may reject papers whose main contents have been published previously in the daily press.

[*Physics Review Letters*, Editorial by Professor Samuel A. Goudsmit, 1 January 1960]

> Gene therapy is the only field in which a researcher can get into the newspapers not only for having done a momentous experiment, but just because of having thought of doing one.

[Peter Rigby, addressing an MRC-sponsored conference on Gene Therapy, London, 1994]

18.1 Introduction

Strategies for gene therapy of neurological disorders differ from gene therapy for any other diseases. The reason for this is that although many important physiological processes are normally carried out by fully differentiated postmitotic cells, those from the brain are the longest lived, and are apparently not replaced in adult life from some type of stem cell precursor, as they are in other somatic tissues. It is estimated that in primates, the total complement of neurons is already present at birth, and at birth these cells have already stopped dividing. Thus, even if neuronal precursors exist, it is unclear if any neuronal turnover occurs during adulthood in primates, based on *in vivo* experiments performed by Pasko Rakic and co-workers (Rakic, 1985). Evolutionarily this does not need to be so, since in birds, neurons in song centres go through a circannual cycle of degeneration and regeneration (Barnea and Nottebohm, 1996). The study of brain stem cells for further understanding of both the ontogeny and pathology of adult mammalian brains is currently a very active field.

Many diseases affect brain structure and function. Each disease affects a particular cell type, the dysfunction of which will correspond to individual pathophysiological manifestations. There are essentially only two general possible avenues to transfer genes into the brain: either they are delivered directly into the target cells of the brain via a gene transfer vector (which could be viral or non-viral, i.e. liposomes) or they are delivered into a platform cell which is then transplanted into the brain. This is important on two accounts. Firstly, different brain diseases affect the function of individual brain cell types and, for therapeutic intervention to be successful, a gene will have to be introduced

Gene Therapy, edited by N.R. Lemoine and D.N. Cooper.
© 1996 BIOS Scientific Publishers Ltd, Oxford.

directly into the affected brain cell type. If neurons are affected, transplanting a gene into the brain glial cells will be of no relevance, and vice versa. Thus, cell type-specific expression and regulation of the levels of transgene expression is an important consideration in the development of gene therapy for brain diseases. Secondly, many disorders of the brain occur because of a breakdown in cell-to-cell communication, which manifests itself as the lack of a factor which is released into the extracellular medium, where it acts on the corresponding receptors located on other brain cells. This occurs in disorders like Parkinson's disease, in which many clinical manifestations are due to a lack of the neurotransmitter dopamine, which is active extracellularly, or in Alzheimer's disease and other neurodegenerative disorders, in which neurodegeneration is thought to be due to a deficiency in a particular neuronal growth factor (reviewed in Lowenstein, 1995).

In many other diseases, ranging from Huntington's disease to the mucopolysaccharidoses, neurons constitute the main cell type affected, and any new treatment will have to target the affected cells, sometimes exclusively. Thus, an increased understanding of pathophysiological mechanisms is crucial to the design of novel therapies. Even in diseases belonging to the same group of disorders; that is, two lysosomal diseases, brain cell types to be targeted by the gene therapy will differ between different diseases. In Tay–Sachs disease, the affected cells are neurons which accumulate unprocessed gangliosides, whereas in metachromatic leukodystrophy, the affected cells are the brain macrophages, which accumulate sulphatides. In both diseases, the function of most other cell types appears unaffected. This is due to the fact that, given the different metabolic routes which are active in different cell types, different gene products are needed to maintain the normal function of different cell types. Genetics can tell us which gene is mutated in which disease, but it cannot tell us in which cell types the function of that gene is absolutely necessary. Cells in which the activity of any particular gene product is dispensable do not show any abnormality when that gene is mutated, and will not need to be targeted by gene therapy. This highlights the important contribution made by pathophysiological studies to our understanding of disease processes and the development of new therapeutic strategies. Thus, whilst in Tay–Sachs disease gene therapy must target neurons, in metachromatic leukodystrophy gene therapy should target the population of brain macrophages.

Although the concepts of gene therapy are relatively straightforward, it will take some time to develop and implement its full potential in the clinic. This is because, in practice, many practical aspects need to be developed to make clinical gene therapy successful (Grossman *et al.*, 1995; Knowles *et al.*, 1995; Mendell *et al.*, 1995). Friedmann (1994, 1996) has emphasized that the clinical implementation of gene therapy may be its greatest challenge. I will now present a logical and rational guide to current developments in neurological gene therapy. Emphasis will be given to new developments; individual vector systems or diseases will not be reviewed in detail, since they are dealt with by other chapters in this book (3–8), and have been reviewed recently (Kaplitt and Loewy, 1995; Lowenstein and Enquist, 1996).

18.2 Six central questions of clinical neurological gene therapy (and some answers)

18.2.1 Which vector to use?

All available vectors (viral and non-viral) have been used to transfer genes into neurons with varying degrees of success. The extensive literature will not be reviewed here, and the interested reader can consult the appropriate book chapters on individual vectors (Chapters 2–8; Kaplitt and Loewy, 1995; Lowenstein and Enquist, 1996). The main conclusion to bear in mind is that all vectors have advantages and disadvantages; all allow for short-term expression at reasonably high levels of transgene expression; further, the immune system is one of the main factors limiting the extent of long-term transgene expression (Kay *et al.*, 1995; Koenig, 1996; Tripathy *et al.*, 1996; and see Section 18.3). Although long-term expression has been shown in some conditions, this is seen in only a restricted population of cells which would probably not be of great significance in a clinical trial of gene therapy. So far, all vectors show almost equal potential.

A major vector development for gene transfer into neurons has been the description of a pseudotyped retroviral system based on human immunodeficiency virus (HIV), which has been demonstrated to express transgenes over a long period of time in the brain *in vivo* (Naldini *et al.*, 1996). The main advantage of a vector that is capable of integrating into postmitotic neurons is that, by allowing sustained transgene expression, it would open up the possibility of treating diseases in which very long-term therapeutic transgene expression is needed, such as Huntington's, Parkinson's and Alzheimer's diseases. An added important feature of retroviral vectors is that the immune response against the vector itself is either very limited or almost non-existent. The normal criticisms directed at retroviral systems still apply; that is, their restricted cloning capacity, their low titres and lack of selective targeting. These are certainly important areas where future research will concentrate.

There have also been reports on the capacity of adeno-associated virus vectors (AAV, a parvovirus vector which depends on adenovirus for growth; see Chapter 5) to infect and express transgenes in postmitotic cells, including neurons (McCown *et al.*, 1996). However, for both retroviral and AAV vectors, formal proof of the integration of vector sequences into the genome of postmitotic neurons is eagerly awaited. The possibility of cytoplasmic expression without genomic integration has been suggested for AAV, and is certainly possible for herpes simplex type 1 (HSV-1) derived vectors. Further modifications to HSV-1 which, during the latent phase of a natural infection, is able to express a limited set of mRNAs episomally over many decades (Efstathiou and Minson, 1995; Fink *et al.*, 1996; Maidment *et al.*, 1996), are also likely to become available in the near future (Mineta *et al.*, 1995).

Another exciting development in brain transplantation has been the implementation of neuronal stem cell transplantation, either derived from early stages of developing brains (Lacorazza *et al.*, 1996; Snyder and Wolfe, 1996) or from adult animals (Fisher and Gage, 1993). Evan Snyder's group has been transplanting genetically engineered brain stem cells into the brains of neonatal and adult lesioned animals and has shown some of the most impressive results in this field. The capacity of such stem cells to integrate into lesioned adult brain areas opens up enormous possibilities, since, for neurodegenerative disorders, transplants would be done into areas of disease-induced lesions, which may thus promote the growth and integration of transplanted stem cells.

Although clinical trials of transplantation using many different cell types is being done in Parkinson's disease (up to 600 patients have received different types of transplants) and is now being proposed for Huntington's disease, progress has been slower than expected (Widner, 1995). The most impressive clinical results have been obtained after transplantation of young patients who suffered acute Parkinsonian symptoms after consuming heroin contaminated with the dopaminergic neurotoxin, 1-methyl-4-phenyl-1,2,3,6-tetrahydropyridine (MPTP). Interestingly, these patients suffer an acute lesion of their nigro-striatal system, similar to the experimental lesion paradigm induced in rodents and monkeys, and that has been shown to respond well to the transplantation of a variety of neural and non-neuronal cells (Dunnett and Bjorklund, 1994).

It is possible that the type of lesion inflicted on nigro-striatal neurons and the pathophysiology of Parkinson's disease determine the outcome and response to neural transplantation. A combination of extensive clinical experience accumulated in transplantation for the treatment of Parkinson's disease, with the possibility of engineering brain stem cells, might be one of the most exciting clinical prospects, if the immune response to stem cells could be limited. Rapid progress in our understanding of immune rejection of transplants might open up many more clinical trials using transplantation of engineered brain stem cells expressing the necessary transgenes.

There are two kinds of limitations imposed by the immune system: (i) the immune response reduces the length of transgene expression, and (ii) it abrogates transgene expression upon the re-administration of the vector. Reduction of transgene expression over time after a single administration of vector is due to cytotoxic T lymphocyte (CTL) responses raised to both vector components and the transgene. A blockade in transgene expression upon re-administration of the vector is due to the development of neutralizing antiviral antibodies. Many different strategies are now being implemented to overcome the unwanted immune responses, for example (i) administration of different vector serotypes on different gene therapy administrations, (ii) the use of vectors deleted in most putative immunogenic genes, (iii) the administration of various immunosuppressants, either drugs such as cyclosporin or immunoregulatory cytokines such as interleukin-12 (IL-12), or (iv) the administration of molecules which interfere with the function of co-stimulatory molecules, such as CTLA4-Ig (Engelhardt *et al.*, 1994; Kay *et al.*, 1995; Yang *et al.*, 1994a,b, 1995).

18.2.2 When to treat?

How many neurons need to be transduced?

These two questions will be treated jointly, because they are intimately linked. Independently of the cause of a neurological disease, neuronal loss depends on the stage of disease progression. As a general concept, one would propose to treat as soon as possible, especially for inherited disorders which offer the added possibility of very early diagnosis. The selection of the proper time to treat thus becomes more problematic when considering treatments for non-inherited neurological diseases, where presymptomatic diagnosis is very difficult, or even impossible, and where the actual onset of the disease constitutes the earliest stage at which therapy could be attempted.

As shown in *Table 18.1*, there are many high incidence neurological diseases, few of which are under consideration as serious clinical gene therapy trial targets. One main stumbling block toward such developments is the decision of when to treat a neurological disease, assuming that one already has evidence that one has developed a viable treatment strategy. The answer to this question should be tackled according to the nature of the disease to be treated; one answer for inherited diseases, and a second for non-inherited diseases. For both, the crucial question is: how early can diagnosis be made?

Table 18.1. Frequency of neurological and neuromuscular inherited and non-inherited disorders

Disease	Incidence (I) or prevalence (P) per 100 000	
Alzheimer's disease	30–59 years:	20 (P)
	60–69 years:	300 (P)
	70–79 years:	3200 (P)
	80–89 years:	10 800 (P)
Parkinson's disease	40–59 years:	100 (P)
	60–69 years:	400 (P)
	70–79 years:	800 (P)
	>80 years:	1000 (P)
Epilepsy		500–1000 (P)
Duchenne muscular dystrophy (in males)		17–244 (I)
		97–390 (P)
Stroke		125 (I)
Hereditary motor and sensory neuropathies (all)		14–280 (P)
Fragile-X, A		93 (P)
Myotonic dystrophy		9–96 (P)
Spinal muscular atrophies (all)		16–66 (I)
		2–40 (P)
Neurofibromatosis		28.6 (P)
Brain tumours		10–20 (I)
Tuberous sclerosis		10 (I)
Huntington's disease		8 (P)

For inherited diseases, the answer ought to be straightforward. In theory, an inherited disease can be diagnosed early or even during the presymptomatic, or at prenatal, stages. When this is possible, one would propose to institute therapy as soon as possible since, as the disease progresses, more and more neurons will be lost, thus irretrievably compromising the capacity of the brain to compensate for the anatomical losses. In support of this concept, it has been shown that in brains of Huntington's disease patients who have inherited a disease-causing expanded allele of the *HD* gene but who have died as a result of other causes during the presymptomatic stages of their disease, neuronal cell death can already be detected. How far presymptomatic treatment should go is difficult to tell, but fetal gene therapy is being developed for a number of diseases, including cystic fibrosis (Coutelle *et al.*, 1995).

In non-inherited diseases, it is more difficult to make an early accurate clinical or even pre-symptomatic diagnosis. However, developing such early pre-symptomatic diagnosis might, in the long run, be crucial in order to develop successful neurological gene therapy. Treatment should then be instituted as soon as the disease is diagnosed. This will be especially important in diseases like Parkinson's disease. It is assumed that no clinical symptoms are evident until more than 90% of basal ganglia neurons have been lost. Importantly, when the disease becomes clinically evident, pharmacological treatment can still be effective for many years, whilst symptomatic treatment does not seem to affect disease progression. If gene therapy is delayed until pharmacological treatments do not work anymore, then the likelihood of success of gene therapy will be rather small, since neuronal damage at this stage might be very extensive and beyond repair. It should be easier to achieve a 20% increase in functional capacity (from 10 to 12% of remaining functional capacity), than a 600% increase when only 2% of neurons are left. In Parkinson's disease, one has to get functionality in up to 12% of nigro-striatal dopaminergic neurons in order to produce symptomatic benefits.

Current concepts of the physiological organization of the central nervous system (CNS) propose the existence of anatomical and functional redundancy within the brain, and are based on the fact that the brain can resist a certain amount of damage to its structure, with relatively little functional impairment. One of the classic examples given to support this idea is the extrapyramidal system which degenerates in Parkinson's disease. Although clear-cut data do not exist, it is apparent from both clinical and experimental paradigms that damage to the system must exceed 80–90% before symptoms of degeneration become clinically apparent. One way of testing this is to block receptor systems known to degenerate in some of these diseases. Thus, it has been suggested that patients with a strong clinical predisposition to Parkinson's disease can, in some cases, show symptoms if dopaminergic blocking agents are administered. Similarly, some patients predisposed to Alzheimer's disease will show signs of memory deficits if low doses of cholinergic antagonists are administered. However, this has not yet led to a generally applicable predictive test to determine which patients will develop any of these diseases.

In support of the idea that presymptomatic neuronal degeneration does continue undetected until an anatomical and functional threshold is crossed, after which symptoms are detected clinically, recent work has shown that presymptomatic Huntington's disease patients, who died before the disease became clinically apparent, already had important signs of neuronal degeneration. All this highlights the importance of early presymptomatic diagnosis to increase the possibilities of success of any novel treatments. If new research on the role of the ApoE $\Sigma4$ and $\Sigma2$ alleles in predisposing towards and/or protecting against the development of Alzheimer's disease is substantiated, these genetic markers could be used in detecting populations at high risk. Even if, in the long run, ApoE alleles prove not to have a very accurate predictive value, the development of further predictive markers is needed, since early diagnosis and early intervention will be one of the main stepping stones in making the new gene therapies clinically effective.

18.2.3 Which neurons need to be transduced?

During the design of viable clinical gene therapy strategies, it is important to identify the appropriate target cell type. The pathological manifestations of individual brain diseases will mostly be due to a specific alteration in the function of particular brain cell types. This is clearly exemplified by the inherited metabolic disorders affecting brain structure and physiology. For example, in diseases like Tay–Sachs or Niemann–Pick, neurons accumulate undegraded metabolites in large lysosomes within cell bodies, which become grossly enlarged. However, non-neuronal cells are less affected or not affected at all. Conversely, in the leukodystrophies (e.g. metachromatic leukodystrophy), macrophages are affected and large numbers of grossly distended macrophages can be seen in perivascular spaces, whereas neurons remain normal during the initial stages of the disease.

In Huntington's disease, whilst neurons in the basal ganglia are one of the main target cells affected by the disease mutation, some striatal neurons (especially those expressing NADPH diaphorase) appear to survive. Similarly, in Parkinson's disease, only dopaminergic cells located within the substantia nigra and which innervate the corpus striatum are selectively affected by the disease, and undergo cell death. This highlights the relevance of an understanding of pathophysiological mechanisms to the development of adequate treatment modalities. This should then lead to the selection of both the right cell type (e.g. neuron or glial cell) and the specific cell subtype [e.g. γ-aminobutyric acid (GABA)ergic neurons in Huntington's disease or dopaminergic neurons in Parkinson's disease; microglial cells in leukodystrophies, etc.].

18.2.4 How much and what levels of transgene expression are needed? Can it be predicted a priori?

Most gene expression is under strictly controlled spatial, developmental and physiological regulation. Thus, different tissues or cells will express different genes at different levels during different stages of development and, in adulthood, gene expression in individual cell types will be, in addition, under physiological regulation. The tight regulation of transcription and translation is the reason why different single gene mutations will affect, for example, neurons but not glial cells, or vice versa. Thus, gene therapy through gene replacement will have to select carefully the appropriate target cell population, and determine the levels and timing of transgene expression needed to correct the deficiency.

To predict the therapeutic level of transgene expression required for the treatment of a particular disease state depends on the answer to two different questions: (i) what is the level of transgene expression needed to maintain a normal phenotype, and (ii) in what percentage of the total target cell population does this level have to be achieved?

For example, in a model metabolic disease affecting neurons, in which a single enzyme gene is mutated (in this case, the Lesch–Nyhan syndrome), the phenotype of the disease varies with the amount of remaining enzyme activity. Patients with less than 1–2% of hypoxanthine guanine phosphoribosyltransferase (HPRT) have the classical phenotype of mental retardation, self-mutilation, choreoathetosis, and hyperuricaemia (classical Lesch–Nyhan syndrome); patients with 2–3% of HPRT activity lack the mental retardation, patients with 3–10% lack the self-mutilation (neurological Lesch–Nyhan), whilst patients with more than approximately 10% are only hyperuricaemic.

How much HPRT enzyme needs to be replaced in the neurons of a patient with no detectable HPRT activity to achieve phenotypic remission? We could suggest a value of at least 10%, which should in principle eliminate all brain symptoms of this devastating disease. However, it is important to consider that in disease-free patients (those with 10% of HPRT enzyme activity remaining), HPRT will be expressed in 100% of cells. Thus, while we might conclude that we would need to replace at least 10% of enzyme activity in our patient, the clinical data do not allow us to predict in what proportion of target cells we will have to achieve that level of activity; that is, at 100% expression level in 10% of the cells, or at 10% expression level in 100% of the cells. To answer this latter question, we must know what percentage of cells need to be transduced in order for that particular brain region to function properly, and this will depend on the physiology of the brain area under examination and the nature of the disease pathology.

In a metabolic disorder not causing neurodegeneration, as the Lesch–Nyhan syndrome is presumed to be, function might be restored by transduction of a low percentage of nigro-striatal neurons (e.g. 10%), because of the known functional redundancy thought to exist in many cerebral circuits as discussed above. It may be that even in Lesch–Nyhan syndrome, close to 100% of neurons might have to be transduced to achieve a phenotypic remission of disease symptoms. In the case of a disorder causing neurodegeneration in cells expressing less than the threshold activity (e.g. Tay–Sachs disease), or expressing high levels of the mutated allele (e.g. Huntington's disease), it will be necessary to transduce all the cells for them to remain alive. While it is possible to define experimentally some of the values in advance of planning clinical gene therapy trials, such experiments are complex to perform even in transgenic animal models (Dorin *et al.*, 1995, and Chapter 9).

18.2.5 For how long does the transgene have to be expressed?

The duration of transgene expression required will depend on the nature of the disease under consideration. In CNS malignancies, where the main therapeutic objective is to kill the tumour cells while sparing surrounding normal cell populations, transgene expression could be limited to 1–2 weeks or even less. However, short-term expression would be of little benefit in the design of a therapy for neurodegenerative diseases or inherited metabolic disorders in which the pathology extends over many decades. In this case, transgene expression would have to be instituted as soon as technically possible, and would have to be sustained for the whole life of the patient.

18.3 Unsolved challenges: limiting the immune response to vectors and transgenes

One of the main challenges to the success of clinical neurological gene therapy is the limitation imposed by the immune system. Although it is sometimes still proposed that the brain is a privileged immunological site, we cannot ignore immune reactions to vector administration in the brain (Medawar, 1948). Even if there is some immune privilege, more aptly described as an extension of graft survival in comparison with allogeneic transplantation to other sites in the body, this appears to be of rather little significance when considering the administration of vectors to the brain.

Recent work has now clearly demonstrated that immune responses do indeed occur in the brain in response to direct intraparenchymal HSV-1 or adenovirus vector administration (Byrnes *et al.*, 1995, 1996; Wood *et al.*, 1994). Moreover, a very strong immune response also occurs when pseudorabies virus enters the brain via the nerve cells from the periphery, a phenomenon that has been used to map neural pathways in the brain (Rinaman *et al.*, 1993). In addition, memory immune responses in the brain which lead to serious tissue damage occur when an adenovirus vector is delivered a second time to the periphery (Byrnes *et al.*, 1996). The implications of this within a context of human gene therapy are all too clear.

To complicate matters further, it has now been demonstrated that immune responses are raised against viral gene products (e.g. those expressed by some first generation adenovirus vectors), but are also generated against the transgene encoded by the viral vector (Riddell *et al.*, 1996; Tripathy *et al.*, 1996). This occurs both if the transgene is encoded by an adenoviral or a retrovirus vector, and also after administration of virus vector to the brain. Even the compromised immune system in acquired immunodeficiency syndrome (AIDS) patients has been able to generate CTL responses against a selectable marker (hygromycin resistance gene) and HSV-1 thymidine kinase (*HSV-tk*) inserted within a retrovirus vector (Koenig, 1996; Riddell *et al.*, 1996). Increased hope in this field is highlighted by striking advances stemming from the field of organ transplantation, which are now inducing selective immuno-suppression through a manipulation of lymphocyte co-stimulatory signals during antigen presentation (Larsen *et al.*, 1996). The recent application of some of these concepts to gene therapy (CTLA4-Ig, a compound blocking the lymphocyte co-stimula-tory molecule CD80) using adenovirus vectors appears to be very encouraging, since transgene expression has now been prolonged for up to 6 months after gene delivery to the liver, without necessitating the induction of generalized non-specific immuno-suppression (Kay *et al.*, 1995; Wilson and Kay, 1995).

18.4 Clinical implementation in gene therapy of neurological disorders

18.4.1 Retroviral transduction of conditionally cytotoxic gene for the treatment of brain tumours

Why are most currently active clinical gene therapy trials in neurology targeted to treat brain tumours (Culver and Blaese, 1994; Fujiwara *et al.*, 1994; see *Table 18.2*)? For three main reasons: (i) there is a great need to develop new treatments for brain tumours given that the mean survival of patients following the diagnosis of brain tumours (6–12 months) has not been improved substantially by developments in neurosurgery, phar-macological treatments and radiotherapy during the last 100 years. (ii) The therapeutic outcome of such therapies can be measured objectively not only as the disappearance of the tumour mass but also as the survival of patients post-treatment. (iii) Tumour cell killing can be achieved using most vectors available, and trials using retrovirus, aden-ovirus and herpes virus vectors are either in progress or will be started in the near future.

Unfortunately, it is difficult to review the current clinical trials in detail, because most of the data of such trials have yet to be published in peer-reviewed publications. Results presented at several scientific meetings have shown a variety of treatment out-comes. In individual cases, it appeared that patient survival had been improved beyond expectations in comparison with historical controls whereas, in other cases, the outcome was not improved at all by gene therapy.

All first generation gene therapy clinical trials for the treatment of brain tumours delivered retroviral vectors encoding HSV-TK, or packaging cell lines releasing retro-viral vectors encoding HSV-TK directly into the tumour mass. In some of the initial experiments, the tumour mass was not removed, either because the tumour mass was inaccessible to surgery, or because the vectors were to be delivered directly to the tumour mass. Multiple injections were performed to cover most of the tumour mass. Even if overall the outcome appeared to be unpredictable, in some cases there seemed to be real positive benefit in terms of patient survival and tumour size regression fol-lowing the treatment with ganciclovir. Such results are extremely encouraging in such a devastating and otherwise rapidly progressing disease.

In some cases, investigators have also reported side-effects, such as intratumoral haemorrhages. It is however difficult to determine at this stage whether this was due to the injections *per se*, or whether it was a result of ganciclovir toxicity towards retrovirally transduced endothelial cells. Another shortcoming of the first generation trials was that a variety of brain tumour types were treated, for example different types of primary and metastatic tumours. This would be expected to make results more difficult to interpret.

Second generation brain tumour gene therapy trials now in progress involve the removal of the tumour mass, followed by injection of vector-producing cells directly into normal brain tissue surrounding the tumour. This new approach was designed to try to target the tumour cells which surround the main tumour mass, infiltrate the surrounding normal brain and give rise to intraparenchymal metastases. These second generation trials have now standardized the tumour cell type and stage of tumour progression being treated by performing a thorough pre-operative histopathological analysis of tumour biopsies. In some cases, radiotherapy is also being included in the trial protocols. Again, there are no published data yet, but news of some promising outcomes have been presented at recent meetings. Much will be learned from large world-wide multicentre trials for gene therapy for brain tumours, which are now examining the effectiveness of standardized procedures in the treatment of large numbers of patients, suffering from similar histological types of brain tumours, and treated with the same retroviral packaging cell line. These studies will provide rigorous comparative data on the effectiveness, power and limitations of current brain tumour gene therapy strategies, and are expected to provide a more stringent assessment of this treatment modality.

18.4.2 Bone marrow transplantation for the treatment of inherited metabolic disorders

The underlying pathophysiology and genetics of many inherited metabolic disorders which affect brain function has been elucidated and, therefore, this group of diseases are potential candidates for neurological gene therapy. CNS dysfunction and neuronal cell death occurs because of an accumulation of toxic intermediates due to an absence of particular lysosomal or other enzymes involved in intermediary metabolism (reviewed and listed in detail in Lowenstein, 1995). Some of these diseases respond well to strict dietary controls, and screening programmes have been introduced to facilitate an early diagnosis and treatment implementation for such cases (e.g. phenylketonuria).

Despite this, while dietary restrictions allow a reduction of toxin build-up and thereby a reduction in cell death in phenylketonuria, this treatment does not actually confer a normal phenotype on the affected brain. For most disorders in this group, however, dietary treatments so far have not been very successful. Two alternative treatments are possible for these metabolic disorders.

(i) enzyme replacement therapy, which has shown excellent results in diseases such as severe combined immunodeficiency syndrome due to the deficiency of the enzyme adenosine deaminase, and in Gaucher disease, in which the gene encoding the enzyme glucocerebrosidase is mutated. In these cases, enzyme is actually taken up by target cells, and thus this treatment can achieve an intracellular decrease in toxic metabolic intermediates. Direct administration of enzyme can also reduce levels of circulating toxic metabolites, but this reduction is only transient and the treatments are usually extremely expensive, especially considering that they need to be administered throughout the lifetime of the patients. Cost

Table 18.2. Clinical neuro-gene therapy protocols submitted or in progress worldwide

Target tumours	Nucleic acid transgene	Method of nucleic acid delivery	Title of gene therapy protocol	Principal investigator(s)	Institution and country
Immunotherapy and antisense					
Brain tumours	*IGF-1* antisense	Lipofection	Gene therapy for human brain tumours using episome-based antisense cDNA transcription of *IGF-1*	J. Ilan	Case Western Reserve Cleveland, OH, USA
Glioblastoma	*IL-2* cDNA	Retrovirus	Injection of a glioblastoma patient with tumour cells and fibroblasts genetically modified to secrete IL-2	R.E. Sobol I. Royston	Sidney Kimmel Cancer Center, San Diego, CA, USA
Glioblastoma	*IL-2* cDNA *TGF-β2* antisense	Retrovirus	Injection of a glioblastoma patients with *TGF-β2* antisense- and *IL-2* gene-modified autologous tumour cells. A phase I study	K.L. Black H. Fakhari	UCLA School of Medicine, Los Angeles, CA, USA
Neuroblastoma	*IFNG* cDNA	Retrovirus	A phase I study of immunization with IFN-γ-transduced neuroblastoma cells	J. Rosenblatt R. Seeger	UCLA, Children's Hospital, Los Angeles, CA, USA
Neuroblastoma	*IL-2* cDNA	Retrovirus	A phase I study of cytokine gene-modified autologous neuroblastoma cells for treatment of relapsed/refractory neuroblastoma	M.K. Brenner	St. Jude Children's Research Hospital, Memphis, TN, USA
Suicide gene therapy					
Brain tumours	*HSV-tk* cDNA	Adenovirus	Treatment of advanced CNS malignancy with the recombinant adenovirus H5.020RSVTK. A phase I trial	S.L. Eck J.B. Alavi	University of Pennsylvania Medical Center, Philadelphia, PA, USA
Brain tumours	*HSV-tk* cDNA	Adenovirus	A phase I study of adenoviral vector delivery of the *HSV-tk* gene and the i.v. administration of ganciclovir in adults with malignant tumour of the CNS	R. Grossman	Baylor College of Medicine, Dallas, TX, USA

Table 18.2. Continued

Target tumours	Nucleic acid transgene	Method of nucleic acid delivery	Title of gene therapy protocol	Principal investigator(s)	Institution and country
Suicide gene therapy					
Glioblastoma	*HSV-tk* cDNA	Retrovirus	Gene therapy for glioblastoma in adult patients: safety and efficacy evaluation of an *in situ* injection of recombinant retrovirus producing cells carrying the *tk* gene of HSV-1, to be followed by ganciclovir	D. Klatzmann	Hôpital de la Pitie, Paris, France
Malignant glioma	*HSV-tk* cDNA	Retrovirus	Stereotaxic injection of HSV-TK producer cells (PA317/GlTkSvNa7) and intravenous ganciclovir for the treatment of recurrent malignant glioma	M. Fetell	Columbia Presbyterian Medical Center, New York, NY, USA
Malignant astrocytomas in children	*HSV-tk* cDNA	Retrovirus	Gene therapy for the treatment of recurrent pediatric malignant astrocytomas with *in vivo* tumour transduction with HSV-TK	C. Raffel K. Culver	Children's Hospital, Los Angeles, CA, and Iowa Methodist Medical Center, Des Moines, IA, USA
Recurrent brain tumours	*HSV-tk* cDNA	Retrovirus	Gene therapy for recurrent pediatric brain tumours	L.E. Kun	St. Jude Children's Research Hospital, Memphis, TN, USA
Brain tumours	*HSV-tk* cDNA	Retrovirus	Gene therapy for the treatment of brain tumours using intratumoral transduction with HSV-TK and i.v. ganciclovir	E.H. Oldfield	DHHS, NINDS, Bethesda, MD, USA
Brain tumours	*HSV-tk* cDNA	Retrovirus	Gene therapy for the treatment of malignant brain tumours with *in vivo* tumour transduction with the HSV-TK–ganciclovir system	K. Culver J. Van Gilder	Iowa Methodist Medical Center, Des Moines, IA, USA
Brain tumours	*HSV-tk* cDNA	Retrovirus	Intratumoral injection of producer cells	Mulder	Groningen, The Netherlands
Brain tumours	*HSV-tk* cDNA	Retrovirus	Intratumoral injection of producer cells	Izquierdo	Madrid, Spain
Brain tumours	*HSV-tk* cDNA	Retrovirus	Intratumoral injection of producer cells	Finocchiaro	Milan, Italy

Marker studies

Brain tumours	β-Galactosidase	Retrovirus	Intratumoral injection of producer cells	Yla-Herttuala Valpalahti Ollikainen	Kuopio, Finland
Neuroblastoma	Neomycin resistance cDNA	Retrovirus	Use of marker genes to investigate the mechanism of relapse and the effect of bone marrow purging in autologous transplantation for stage D neuroblastoma	M.K. Brenner	Baxter Health Care Corp., and St. Jude Children's Hospital, Memphis, TN, USA
Neuroblastoma	Neomycin resistance cDNA	Retrovirus	Use of gene transfer to determine the role of tumour cells in bone marrow used for autologous transplantation and the efficiency of immunomagnetic purging of the bone marrow	J. Kemshead	Pediatric and Neuro-Oncology Group, Frenchay Hospital, Bristol, UK

Modified from the data published and updated regularly in *Cancer Gene Therapy* and *Human Gene Therapy* (1996). Full clinical trials have been published in *Human Gene Therapy* and can be found with reference to the principal investigators shown in this table.

Abbreviations: IGF-1, insulin-like growth factor-1; IL-2, interleukin-2; TGF-β, transforming growth factor-β.

considerations have also been raised during recent debates on the implications of high technology health costs for relatively few patients. In addition, peripheral administration of enzyme does not cross the blood–brain barrier, and its administration is thus ineffective to treat the neurological symptoms.

(ii) Bone marrow transplantation (BMT) is a proven therapeutic intervention which has already been tested over several years within the setting of clinical trials (see *Table 18.3*). The presumed effectiveness of BMT in the treatment of the neurological symptoms of metabolic lysosomal disorders is thought to be due to the fact that the accumulation of toxic intermediate metabolites could be reversed by the expression of the missing metabolic enzyme activity in brain macrophages or microglia derived from bone marrow transplants. Studies in experimental models have shown that several lysosomal enzymes, such as α-L-iduronidase, β-glucuronidase or α-mannosidase, can be taken up by brain cells, including neurons, and stored, both *in vivo* and *in vitro* (P.R. Lowenstein *et al.*, unpublished observations; Walkely *et al.*, 1996).

The theoretical possibility that neurons take up enzymes released by microglia appears to be possible in principle. Nevertheless, many questions remain regarding the use of BMT for the treatment of inherited neurological metabolic disease, and these are currently being pursued by several research teams throughout the world (see *Table 18.3*; reviewed recently in Walkley *et al.*, 1996). The recent achievement of long-term expression lasting several months *in vitro* following retroviral transduction of bone marrow lends support to the fact that such techniques could be used within a clinical setting in the near future (Fairbairn *et al.*, 1996).

Table 18.3. Clinical BMT studies for the treatment of neurological disease (modified from Walkley *et al.*, 1996)

Diseases treated in humans	Diseases treated in experimental animals
Lysosomal enzyme deficiencies characterized by significant intraneuronal storage	
Hurler's disease [mucopolysaccharidosis (MPS) type I]	Hurler's disease (dogs)
Hunter's disease (MPS type II)	Maroteaux–Lamy disease (MPS type VI) (cats)
Sanfilippo disease (MPS type III)	Sly syndrome (MPS type VII)
Niemann–Pick disease type A	GM_1 gangliosidosis (cats)
GM_1 gangliosidosis	GM_2 gangliosidosis (cats)
α mannosidosis	α mannosidosis (cats)
Pompe disease (glycogenosis type III)	Fucosidosis (dogs)
Lysosomal enzyme deficiencies characterized by major glial cell involvement	
Gaucher's disease (glucosylceramide lipidosis)	Krabbe disease (mice)
Krabbe disease (globoid cell leukodystrophy)	
Metachromatic leukodystrophy	
Adrenoleukodystrophy	
Other metabolic diseases of the nervous system	
Batten disease	Batten disease (dogs)

18.5 The future of neurological (gene) therapy

While evaluating the future of neurological gene therapy, it is important to keep a balanced view that does recognize and consider progress in therapy stemming from more traditional approaches to neurological therapeutics. Neural transplantation especially has made real advances in the last few years, and as results from more transplantation trials become available it becomes easier to evaluate more objectively the potential and shortcomings of this approach, especially for the treatment of Parkinson's disease. Also, the transplantation work which has been proceeding for over two decades will provide much background information useful for the development of novel strategies, such as the use of transduced brain stem cells which can integrate into the host brain (Lacorazza *et al.*, 1996; Snyder and Wolfe, 1996). Also, improved surgical techniques such as pallidotomy are again gaining favour among neurosurgeons attempting to improve on the treatment of Parkinson's disease (Goetz and Diederich, 1996).

Another avenue for new discoveries is provided by an advance in the basic understanding of disease pathophysiology and some novel effects of available drugs. For example, recent work demonstrated that aminoglycosides can help to correct the lack of activity of cystic fibrosis transmembrane regulator (CFTR) due to the $\Delta F508$ mutation by improving the folding of the mutated protein in such a way that allows the mutated protein to be inserted in the plasma membrane and thus become functional (Delaney and Wainwright, 1996; Howard *et al.*, 1996). Similar progress has been achieved recently through the demonstration that mutated proteins synthesized from genes containing expanded trinucleotides (e.g. Huntington's disease) bind glyceraldehyde-3-phosphate dehydrogenase (GPDH) (Burke *et al.*, 1996; Roses, 1996). This suggests that treatments targeting neuronal energy metabolism offer new avenues for the treatment of neurological diseases caused by trinucleotide expansion.

Along the same lines, oxidative stress is once again being considered a major pathophysiological factor inducing cell death. Oxidative stress recently has been linked to neuronal cell death and apoptosis, and anti-oxidant administration appeared to have an effect in slowing the progression of early stage Huntington's disease, and a transgenic animal model of amyotrophic lateral sclerosis (ALS) (Gurney *et al.*, 1996; Peyser *et al.*, 1995). Interestingly, the over-expression of Cu/Zn superoxide dismutase (SOD) increases the survival of neurons transplanted into lesioned rodent basal ganglia (Nakao *et al.*, 1995), providing evidence that ongoing oxidative damage might limit the survival and therapeutic effect of transplants. Some familial cases of ALS display mutations in the *SOD1* gene encoding the anti-oxidant enzyme SOD, and changes in the levels of this enzyme have also been found in sporadic ALS (Bergeron *et al.*, 1994). A transgenic mouse model has been produced which has demonstrated that the mutation confers a novel function upon SOD, thus suggesting that it is not SOD loss-of-function that causes ALS (Gurney *et al.*, 1994). Indeed, mutated SOD has an increased capacity to generate free radicals (Yim *et al.*, 1996). Also, adenovirus vectors expressing SOD were effective in protecting striatal neurons from excitotoxic damage (Barkats *et al.*, 1996). Thus, more traditional pharmacological agents (i.e. antioxidants) may prove important in the treatment of diseases about which much information on genetics and hopefully on pathophysiological mechanisms will continue to be generated. The administration of growth factors [e.g. ciliary neurotrophic factor (CNTF)] through cell implants intrathechally implanted in human patients is also being explored for the treatment of ALS. Previously, serious adverse effects hampered the peripheral administration of high doses of the growth factor CNTF. The use of cell implants now allows CNTF to be delivered directly to the brain in much lower doses which do not show the side-effects of peripheral administration (Aebisher *et al.*, 1996).

18.6 Ethical issues. When is it appropriate to start?

Some neurological diseases are genetically dominant. For such disorders, exemplified by Parkinson's disease, patients could ask to be treated by germ-line gene therapy to rid their descendants of the disease. Whilst alternatives are available for the range of other inherited disorders, like pre-implantation diagnosis, those alternatives are not applicable in the case of truly dominant disorders. Germ-line gene therapy is currently banned by most gene therapy regulatory bodies world-wide. However, Ralph Brinster recently has designed a method for gene therapy of the male germ line (Brinster and Avarbock, 1994; Brinster and Zimmermann, 1994). Although this technique could not yet be used on human patients, the basic methodology is being developed very quickly. It will be important in the coming years for societies to open up the debate and decide on where they stand with respect to these new technologies. Ideally, this should be done years in advance of the technique being actually capable of being implemented on human patients. The same arguments could be applied for the development of fetal gene therapy for neurological disorders, as for gene transfer into fetal lungs for the prevention of cystic fibrosis.

In thinking about the ethical issues involved, it is also important to keep an open mind on when a given experiment can start; that is, how much reasonable information on the likelihood of success of a new experiment should be available before the investigators are allowed to go ahead. I think an imaginary encounter can be brought to mind: this is a meeting of an ethical committee convened in 1796 in England to decide on whether to authorize an application put forward by a Dr Jenner on the implementation of a novel treatment for the prevention of smallpox (Jenner, 1798). Would Dr Jenner be able to pass the standards of the Recombinant DNA Advisory Committee or the Gene Therapy Advisory Commitee, or would he be required to present properly controlled experimental results only available centuries later? In retrospect, it is easy to see that Dr Jenner should have gone ahead, but how could an ethical committe have known that the initial experiments done by Dr Jenner would prove so successful?

Acknowledgements

P.R.L. is a Research Fellow of The Lister Institute of Preventive Medicine. The author wishes to thank Drs M. Castro, A. Larregina and A. Morelli for challenging discussions on the future directions of neurological gene therapy. Funding for the work on gene therapy for neurological diseases and related work in the Molecular Medicine Unit has been kindly provided by Cancer Research Campaign, Parkinson's Disease Society, The Lister Institute of Preventive Medicine, The Wellcome Trust, The Sandoz Foundation for Gerontological Research, BBSRC, The Sir Halley Stewart Trust, REMEDI, The Faculty of Medicine (University of Manchester) Bequest Fund and The Royal Society.

References

Aebischer P, Schluep M, Deglon N, Joseph JM, Hirt L, Heyd B, Goddard M, Hammang JP, Zurn AD, Kato AC, Regli F, Baetge EE. (1996) Intrathecal delivery of CNTF using encapsulated genetically modified xenogeneic cells in amyotrophic lateral sclerosis patients. *Nature Med.* 2: 696–699.

Barkats M, Bemelmans AP, Geoffroy MC, Robert JJ, Loquet I, Horellou P, Revah F, Mallet J. (1996) An adenovirus encoding CuZn-SOD protects cultured striatal neurons against glutamate toxicity. *NeuroReport* 7: 497–501.

Barnea A, Nottebohm F. (1996) Recruitment and replacement of hippocampal neurons in young and adult chickadees. An addition to the theory of hippocampal learning. *Proc. Natl Acad. Sci. USA* 93: 714–718.

Bergeron C, Muntasser S, Somerville MJ, Weyer L, Percy ME. (1994) Copper/zinc superoxide dismutase messenger RNA levels are increased in sporadic amyotrophic lateral sclerosis motor-neurons. *Brain Res.* 659: 272–276.

Brinster RL, Avarbock MR. (1994) Germline transmission of donor haplotype following spermatogonial transplantation [see Comments]. *Proc. Natl Acad. Sci. USA* 91: 11303–11307.

Brinster RL, Zimmermann JW. (1994) Spermatogenesis following male germ-cell transplantation [see Comments]. *Proc. Natl Acad. Sci. USA* 91: 11298–11302.

Burke JR, Enghild JJ, Martin ME, Jou Y, Myers RM, Roses AD, Vance JM, Strittmatter WJ. (1996) Huntington and DRPLA protein selectively interact with the enzyme GAPDH. *Nature Med.* 2: 347–350.

Byrnes AP, Rusby JE, Wood MJA, Charlton HM. (1995) Adenovirus gene transfer causes inflammation in the brain. *Neuroscience.* 66: 1015–1024.

Byrnes AP, MacLaren RE, Charlton HM. (1996) Immunological instability of persistent adenovirus vectors in the brain: peripheral exposure to vector leads to renewed inflammation, reduced gene expression, and demyelination. *J. Neurosci.* 16: 3045–3055.

Coutelle C, Douar A, Colledge WH, Froster U. (1995). The challenge of fetal gene therapy. *Nature Med.* 1: 864–866.

Culver KW, Blaese RM. (1994) Gene therapy for cancer. *Trends Genet.* 10: 174–178.

Delaney SJ, Wainwright BJ. (1996) New pharmaceutical approaches to the treatment of cystic fibrosis. *Nature Med.* 2: 392–393.

Dunnett SB, Bjorklund A. (ed) (1995) *Functional Neural Transplantation.* Raven Press, New York.

Efstathiou S, Minson AC. (1995) Herpesvirus-based vectors. *Br. Med. Bull.* 51: 45–55.

Engelhardt JF, Ye XH, Doranz B, Wilson JM. (1994) Ablation of E2A in recombinant adenoviruses improves transgene persistence and decreases inflammatory response in mouse liver. *Proc. Natl Acad. Sci. USA* 91: 6196–6200.

Fairbairn LJ, Lashford LS, Spooncer E, McDermott RH, Lebens G, Arrand JE, Arrand JR, Bellantuono I, Holt R, Hatton CE, Cooper A, Besley GTN, Wraith JE, Anson DS, Hopwood JJ, Dexter TM. (1996) Long-term *in vitro* correction of α-L-iduronadase deficiency (Hurler syndrome) in human bone marrow. *Proc. Natl Acad. Sci. USA* 93: 2025–2030.

Fink DJ, DeLuca NA, Goins WF, Glorioso JC. (1996) Gene transfer to neurons using herpes simplex virus based vectors. *Annu. Rev. Neurosci.* 19: 265–287.

Fisher LJ, Gage FH. (1993) Grafting in the mammalian central nervous system. *Physiol. Rev.* 73: 583–616.

Friedmann T. (1994) Editorial: the promise and overpromise of human gene therapy. *Gene Ther.* 1: 217–218.

Friedmann T. (1996) Human gene therapy. An immature genie, but certainly out of the bottle. *Nature Med.* 2: 144–147.

Fujiwara T, Grimm EA, Roth JA. (1994) Gene therapeutics and gene therapy for cancer. *Curr. Opin. Oncol.* 6: 96–105.

Goetz CG, Diederich NJ. (1996) There is a renaissance of interest in pallidotomy for Parkinson's disease. *Nature Med.* 2: 510–514.

Grossman M, Rader DJ, Muller DWM, Kolansky DM, Kozarsky K, Clark BJ, III, Stein EA, Lupien PJ, Brewer HB, Jr, Raper SE, Wilson JM. (1995) A pilot study of *ex vivo* gene therapy for homozygous familial hypercholesterolaemia. *Nature Med.* 1: 1148–1154.

Gurney ME, Cutting FB, Zhai P, Doble A, Taylor CP, Andrus, PK, Hall ED. (1996) Benefit of vitamin-E, riluzole, and gabapentin in a transgenic model of familiar amyotrophic–lateral sclerosis. *Ann. Neurol.* 39: 147–157.

Gurney ME, Pu HF, Chiu AY, Dalcanto MC, Polchow CY, Alexander DD, Caliendo J, Hentati A, Kwon YW, Deng HX, Chen WJ, Zhai P, Sufit RL, Siddique T. (1994) Motor neuron degeneration in mice that express a human Cu, Zn superoxide dismutase mutation. *Science* 264: 1772–1775.

Howard M, Frizzell RA, Bedwell DM. (1996) Aminoglycoside antibiotics restore CFTR function by overcoming premature stop mutations. *Nature Med.* 2: 467–469.

Jenner E. (1798) An inquiry into the causes and effects of the variolae vaccinae, a disease discovered in some of the western counties of England, particularly Gloucestershire, and known by the name of the cow pox. Printed, for the author, by Sampson Low, No. 7, Berwick Street, Soho, London, England. Reprinted in: (1964) *Selected Papers on Virology* (ed. N Hahon). Prentice Hall, Englewood Cliffs, NJ, pp. 1–36.

Kaplitt MG, Loewy AD. (eds) (1995) *Viral Vectors: GeneTherapy and Neuroscience Applications.* Academic Press, London.

Kay MA, Holterman A, Meuse L, Gown A, Ochs HD, Linsley PS, Wilson CB. (1995) Long-term hepatic adenovirus-mediated gene expression in mice following CTLA4Ig administration. *Nature Genet.* **11**: 191–197.

Knowles MR, Hohneker KW, Zhou Z, Olsen JC, Noah TL, Hu P, Leigh MW, Engelhardt JF, Edwards LJ, Jones KR, Grossman M, Wilson JM, Johnson LG, Boucher RC. (1995) A controlled study of adenoviral-vector-mediated gene transfer in the nasal epithelium of patients with cystic fibrosis. *New Engl. J. Med.* **333**: 823–831.

Koenig S. (1996) A lesson from the HIV patient: the immune response is still the bane (or promise) of gene therapy. *Nature Med.* **2**: 165–167.

Lacorazza D, Flax JD, Snyder EY, Jendoubi M. (1996) Expression of human β-hexosaminidase α-subunit gene (the gene defect of Tay–Sachs disease) in mouse brains upon engraftment of transduced progenitor cells. *Nature Med.* **2**: 424–429.

Larsen CP, Elwood ET, Alexander DZ, Ritchie SC, Hendrix R, Tucker-Burden C, Cho HR, Aruffo A, Hollenbaugh D, Linsley PS, Winn KJ, Pearson TC. (1996) Long-term acceptance of skin and cardiac allografts after blocking CD40 and CD28 pathways. *Nature* **381**: 434–438.

Lowenstein PR. (1995) Degenerative and inherited neurological disorders. In: *Molecular and Cell Biology of Human Gene Therapeutics* (ed. G Dickson). Chapman & Hall, London, pp. 301–349.

Lowenstein PR, Enquist LW. (Eds) (1996) *Protocols for Gene Transfer in Neuroscience: Towards Gene Therapy of Neurological Disorders.* John Wiley & Sons, Chichester.

Maidment NT, Tan AM, Bloom DC, Anton B, Feldman LT, Stevens JG. (1996) Expression of the *lacZ* reporter gene in the rat basal forebrain, hippocampus, and nigrostriatal pathway using a nonreplicating herpes simplex vector. *Exp. Neurol.* **139**: 107–114.

Maron A, Gustin T, Leroux A, Mottet I, Dedieu JF, Brion JP, Demeure R, Perricaudet M, Octave JN. (1996) Gene therapy of rat C6 glioma using adenovirus mediated transfer of the herpes simplex virus thymidine kinase gene. Long term follow up by magnetic resonance imaging. *Gene Ther.* **3**: 315–322.

McCown TJ, Xiao X, Li J, Breese GR, Samulski RJ. (1996) Differential and persistent expression patterns of CNS gene transfer by an adeno-associated virus (AAV) vector. *Brain Res.* **713**: 99–107.

Medawar PB. (1948) Immunity of homologous grafted skin. III. The fate of skin homografts transplanted to the brain, subcutaneous tissue and to the anterior chamber of the eye. *Br. J. Exp. Pathol.* **29**: 58–69.

Mendell JR, Kissel JT, Amato AA, King W, Signore L, Prior TW, Sahenk Z, Benson S, McAndrew PE, Rice R, Nagaraja H, Stephens R, Lantry L, Morris GE, Burghes AHM. (1995) Myoblast transfer in the treatment of Duchenne's muscular dystrophy. *New Engl. J. Med.* **333**: 832–838.

Mineta T, Rabkin SD, Yazaki T, Hunter WD, Martuza RL. (1995) Attenuated multi-mutated herpes simplex virus-1 for the treatment of malignant gliomas. *Nature Med.* **1**: 938–943.

Nakao N, Frodl EM, Widner H, Carlson E, Eggerding FA, Epstein CJ, Brundin P. (1995) Overexpressing Cu/Zn superoxide-dismutase enhances survival of transplanted neurons in a rat model of Parkinson's disease. *Nature Med.* **1**: 226–231.

Naldini L, Blomer U, Gallay P, Ory D, Mulligan R, Gage, FH, Verma IH. (1996) *In vivo* gene delivery and stable transduction of nondividing cells by a lentiviral vector. *Science* **272**: 263–267.

Peyser CE, Folstein M, Chase GA, Starkstein S, Brandt J, Cockrell JR, Bylsma F, Coyle JT, McHugh PR, Folstein SE. (1995) Trial of D-alpha-tocopherol in Huntington's Disease. *Am. J. Psychiatr.* **152**: 1771–1775.

Rakic P. (1985) DNA synthesis and cell division in the adult primate brain. *Ann. NY Acad. Sci.* **457**: 193–211.

Riddell SR, Elliott M, Lewinsohn DA, Gilbert MJ, Wilson L, Manley SA, Lupton SD, Overell RW, Reynolds TC, Corey L, Greenberg PD. (1996) T-cell mediated rejection of gene-modified HIV-specific cytotoxic T lymphocytes in HIV-infected patients. *Nature Med.* **2**: 216–223.

Rinaman L, Card JP, Enquist LW. (1993) Spatiotemporal responses of astrocytes, ramified microglia, and brain macrophages to central nervous system infection with pseudorabies virus. *J. Neurosci.* **13**: 685–702.

Roses A. (1996) From genes to mechanisms to therapies: lessons to be learned from neurological disorders. *Nature Med.* **2**: 267–269.

Snyder EY, Wolfe JH. (1996) Central nervous system cell transplantation. A novel therapy for storage diseases. *Curr. Opin. Neurol.* **9**: 126–136.

Tripathy SK, Black HB, Goldwasser E, Leiden JM. (1996) Immune responses to transgene-encoded proteins limit the stability of gene expression after injection of replication-defective adenovirus vectors. *Nature Med.* **2**: 545–550.

Walkley SU, Thrall MA, Dobrenis K. (1996) Targeting gene products to the brain and neurons using bone marrow transplantation: a cell mediated delivery system for therapy of inherited metabolic human disease. In: *Protocols for Gene Transfer in Neuroscience: Towards Gene Therapy of Neurological Disorders* (eds PR Lowenstein, LW Enquist). John Wiley & Sons, Chichester, pp. 275–304.

Widner H. (1995) Transplantation of neuronal and nonneuronal cells into the brain. In: *Immune Responses in the Nervous System* (ed. NJ Rothwell). BIOS Scientific Publishers, Oxford, UK, pp. 189–227.

Wilson C, Kay MA. (1995) Immunomodulation to enhance gene therapy. *Nature Med.* **1**: 887–889.

Wood MJA, Byrnes AP, Pfaff DW, Rabkin SD, Charlton HM. (1994) Inflammatory effects of gene transfer into the CNS with defective HSV-1 vectors. *Gene Ther.* **1**: 283–291.

Yang YP, Ertl HCJ, Wilson JM. (1994a) MHC class I restricted cytotoxic T-lymphocytes to viral antigens destroy hepatocytes in mice infected with E1 deleted recombinant adenoviruses. *Immunity* **1**: 433–442.

Yang YP, Nunes FA, Berencsi K, Gonczol E, Engelhardt JF, Wilson JM. (1994b) Inactivation of E2A in recombinant adenoviruses improves the prospect for gene therapy in cystic fibrosis. *Nature Genet.* **7**: 362–369.

Yang YP, Trinchieri G, Wilson JM. (1995) Recombinant IL12 prevents formation of blocking IgA antibodies to recombinant adenovirus and allows repeated gene therapy to mouse lung. *Nature Med.* **1**: 890–893.

Yim MB, Kang JH, Yim HS, Kwak HS, Chock PB, Stadtman ER. (1996) A gain of function of an amyotrophic lateral sclerosis associated Cu,Zn superoxide dismutase mutant. An enhancement of free radical formation due to a decrease in KM for hydrogen peroxide. *Proc. Natl Acad. Sci. USA* **93**: 5709–5714.

Index

ORDERING DETAILS

Main address for orders

BIOS Scientific Publishers Ltd
9 Newtec Place, Magdalen Road,
Oxford OX4 1RE, UK
Tel: +44 1865 726286
Fax: +44 1865 246823

Australia and New Zealand
DA Information Services
648 Whitehorse Road, Mitcham, Victoria 3132, Australia
Tel: (03) 9210 7777
Fax: (03) 9210 7788

India
Viva Books Private Ltd
4325/3 Ansari Road, Daryaganj, New Delhi 110 002, India
Tel: 11 3283121
Fax: 11 3267224

Singapore and South East Asia
(Brunei, Hong Kong, Indonesia, Korea, Malaysia, the Philippines,
Singapore, Taiwan, and Thailand)
Toppan Company (S) PTE Ltd
38 Liu Fang Road, Jurong, Singapore 2262
Tel: (265) 6666
Fax: (261) 7875

USA and Canada
BIOS Scientific Publishers
PO Box 605, Herndon, VA 20172-0605, USA
Tel: (703) 661 1500
Fax: (703) 661 1501

Payment can be made by cheque or credit card (Visa/Mastercard, quoting number
and expiry date). Alternatively, a *pro forma* invoice can be sent.

Prepaid orders must include £2.50/US$5.00 to cover postage and packing
(two or more books sent post free)